Nanotechnology in the Life Sciences

Series Editor

Ram Prasad
School of Environmental Science and Engineering,
Sun Yat-sen University, Guangzhou, China

Amity Institute of Microbial Technology,
Amity University, Noida, Uttar Pradesh, India

Nano and biotechnology are two of the 21st century's most promising technologies. Nanotechnology is demarcated as the design, development, and application of materials and devices whose least functional make up is on a nanometer scale (1 to 100 nm). Meanwhile, biotechnology deals with metabolic and other physiological developments of biological subjects including microorganisms. These microbial processes have opened up new opportunities to explore novel applications, for example, the biosynthesis of metal nanomaterials, with the implication that these two technologies (i.e., thus nanobiotechnology) can play a vital role in developing and executing many valuable tools in the study of life. Nanotechnology is very diverse, ranging from extensions of conventional device physics to completely new approaches based upon molecular self-assembly, from developing new materials with dimensions on the nanoscale, to investigating whether we can directly control matters on/in the atomic scale level. This idea entails its application to diverse fields of science such as plant biology, organic chemistry, agriculture, the food industry, and more.

Nanobiotechnology offers a wide range of uses in medicine, agriculture, and the environment. Many diseases that do not have cures today may be cured by nanotechnology in the future. Use of nanotechnology in medical therapeutics needs adequate evaluation of its risk and safety factors. Scientists who are against the use of nanotechnology also agree that advancement in nanotechnology should continue because this field promises great benefits, but testing should be carried out to ensure its safety in people. It is possible that nanomedicine in the future will play a crucial role in the treatment of human and plant diseases, and also in the enhancement of normal human physiology and plant systems, respectively. If everything proceeds as expected, nanobiotechnology will, one day, become an inevitable part of our everyday life and will help save many lives.

More information about this series at http://www.springer.com/series/15921

Indu Bhushan • Vivek Kumar Singh
Durgesh Kumar Tripathi
Editors

Nanomaterials and Environmental Biotechnology

Springer

Editors
Indu Bhushan
School of Biotechnology
Shri Mata Vaishno Devi University
Katra, Jammu and Kashmir, India

Vivek Kumar Singh
School of Physics
Shri Mata Vaishno Devi University
Katra, Jammu and Kashmir, India

Durgesh Kumar Tripathi
Amity Institute of Organic Agriculture
Amity University
Noida, Uttar Pradesh, India

ISSN 2523-8027 ISSN 2523-8035 (electronic)
Nanotechnology in the Life Sciences
ISBN 978-3-030-34543-3 ISBN 978-3-030-34544-0 (eBook)
https://doi.org/10.1007/978-3-030-34544-0

This Springer imprint is published by the registered company Springer Nature Switzerland AG
The registered company address is: Gewerbestrasse 11, 6330 Cham, Switzerland

Preface

Nanoscience is the science of small, i.e., nanoscale particles (1–100 nm), which have a distinctive place not only because of their specific properties and reduced dimensions but also due to their potential role as building blocks for the synthesis of many nanostructure compounds. Nanoscience has added altogether another dimension of change in the direction of the revolutionary advancement in scientific understanding and it is applied across many other new technologies. From the time when nanoscience was introduced several years ago, it has changed our lives significantly – and will continue to do so! It all started in the field of physics "especially in electronics" and later followed in pharmaceutical and chemical sciences. Although the progress in the field of nanotechnology is a contemporary multidisciplinary science which involves the fields of chemistry, physics, engineering, and biology, the production of nanoparticles (NPs), both by humans and in nature, dates back to the pre-Christian era. In the present times, nanoscience and nanotechnology have taken major roles in branches of research and industry which impact human lives and yield unexpected results. However, the core aspect of nanoscience has always been in the betterment of technological applications in various arenas, thus making our daily lives easy and better. It enables quick advancements in a wide array of fields, i.e., economical, environmental, pharmaceutical, industrial, and several others. This book presents a fresh insight into the recent developments taking place into various sectors, consolidating all the knowledge into one place. It is aimed at an extensive array of book lovers with backgrounds in biology, chemistry, medicine, and physics. This book emphasizes that nanotechnologies are now extensively believed to have the capability to benefit diverse areas such as water decontamination, drug development, communication technologies and information, and manufacture lighter and stronger equipment. Each chapter in it gives an account of the application of nanotechnology in basic sciences and manufacturing, along with the commercial potential of nanomaterials in various industries and environment.

Katra, Jammu and Kashmir, India — Indu Bhushan
Katra, Jammu and Kashmir, India — Vivek Kumar Singh
Noida, Uttar Pradesh, India — Durgesh Kumar Tripathi

Contents

Chapter 1
Nanoparticles and Plant Interaction with Respect to Stress Response

Mohammed Shariq Iqbal, Akhilesh Kumar Singh, Satarudra Prakash Singh, and Mohammad Israil Ansari

1.1 Introduction

Nanotechnology is an incipient multidirectional technique with extensive applications in cancer remedies, drug delivery, microelectronics, biosensors, and cosmetic production, and also in agricultural fields (Nel et al. 2006; Singh et al. 2016, 2019; Arif et al. 2018; Shweta et al. 2018, 2017; Vishwakarma et al. 2018; Rastogi et al. 2019a, b). However, unspecified discharge of nanoparticles (metallic) into ecological communities has increased worldwide apprehension about their probable toxicity. The length of nanoparticles generally ranges from 1 to 100 nm in at least two dimensions, so that these are extremely fine particles with more surface area (Nowack and Bucheli 2007). Compared to molecules and bulk materials, nanoparticles are intermediate in size. The distinctive physical as well as chemical properties of nanoparticles result from their detachment in bulk material into reduced and smaller pieces (Jefferson 2000). However, because of their nanoscale size, their surface area thus increases, which makes them extremely catalytic or reactive.

The haphazard release of nanoparticles into natural environments from industrial effluents leads to their bulk production (Brunner et al. 2006; Owen and Handy 2007). Thus, most produced nanoparticles consist of heavy metals, toxifying water and soil with metallic nanoparticles, now a concern in environmental degradation. Plant interaction with excess nanoparticles in water and soil may cause the uptake and accumulation of the nanoparticles in the plant, leading to their ultimate conveyance to the ecosystem. However, nanoparticles may persist and be accumulated

M. S. Iqbal · A. K. Singh · S. P. Singh
Amity Institute of Biotechnology, Amity University Uttar Pradesh, Lucknow Campus, Lucknow, India

M. I. Ansari (✉)
Department of Botany, University of Lucknow, Lucknow, India
e-mail: ansari_mi@lkouniv.ac.in

I. Bhushan et al. (eds.), *Nanomaterials and Environmental Biotechnology*, Nanotechnology in the Life Sciences,
https://doi.org/10.1007/978-3-030-34544-0_1

within the plant, thus causing physical as well as chemical damage to various parts of the plant. Generally, nanoparticles pass into the plant root system through junctions of the lateral roots, thereby reaching the xylem tissue through the pericycle and the cortex (Dietz and Herth 2011). Moreover, although the cell wall restricts the entrance of nanoparticles into the plant body, cell walls have a definite pore size, allowing the transference of nanoparticles smaller than the pore size of the cell wall (Fleischer et al. 1999; Navarro et al. 2008). The degree of penetration is subject to the surface characteristics and dimensions of the nanoparticles. Undeniably, the smaller nanoparticles can pass through the cell wall easily. Larger nanoparticles are unable to pass through the cell wall and thus cannot disturb the metabolic pathways of the cell (Verano-Braga et al. 2014). However, flower stigmas, hydathodes, and stomata possess larger cell-wall openings, so that larger-size nanoparticles can pass through and possibly affect the plant. Chemical and physical interactions of nanoparticles with plants could be a natural or induced phenomenon. Chemical interfaces encompass the production of reactive oxygen or nitrogen species (ROS/RNS) (Nel et al. 2006), disruption in cell membrane transport activity of ions (Auffan et al. 2008), injury from oxidative stress (Foley et al. 2002; Jalil et al. 2017), and peroxidation of lipids present in the cells of the plant (Kamat et al. 2000). The ensuing admittance into the body of the plant through the cell wall subsequently allows nanoparticles to associate and work as metallic ions, reacting with carboxyl and sulfhydryl groups and eventually modifying protein activity. The effect of nanoparticle applications in shown in Fig. 1.1.

However, while studying engineered nanomaterials that arbitrate ecotoxicity, several artifacts that often lead to misconceptions of outcomes should be considered (Petersen et al. 2014). These probable aspects consist of noxious scum in engineered nanomaterials, their apposite storage and dissemination in a particular medium. Furthermore, engineered nanomaterials imply unintended effects on plant

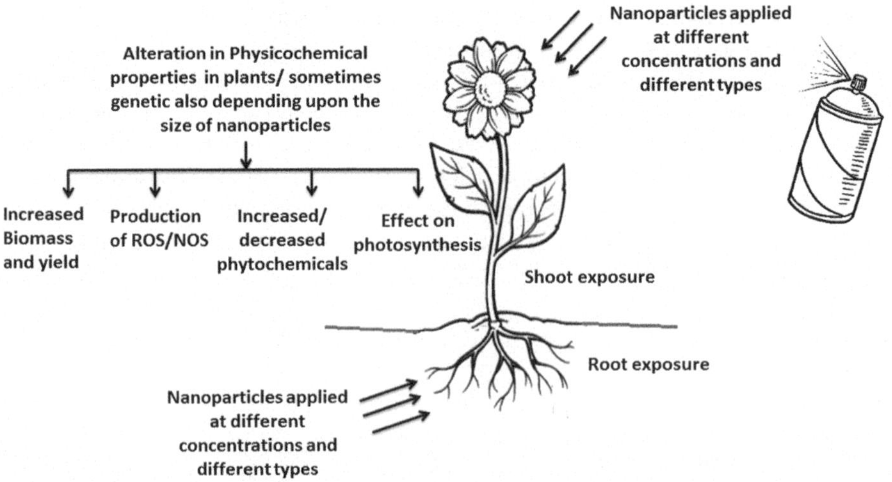

Fig. 1.1 Effect of nanoparticles on plants, leading to stress conditions

growth and development by depletion of nutrients with time and the diffusion of the engineered nanomaterials in organisms. Additionally, the dissolution, settling, and agglomeration properties of engineered nanomaterials cause diverse variations all through the study period that are difficult to evaluate accurately. Characteristics such as greater superficial area with distinct physical as well as chemical properties allow engineered nanomaterials to freely interact with ions physicochemically within the nutrient medium, thereby causing unintended toxic reactions such as wilting and chlorosis (Slomberg and Schoenfisch 2012; Begum and Fugetsu 2012; Jalil et al. 2018). Furthermore, as engineered nanomaterials interact with organic acids in the roots of the plant, the pH of the media falls, thereby modifying the nutrient quantity and properties of the engineered materials (Marschner 1995). The ineffectiveness of such interactions and their effects can be observed as an incongruous elucidation of phytotoxicity, and eventually a fictitious impression of engineered nanomaterials is understood (Ma et al. 2010).

1.2 The Nanoparticle and Its Role in Plant Stress

It is been reported that nanoparticles generally possess desired and undesired outcomes on various plant species because of their configuration, dissimilar size, altered physicochemical properties, and the concentrations of different nanoparticles used (Ma et al. 2010; Tripathi et al. 2017). It was reported that carbon nanotubes (multi-walled) substantially exaggerated the upregulation of stress-related gene expression in tomato seed germination and further affected the progress of the seedlings (Khodakovskaya et al. 2009; Singh et al. 2016; Koul et al. 2018). In a finding by Lee et al. (2010), Al_2O_3 nanoparticles were found to be less toxic than ZnO, SiO_2, or Fe_2O_3 nanoparticles when used to treat *Arabidopsis thaliana*. Earlier investigations on algae also underscored the noxious effects of nanoparticles (Arouja et al. 2009). Several nanoparticles such as CeO_2, ZnO, TiO_2, and Ag nanoparticles were observed to be deposited on the cell-wall surface as well as on the surface of the organelles, which causes stimulation in the form of stress response from oxidative stress within the cell (Buzea et al. 2007). In another study of *Cucurbita pepo*, the outcome of showed that seed germination was not altered by treatment with Cu, Si, Ag, and ZnO nanoparticles. On the other hand, their complementary bulk constituents and Cu nanoparticles cause alteration in root length when estimated against the control and a powder of bulk copper (Stampoulis et al. 2009). Treatment with ZnO nanoparticles affected root length in rice plants, but TiO_2 nanoparticles had no effect on the roots (Boonyanitipong et al. 2011). In another study, Riahi-Madvar et al. (2012) revealed effects on the roots of *Triticum aestivum* by treating the plant withAl_2O_3 nanoparticles at different concentrations, although the nanoparticles did not influence seed germination, shoot length, or fresh weight/dry weight ratio. Treatment of rice seedlings with CuO nanoparticles had an effect on enzymatic activity. Enzymatic antioxidant values were shown to be elicited (Shaw and Hossain 2013). An analogous experiment was conducted by Shaw et al. (2014)

wherein treatment of *Hordeum vulgare* with CuO nanoparticles caused an effect on photosynthetic activity and also on the antioxidants. It was elucidated that the decreased growth of shoot and root length caused poor photosynthetic activity. Furthermore, Atha et al. (2012) showed CuO nanoparticles cause decreased growth and DNA damage in *Raphanus sativus*, *Lolium perenne*, and *Lolium rigidum* plants. Rico et al. (2013) revealed effects on photosynthetic activity, enzymatic activity, and ascorbate and thiol levels when CeO_2 nanoparticles were used to treat rice seedlings. Formation of ROS/RNS and H_2O_2 when *Spirodela punctata* plants were treated with Ag and ZnO nanoparticles showed the significant toxicity of these nanoparticles (Thwala et al. 2013). Among the numerous metallic nanoparticles, ample consideration has been paid to Ag nanoparticles because of their distinctive biological and physicochemical properties when compared to various other large-size nanoparticles (Sharma et al. 2009). The fungicidal and bactericidal properties of Ag nanoparticles have widespread application as an indispensable constituent in various products at domestic, nutritional, and industrial levels (Tran et al. 2013). Silver nanoparticles when compared to silver-based compound nanoparticles possess an enlarged surface area accessible for microbe interface; further, it was reported that such are more noxious to various fungi and bacteria and several viruses as well. Similar to alternate metallic nanoparticles, Ag nanoparticles possess more impact on prompt stress reactions (ROS/RNS) in various microorganisms, algae, animals, and plants (Jiang et al. 2012). Conversely, the stimulus of Ag nanoparticles on plants is basically governed by such factors as plant species, growth and developmental stage, type and concentration of nanoparticles used, and experimental parameters such as period of treatment, humidity and temperature, and method of nanoparticle treatment (Vannini et al. 2013). Ag nanoparticles are one of the most widely examined nanoparticles whose toxic effects have been investigated in numerous plant crops (Stampoulis et al. 2009; Jiang et al. 2012; Kumari et al. 2009). Several studies have revealed that Ag nanoparticles were disadvantageous for plant growth as compared to the growth-enhancing possession of Ag nanoparticles in wetland plants (Yin et al. 2012), *Proteus vulgaris* and *Zea mays* (Salama 2012), *Brassica juncea* (Sharma et al. 2012), and *Eruca sativa* (Vannini et al. 2013). Ag nanoparticles showed a chromotoxic outcome on mitotic cell division in *Allium cepa* (Kumari et al. 2009). Further, Ag nanoparticles intermingle with proteins (are membrane bound) and stimulate various metabolic pathways, which restricts cell propagation (Roh et al. 2012; Gopinath et al. 2010). Some of the numerous effects of nanoparticles on plants are summarized in Table 1.1.

1.3 Mechanistic Interaction of Nanoparticles in Plant Stress

Recent studies showed that all the interactions of nanoparticles may be determined and noted to be affected by plant species, nanoparticle type and size, and the chemical structure, constancy, and functional aspects of the nanoparticles. The interaction of nanoparticles with plants leading to stress can be classified into different phases of nanoparticle uptake, translocation within the plant, accumulation in different

Table 1.1 Response of different plants toward different kinds of nanoparticles

Plants	Types of nanoparticles applied	Response	References
Boswellia ovalifoliolata, Egeria densa, Juncus effusus, Quercus robur	Silver	Improved germination rate; enhanced enzymatic antioxidants; no effect on chlorophyll; no phytotoxicity	Savithramma et al. (2012); Yuan et al. (2018b); Olchowik et al. (2017)
Gloriosa superba, Arabidopsis thaliana	Cerium oxide	Toxic; Increased plant growth	Arumugama et al. (2015); Ma et al. (2013)
Gum karaya, *Phyllanthus amarus, Cassia alata*	Copper oxide	Therapeutic applications; Positive antimicrobial activity	Jayalakshmi and Yogamoorthi (2014); Vellora et al. (2013); Acharyulu et al. (2014)
Quercus robur	Copper	Positive antimicrobial activity; no phytotoxicity	Olchowik et al. (2017)
Euphorbia condylocarpa	Palladium	As a catalyst	Nasrollahzadeha et al. 2015
Oryza sativa, Triticum aestivum, Lycopersicon esculentum	Titanium dioxide	Phytocatalyst Elicitation of chlorophyll	Ramimoghadam et al. (2014); Mahmoodzadeh et al. (2013); Qi et al. (2013)
Glycine max, Vigna radiata	Iron oxide	Improved productivity and quantity	Dhoke et al. (2013); Sheykhbaglou et al. (2010)
Capsicum annuum	Iron	Promoted plant growth; alleviated iron deficiency	Yuan et al. (2018a, b)
Arachis hypogaea	Zinc oxide	Improved productivity	Prasad et al. (2012)
Cucumis sativus	Gold	Improved germination rate	Barrena et al. (2009)
Lycopersiconm esculentum	Carbon nanotubes	Improved germination rate	Morla et al. (2011)
Solanum lycopersicon	Nickel oxide	Induced apoptosis in roots; enhanced antioxidants	Faisal et al. (2013); Soares et al. (2016)
Lemna minor	Alumina	Enlarged plant growth; improved productivity and quantity	Juhel et al. (2011)

cells and tissues, and desired or undesired outcomes. Some of the studies revealing such conditions are discussed next.

1.3.1 Phytotoxicity Mechanism of Nanoparticles

Investigation related to phytotoxicity in higher plants is imperative for elucidating the toxic effect of nanoparticles. Both undesirable and desirable or insignificant effects have been elucidated concerning the potential noxiousness of nanoparticles to various plants (Bystrzejewska-Piotrowska et al. 2009; Sohaebuddin et al. 2010; Muller et al. 2005; Jalil et al. 2018). Various studies point toward nanoparticle toxicity (Ghodake et al. 2010; Stampoulis et al. 2009). A noticeable variation in germination rate and growth was detected in the seeds of rice when exposed to carbon nanomaterials, mainly carbon nanotubes (Wang et al. 2012; Smirnova et al. 2012; Tan and Fugetsu 2007). In the experiment, water content in carbon nanotube-treated seeds was compared to that in control seeds, with better water content observed in the treated seeds. The germinating seeds were supplemented with carbon nanotubes to elucidate the effect on further developmental stages. The findings show significant use of carbon nanotubes to improve the growth of rice seedlings (Smirnova et al. 2012). In another example using Al_2O_3 nanoparticles, root length elongation was hindered in soybean, carrot, cabbage, cucumber, and corn (Kollmeier et al. 2000; Yamamoto et al. 2001; Tian et al. 2007; Ryan et al. 1992), whereas nanoparticles of ZnO were found to be maximally toxic, impeding the growth rate of the roots in various plants (Stella et al. 2010; Ma et al. 2009; Huang et al. 2002). Nanoparticles of ZnO were observed to be noxious in high concentrations when used to treat *Arabidopsis thaliana* plants, where a decreased germination rate was observed, with Al_2O_3, SiO_2, and Fe (II, III) oxide nanoparticles showing a moderate effect (Bin Hussein et al. 2002; Wang et al. 2004; Dwivedi and Randhawa 1974). With consideration for the toxicological aspect, the ratio of particle size to surface area is a vital physical property of a nanoparticle; the lesser the particle size, the greater the surface area. Hence, more of its atoms or molecules are exhibited externally than internally (Fugetsu and Parvin 2011; Begum et al. 2011). Several studies revealed surface properties of nanoparticles to be more toxic with a higher toxicity level than finer particles of the same material (Clarke and Brennan 1989; Kashem and Kawai 2007): this has been experimentally demonstrated by the use of diverse types of nanoparticles, such as cobalt, nickel, titanium dioxide, and carbon black. It was observed that TiO_2 nanoparticles with a minimum size less than 30 nm consequently are 43 fold more inflammatory than nanoparticles larger than 200 nm (Feizi et al. 2012; Castiglione et al. 2011; Qiu et al. 2013). Numerous investigations revealed that nano-sized particles are somewhat more toxic than micro-sized particles (Currie and Perry 2007). It was elucidated that one of the important parameters of noxiousness of nanoparticles is surface area. For instance, crystalline TiO_2 did not exhibited more severe toxicity than shown by TiO_2 nanoparticles (Stephen et al. 2012; Han et al. 2010). Generally, the present phytotoxic outline of nanoparticles is somewhat hypothetical; initially, the effects of nanoparticle properties are not well understood and further investigation on toxic effects is necessary, particularly on valuable food crops (Groppa et al. 2008). Thus, it can be concluded that various investigations have revealed that direct exposure to a particular kind of nanoparticles instigated a noteworthy phytotoxic effect, underscoring the necessity for environmental accountability for discarding wastes containing nanoparticles. Further studies on the influence of nanoparticles on valuable food crops and on the environment are required.

1.3.2 Uptake Mechanism of Nanoparticles

Investigations on the mechanism of uptake of nanoparticles in plants lack reliable and widely acceptable information (Nevius et al. 2012). Previous observations showed that nanoparticles may adhere to the plant root system and cause physico-chemical changes during uptake within the plant (Hartley and Lepp 2008; Taylor and Foy 1985). Currently, various investigations emphasize revealing the interface mechanism of nanoparticles toward plants (Besson-Bard et al. 2009; Zhang et al. 2017; Ma et al. 2018). Nanoparticle uptake and accumulation may differ depending on the difference in size and type of nanoparticle within the plant. Undeniably, validation of the mechanism of nanoparticle uptake is very restricted, and it depends on the concentration applied (Smirnova et al. 2012). Thus, most of the investigations reported do not yield similar outcomes for diverse forms (shapes and sizes) of nanoparticles (John et al. 1972). The majority of information is related to metal-based nanoparticles such as TiO_2, ZnO, Ag, Au, or Fe that correspond to a particular germination stage of the plant. Various possibilities have been suggested for nanoparticle uptake by the cells of the plants. Studies have suggested that nanoparticles move in plant cells by binding with protein biomolecules or ion channels, or through the process of endocytosis by means of new pores formed, finally binding to some organic molecule (Maine et al. 2001; Kurepa et al. 2010). For such investigations, carbon nanotubes have been preferred over other nanoparticles (Smirnova et al. 2012). However, it was reported that nanoparticles when compared to the bulk metals cause more reactivity by the greater surface area to mass ratio (Yuan et al. 2011). Subsequently, the nanoparticles might align with membrane transporters to form complexes as the root absorbs these and transports the particles into the plants. Thus, nanoparticles have been identified that can recognize ion transporters and be readily taken up by the plant (Tani and Barrington 2005). Selectivity among types of plants and the uptake of nanoparticles, which is still not very clear, is an area of further investigation.

1.3.3 Translocation Mechanism of Nanoparticles

Several investigations supported that the translocation of nanoparticles is determined by the quantity delivered and the species of plant (Yang and Ma 2010). Specific nanoparticles move swiftly within the plant, forming interactions with other biomolecules. Thus, the other nutrients are estimated according to the translocation of the nanoparticles applied (Zhu et al. 2008). The mechanism of translocation is instigated by the permeation of nanoparticles, into first the cell walls and then the plasma membrane of the cells. Through conduction by the plant xylem, the uptake mechanism and nanoparticle transferences take place in the shoot system

(Pola et al. 2012; Birbaum et al. 2010). The pore size of the cell wall is a vital criterion for the selection of nanoparticles, determining which nanoparticles can penetrate. As was investigated in *Allium porrum*, nanoparticle penetration was swifter in stomata than in the leaf (Birbaum et al. 2010).

1.3.4 Interaction Mechanism of Nanoparticles Leading to Stress

For the past few decades, the phytotoxicity of nanoparticles has been extensively investigated in several plant species, mainly focusing morpho-physiologically and biochemically. Nevertheless, only a few experiments have been focused on nanoparticle interaction with biomolecules with consideration of proteomics and causes of stress in the plant. Mirzajani et al. (2014) revealed by proteomic technique (gel-based) that the interaction of Ag nanoparticles on *Oryza sativa* causes toxicity. This investigation, based on root proteomics, elucidated that Ag nanoparticle-associated proteins were mainly related to the oxidative stress pathway, transcription, cell-wall synthesis, ion signaling and its regulation, division of cells, and degradation of protein. The effect of nanoparticles on the cell leading to such alterations is shown in Fig. 1.2. It was further observed that elicitation of enzymatic antioxidants such as peroxidases, glutathione-*S*-transferase, L-ascorbate, and superoxide dismutase induces enhanced formation of ROS under Ag nanoparticle treatment stress (Vannini et al. 2013). When Ag nanoparticles and $AgNO_3$ compounds were applied to *Erruca sativa* roots, both forms of silver produced alterations in the proteins associated with cellular homeostasis and redox regulation. These outcomes showed that the noxiousness of Ag nanoparticles mainly derives from its distinctive physicochemical characteristics (Vannini et al. 2013). Under flooding stress, the toxicity mechanism of Ag nanoparticles was studied in early stages of *Glycine max* plants, showing that proteins associated with signaling pathways, and the metabolism of cells and stress response, were altered. Furthermore, glyoxalase, an enzyme related to the detoxification pathway, was also degraded by Ag nanoparticle treatment (Mustafa et al. 2015a, b). In another study by Mustafa et al. (2015a, b), wherein the effects of Ag, ZnO, and Al_2O_3 nanoparticles were compared for treating *Glycine max* plant under flooding stress, protein synthesis was degraded, and glycolysis and lipid metabolism were also affected. Such investigations exhibit the interaction of nanoparticles with plants, but more research is needed to fully elucidate such effects. Thus, the desired as well as undesirable effects of nanoparticles can be observed on plants depending upon the requirements. Future investigations on interaction of nanoparticles with plants would help elucidate better understanding, implying specific nanoparticles can be applied to plants for desired outcomes, which may lead to better agricultural yields.

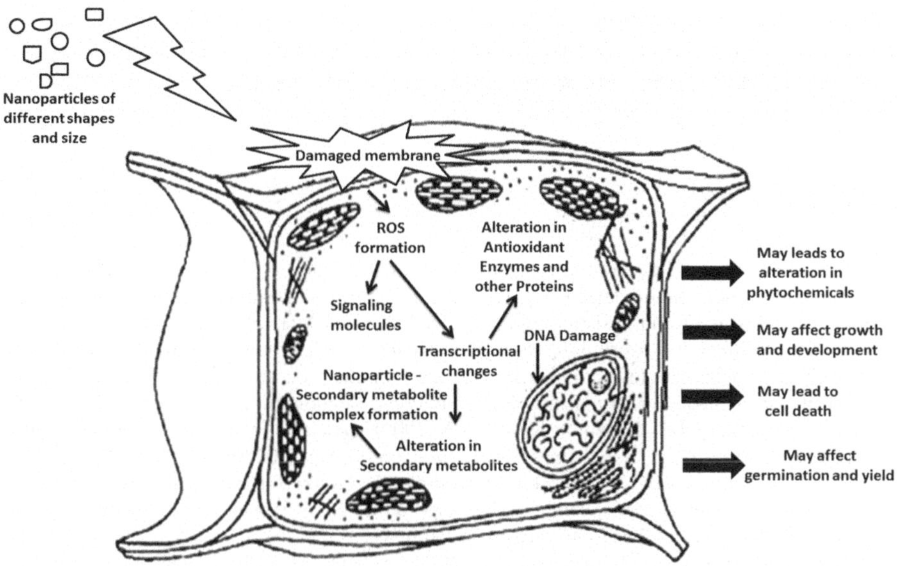

Fig. 1.2 Systematic representation of effect of nanoparticles within the plant cell leading to stress conditions

1.4 Conclusions and Future Prospects

The investigations conducted so far mostly concern plant reactions to a particular nanoparticle stress displaying an abundance of proteins associated with ROS, signaling caused by stress, pathways related to plant hormones, oxidation-reduction within the cell, and detoxification. Investigations on nanoparticles causing phytotoxicity showed that nanoparticle size is an important aspect in the type and degree of response within the plant cell. Further investigations are needed to fully elucidate whether metallic nanoparticles wield their noxious effects because of their distinctive characteristics or the loose metallic ions. Furthermore, exploration intended to recognize and illustrate subcellular organelles for elucidating the detailed alterations within the cell helps to understand the stress mechanism caused by nanoparticles. Additionally, metabolomics and transcriptomics techniques can have great prospects to fully elucidate the stress response toward nanoparticles. All this information would give us a wide explanation of the response mechanism of plants to stress caused by nanoparticles. Such investigations on plant stress tolerance mechanisms toward nanoparticles can lead to better plant yields for the production of specific valuable phytochemicals. These data, based on the interaction of nanoparticles with plants, would help elucidate improved understanding about the plant responses, which would suggest particular nanoparticles for plants for anticipated outcomes. This information can be advantageous for the agricultural perspective in improved yields and increased production of secondary metabolites that are beneficial in the pharmaceutical and nutraceutical industries.

References

Acharyulu NPS, Dubey RS, Swaminadham V, Kalyani RL, Kollu P, Pammi SVN (2014) Green synthesis of CuO nanoparticles using *Phyllanthus amarus* leaf extract and their antibacterial activity against multidrug resistance bacteria. Int J Eng Res Technol 3(4):639–641

Arif N, Yadav V, Singh S, Tripathi DK, Dubey NK, Chauhan DK, Giorgetti L (2018) Interaction of copper oxide nanoparticles with plants: uptake, accumulation, and toxicity. In: Nanomaterials in plants, algae, and microorganisms. Academic Press, London, pp 297–310

Arouja V, Dubourguier HC, Kasemets K, Kahru A (2009) Toxicity of nanoparticles of CuO, ZnO, and TiO2 to microalgae *Pseudokirchneriella subcapitata*. Sci Total Environ 407:1461–1468

Arumugama A, Karthikeyan C, Hameed ASH, Gopinath K, Gowri S, Karthika V (2015) Synthesis of cerium oxide nanoparticles using *Gloriosa superba* L. leaf extract and their structural, optical and antibacterial properties. Mater Sci Eng 49:408–415

Atha DH, Wang H, Petersen EJ, Cleveland D, Holbrook RD, Jaruga P, Dizdaroglu M, Xing B, Nelson BC (2012) Copper oxide nanoparticle mediated DNA damage in terrestrial plant models. Environ Sci Technol 46:1819–1827

Auffan M, Achouak W, Rose J, Roncato MA, Chanéac C, Waite DT, Masion A, Woicik JC, Wiesner MR, Bottero JY (2008) Relation between the redox state of iron-based nanoparticles and their cytotoxicity toward Escherichia coli. Environ Sci Technol 42:6730–6735

Barrena R, Casals E, Colón J, Font X, Sánchez A, Puntes V (2009) Evaluation of the ecotoxicity of model nanoparticles. Chemosphere 75(7):850–857

Begum P, Fugetsu B (2012) Phytotoxicity of multi-walled carbon nanotubes on red spinach (*Amaranthus tricolor* L) and the role of ascorbic acid as an antioxidant. J Hazard Mater 243:212–222

Begum P, Ikhtiari R, Fugetsu B (2011) Graphene phytotoxicity in the seedling stage of cabbage, tomato, red spinach, and lettuce. Carbon 49(12):3907–3919

Besson-Bard A, Gravot A, Richaud P et al (2009) Nitric oxide contributes to cadmium toxicity in arabidopsis by promoting cadmium accumulation in roots and by up-regulating genes related to iron uptake. Plant Physiol 149(3):1302–1315

Bin Hussein MZ, Zainal Z, Yahaya AH, Foo DWV (2002) Controlled release of a plant growth regulator, ũFC;- naphthaleneacetate from the lamella of Zn-Al-layered double hydroxide nanocomposite. J Control Release 82(2–3):417–427

Birbaum K, Brogioli R, Schellenberg M et al (2010) No evidence for cerium dioxide nanoparticle translocation in maize plants. Environ Sci Technol 44(22):8718–8723

Boonyanitipong P, Kositsup B, Kumar P, Baruah S, Dutta J (2011) Toxicity of ZnO and TiO2 nanoparticles on germinating rice seed *Oryza sativa* L. Int J Biosci Biochem Bioninform 1:282–285

Brunner TJ, Wick P, Manser P, Spohn P, Grass RN, Limbach LK, Bruinink A, Stark WJ (2006) *In vitro* cytotoxicity of oxide nanoparticles: comparison to asbestos, silica, and the effect of particle solubility. Environ Sci Technol 40:4374–4381

Buzea C, Pacheco II, Robbie K (2007) Nanomaterials and nanoparticles: sources and toxicity. Biointerphases 2:17–71

Bystrzejewska-Piotrowska G, Golimowski J, Urban PL (2009) Nanoparticles: their potential toxicity, waste and environmental management. Waste Manag 29(9):2587–2595

Castiglione MR, Giorgetti L, Geri C, Cremonini R (2011) The effects of nano-TiO2 on seed germination, development and mitosis of root tip cells of *Vicia narbonensis* L. and *Zea mays* L. J Nanopart Res 13(6):2443–2449

Clarke BB, Brennan E (1989) Differential cadmium accumulation and phytotoxicity in sixteen tobacco cultivars. J Air Waste Manage Assoc 39(10):1319–1322

Currie HA, Perry CC (2007) Silica in plants: biological, biochemical and chemical studies. Ann Bot 100(7):1383–1389

Dhoke SK, Mahajan P, Kamble R, Khanna A (2013) Effect of nanoparticles suspension on the growth of mung (*Vigna radiata*) seedlings by foliar spray method. Nanotechnol Dev 3:1

Dietz KJ, Herth S (2011) Plant nanotoxicology. Trends Plant Sci 16:582–589

Dwivedi RS, Randhawa NS (1974) Evaluation of a rapid test for the hidden hunger of zinc in plants. Plant Soil 40(2):445–451

Faisal M, Saquib Q, Alatar AA, Al-Khedhairy AA, Hegazy AK, Musarrat J (2013) Phytotoxic hazards of NiO-nanoparticles in tomato: a study on mechanism of cell death. J Hazard Mater 15:250–251, 318–32

Feizi H, Rezvani MP, Shahtahmassebi N, Fotovat A (2012) Impact of bulk and nanosized titanium dioxide (TiO_2) on wheat seed germination and seedling growth. Biol Trace Elem Res 146(1):101–106

Fleischer A, O'Neill MA, Ehwald R (1999) The pore size of non-graminaceous plant cell walls is rapidly decreased by borate ester cross-linking of the pectic polysaccharide rhamnogalacturonan II. Plant Physiol 121:829–838

Foley S, Crowley C, Smaihi M, Bonfils C, Erlanger BF, Seta P, Larroque C (2002) Cellular localisation of a water-soluble fullerene derivative. Biochem Biophys Res Commun 294:116–119

Fugetsu B, Parvin B (2011) Graphene phytotoxicity in the seedling stage of cabbage, tomato, red spinach, and lettuce. In: Bianco S (ed) Carbon nanotubes—from research to applications. CC BY-NC-SA

Ghodake G, Seo YD, Park D, Lee DS (2010) Phytotoxicity of carbon nanotubes assessed by *Brassica juncea* and *Phaseolus mungo*. J Nanoelectron Optoelectron 5(2):157–160

Gopinath P, Gogoi SK, Sanpui P, Paul A, Chattopadhyay A, Ghosh SS (2010) Signaling gene cascade in silver nanoparticle induced apoptosis. Colloids Surf. B 77:240–245

Groppa MD, Rosales EP, Iannone MF, Benavides MP (2008) Nitric oxide, polyamines and cd-induced phytotoxicity in wheat roots. Phytochemistry 69(14):2609–2615

Han T, Fan T, Chow S, Zhang D (2010) Biogenic N-PcodopedTiO_2: synthesis, characterization and photocatalytic properties. Bioresour Technol 101(17):6829–6835

Hartley W, Lepp NW (2008) Remediation of arsenic contaminated soils by iron-oxide application, evaluated in terms of plant productivity, arsenic and phytotoxic metal uptake. Sci Total Environ 390(1):35–44

Huang FM, Tai KW, Chou MY, Chang YC (2002) Cytotoxicity of resin-, zinc oxide-eugenol-, and calcium hydroxidebased root canal sealers on human periodontal ligament cells and permanent V79 cells. Int Endod J 35(2):153–158

Jalil SU, Ahmed I, Ansari MI (2017) Functional loss of GABA transaminase (GABA-T) expressed early leaf senescence under various stress conditions in *Arabidopsis thaliana*. Curr Plant Biol 9–10:11–22

Jalil SU, Zahera M, Khan MS, Ansari MI (2018) Biochemical synthesis of gold nanoparticles from leaf protein of *Nicotiana tabacum* L. cv. xanthi and their physiological, developmental and ROS scavenging responses on tobacco plant under stress conditions. IET Nanobiotechnol. https://doi.org/10.1007/978-3-319-95480-6_14

Jayalakshmi, Yogamoorthi A (2014) Green synthesis of copper oxide nanoparticles using aqueous extract of *Cassia alata* and particles characterization. Int. J Nanomat Biostuct 4(4):66–71

Jefferson DA (2000) The surface activity of ultrafine particles. Philos Trans R Soc Lond Ser A 358:2683–2692

Jiang HS, Li M, Chang FY, Li W, Yin LY (2012) Physiological analysis of silver nanoparticles and $AgNO_3$ toxicity to Spirodela polyrrhiza. Environ Toxicol Chem 31:1880–1886

John MK, VanLaerhoven CJ, Chuah HH (1972) Factors affecting plant uptake and phytotoxicity of cadmium added to soils. Environ Sci Technol 6(12):1005–1009

Juhel G, Batisse E, Hugues Q, Daly D, van Pelt FN, O'Halloran J, Jansen MA (2011) Alumina nanoparticles enhance growth of Lemna minor. Aquat Toxicol 105(3): 328–336

Kamat JP, Devasagayam TP, Priyadarsini KI, Mohan H (2000) Reactive oxygen species mediated membrane damage induced by fullerene derivatives and its possible biological implications. Toxicology 155:55–61

Kashem MA, Kawai S (2007) Alleviation of cadmium phytotoxicity by magnesium in Japanese mustard spinach. Soil Sci Plant Nutr 53(3):246–251

Khodakovskaya M, Dervishi E, Mahmood M, Xu Y, Li Z, Watanabe F, Biris AS (2009) Carbon nanotubes are able to penetrate plant seed coat and dramatically affect seed germination and plant growth. ACS Nano 3:3221–3227

Kollmeier M, Felle HH, Horst WJ (2000) Genotypical differences in aluminum resistance of maize are expressed in the distal part of the transition zone. Is reduced basipetal auxin flow involved in inhibition of root elongation by aluminum? Plant Physiol 122(3):945–956

Koul A, Kumar A, Singh VK, Tripathi DK, Mallubhotla S (2018) Exploring plant-mediated copper, iron, titanium, and cerium oxide nanoparticles and their impacts. In: Nanomaterials in plants, algae, and microorganisms. Academic Press, San Diego, pp 175–194

Kumari M, Mukherjee A, Chandrasekaran N (2009) Genotoxicity of silver nanoparticles in Allium cepa. Sci Total Environ 407:5243–5246

Kurepa J, Paunesku T, Vogt S et al (2010) Uptake and distribution of ultrasmall anatase TiO_2 alizarin red S nanoconjugates in *Arabidopsis thaliana*. Nano Lett 10(7):2296–2302

Lee CW, Mahendra S, Zodrow K, Li D, Tsai YC, Braam J, Alvarez PJ (2010) Developmental phytotoxicity of metal oxide nanoparticles to *Arabidopsis thaliana*. Environ Toxicol Chem 29:669–675

Ma H, Bertsch PM, Glenn TC, Kabengi NJ, Williams PL (2009) Toxicity of manufactured zinc oxide nanoparticles in the nematode *Caenorhabditis elegans*. Environ Toxicol Chem 28(6):1324–1330

Ma X, Geisler-Lee J, Deng Y, Kolmakov A (2010) Interactions between engineered nanoparticles (ENPs) and plants: phytotoxicity, uptake and accumulation. Sci Total Environ 408:3053–3061

Ma C, Chhikara S, Xing B, Musante C, White JC, Dhankher OP (2013) Physiological andmolecular response of *Arabidopsis thaliana* (L.) to nanoparticle cerium and indium oxide exposure. ACS Sustainable Chem Eng 1(7):768–778

Ma C, White JC, Zhao J, Zhao Q, Xing B (2018) Uptake of engineered nanoparticles by food crops: characterization, mechanisms, and implications. Annu Rev Food Sci Technol 9:129–153

Mahmoodzadeh H, Nabavi M, Kashefi H (2013) Effect of nanoscale titanium dioxide particles on the germination and growth of canola (*Brassica napus*). J Ornamental Hortic Plants 3:25–32

Maine MA, Duarte MV, Su˜n'c NL (2001) Cadmium uptake by floating macrophytes. Water Res 35(11):2629–2634

Marschner HM (1995) Nutrition of higher plants, 2nd edn. Academic Press, San Diego

Morla S, Ramachandra Rao CSV, Chakrapani R (2011) Factors affecting seed germination and seedling growth of tomato plants cultured in vitro conditions. J Chem Bio Phys Sci B 1: 328–334.

Mirzajani F, Askari H, Hamzelou S, Schober Y, Römpp A, Ghassempour A, Spengler B (2014) Proteomics study of silver nanoparticles toxicity on *Oryza sativa* L. Ecotoxicol Environ Saf 108:335–339

Muller J, Huaux F, Moreau N et al (2005) Respiratory toxicity of multi-wall carbon nanotubes. Toxicol Appl Pharmacol 207(3):221–231

Mustafa G, Sakata K, Komatsu S (2015a) Proteomic analysis of flooded soybean root exposed to aluminumoxide nanoparticles. J Proteome 128:280–297

Mustafa G, Sakata K, Hossain Z, Komatsu S (2015b) Proteomic study on the effects of silver nanoparticles onsoybean under flooding stress. J Proteome 122:100–118

Nasrollahzadeha M, Sajadib SM, Vartoonia AR, Khalajc M (2015) Green synthesis of Pd/Fe_3O_4 nanoparticles using *Euphorbia condylocarpa* M. bieb root extract and their catalytic applications as magnetically recoverable and stable recyclablecatalysts for the phosphine-free Sonogashira and Suzuki coupling reactions. J Mol Catal A Chem 396:31–39

Navarro E, Baun A, Behra R, Hartmann NB, Filser J, Miao AJ, Quigg A, Santschi PH, Sigg L (2008) Environmental behavior and ecotoxicity of engineered nanoparticles to algae, plants, and fungi. Ecotoxicology 17:372–386

Nel A, Xia T, Mädler L, Li N (2006) Toxic potential of materials at the nanolevel. Science 311:622–627

Nevius BA, Chen YP, Ferry JL, Decho AW (2012) Surface functionalization effects on uptake of fluorescent polystyrene nanoparticles by model biofilms. Ecotoxicology 21(8):2205–2213

Nowack B, Bucheli TD (2007) Occurrence, behavior and effects of nanoparticles in the environment. Environ Pollut 150:5–22

Olchowik J, Bzdyk RM, Studnicki M, Bederska-Błaszczyk M, Urban A, Aleksandrowicz-Trzcinska M (2017) The effect of silver and copper nanoparticles on the condition of English Oak (*Quercus robur* L.) seedlings in a container nursery experiment. Forests 8:310

Owen R, Handy R (2007) Formulating the problems for environmental risk assessment of nanomaterials. Environ Sci Technol 41:5582–5588

Petersen EJ, Henry TB, Zhao J, MacCuspie RI, Kirschling TL, Dobrovolskaia MA, Hackley V, Xing B, White JC (2014) Identification and avoidance of potential artifacts and misinterpretations in nanomaterial ecotoxicity measurements. Environ Sci Technol 48:4226–4246

Pola M, Tamara LC, Andrew TH (2012) Toxicity, uptake, and translocation of engineered nanomaterials in vascular plants. Environ Sci Technol 46(17):9224–9239

Prasad TNVKV, Sudhakar P, Sreenivasulu Y, Latha P, Munaswamy V, Reddy KR, Sreeprasad TSP, Sajanlal R, Pradeep T (2012) Effect of nanoscale zinc oxide particles on the germination, growth and yield of peanut. J Plant Nutr 35(6):905–927

Qi M, Liu Y, Li T (2013) Nano-TiO2 improve the photosynthesis of tomato leaves under mild heat stress. Biol Trace Elem Res 156(1–3):323–328

Qiu Z, Yang Q, Liu W (2013) Photocatalytic degradation of phytotoxic substances in waste nutrient solution by various immobilized levels of nano-TiO_2. Water Air Soil Pollut 224(3):1–10

Ramimoghadam D, Bagheri S, Bee S, Hamid A (2014) Biotemplated synthesis of anatase titanium dioxide nanoparticles via lignocellulosic waste material. Biomed Res Int 2014:205636

Rastogi A, Tripathi DK, Yadav S, Chauhan DK, Živčák M, Ghorbanpour M, El-Sheery NI, Brestic M (2019a) Application of silicon nanoparticles in agriculture. 3 Biotech 9(3):90

Rastogi A, Zivcak M, Tripathi DK, Yadav S, Kalaji HM, Brestic M (2019b) Phytotoxic effect of silver nanoparticles in Triticum aestivum: improper regulation of photosystem I activity as the reason for oxidative damage in the chloroplast. Photosynthetica 57(1):209–216

Riahi-Madvar A, Rezaee F, Jalili V (2012) Effects of alumina nanoparticles on morphological properties and antioxidant system of *Triticum aestivum*. Iran J Plant Physiol 3:595–603

Rico CM, Hong J, Morales MI, Zhao L, Barrios AC, Zhang JY, Peralta-Videa JR, Gardea-Torresdey JL (2013) Effect of cerium oxide nanoparticles on rice: a study involving the antioxidant defense system and in vivo fluorescence imaging. Environ Sci Technol 47:5635–5642

Roh JY, Eom HJ, Choi J (2012) Involvement of Caenorhabditis elegans MAPK signaling pathways in oxidative stress response induced by silver nanoparticles exposure. Toxicol Res 28:19–24

Ryan PR, Shaff JE, Kochian LV (1992) Aluminum toxicity in roots: correlation among ionic currents, ion fluxes, and root elongation in aluminum-sensitive and aluminum-tolerant wheat cultivars. Plant Physiol 99(3):1193–1200

Salama HMH (2012) Effects of silver nanoparticles in some crop plants, common bean (*Phaseolus vulgaris* L.) and corn (*Zea mays* L.). Int. res. J. Biotechnol 3:190–197

Savithramma N, Ankanna S, Bhumi G (2012) Effect of nanoparticles on seed germination and seedling growth of *Boswellia ovalifoliolata* an endemic and endangered medicinal tree taxon. Nano Vision 2:61–68

Sharma VK, Yngard RA, Lin Y (2009) Silver nanoparticles: green synthesis and their antimicrobial activities. Adv Colloid Interf Sci 145:83–96

Sharma P, Bhatt D, Zaidi MG, Saradhi PP, Khanna PK, Arora S (2012) Silver naoparticle-mediated enhancement in growth and antioxidant status of Brassica juncea. Appl Biochem Biotechnol 167:2225–2233

Shaw AK, Hossain Z (2013) Impact of nano-CuO stress on rice (*Oryza sativa* L.) seedlings. Chemosphere 93:906–915

Shaw AK, Ghosh S, Kalaji HM, Bosa K, Brestic M, Zivcak M, Hossain Z (2014) Nano-CuO stress induced modulation of antioxidative defense and photosynthetic performance of syrian barley (*Hordeum vulgare* L.). Environ Exp Bot 102:37–47

Sheykhbaglou R, Sedghi M, Shishevan MT, Sharifi RS (2010) Effects of nano-iron oxide particles on agronomic traits of soybean. Not Sci Biol 2:112–113

Shweta, Vishwakarma K, Sharma S, Narayan RP, Srivastava P, Khan AS, Dubey NK, Tripathi DK, Chauhan DK (2017) Plants and carbon nanotubes (CNTs) interface: present status and future prospects. In: Nanotechnology. Springer, Singapore, pp 317–340

Shweta, Tripathi DK, Chauhan DK, Peralta-Videa JR (2018) Availability and risk assessment of nanoparticles in living systems: a virtue or a peril? In: Nanomaterials in plants, algae, and microorganisms. Academic Press, San Diego, pp 1–31

Singh S, Tripathi DK, Dubey NK, Chauhan DK (2016) Effects of nano-materials on seed germination and seedling growth: striking the slight balance between the concepts and controversies. Mater Focus 5(3):195–201

Singh J, Vishwakarma K, Ramawat N, Rai P, Singh VK, Mishra RK, Kumar V, Tripathi DK, Sharma S (2019) Nanomaterials and microbes' interactions: a contemporary overview. 3 Biotech 9(3):68

Slomberg DL, Schoenfisch MH (2012) Silica nanoparticle phytotoxicity to *Arabidopsis thaliana*. Environ Sci Technol 46:10247–10254

Smirnova E, Gusev A, Zaytseva O et al (2012) Uptake and accumulation of multiwalled carbon nanotubes change themorphometric and biochemical characteristics of *Onobrychis arenaria* Seedlings. Front Chem Sci Eng 6(2):132–138

Soares C, Branco-Neves S, de Sousa A, Pereira R, Fidalgo F (2016) Ecotoxicological relevance of nano-NiO and acetaminophen to Hordeum vulgare L.: Combining standardized procedures and physiological endpoints. Chemosphere 165:442–452

Sohaebuddin SK, Thevenot PT, Baker D, Eaton JW, Tang L (2010) Nanomaterial cytotoxicity is composition, size, and cell type dependent. Part Fibre Toxicol 7:22

Stampoulis D, Sinha SK, White JC (2009) Assay-dependent phytotoxicity of nanoparticles to plants. Environ Sci Technol 43(24):9473–9479

Stella WYW, Priscilla TYL, Djuriˇsi'c AB, MYL K (2010) Toxicities of nano zinc oxide to five marinenorganisms: influences of aggregate size and ion solubility. Anal Bioanal Chem 396(2):609–618

Stephen GW, Li H, Jennifer H, Da-Ren C, In-Chul K, Yinjie JT (2012) Phytotoxicity of metal oxide nanoparticles is related to both dissolved metals ions and adsorption of particles on seed surfaces. J Pet Environ Biotechnol 3:4

Tan XM, Fugetsu B (2007) Multi-walled carbon nanotubes interact with cultured rice cells: evidence of a self-defense response. J Biomed Nanotechnol 3(3):285–288

Tani FH, Barrington S (2005) Zinc and copper uptake by plants under two transpiration rates. Part I.Wheat (*Triticum aestivum* L). Environ Pollut 138(3):538–547

Taylor GJ, Foy CD (1985) Differential uptake and toxicity of ionic and chelated copper in *Triticum aestivum*. Can J Bot 63(7):1271–1275

Thwala M, Musee N, Sikhwivhilu L, Wepener V (2013) The oxidative toxicity of ag and ZnO nanoparticles towards the aquatic plant Spirodela punctuta and the role of testing media parameters. Environ Sci Process Impacts 15:1830–1843

Tian Q, Sun D, Zhao M, Zhang W (2007) Inhibition of nitric oxide synthase (NOS) underlies aluminum-induced inhibition of root elongation in *Hibiscus moscheutos*. New Phytol 174(2):322–331

Tran QH, Nguyen VQ, Le AT (2013) Silver nanoparticles: synthesis, properties, toxicology, applications and perspectives. Adv Nat Sci 4:033001

Tripathi DK, Ahmad P, Sharma S, Chauhan DK, Dubey NK (eds) (2017) Nanomaterials in plants, algae, and microorganisms: concepts and controversies (vol. 1). Academic Press, London

Vannini C, Domingo G, Onelli E, Prinsi B, Marsoni M, Espen L, Bracale M (2013) Morphological and proteomic responses of Eruca sativa exposed to silver nanoparticles or silver nitrate. PLoS One 8:e68752

Vellora V, Padil T, Černík M (2013) Green synthesis of copper oxide nanoparticles using *Gum karaya* as a biotemplate and their antibacterial application. Int J Nanomedicine 8:889–898

Verano-Braga T, Miethling-Graff R, Wojdyla K, Rogowska-Wrzesinska A, Brewer JR, Erdmann H, Kjeldsen F (2014) Insights into the cellular response triggered by silver nanoparticles using quantitative proteomics. ACS Nano 8:2161–2175

Vishwakarma K, Upadhyay N, Kumar N, Tripathi DK, Chauhan DK, Sharma S, Sahi S (2018) Potential applications and avenues of nanotechnology in sustainable agriculture. In: Nanomaterials in plants, algae, and microorganisms. Academic Press, London, pp 473–500

Wang X, Summers CJ, Wang ZL (2004) Large-scale hexagonal-patterned growth of aligned ZnO nanorods for nano-optoelectronics and nanosensor arrays. Nano Lett 4(3):423–426

Wang X, Han H, Liu X, Gu X, Chen K, Lu D (2012) Multiwalled carbon nanotubes can enhance root elongation of wheat (*Triticum aestivum*) plants. J Nanopart Res 14(6):841

Yamamoto Y, Kobayashi Y, Matsumoto H (2001) Lipid peroxidation is an early symptom triggered by aluminum, but not the primary cause of elongation inhibition in pea roots. Plant Physiol 125(1):199–208

Yang K, Ma Y (2010) Computer simulation of the translocation of nanoparticles with different shapes across a lipid bilayer. Nat Nanotechnol 5(8):579–583

Yin L, Colman BP, McGill BM, Wright JP, Bernhardt ES (2012) Effects of silver nanoparticle exposure on germination and early growth of eleven wetland plants. PLoS One 7:e47674

Yuan H, Hu S, Huang P et al (2011) Single walled carbon nanotubes exhibit dual-phase regulation to exposed *Arabidopsis* mesophyll cells. Nanoscale Res Lett 6(1):1–9

Yuan J, Chen Y, Li H, Lu J, Zhao H, Liu M, Nechitaylo GS, Glushchenko NN (2018a) New insights into the cellular responses to iron nanoparticles in *Capsicum annuum*. Sci Rep 8:3228

Yuan L, Richardson CJ, Ho M, Willis CW, Colman BP, Wiesner MR (2018b) Stress responses of aquatic plants to silver nanoparticles. Environ Sci Technol 52(5):2558–2565

Zhang W, Dan Y, Shi H, Ma X (2017) Elucidating the mechanisms for plant uptake and *in-planta* speciation of cerium in radish (*Raphanus sativus* L.) treated with cerium oxide nanoparticles. J Environ Chem Eng 5(1):572–577

Zhu H, Han J, Xiao JQ, Jin Y (2008) Uptake, translocation, and accumulation of manufactured iron oxide nanoparticles by pumpkin plants. J Environ Monit 10(6):713–717

Chapter 2
Nanoencapsulation Technology: Boon to Food Packaging Industries

Somenath Das, Anand Kumar Chaudhari, Abhishek Kumar Dwivedy, Neha Upadhyay, Vipin Kumar Singh, Akanksha Singh, and Nawal Kishore Dubey

2.1 Introduction

Global food security is one of the biggest issues nowadays. Various food-borne pathogens like bacteria, fungi, and insects not only invade the food but also alter their nutritional profile by producing toxic metabolites (Scallan et al. 2011; Marin et al. 2013; Kim et al. 2003; Park et al. 2003; in't Veld 1996). Fungi, most importantly belonging to the species of *Aspergillus*, *Penicillium*, and *Fusarium*, are the major contributors for food and feedstuff spoilage. They produce mycotoxins, viz., aflatoxins, fumonisins, ochratoxins, deoxynivalenol, and zearalenone. Among these mycotoxins, aflatoxins in particular aflatoxin B_1 (AFB_1) produced mainly by different species of *Aspergillus* are a severe concern as it pose carcinogenic, teratogenic, mutagenic, and immunosuppressive potential. Based on the toxicity, it has also been recognized as class I human carcinogen (IARC 1993). The consumption of these contaminated products may have a direct effect on the consumer's health in the form of chronic and acute toxicity. It is estimated that annually around 200 deaths of adults and 40% of children's death up to 5 years are due to illness caused by consumption of contaminated foods (WHO 2017). In order to reduce these harmful contaminants from food, several physical methods like heating, UV or ionizing radiation, membrane filtrations (Farkas 2007), and use of synthetic chemical preservatives, viz., potassium bromate, butylated hydroxyanisole (BHA), and butylated hydroxytoluene (BHT) (Inetianbor et al. 2015), are widely used; however their irrational use may cause pest resistance and adverse effect on humans as well as on the environment, which has enforced the scientists to look toward natural food preservatives in order to meet sustainable food security (Damalas and Eleftherohorinos 2011)

S. Das · A. K. Chaudhari · A. K. Dwivedy · N. Upadhyay · V. K. Singh · A. Singh · N. K. Dubey (✉)
Department of Botany, Center of Advanced Study, Institute of Science, Banaras Hindu University, Varanasi, India

I. Bhushan et al. (eds.), *Nanomaterials and Environmental Biotechnology*, Nanotechnology in the Life Sciences,
https://doi.org/10.1007/978-3-030-34544-0_2

Since antiquity, plant products in different forms like powder or extracts have been used as preferred alternatives for the protection of food commodities from storage contaminants; however plant products in the form of essential oils are of great concerns as they are volatile, biodegradable, with no residual toxicity, safe to mammalian system, and eco-friendly (Isman 2000; Burt 2004). Many literatures also claim the importance of essential oils as the source of natural preservatives in view of their strong antimicrobial and antioxidant potential (Tripathi et al. 2009; Sánchez-González et al. 2011). Although, essential oils have merit over chemical preservatives, they often fail to exert their full preservative potential, due to oxidation and instability in presence of heat, light and other abiotic factors (Prakash et al. 2018).

In view of these obstacles, introduction of nanotechnology has revolutionized the food industry for the progress of nano based formulation of essential oils and other bioactive components, to achieve major goal of sustainable food preservation and security. Nanoparticles enclosing active principles are claimed to enhance the efficacy of bioactive components, as well as the stability, texture, taste and flavor of food products.

Based on the above background, the present chapter provides an overview of currently employed nanotechnologies and their application in food sectors to develop an approach for the preservation of food during storage in order to achieve sustainable food security.

2.2 Nanomaterials Used for Food Packaging

Recent development in EO formulations based on nanoencapsulation technology has gained much attention in food sectors. These formulations can take display several advantages over conventional methods of food preservation. They confer the controlled release of essential oils and protect them from thermal and photodegradation which assures their improved stability, organoleptic impacts, flavor, and function, consequently extending the final product shelf life (Liang et al. 2012; Madene et al. 2006; Lakkis 2016). There are several methods which have been used for the preparation of nanoparticles using various coating materials. Some of the important methods including different types of polymer used for encapsulation of essential oils or plant bioactive components are discussed and listed in Table 2.1.

2.2.1 Lipid-Based Encapsulation of Essential Oils

Lipids are an important group of natural molecules supporting vital life processes. Lipids give desired qualities like texture, aroma, color, and flavor and mouth-filling properties to food products (Bourne 2002). Currently, due to their nontoxicity and

Table 2.1 List of key nanoparticle types enclosing active materials and their biological properties

S no.	Nanoparticle type	EO/bioactive compounds	Inference	References
1.	Nanoemulsion	*Eugenia brejoensis*	Antimicrobial	Mendes et al. (2018)
		Pimpinella anisum	Insecticidal	Hashem et al. (2018)
		Lemongrass	Antimicrobial	Guerra-Rosas et al. (2017)
		Thyme oil	Antimicrobial	Ryu et al. (2018)
		Lemon myrtle	Antimicrobial	Buranasuksombat et al. (2011)
		Cinnamomum zeylanicum	Antimicrobial	Tian et al. (2016)
2.	Solid lipid NPs	*Eugenia caryophyllata*	Antimicrobial	Fazly Bazzaz et al. (2018)
		Melaleuca alternifolia	Insecticidal/ repellent	Clerici et al. 2018
		Frankincense and myrrh	Anticancer	Shi et al. (2012)
		Miconazole (drug)	Antifungal	Aljaeid and Hosny (2016)
3.	Liposome-mediated NPs	*Geranium maculatum*	Insecticidal	González et al. (2017)
		Nigella sativa	Anticancer	Shanmugam et al. (2017)
		Torreya grandis	Antibacterial	Wu et al. (2018)
4.	Micelles	Tea tree	Antimicrobial	Ganguly et al. (2018)
		Carvacrol and eugenol	Antibacterial	Gaysinsky et al. (2005)
5.	Chitosan-based NPs	*Gaultheria procumbens*	Antifungal	Kujur et al. (2017)
		Zataria multiflora	Antifungal	Mohammadi et al. (2015)
		Garlic	Antifungal	Yang et al. (2009)
6.	Strach-based NPs	Menthone, citral, and lavender oil	Antimicrobial	Qiu et al. (2017)
		Oregano oil	Antimicrobial	Pelissari et al. (2009)
		Cinnamon oil	Antimicrobial	Souza et al. (2013)
7.	Gum-based NPs	Carvacrol and thymol	Antibacterial	Guarda et al. (2011)
		Lemongrass and cinnamon oil	Antifungal	Maqbool et al. (2011)

high acceptability in food products, lipid-based encapsulation of EOs is gaining high attention among scientists worldwide for their use in the food system (Aditya and Ko 2015). Generally lipid-based encapsulation includes emulsions (nano- and micro-emulsions), solid lipid-based nanoparticles (SLBNs), liposome-mediated nanoparticles, micelles, etc.

2.2.1.1 Emulsions

The emulsions can be grouped into three major categories, viz., macro-, micro- and nanoemulsions, which are of major focus for food industries employing bioactives exhibiting low water solubility including essential oils (Xue 2015). Generally, surfactants are used for the preparation of emulsions; however, proteins and lipids have also been used. Preparation and characterization of essential oil loaded nanoemulsion incorporates different modern technologies such as homogenization, sonication, X-ray diffraction and scanning and transmission electron microscopy (Gupta et al. 2016; Solans et al. 2005).

2.2.1.2 Solid Lipid Nanoparticle (SLNs)

These are novel colloidal carriers developed as an alternative material to the polymers, where the solid lipids were involved. Solid-phase lipid nanoparticles are one of the most popular approaches in the field of food sciences for improvement of essential oils biocompatibility (Mukherjee et al. 2009; Saupe and Rades 2006). So far, several methods have been used for the preparation of SLNs, among which high-pressure homogenization, ultrasonication, solvent evaporations, spray drying, and emulsion-based preparations are the most common (Ekambaram et al. 2012).

2.2.1.3 Liposome as Nanocarriers of Bioactive Molecules

Liposome is one of the important systems formed by one or more phospholipids bilayer with an internal aqueous compartment. Due to their amphipathic nature, phospholipids are able to self-organize in the aqueous phase. On account of these properties, liposome-mediated encapsulation is useful for both the hydrophobic and hydrophilic components (El Asbahani et al. 2015). They are efficient carrier molecules for incorporating EOs and components, by improving their physicochemical properties, bioavailability, and solubility (Coimbra et al. 2011). Thin film evaporation, extrusions, sonication, freeze thaw, and saturation of gas-phase solution were the most commonly used methods in the preparation of liposome-mediated encapsulation (Sherry et al. 2013).

2.2.1.4 Micelles

Micelles are the supramolecular complexes consisting of polar and nonpolar materials containing different interwoven gelling agents (Rodríguez et al. 2016). Incorporation of essential oils with anticancerous activities into micelles matrix may be an important strategy to combat the rising incidence of cancers and other untreated diseases. Micelle sizes vary in the dimension from 2 to 20 nanometers based on experimental procedures. Commonly used techniques for micelle preparation are suspension, solvent removal through appropriate methods, dispersion, and dialysis (Hoar and Schulman 1943; Jones and Leroux 1999). An outline of preparation of different lipid-based nanoparticle has been presented in Fig. 2.1.

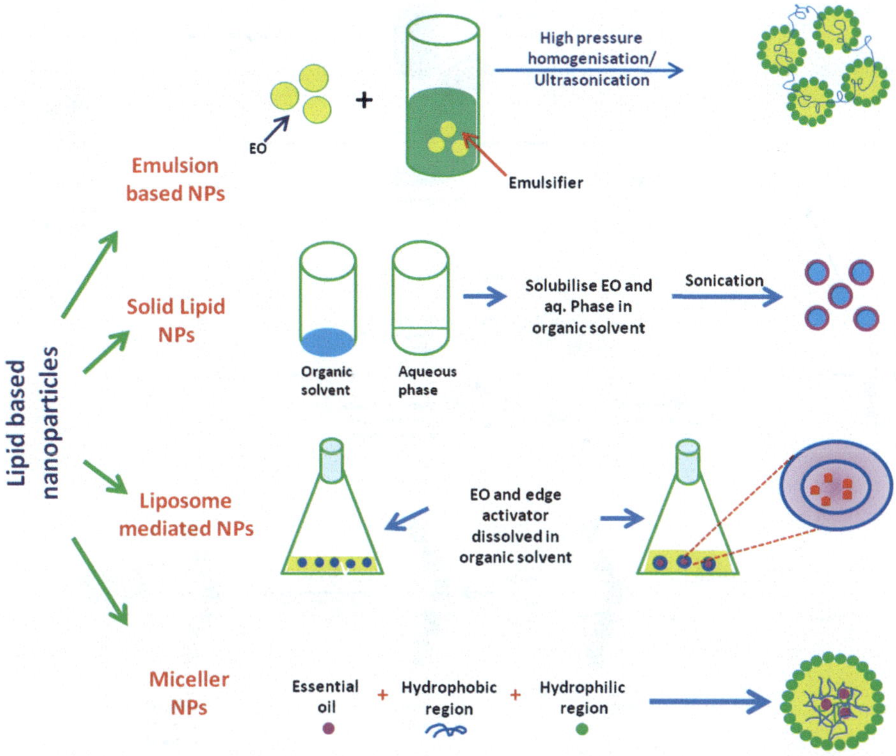

Fig. 2.1 An illustrated diagram showing methodology for the preparation of lipid-based nanoparticles

2.2.2 Polymer-Based Encapsulation of Essential Oils

Polymers including chitosan, starch, cellulose, plant gums, dextrin, and alginates are some of the main wall materials used to entrap essential oils into micro- or nano-sized matrix to prevent them from physical degradation and to enhance their antimicrobial efficiencies.

Starch is a natural and biodegradable polymer used as energy source by many plants. It is also an important polymer in nature. Starch and its modified forms maltodextrins, cyclodextrins, and dextrin whites are being used in the formation of essential oil loaded nanoparticle owing to their abundance, emulsifier property, high stability, and biocompatibility. To date, several starch-based nanoparticles have been developed to enhance the efficacy of bioactive components as evidenced by recent literatures (Shao et al. 2018; Hasani et al. 2018).

Chitosan is an important biopolymer mostly used for active encapsulation of essential oils and their components to enhance activity toward a wide range of pathogens and storage pests (Hosseini et al. 2013). In general, chitosan is isolated

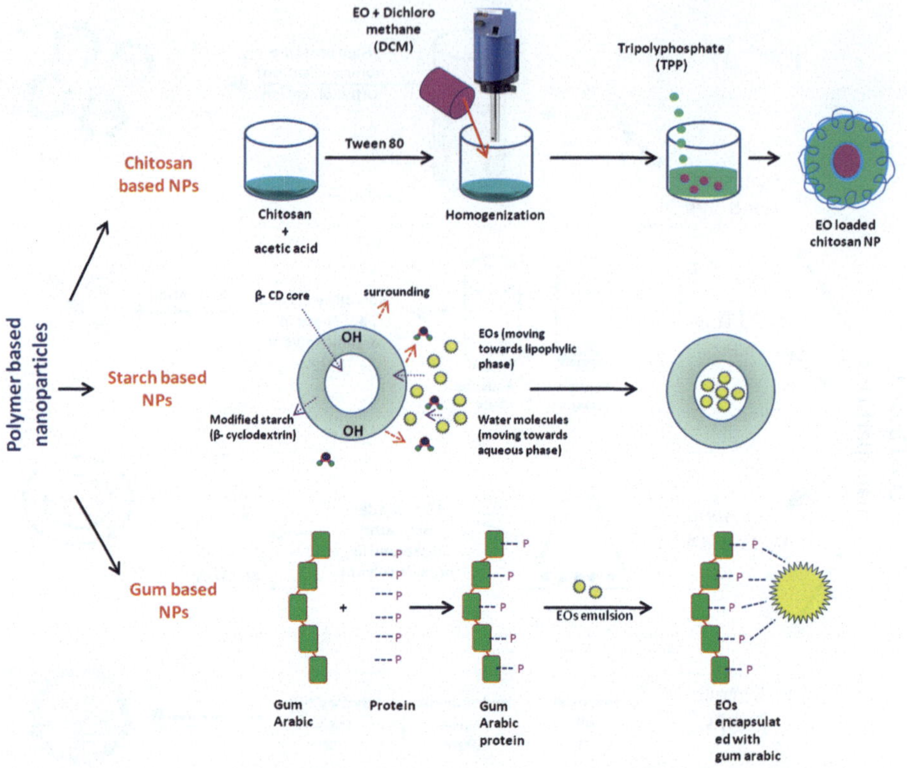

Fig. 2.2 Important polymer based nanoencapsluation methods for bioactive materials

from outer skeletal part of crustacean members, which contains different amines along with primary and secondary alcoholic groups for active binding with essential oil (Vishwakarma et al. 2016).

Gums are natural plant products serving as excellent wall materials for the encapsulation of important plant-derived bioactive components. The semipermeable character of gums helps in stabilization during the encapsulation process (Sarkar et al. 2012; Hamid Akash et al. 2015). In addition, some other polymers like cellulose, wax, gluten, alginate, xanthan, whey, etc. have also been utilized for essential oil encapsulation. A systemic representation on polymer-based nanoencapsulation process for bioactive materials is presented in Fig. 2.2.

2.3 Active Packaging of EO Nanoparticles as Food Protectant

Active packaging of EOs and their bioactive components is one of the important approaches used by food industries to protect food items from microbial and toxic metabolites contamination thereby maintaining their nutritional status (Coles et al.

2003). Utilization of essential oils through packaging provides protection against three major classes of external factors, viz., chemical, physical, and biological. Therefore, current researchers are focussing on modern delivery systems for encapsulation and release of essential oil vapors in a controlled manner. Since, food industries are facing numerous challenges in developing healthy, safe, and environmentally acceptable foods (Neethirajan and Jayas 2011), nanotechnology may be exploited successfully as one of the important tools to improve production and preservation of food, for extending shelf life (Roco 2002). Nanoencapsulation deals with the reduction in size of the coating materials up to 1 μm having the bioactive components enclosed in a tight matrix. The involvement of essential oil based nanotechnology for efficient food packaging supports the preservative potential in terms of controlled release, greater surface area to volume ratio and optimized interaction with food components. Another important point of food packaging is the selection of suitable active packaging polymers, which do not interact with the organoleptic properties of food viz. flavor, color, taste and texture. The most commonly used matrix materials for edible thin film packaging of food products are chitosan, gelatin, carrageenan, polylactic acid, and alginate. These edible films are commonly developed by carbohydrates, proteins, or lipids acting as a barrier of oxygen, carbon dioxide, and other soluble gases providing off-flavors and odors. Among different polymeric matrices, montmorillonite (MMT) has been used by some workers due to associated benefits of potential release of antimicrobial compounds and controlled permeability of the gas and water vapor inside the packaging matrix (Fabra et al. 2009).

Polymeric particles are generally of two types, i.e., nanocapsules and nanospheres. Incorporation of essential oils into the polymeric matrix provides significant antimicrobial and antioxidant properties for control of microbial contamination in the food products. Essential oils get conjugated with these polymeric materials and constitute a biocompatible nanoparticle. The antimicrobial property of developed packaging film greatly rely over the ratio of essential oil and polymeric matrix utilized. Encapsulation of essential oil enhances effective concentration of bioactive components in the food system especially in water-rich phases or solid-liquid interphases where microbial contamination occurs. For instance, nanoencapsulated terpenes and D-limonene showed significant efficacy against *Lactobacillus delbrueckii* and *Escherichia coli* (Donsì et al. 2011). Gaysinsky et al. (2005) performed nanoencapsulation of two important essential oil components eugenol and carvacrol with different nano-based surfactant for enhancing antimicrobial efficacy against pathogenic microbes. Citronellal, one of the major components of the *Eucalyptus citriodora* essential oil encapsulated in chitosan matrix, has shown better antifungal, antioxidant, and insect repellant activity with significant distribution and polydispersity (Ribeiro et al. 2014). Encapsulation of *Zataria multiflora* essential oil within chitosan nanoparticle significantly inhibited the growth and sporulation of *Botrytis cinerea* in the storage condition at 4 °C (Mohammadi et al. 2015). Essential oil loaded nanoemulsion delivery system is widely acceptable for practical application being stable, transparent and colloidal in nature. The use of oregano essential oil nanoemulsion with active preservative potential against

Listeria monocytogenes, *Salmonella typhimurium*, and *Escherichia coli* on fresh lettuce has recently been reported by Bhargava et al. (2015). Recent findings of mandarin essential oil enclosed within modified chitosan-based coating for beans have been investigated for large-scale future application (Severino et al. 2014; Donsì et al. 2015). The schematic representation of preparation and role of nanoparticle-based delivery system in food industries is represented in Fig. 2.3.

There are different delivery systems used to achieve maximum oil concentration and efficient antimicrobial activity in food system. The most common delivery systems are emulsions, suspensions, and liposomes which are colloidal particles associated with essential oils as internal core materials, acting as good delivery system with potential sustained release properties and significant activity against fungi and other microbes (Gomes et al. 2011). Essential oil loading within different biopolymeric matrices following a variety of fabrication techniques, viz., interfacial polymerization, solvent evaporation, and emulsification, has been demonstrated to maintain good quality and stability of nanoparticles in storage practices (Akrachalanont 2008).

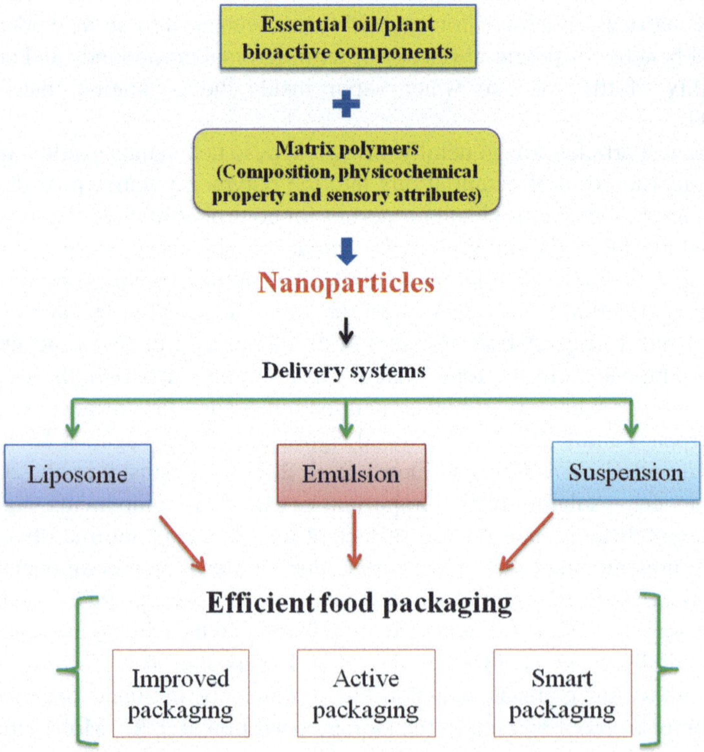

Fig. 2.3 Brief outline of synthesis and application of essential oil-based nanoparticles in food sector

2.4 Mode of Action of Nanoparticles

Recently, there has been an increasing interest for the use of plant essential oils as fungicide, bactericide, insecticides, and rodenticides to control fungal, bacterial, and insect- and rodent-mediated deterioration of food items. However, volatile nature of essential oils and physico-chemical degradation by abiotic factors limits their long term application. Therefore, use of nano based delivery system may provide the efficient way to control the pests by involving different mode of action on cells leading to cell lysis and subsequent death. The smaller size of nanoparticles allows them to traverse within the cells culminating into loss of cellular activity. Furthermore, the lipophilic chemical components of essential oils can also easily cross the plasma membrane of fungi and insects and cause different enzymatic alterations and biochemical dysfunction inside their body (Lee et al. 2004). Inhibition of fungal membrane ergosterol by different essential oils has been actively reported by Tian et al. (2012). The active ingredients of essential oils can be used as potent repellant, fumigant, antifeedant, and insecticidal agent. In general, the terpenoids and phenolic components of essential oils target nervous system of insect by interfering with functions of enzymes such as acetyl choline esterase (AChE) or neurotransmitters such as octopamine and γ-aminobutyric acid (GABA). In insects, the toxic effects on octopamine neurotransmitter, GABA gated chloride channel and cytochrome P_{450} monooxygenase may ultimately cause hypersensitivity, convulsions and paralysis of organs (Enan 2001). Besides their lipophilic nature, the bioactivity of essential oil also depends on different functional groups (nature and position), volatility, and molecular weight of compounds. Acetate-derived methyl groups and double-bonded benzene ring of eugenol, safrole, and limonene actively take part in knockdown of different enzymes and fumigant toxicity of *Spodoptera litura*, *Callosobruchus chinensis*, *Tribolium castaneum*, and other storage insects. Lower solubility, higher volatility and oxidation tendency, are the major problem in practical application of essential oil for post harvest pest management. These problems may be effectively resolved to a greater extent by nanoencapsulated essential oil-based pesticides (nanopesticides) or nanoformulated essential oil. Christofoli et al. (2015) reported on *Zanthoxylum rhombifolium* essential oil loaded nanospheres against *Bemisia tabaci.* Silver nanoparticles loaded with bioactive principles of *Aristolochia indica* showed maximum larvicidal, repellant, and cytotoxic effect on *Helicoverpa armigera* (Siva and Kumar 2015). Lead and silver nanoparticle of *Avicennia marina* plant extract exhibited prominent toxicity against *Sitophilus oryzae* (Sankar and Abideen 2015). The list of chemical and biological nanoparticles used for broad management of insect pests is described in Table 2.2.

Table 2.2 Important chemical and biological nanoparticles against fungi, bacteria, insects, and rodents

Types of nanoparticles	Targeted pests	References
Silver nanoparticles	*Spodoptera litura* and *Achaea janata*	Yasur and Rani (2015)
Zinc nanoparticle	*Aphis nerii*	Rouhani et al. (2012)
Silver nanoparticle	*Pseudomonas aeruginosa*	Salomoni et al. (2017)
Chitosan nanoparticle	*Spodoptera litura*	Chandra et al. (2013)
Silver nanoparticle with *Aristolochia indica*	*Helicoverpa armigera*	Siva and Kumar (2015)
PEG-based garlic essential oil nanoparticle	*Tribolium castaneum*	Yang et al. (2009)
Chitosan nanoparticle	*Callosobruchus maculatus*	Sahab et al. (2015)
Silver nanoparticle	*Sitophilus oryzae*	Zahir et al. (2012)
Illicium verum essential oil loaded chitosan nanoparticle	*Aspergillus flavus* and other storage fungi and aflatoxin B_1	Dwivedy et al. (2018)
Mentha piperita essential oil loaded in chitosan-cinnamic acid nanogel	*Aspergillus flavus*	Beyki et al. (2014)
Pectin/papaya puree/ cinnamaldehyde nanoemulsion	*Escherichia coli*, *Staphylococcus aureus*, *Listeria monocytogenes*, and *Salmonella enterica*	Otoni et al. (2014)
Clove/cinnamon loaded nanoemulsion	*Bacillus subtilis*, *Salmonella typhimurium*, and *Staphylococcus aureus*	Zhang et al. (2017)
Mustard oil microemulsion	*Escherichia coli*	Gaysinsky et al. (2008)
Eugenol loaded nanoemulsion	*Staphylococcus aureus*	Ghosh et al. (2014)
Encapsulated terpene mixture and D-limonene nanoemulsion	*Lactobacillus delbrueckii*, *Saccharomyces cerevisiae*, *Escherichia coli*	Donsì et al. (2011)
Santolina insularis essential oil encapsulated with liposome	Herpes simplex virus type-1	Valenti et al. (2001)

2.5 Factors Controlling the Stability of Nanoparticles in Food System

The rising disciplines of nanotechnology and their applications in food and agricultural industries are of major focus in the current generation. Varied applications of nanotechnology are expanding rapidly with a multitude of potential factor for nanoparticle stability and enhancement of food shelf life by maintaining the flavor properties and ultimately improved nutritional value of food products during storage (Chellaram et al. 2014). Nanoencapsulation provides protection to bioactive food ingredients by eradication of incompatibilities and masking the unpleasant taste or odor. Besides having effective applications, there are some controlling factors which affect their physical and chemical stability.

2.5.1 Free Energy of Different Phases

Nonpolar interaction of lipophilic oil with hydrophilic component increases the thermodynamic instability of nanoemulsion depending upon the free energy status (Wooster et al. 2016). Therefore, changes in physical characteristics of nanoemulsions ultimately result in phase alteration (Chung and McClements 2018). Thus, nanoemulsion characterization is a major aspect not only in the early stage but also during long-term storage. Generally, storage practices occur in varied environmental conditions that may be quite useful to enhance the stability and activity of the nanoemulsions. Furthermore, under certain conditions, nanoemulsions may become unstable due to gravitational separations resulting primarily in the differences in densities as observed by densitometric studies (Chung and McClements 2018).

2.5.2 Droplet Aggregation and Particle Size

Nano-sized droplets in emulsions are always in constant collision reaction because of gravitational force, Brownian motion, and other mechanical forces (McClements 2015), favoring the aggregations of nano-droplets. Droplets may get associated with each other without interfering their original size, and this process is known as flocculation. In addition, a number of nano-droplets may come across each other and form a large droplet by the process known as coalescence.

However, for effective commercial applications, it is necessary to avoid nano-droplet aggregation as much as possible. For this, attractive forces, viz., hydrophobic and van der Waals forces, may be compensated by the repulsive forces, viz., electrostatic and steric actions between the nano-droplets. In addition, the decreased nanoparticle size is correlated with reduction of nano-droplet aggregation (Degner et al. 2014). According to McClements (2004), along homogenization, differences in soluble phases and emulsifier concentration and types are the governing factors regulating the size of the droplets. The stability of nanoencapsule suspension can also be achieved by drying the nano-suspension (Nakagawa et al. 2011).

2.5.3 Emulsifier Type

The stability and size of nanoemulsion suspension formed after homogenization are greatly influenced by the nature of emulsifier used. McClements (2004) studied protein and lipid droplet interaction which are very much susceptible to changes in ionic strength and pH as their major stabilization by repulsing forces. In contrast to this, Qian et al. (2011) have described the coating of lipid droplets into polysaccharide and non-ionic surfactants that are not much susceptible to these parameters such as steric repulsion.

2.5.4 *Ionic Strength and pH*

Ionic strength and pH of aqueous phase are two other key factors which have leading impact over stability as well as properties of nanoemulsion by influencing their electrostatic interactions. Ionic concentration and its type present in the aqueous phase can influence the strength of electrostatic interactions by influencing the effect of ion binding, ion bridging, and electrostatic screening. Besides ionic concentration, pH of nano-suspension also has considerable impact over strength of electrostatic interactions depending on charge potential on nanoparticle surface (McClements 2004).

Nanoparticles consisting of protein-coated fat droplets have strong affinity toward flocculation when ionic strength of nano-suspension exceeds beyond a certain level because of electrostatic screening and ion binding effects (De La Fuente et al. 1998). Moreover, addition of salt to the nanoemulsion prior to or after homogenization may influence the stability and aggregation of nanoparticles (Kim et al. 2005). Similar effects are also observed when pH value of nanoemulsion is near to their isoelectric point (Demetriades and McClements 1998). As mentioned above, ionic strength and pH value of nano-suspension influence the aggregation and stability of nano-droplets. Besides this, some other properties like texture of emulsion and their optical properties may also get affected.

2.5.5 *Thermal Processing*

Thermal processing is another factor that governs nanoemulsion stability by influencing the droplet aggregation over and above visual and textural properties of nano-suspension. Croguennec et al. (2004) have reported that adsorbed globular proteins may experience an irreversible change in their structure and chemical reactivity on heating beyond their thermal denaturation. Moreover, Donato et al. (2007) have reported the formation of protein aggregates after heat treatments.

2.6 Nanoparticles as Active Biosensor for Detection of Food Contaminants (Chemicals and Food-Borne Pathogens)

A sensor is a measurement system consisting of specific probe to detect particular species or element at trace levels. Biosensor have been used for detection of different biological organisms such as viruses, bacteria, proteins, nucleic acids and pathogens (Pérez-López and Merkoçi 2011). Biosensors contain a bioreceptor and transducer which interact with specific analyte molecule, generating an electrical signal. Bioreceptor consists of living systems viz. cells, tissue, enzymes, proteins, nucleic acids and antibodies which are utilized for the biochemical adhesion property. Nowadays, nanobiosensing technology is being utilized in food safety analysis to detect various sensitive contaminants in food materials. A nanobiosensor is also called as a second-generation biosensor due to the involvement of ultrasensitive nanoparticles and their effective transducers (Sagadevan and Periasamy 2014). The major advantages of using nanomaterials in biosensing are the combined effect on biological receptor molecules and higher surface to volume ratio (Kumar et al. 2012). Among different synthesized nanoparticles, metal nanoparticles, carbon nanoparticles, semiconductor quantum nanoparticles, and magnetic nanoparticles have been used for detection of food allergens. Metal nanoparticles, basically colloidal gold, have been successfully used in DNA biosensing based on their amount of immobilized biomolecules in an active sensor (Cai et al. 2002). There are different types of nanotubes, viz., one-dimensional nanostructures (1-D), nanotubes with single or multichannel carbon tape, and carbon nanowires, based on the conducting polymer and biochemical reaction routes. The organic nanosensors have been actively applied for detection of organophosphate pesticide contamination in fruits and other stored products as well as their solubility and residual toxicity in different food products. Interestingly these nanobiosensors have marked advancement for detection of mycotoxins and different mycotoxigenic fungi in different food products (Li et al. 2012). Aflatoxin contamination can be measured on pre-harvested and post-harvested food products at a very low concentration, i.e., 0.3 pg/mL by different nanocomposites (Gan et al. 2013). Yotova et al. (2013) have developed a nanosensor with modified tyrosinase enzyme for detection of toxigenic fungi. Gold nanorods (GNRs) have active sensing capacity for AFB_1 detection up to 0.04 ppb in food samples (Xu et al. 2013). Modified gold nanoparticles with electrochemiluminescent aptamers have been used for detection of ochratoxin A. AF-oxidase enzyme linked with modified carbon nanotubes and poly-o-phenylenediamine integrated with gold electrode has the ability to detect sterigmatocystin concentration up to 3 ng/mL (Chen et al. 2009). Similarly, immuno-chromatographic biosensors with attached monoclonal antibodies have been used for rapid detection of zearalenone in corn samples (Shim et al. 2009). Maragos (2012) has developed colloidal gold nanoparticles with interferometry principle for detection of deoxynivalenol in wheat.

2.7 Application of Nanoparticles in Different Food Sectors

In today's world, the applications of nanoscience and nanotechnology in food materials have contributed a major status for nutrition and health of consumers. Most of the nanoparticles are colloidal in nature, and stability is achieved by van der Waals forces and steric stabilization provided by core materials and adsorbing matrix polymers. In the food industry, nanotechnology have both "top-down" and "bottom-up" approaches with smart delivery of nutrients and proteins, rapid identification of contaminants, and encapsulation of essential oil with adsorbing matrices (Verma et al. 2009). Globally, the overall sales of nanotechnology products have been observed to increase from US$ 150 million to US$ 860 million in between 2002 and

2004. Application of essential oil in food matrices as biopreservatives and natural antioxidant is one of the most important aspects in the current generation. This rapidly developing encapsulation technology have an ultimate focus on food processing, packaging, transportation, shelf life enhancement, and bioavailability of nutrient compounds. The microencapsulated essential oils having antimicrobial agents are used in efficient smart packaging, an effective approach in food industries (de Barros Fernandes et al. 2014). Use of nanoparticles for the improvement of food processing involves different biological aspects such as effective preservative potential, targeted and controlled release of bioactive components, high absorption rate related to greater surface area, high availability, and improved organoleptic properties. Active antimicrobial effect, improved barrier properties and biosensing properties have been applied in nanopackaging of food materials. The active use of nanosensors for pathogenic microbes and toxin detection in stored foods is an important aspect of nanotechnology during transportation. Nanoparticles are efficient due to selective binding and easy removal of pathogens from food surface (Duncan 2011). Industrial application of encapsulated essential oil in food systems is a very limited approach until now. Wu et al. (2012) have reported efficient antimicrobial property of zein nanoparticle loaded with thymol and carvacrol without affecting organoleptic properties of food commodities. Nano-formulation of *Artemisia arborescens* essential oil with solid lipid nanoparticles allowed controlled release of vapor and served as an efficient ecological pesticide (Lai et al. 2006). A number of essential oils and their bioactive components have been incorporated into edible film exhibiting better efficacy against food-borne pests and pathogens. Khalili et al. (2015) confirmed potential efficacy of thyme essential oil encapsulated within chitosan-benzoic acid nanogel against growth and aflatoxin secretion by *Aspergillus flavus* in stored foods. Hence efficient food packaging, formulation, and delivery can be assisted by several compatible nanomaterials. A comprehensive diagram representing the application of nanoparticles in different food sector is shown in Fig. 2.4.

2.8 Safety Issues Associated with Application of Nanotechnology in Food Packaging/Food Preservation

Before discussing the possible consequences of nanomaterial ingestion via food, it is very important to recall that our food naturally comprises of several nanorange components. Several forms of biopolymers such as proteins, carbohydrates, and fats are present in the nanorange in variety of food items (Magnuson et al. 2011; Cockburn et al. 2012). Even the digestive processes (mechanical/chemical) performed in the gastrointestinal tract leads to the conversion of complex food structures into simple forms of mono−/disaccharides, fatty acids, or amino acids/peptides, which can be included under nano−/micro-material range (Martirosyan and Schneider 2014). Such nanomaterials formed in the gastrointestinal tract are subjected to absorption or translocation into the blood/lymph through the gut epi-

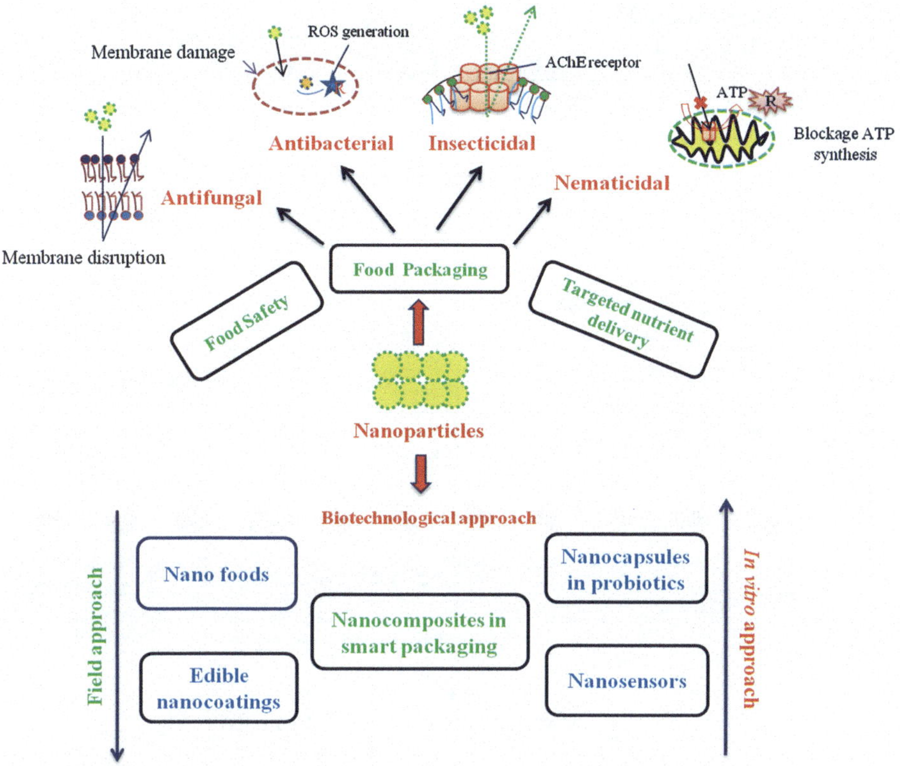

Fig. 2.4 Application of nanoparticles in different food sectors

thelia (Powell et al. 2010). Thus, it should be made clear that the gut lumen and epithelia are well exposed to various nanomaterials ingested orally via food.

The major concern of nanomaterial usage in food packaging is “leaching out” of active ingredients from the packaging materials into the food items. The incorporation of nanomaterials in the food as stabilizers, anticaking agents, in order to improve the shelf life, odour, taste and aroma, will also increase the chances of oral entrance (Laloux et al. 2017). In this context, there are several issues, most importantly the very small size of nanomaterials favoring the greater absorption through the intestinal/gut epithelia (Martirosyan and Schneider 2014). If there is a possibility of internalization and retention in cells or tissues, the nanomaterials could enter into the food chain. After entry, the nanomaterials may get coated with lipids/proteins as such forming a “corona” which facilitates the absorption, thus making the nanomaterial biofunctional (Bellmann et al. 2015). This corona stage is not stable and is subjected to alterations under changing macromolecules and environmental conditions it encounters. This property complicates the expected behavior of nanomaterials as well as the absorption/metabolic profile of food components inside the gastrointestinal tract. Such unknown effects associated with the nanomaterial behavior inside the gut have been termed as the “Trojan horse effect” (Park et al.

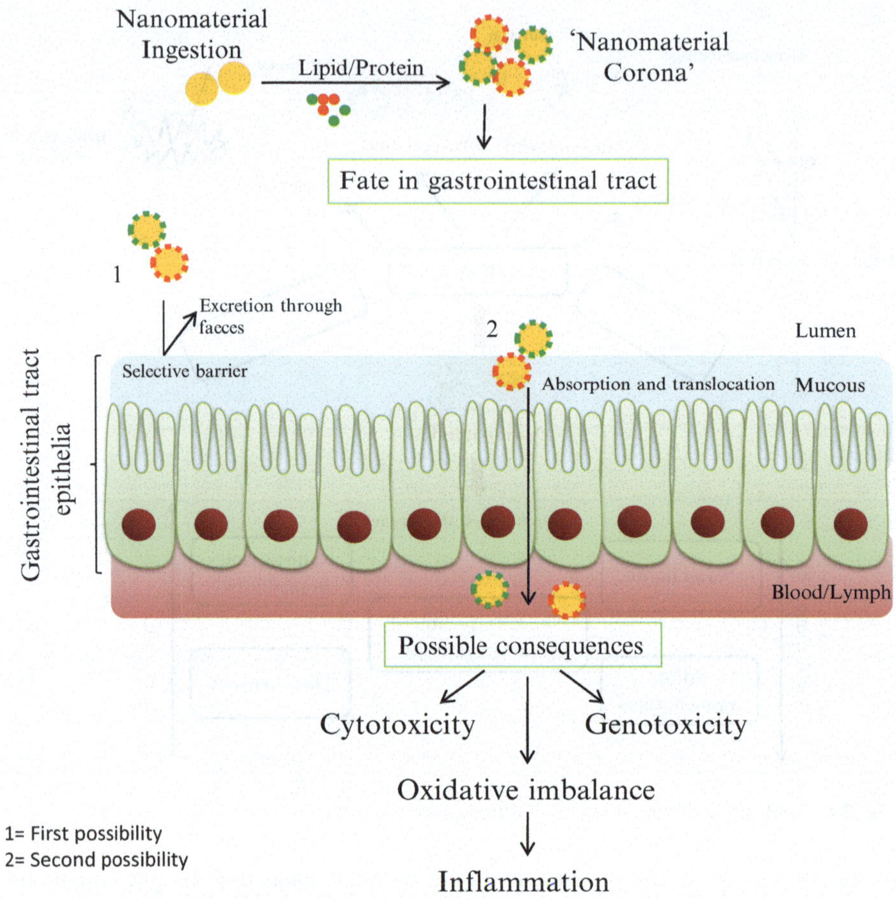

Fig. 2.5 The fate and possible effects of nanomaterial ingestion

2010; Souza et al. 2018). The fate and the potential consequences of nanomaterial ingestion via food have been represented in Fig. 2.5.

The gastrointestinal tract has a mucous layering which functions as mechanical barrier to foreign particles entering the tract. The nanomaterial may get excreted via feces if not allowed to pass through the epithelia. However, if it gets absorbed/translocated into the blood/lymph, it has the passage to stream through different compartments of the body (Bellmann et al. 2015). The consequences of nanomaterial acquisition in different body parts have been demonstrated in vitro (Owens III and Peppas 2006; Corbalan-Penas 2010; Athinarayanan et al. 2014). Factors governing the fate of nanomaterials inside the body include dose, particle size, shape, charge, solubility, reactivity, and surface coating. The lipid bilayer of the cell membrane has been speculated as the primary target of nanorange materials. Pores formed in the membrane due to nanomaterials lead to cytotoxicity. The nanomaterials have been shown to induce ROS (reactive oxygen species) and RNS (reactive nitrogen species; nitrosative stress) production (Athinarayanan et al. 2014). The oxidative imbalance inside the cell has also been shown to activate the pro-inflammatory cytokines and inflammasome-mediated inflammation (Dankovic et al. 2007; Moos et al. 2011). Genotoxicity induced by nanomaterial interaction could result in direct interaction with genomic DNA or indirectly through ROS generation (Singh et al. 2017).

In conclusion, it is essential to perform safety assessment of nanomaterials before their application in food packaging or processing. Cockburn et al. (2012) have provided a five-step systematic approach in this connection. Therefore, definite description as well as the comprehensive characterization of the nanomaterials is vital, followed by risk evaluation through utilization of "decision tree," whether the properties favor the need of detailed investigations on toxicological testing or not.

2.9 Future Prospective

In the recent past, there is an increasing interest toward the use and incorporation of nanometric structures in food industries for active packaging, processing, or preservation of quality. Therefore, the scientific community has focused greatly on the development of nanoscale materials that could fulfill changing world food demands. Nanobiotechnology in food packaging has immense potential to preserve food quality and nutritional profile by serving as antimicrobials and barriers against UV rays, gas, and moisture. To date, various methods have been proposed to encapsulate the bioactive compounds in an array of nano-vehicles to facilitate the stability as well as efficacy. Although research toward nanopackaging has gained momentum, still a lot of questions remained unanswered regarding their fate in the biological systems, environmental impacts, consumer acceptance, effects over sensory profile, and the cost to benefit ratio. Moreover, consumers face hesitation in accepting the incorporation of nanotechnology in the food system. Nanosensors incorporated in plastic packaging could be very much effective against detection of food contamination and spoilage (Thiruvengadam et al. 2018). Rigorous research over safety evaluation together with efforts for increasing the consumer's acceptance should be in priority to further explore the application of nanotechnology in food industries.

Acknowledgments SD and AKC are thankful to the Council of Scientific and Industrial Research (CSIR), New Delhi, for providing research fellowship. Authors are thankful to the Head, CAS in Botany and DST-FIST, Banaras Hindu University, for laboratory facilities.

References

Aditya NP, Ko S (2015) Solid lipid nanoparticles (SLNs) delivery vehicles for food bioactives. RSC Adv 5(39):30902–30911

Akrachalanont P (2008) Preparation and evaluation of liposome containing clove oil. Doctoral Dissertation, Thesis at Silpakorn University, 2008. http://www.thapra.lib.su.ac.th/objects/thesis/fulltext/snamcn/Pilaslak—Akrachalanont/Fulltext.pdf, XP002669247

Aljaeid BM, Hosny KM (2016) Miconazole-loaded solid lipid nanoparticles: formulation and evaluation of a novel formula with high bioavailability and antifungal activity. Int J Nanomedicine 11:441

Athinarayanan J, Periasamy VS, Alsaif MA, Al-Warthan AA, Alshatwi AA (2014) Presence of nanosilica (E551) in commercial food products: TNF-mediated oxidative stress and altered cell cycle progression in human lung fibroblast cells. Cell Biol Toxicol 30:89–100. https://doi.org/10.1007/s10565-014-9271-8

Bellmann S, Carlander D, Fasano A, Momcilovic D, Scimeca JA, Waldman WJ, Gombau L, Tsytsikova L, Canady R, Pereira DI, Lefebvre DE (2015) Mammalian gastrointestinal tract parameters modulating the integrity, surface properties, and absorption of food-relevant nanomaterials. Wiley Interdiscip Rev Nanomed Nanobiotechnol 7(5):609–622

Beyki M, Zhaveh S, Khalili ST, Rahmani-Cherati T, Abollahi A, Bayat M, Mohsenifar A (2014) Encapsulation of *Mentha piperita* essential oils in chitosan–cinnamic acid nanogel with enhanced antimicrobial activity against Aspergillus flavus. Ind Crop Prod 54:310–319

Bhargava K, Conti DS, da Rocha SR, Zhang Y (2015) Application of an oregano oil nanoemulsion to the control of foodborne bacteria on fresh lettuce. Food Microbiol 47:69–73

Bourne M (2002) Food texture and viscosity: concept and measurement. Academic Press, New York

Buranasuksombat U, Kwon YJ, Turner M, Bhandari B (2011) Influence of emulsion droplet size on antimicrobial properties. Food Sci Biotechnol 20(3):793–800

Burt S (2004) Essential oils: their antibacterial properties and potential applications in foods—a review. Int J Food Microbiol 94(3):223–253

Cai H, Wang Y, He P, Fang Y (2002) Electrochemical detection of DNA hybridization based on silver-enhanced gold nanoparticle label. Anal Chim Acta 469(2):165–172

Chandra JH, Raj LA, Namasivayam SKR, Bharani RA (2013) Improved pesticidal activity of fungal metabolite from Nomuraea rileyi with chitosan nanoparticles. In: Advanced Nanomaterials and Emerging Engineering Technologies (ICANMEET), September 2013, Chennai, India 387–390

Chellaram C, Murugaboopathi G, John AA, Sivakumar R, Ganesan S, Krithika S, Priya G (2014) Significance of nanotechnology in food industry. APCBEE Proc 8:109–113

Chen J, Li S, Yao D, Liu D (2009) Detection of sterigmatocystin based on the novel aflatoxin-oxidase/chitosan-single-walled carbon nanotubes/poly-o-phenylenediamine modified electrode. Chinese J Biotechnol 25(12):2029–2035

Christofoli M, Costa ECC, Bicalho KU, de Cássia Domingues V, Peixoto MF, Alves CCF, Araújo WL, de Melo Cazal C (2015) Insecticidal effect of nanoencapsulated essential oils from *Zanthoxylum rhoifolium* (Rutaceae) in *Bemisia tabaci* populations. Ind Crop Prod 70:301–308

Chung C, McClements DJ (2018) Characterization of physicochemical properties of nanoemulsions: appearance, stability, and rheology. In: Nanoemulsions. Academic Press, Cambridge, pp 547–576

Clerici DJ, de Souza ME, Quatrin PM, Vianna Santos RC, Costa-Leonardo AM, Ourique AF, Raffin RP, Marques CT, de Camargo Dietrich CRR, de Silva Gündel S, Klein B (2018) Effects of nanostructured essential oils against subterranean termites (Coptotermes gestroi). J Appl Entomol Z 142(4):406–412

Cockburn A, Bradford R, Buck N, Constable A, Edwards G, Haber B, Hepburn P, Howlett J, Kampers F, Klein C, Radomski M (2012) Approaches to the safety assessment of engineered nanomaterials (ENM) in food. Food Chem Toxicol 50(6):2224–2242

Coimbra M, Isacchi B, van Bloois L, Torano JS, Ket A, Wu X, Bilia R (2011) Improving solubility and chemical stability of natural compounds for medicinal use by incorporation into liposomes. Int J Pharm 416(2):433–442

Coles R, McDowell D, Kirwan MJ (2003) Food packaging technology (vol. 5). CRC Press, Florida

Corbalan-Penas JJ (2010) Synthetic amorphous silica nanoparticles induced human endothelial cell inflammation and platelet aggregation: role of NF-jB and reactive nitrogen species. Doctoral Dissertation, University of Dublin, Trinity College

Croguennec TT, O'Kennedy B, Mehra R (2004) Heat-induced denaturation/aggregation of β-lactoglobulin A and B: kinetics of the first intermediates formed. Int Dairy J 14(5):399–409
Damalas CA, Eleftherohorinos IG (2011) Pesticide exposure, safety issues, and risk assessment indicators. Int J Environ Res Pub Health 8(5):1402–1419
Dankovic D, Kuempel E, Wheeler M (2007) An approach to risk assessment for TiO2. Inhal Toxicol 19:205–212
de Barros Fernandes RV, Borges SV, Botrel DA (2014) Gum arabic/starch/maltodextrin/inulin as wall materials on the microencapsulation of rosemary essential oil. Carbohydr Polym 101:524–532
Degner BM, Chung C, Schlegel V, Hutkins R, McClements DJ (2014) Factors influencing the freeze-thaw stability of emulsion-based foods. Compr Rev Food Sci Food Saf 13(2):98–113
De La Fuente J, Rodríguez M, Redondo M, Montero C, García-García J, Méndez L, Ramos E (1998) Field studies and cost-effectiveness analysis of vaccination with Gavac™ against the cattle tick Boophilus microplus. Vaccine 16(4):366–373
Demetriades K, McClements DJ (1998) Influence of pH and heating on physicochemical properties of whey protein-stabilized emulsions containing a nonionic surfactant. J Agric Food Chem 46(10):3936–3942
Donato L, Guyomarc'h F, Amiot S, Dalgleish DG (2007) Formation of whey protein/κ-casein complexes in heated milk: preferential reaction of whey protein with κ-casein in the casein micelles. Int Dairy J 17(10):1161–1167
Donsì F, Annunziata M, Sessa M, Ferrari G (2011) Nanoencapsulation of essential oils to enhance their antimicrobial activity in foods. LWT - Food Sci Technol 44(9):1908–1914
Donsì F, Marchese E, Maresca P, Pataro G, Vu KD, Salmieri S, Lacroix M, Ferrari G (2015) Green beans preservation by combination of a modified chitosan based-coating containing nanoemulsion of mandarin essential oil with high pressure or pulsed light processing. Postharvest Biol Tecnol 106:21–32
Duncan TV (2011) Applications of nanotechnology in food packaging and food safety: barrier materials, antimicrobials and sensors. J Colloid Interface Sci 363(1):1–24
Dwivedy AK, Singh VK, Prakash B, Dubey NK (2018) Nanoencapsulated *Illicium verum* Hook. f. essential oil as an effective novel plant-based preservative against aflatoxin B1 production and free radical generation. Food Chem Toxicol 111:102–113
Ekambaram P, Sathali AAH, Priyanka K (2012) Solid lipid nanoparticles: a review. Sci Rev Chem Commun 2(1):80–102
El Asbahani A, Miladi K, Badri W, Sala M, Addi EA, Casabianca H, El Mousadik A, Hartmann D, Jilale A, Renaud FNR, Elaissari A (2015) Essential oils: from extraction to encapsulation. Int J Pharm 483(1–2):220–243
Enan E (2001) Insecticidal activity of essential oils: octopaminergic sites of action. Comp Biochem Physiol 130(3):325–337
Fabra MJ, Jiménez A, Atarés L, Talens P, Chiralt A (2009) Effect of fatty acids and beeswax addition on properties of sodium caseinate dispersions and films. Biomacromolecules 10(6):1500–1507
Farkas J (2007) Physical methods of food preservation. In: Food microbiology: fundamentals and frontiers, 3rd edn. American Society of Microbiology, Washington, DC, pp 685–712
Fazly Bazzaz BS, Khameneh B, Namazi N, Iranshahi M, Davoodi D, Golmohammadzadeh S (2018) Solid lipid nanoparticles carrying Eugenia caryophyllata essential oil: the novel nanoparticulate systems with broad-spectrum antimicrobial activity. Lett Appl Microbiol 66(6):506–513
Gan N, Zhou J, Xiong P, Hu F, Cao Y, Li T, Jiang Q (2013) An ultrasensitive electrochemiluminescent immunoassay for Aflatoxin M1 in milk, based on extraction by magnetic graphene and detection by antibody-labeled CdTe quantum dots-carbon nanotubes nanocomposite. Toxins 5(5):865–883
Gaysinsky S, Davidson PM, Bruce BD, Weiss J (2005) Growth inhibition of Escherichia coli O157: H7 and *Listeria monocytogenes* by carvacrol and eugenol encapsulated in surfactant micelles. J Food Prot 68(12):2559–2566

Gaysinsky S, Davidson PM, McClements DJ, Weiss J (2008) Formulation and characterization of phytophenol-carrying antimicrobial microemulsions. Food Biophys 3(1):54–65

Ghosh V, Mukherjee A, Chandrasekaran N (2014) Eugenol-loaded antimicrobial nanoemulsion preserves fruit juice against, microbial spoilage. Colloid Surface B Biointerfaces 114:392–397

Gomes C, Moreira RG, Castell-Perez E (2011) Poly (DL-lactide-co-glycolide)(PLGA) nanoparticles with entrapped trans-cinnamaldehyde and eugenol for antimicrobial delivery applications. J Food Sci 76(2):N16–N24

González JOW, Jesser EN, Yeguerman CA, Ferrero AA, Band BF (2017) Polymer nanoparticles containing essential oils: new options for mosquito control. Environ Sci Pollut Res 24(20):17006–17015

Guarda A, Rubilar JF, Miltz J, Galotto MJ (2011) The antimicrobial activity of microencapsulated thymol and carvacrol. Int J Food Microbiol 146(2):144–150

Guerra-Rosas MI, Morales-Castro J, Cubero-Márquez MA, Salvia-Trujillo L, Martín-Belloso O (2017) Antimicrobial activity of nanoemulsions containing essential oils and high methoxyl pectin during long-term storage. Food Control 77:131–138

Gupta A, Eral HB, Hatton TA, Doyle PS (2016) Nanoemulsions: formation, properties and applications. Soft Matter 12(11):2826–2841

Hamid Akash MS, Rehman K, Chen S (2015) Natural and synthetic polymers as drug carriers for delivery of therapeutic proteins. Polym Rev 55(3):371–406

Hasani S, Ojagh SM, Ghorbani M (2018) Nanoencapsulation of lemon essential oil in Chitosan-Hicap system. Part 1: study on its physical and structural characteristics. Int J Biol Macromol 115:143–151

Hashem AS, Awadalla SS, Zayed GM, Maggi F, Benelli G (2018) *Pimpinella anisum* essential oil nanoemulsions against *Tribolium castaneum*—insecticidal activity and mode of action. Environ Sci Pollut Res 25(19):18802–18812

Hoar TP, Schulman JH (1943) Transparent water-in-oil dispersions: the oleopathic hydro-micelle. Nature 152(3847):102

Hosseini SF, Zandi M, Rezaei M, Farahmandghavi F (2013) Two-step method for encapsulation of oregano essential oil in chitosan nanoparticles: preparation, characterization and in vitro release study. Carbohydr Polym 95(1):50–56

IARC Working Group on the Evaluation of the Carcinogenic Risk of Chemicals to Humans. (1993) IARC monographs on the evaluation of the carcinogenic risk of chemicals to humans, vol 35, World Health Organization

in't Veld JH (1996) Microbial and biochemical spoilage of foods: an overview. Int J Food Microbiol 33(1):1–18

Inetianbor JE, Yakubu JM, Ezeonu SC (2015) Effects of food additives and preservatives on man–a review. Asian J Sci Technol 6(2):1118–1135

Isman MB (2000) Plant essential oils for pest and disease management. Crop Prot 19(8–10):603–608

Jones MC, Leroux JC (1999) Polymeric micelles–a new generation of colloidal drug carriers. Eur J Pharm Biopharm 48(2):101–111

Khalili ST, Mohsenifar A, Beyki M, Zhaveh S, Rahmani-Cherati T, Abdollahi A, Bayat M, Tabatabaei M (2015) Encapsulation of Thyme essential oils in chitosan-benzoic acid nanogel with enhanced antimicrobial activity against *Aspergillus flavus*. LWT-Food Sci Technol 60(1):502–508

Kim SI, Roh JY, Kim DH, Lee HS, Ahn YJ (2003) Insecticidal activities of aromatic plant extracts and essential oils against *Sitophilus oryzae* and *Callosobruchus chinensis*. J Stored Prod Res 39(3):293–303

Kim HJ, Decker EA, McClements DJ (2005) Influence of protein concentration and order of addition on thermal stability of β-lactoglobulin stabilized n-hexadecane oil-in-water emulsions at neutral pH. Langmuir 21(1):134–139

Kujur A, Kiran S, Dubey NK, Prakash B (2017) Microencapsulation of *Gaultheria procumbens* essential oil using chitosan-cinnamic acid microgel: improvement of antimicrobial activity, stability and mode of action. LWT-Food Sci Technol 86:132–138

Kumar S, Dilbaghi N, Barnela M, Bhanjana G, Kumar R (2012) Biosensors as novel platforms for detection of food pathogens and allergens. BioNanoScience 2(4):196–217

Lai F, Wissing SA, Müller RH, Fadda AM (2006) *Artemisia arborescens* L essential oil-loaded solid lipid nanoparticles for potential agricultural application: preparation and characterization. AAPS PharmSciTech 7(1):E10

Lakkis JM (2016) Encapsulation and controlled release technologies in food systems. Wiley, Chichester

Laloux L, Polet M, Schneider YJ (2017) Interaction between ingested-engineered nanomaterials and the gastrointestinal tract: in vitro toxicology aspects. In: Nanotechnology in agriculture and food science

Lee BH, Annis PC, Choi WS (2004) Fumigant toxicity of essential oils from the Myrtaceae family and 1, 8-cineole against 3 major stored-grain insects. J Stored Prod Res 40(5):553–564

Li Y, Liu X, Lin Z (2012) Recent developments and applications of surface plasmon resonance biosensors for the detection of mycotoxins in foodstuffs. Food Chem 132(3):1549–1554

Liang R, Xu S, Shoemaker CF, Li Y, Zhong F, Huang Q (2012) Physical and antimicrobial properties of peppermint oil nanoemulsions. J Agric Food Chem 60:7548–7555

Madene A, Jacquot M, Scher J, Desobry S (2006) Flavour encapsulation and controlled release–a review. Int J Food Sci Technol 41(1):1–21

Magnuson BA, Jonaitis TS, Card JW (2011) A brief review of the occurrence, use, and safety of food-related nanomaterials. J Food Sci 76(6):126–133

Maqbool M, Ali A, Alderson PG, Mohamed MTM, Siddiqui Y, Zahid N (2011) Postharvest application of gum Arabic and essential oils for controlling anthracnose and quality of banana and papaya during cold storage. Postharvest Biol Technol 62(1):71–76

Maragos CM (2012) Signal amplification using colloidal gold in a biolayer interferometry-based immunosensor for the mycotoxin deoxynivalenol. Food Addit Contam Part A 29(7):1108–1117

Marin S, Ramos AJ, Cano-Sancho G, Sanchis V (2013) Mycotoxins: occurrence, toxicology, and exposure assessment. Food Chem Toxicol 60:218–237

Martirosyan A, Schneider YJ (2014) Engineered nanomaterials in food: implications for food safety and consumer health. Int J Environ Res Pub Health 11(6):5720–5750

McClements DJ (2004) Protein-stabilized emulsions. Curr Opin Colloid Interface Sci 9(5):305–313

McClements DJ (2015) Food emulsions: principles, practices, and techniques. CRC Press, Boca Raton

Mendes JF, Martins HHA, Otoni CG, Santana NA, Silva RCS, Da Silva AG, Silva MV, Correia MTS, Machado G, Pinheiro ACM, Piccoli RH (2018) Chemical composition and antibacterial activity of *Eugenia brejoensis* essential oil nanoemulsions against *Pseudomonas fluorescens*. LWT-Food Sci Technol 93:659–664

Mohammadi A, Hashemi M, Hosseini SM (2015) Nanoencapsulation of *Zataria multiflora* essential oil preparation and characterization with enhanced antifungal activity for controlling *Botrytis cinerea*, the causal agent of gray mould disease. Innov Food Sci Emerg Technol 28:73–80

Moos PJ, Olszewski K, Honeggar M, Cassidy P, Leachman S, Woessner D, Cutler NS, Veranth JM (2011) Responses of human cells to ZnO nanoparticles: a gene transcription study. Metallomics 3(11):1199–1211

Mukherjee S, Ray S, Thakur RS (2009) Solid lipid nanoparticles: a modern formulation approach in drug delivery system. Indian J Pharm Sci 71(4):349

Nakagawa K, Surassmo S, Min SG, Choi MJ (2011) Dispersibility of freeze-dried poly (epsilon-caprolactone) nanocapsules stabilized by gelatin and the effect of freezing. J Food Eng 102(2):177–188

Neethirajan S, Jayas DS (2011) Nanotechnology for the food and bioprocessing industries. Food Bioprocess Technol 4(1):39–47

Otoni CG, Pontes SF, Medeiros EA, Soares NDF (2014) Edible films from methylcellulose and nanoemulsions of clove bud (*Syzygium aromaticum*) and oregano (*Origanum vulgare*) essential oils as shelf life extenders for sliced bread. J Agric Food Chem 62(22):5214–5219

Owens DE III, Peppas NA (2006) Opsonization, biodistribution, and pharmacokinetics of polymeric nanoparticles. Int J Pharm 307:93–102

Park IK, Lee SG, Choi DH, Park JD, Ahn YJ (2003) Insecticidal activities of constituents identified in the essential oil from leaves of *Chamaecyparis obtusa* against *Callosobruchus chinensis* (L.) and *Sitophilus oryzae* (L.). J Stored Prod Res 39(4):375–384

Park E-J et al (2010) Silver nanoparticles induce cytotoxicity by a Trojan-horse type mechanism. Toxicol In Vitro 24(3):872–878

Pelissari FM, Grossmann MV, Yamashita F, Pineda EAG (2009) Antimicrobial, mechanical, and barrier properties of cassava starch–chitosan films incorporated with oregano essential oil. J Agric Food Chem 57(16):7499–7504

Pérez-López B, Merkoçi A (2011) Nanomaterials based biosensors for food analysis applications. Trends Food Sci Technol 22(11):625–639

Powell JJ, Faria N, Thomas-McKay E, Pele LC (2010) Origin and fate of dietary nanoparticles and microparticles in the gastrointestinal tract. J Autoimmun 34:J226–J233

Prakash B, Kujur A, Yadav A, Kumar A, Singh PP, Dubey NK (2018) Nanoencapsulation: an efficient technology to boost the antimicrobial potential of plant essential oils in food system. Food Control 89:1–11

Qian C, Decker EA, Xiao H, McClements DJ (2011) Comparison of biopolymer emulsifier performance in formation and stabilization of orange oil-in-water emulsions. J Am Oil Chem Soc 88(1):47–55

Qiu C, Chang R, Yang J, Ge S, Xiong L, Zhao M, Li M, Sun Q (2017) Preparation and characterization of essential oil-loaded starch nanoparticles formed by short glucan chains. Food Chem 221:1426–1433

Ribeiro JC, Ribeiro WLC, Camurça-Vasconcelos ALF, Macedo ITF, Santos JML, Paula HCB, Bevilaqua CML (2014) Efficacy of free and nanoencapsulated *Eucalyptus citriodora* essential oils on sheep gastrointestinal nematodes and toxicity for mice. Vet Parasitol 204(3–4):243–248

Roco MC (2002) Government nanotechnology funding: an international outlook. JOM 54(9):22–23

Rodríguez J, Martín MJ, Ruiz MA, Clares B (2016) Current encapsulation strategies for bioactive oils: from alimentary to pharmaceutical perspectives. Food Res Int 83:41–59

Rouhani M, Samih MA, Kalantari S (2012) Insecticide effect of silver and zinc nanoparticles against Aphis nerii Boyer De Fonscolombe (Hemiptera: Aphididae). Chil J Agric Res 72(4):590

Ryu V, McClements DJ, Corradini MG, McLandsborough L (2018) Effect of ripening inhibitor type on formation, stability, and antimicrobial activity of thyme oil nanoemulsion. Food Chem 245:104–111

Sagadevan S, Periasamy M (2014) Recent trends in nanobiosensors and their applications-a review. Rev Adv Mater Sci 36:62–69

Sahab AF, Waly AI, Sabbour MM, Nawar LS (2015) Synthesis, antifungal and insecticidal potential of Chitosan (CS)-g-poly (acrylic acid)(PAA) nanoparticles against some seed borne fungi and insects of soybean. Int J Chem Tech Res 8:589–598

Salomoni R, Léo P, Montemor AF, Rinaldi BG, Rodrigues MFA (2017) Antibacterial effect of silver nanoparticles in Pseudomonas aeruginosa. Nanotechnol Sci Appl 10:115

Sánchez-González L, Vargas M, González-Martínez C, Chiralt A, Cháfer M (2011) Use of essential oils in bioactive edible coatings: a review. Food Eng Rev 3(1):1–16

Sankar MV, Abideen S (2015) Pesticidal effect of green synthesized silver and lead nanoparticles using Avicennia marina against grain storage pest Sitophilus oryzae. Int J Nanomater Biost 5(3):32–39

Sarkar S, Gupta S, Variyar PS, Sharma A, Singhal RS (2012) Irradiation depolymerized guar gum as partial replacement of gum Arabic for microencapsulation of mint oil. Carbohydr Polym 90(4):1685–1694

Saupe A, Rades T (2006) Solid lipid nanoparticles. In: Nanocarrier technologies. Springer, Dordrecht, pp 41–50

Scallan E, Hoekstra RM, Angulo FJ, Tauxe RV, Widdowson MA, Roy SL, Griffin PM (2011) Foodborne illness acquired in the United States—major pathogens. Emerg Infect Dis 17(1):7

Severino R, Vu KD, Donsì F, Salmieri S, Ferrari G, Lacroix M (2014) Antibacterial and physical effects of modified chitosan based-coating containing nanoemulsion of mandarin essential oil and three non-thermal treatments against Listeria innocua in green beans. Int J Food Microbiol 191:82–88

Shanmugam MK, Arfuso F, Kumar AP, Wang L, Goh BC, Ahn KS, Sethi G (2017) Modulation of diverse oncogenic transcription factors by thymoquinone, an essential oil compound isolated from the seeds of *Nigella sativa* Linn. Pharmacol Res

Shao P, Zhang H, Niu B, Jin W (2018) Physical stabilities of taro starch nanoparticles stabilized Pickering emulsions and the potential application of encapsulated tea polyphenols. Int J Biol Macromol 118:2032–2039

Sherry M, Charcosset C, Fessi H, Greige-Gerges H (2013) Essential oils encapsulated in liposomes: a review. J Liposome Res 23(4):268–275

Shi F, Zhao JH, Liu Y, Wang Z, Zhang YT, Feng NP (2012) Preparation and characterization of solid lipid nanoparticles loaded with frankincense and myrrh oil. Int J Nanomedicine 7:2033

Shim WB, Dzantiev BB, Eremin SA, Chung DH (2009) One-step simultaneous immunochromatographic strip test for multianalysis of ochratoxin A and zearalenone. J Microbiol Biotechnol 19(1):83–92

Singh T, Shukla S, Kumar P, Wahla V, Bajpai VK, Rather IA (2017) Application of nanotechnology in food science: perception and overview. Front Microbiol 8:1501

Siva C, Kumar MS (2015) Pesticidal activity of eco-friendly synthesized silver nanoparticles using Aristolochia indica extract against *Helicoverpa armigera* Hubner (Lepidoptera: Noctuidae). Int J Adv Sci Tech Res 2:197–226

Solans C, Izquierdo P, Nolla J, Azemar N, Garcia-Celma MJ (2005) Nano-emulsions. Curr Opin Colloid Interface Sci 10(3–4):102–110

Souza AC, Goto GEO, Mainardi JA, Coelho ACV, Tadini CC (2013) Cassava starch composite films incorporated with cinnamon essential oil: antimicrobial activity, microstructure, mechanical and barrier properties. LWT-Food Sci Technol 54(2):346–352

Souza LRR, da Silva VS, Franchi LP, de Souza TAJ (2018) Toxic and beneficial potential of silver nanoparticles: the two sides of the same coin. In: Cellular and molecular toxicology of nanoparticles. Springer, Cham, pp 251–262

Thiruvengadam M, Rajakumar G, Chung IM (2018) Nanotechnology: current uses and future applications in the food industry. 3 Biotech 8(1):74

Tian J, Ban X, Zeng H, He J, Chen Y, Wang Y (2012) The mechanism of antifungal action of essential oil from dill (*Anethum graveolens* L.) on *Aspergillus flavus*. PLoS One 7(1):e30147

Tian WL, Lei LL, Zhang Q, Li Y (2016) Physical stability and antimicrobial activity of encapsulated cinnamaldehyde by self-emulsifying nanoemulsion. J Food Process Eng 39(5):462–471

Tripathi AK, Upadhyay S, Bhuiyan M, Bhattacharya PR (2009) A review on prospects of essential oils as biopesticide in insect-pest management. J Pharmacogn Phytother 1(5):052–063

Valenti D, De Logu A, Loy G, Sinico C, Bonsignore L, Cottiglia F, Garau D, Fadda AM (2001) Liposome-incorporated Santolina insularis essential oil: preparation, characterization and in vitro antiviral activity. J Liposome Res 11(1):73–90

Verma S, Gokhale R, Burgess DJ (2009) A comparative study of top-down and bottom-up approaches for the preparation of micro/nanosuspensions. Int J Pharm 380(1–2):216–222

Vishwakarma GS, Gautam N, Babu JN, Mittal S, Jaitak V (2016) Polymeric encapsulates of essential oils and their constituents: a review of preparation techniques, characterization, and sustainable release mechanisms. Polym Rev 56(4):668–701

Wooster TJ, Labbett D, Sanguansri P, Andrews H (2016) Impact of microemulsion inspired approaches on the formation and destabilisation mechanisms of triglyceride nanoemulsions. Soft Matter 12(5):1425–1435

World Health Organization (2017) Noncommunicable diseases: progress monitor. WHO, Geneva

Wu Y, Luo Y, Wang Q (2012) Antioxidant and antimicrobial properties of essential oils encapsulated in zein nanoparticles prepared by liquid–liquid dispersion method. LWT-Food Sci Technol 48(2):283–290

Wu J, Shu Q, Niu Y, Jiao Y, Chen Q (2018) Preparation, characterization, and antibacterial effects of chitosan nanoparticles embedded with essential oils synthesized in an ionic liquid containing system. J Agric Food Chem 66(27):7006–7014

Xu X, Liu X, Li Y, Ying Y (2013) A simple and rapid optical biosensor for detection of aflatoxin B1 based on competitive dispersion of gold nanorods. Biosens Bioelectron 47:361–367

Xue J (2015) Essential oil nanoemulsions prepared with natural emulsifiers for improved food safety. Dissertation, University of Tennessee, Knoxville

Yang FL, Li XG, Zhu F, Lei CL (2009) Structural characterization of nanoparticles loaded with garlic essential oil and their insecticidal activity against *Tribolium castaneum* (Herbst) (Coleoptera: Tenebrionidae). J Agric Food Chem 57(21):10156–10162

Yasur J, Rani PU (2015) Lepidopteran insect susceptibility to silver nanoparticles and measurement of changes in their growth, development and physiology. Chemosphere 124:92–102

Yotova L, Yaneva S, Marinkova D, Serfaty S (2013) Co-immobilization of peroxidase and tyrosinase onto hybrid membranes obtained by the sol-gel method for the construction of an optical biosensor. Biotechnol Biotechnol Equip 27(3):3885–3889

Zahir AA, Bagavan A, Kamaraj C, Elango G, Rahuman AA (2012) Efficacy of plant-mediated synthesized silver nanoparticles against Sitophilus oryzae. J Biop 5:95

Zhang S, Zhang M, Fang Z, Liu Y (2017) Preparation and characterization of blended cloves/cinnamon essential oil nanoemulsions. LWT-Food Sci Technol 75:316–322

Chapter 3
Ecotoxicity of Metallic Nanoparticles and Possible Strategies for Risk Assessment

Ifra Zoomi, Harbans Kaur Kehri, Ovaid Akhtar, Dheeraj Pandey, Pragya Srivastava, and Raghvendra Pratap Narayan

3.1 Introduction

Nanoparticles are defined as particulate matter, usually with nanoscale dimensions (1–100 nm), whose properties are diverse from their bulk form (Auffan et al. 2009). Various distinctive properties of nanoparticles such as electronic (Kim et al. 2007), optical/photonic (Chan et al. 2013), magnetic (Mornet et al. 2006), and catalytic (Nasrollahzadeh et al. 2015) have significant roles in daily human life. Fundamentally, nanoparticles are categorized in two groups: (i) carbon-containing nanoparticles and (ii) metal-containing nanoparticles. Carbon-containing nanoparticles are made of carbon nanotubes and fullerenes. However, most of the metal-containing nanoparticles are made from metals such as gold (Au), iron (Fe), silver (Ag), copper (Cu), and metal oxides such as titanium dioxide (TiO_2), antimony oxide (Sb_2O_3), cerium dioxide (CeO_2), copper oxide (CuO), nickel oxide (NiO), iron oxide (FeO), and zinc oxide (ZnO).

Metallic nanoparticles are important because of their physical, chemical, and optoelectronic properties. Metallic nanoparticles have been used in various products with different purposes, such as sensors (Li et al. 2007), as catalysts in various processes (Carnes and Klabunde 2003), drug delivery (Hola et al. 2015), sunscreens (Gulson et al. 2015), solar-driven energy production (Sau et al. 2010), and in pollutant remediation (Kamat and Meisel 2003; Choopun et al. 2009; Raman and Kanmani

I. Zoomi · H. K. Kehri · D. Pandey · P. Srivastava
Sadasivan Myco-pathology Laboratory, Department of Botany,
University of Allahabad, Allahabad, India

O. Akhtar
Department of Botany, Kamla Nehru Institute of Physical and Social Sciences,
Sultanpur, India

R. P. Narayan (✉)
Netaji Subhash Chandra Bose Government Girls P.G. College, Lucknow, India

I. Bhushan et al. (eds.), *Nanomaterials and Environmental Biotechnology*,
Nanotechnology in the Life Sciences,
https://doi.org/10.1007/978-3-030-34544-0_3

2016). Additionally, nanoparticles have been shown to inhibit microorganism growth by exhibiting antibacterial, antiviral, and antifungal properties (Padmavathy and Vijayaraghavan 2008; Khatami et al. 2015; Rai et al. 2016). As metallic nanoparticles are widely used, their exposure is likely to increase substantially, and they interact negatively with microorganisms, green plants, animals, human beings, and their surrounding environment (Navarro et al. 2008; Wise et al. 2010; Singh et al. 2016, 2019; Shweta et al. 2017, 2018; Arif et al. 2018; Vishwakarma et al. 2018). Therefore, detailed understanding of their synthesis, interaction, and possible risk valuation would offer a foundation for harmless use of nanoparticles.

3.2 Synthesis of Metallic Nanoparticles

Synthesis of metallic nanoparticles is a timely area of research because their use is widespread. Various physical and chemical methods are used for the production of metallic nanoparticles (Tripathi et al. 2017; Koul et al. 2018). For the synthesis of metal oxide nanoparticles, stabilized precipitation and flame pyrolysis are commonly used methods (Christian et al. 2008). Monodisperse nickel phosphide (Ni_2P) nanorods and nanoparticles are made by a one-step solution-phase route (Li et al. 2015). Moreover, there is growing interest in the controlled synthesis of metallic nanoparticles (Wiley et al. 2005; Xia et al. 2009). However, these methods have certain drawbacks because of the use of poisonous chemicals and radiation and are an expensive process. Therefore, academic research is shifting toward biological synthesis of metallic nanoparticles, as it is rapid, feasible, and more productive relative to its cost. In this context, microorganisms have an increasingly critical role as they can provide inorganic materials either intra- or extracellularly for the synthesis of nanoparticles (Simkiss and Wilbur 1989; Mann 1996). Numerous microorganisms such as bacteria (Shahverdi et al. 2007; Saifuddin et al. 2009; Pantidos and Horsfall 2014), actinomycetes (Abdeen et al. 2014; Golinska et al. 2014), algae (Singaravelu et al. 2007; Aruoja et al. 2009; Abboud et al. 2014), and fungi (Mukherjee et al. 2001; Ahmad et al. 2003; Yadav et al. 2015) are used for nanoparticle synthesis. Also, peptides (Tomczak et al. 2007), starches (Kumar et al. 2014), and almost all parts of plants have been used for the synthesis of metallic nanoparticles.

3.3 Application of Nanoparticles

Nowadays, metal-based nanoparticles have become one of the main and increasing aims of nanotechnology, as these particles are usually used in cosmetics, antibacterial agents, tires, stain-resistant clothing, optic devices, toothpaste, sensors, food additives, and data storage (Fig. 3.1).

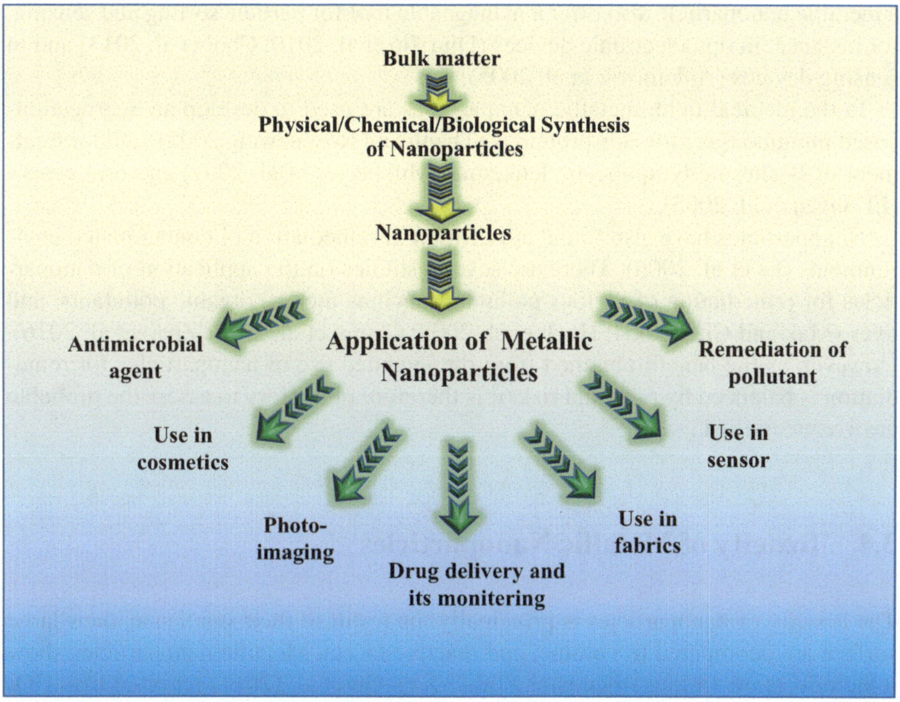

Fig. 3.1 Application of metallic nanoparticles

As antibacterial agents, these particles are widely used in medical applications, food storage, and water treatment (Bosetti et al. 2002; Cho et al. 2005; Singh et al. 2008; Espitia et al. 2012). In the textile industry, silver nanoparticles are used to prepare cotton fibers that exhibit antibacterial activity (Durán et al. 2007). It has been reported that Ag or Au nanoparticles extracellularly produced from *Fusarium oxysporum* can be used to prevent or to reduce the infection of *Staphylococcus aureus* (Durán et al. 2007). Moreover, metallic nanoparticles are of great scientific importance regarding their catalytic activity in various metal-based reactions (Hvolbæk et al. 2007). Metallic nanoparticles also show superior catalytic activity in the reduction and removal of dye. For instance, gold nanoparticles are reported to catalyze the reduction of dye in the presence of stannous chloride (Gupta et al. 2010). Silver and gold nanoclusters were reported to catalyze the reduction of methylene blue dye ($C_{16}H_{18}ClN_3S$) by arsine in micellar medium (Kundu et al. 2002). Köhler et al. (2008) reported that the catalytic activities of nanoparticles enhanced the bleaching of the organic dyes. However, catalytic activity of metallic nanoparticles also varies from metal to metal. For instance, the Ag nanoparticle was found to be superior to Au and Pt colloid in catalyzing chemiluminescence from the luminol–hydrogen peroxide system (Guo et al. 2008). Despite the aforementioned antibacterial and catalytic activity of metallic nanoparticles, the optical properties of a metallic nanoparticle also offer a manageable tool for particle sorting and sensing, for instance, in optoelectronic devices (Djurišić et al. 2010; Choi et al. 2013) and in sensing devices (Ankamwar et al. 2005).

In the medical field, metallic nanoparticles are used to develop an aggregation-based immunoassay for anti-protein A (Thanh and Rosenzweig 2002), and for treatment of B-chronic lymphocytic leukemia (Mukherjee et al. 2007) and oral cancer (El-Sayed et al. 2005).

Nanoparticles have also found application in remediation of contaminated environments (Li et al. 2006). There are several studies on the application of nanoparticles for remediation of various pollutants such as metals, organic pollutants, and dyes (Mak and Chen 2004; Hoch et al. 2008; Cheng et al. 2013; Zhao et al. 2016). However, as the benefit obtained from the intended use of nanoparticles for remediation is balanced by potential risk, it is therefore obligatory to assess the probable environmental risk.

3.4 Toxicity of Metallic Nanoparticles

The toxicity of nanoparticles is principally the result of their small size, their large surface area compared to volume, and reactive facets. Metallic nanoparticles show toxic effects on various organisms (Table 3.1). Ge et al. (2011) reported that TiO_2 and ZnO nanoparticles reduced the biomass of a microbial community. Among microbial communities, nitrogen-fixing bacteria are an important component of the soil ecosystem as they maintain soil health and fertility. Cherchi et al. (2011) reported bactericidal effects of TiO_2 nanoparticles in *Anabaena variabilis*. Toxicity of nanometal oxides in aquatic ecosystems has also been studied by various research groups (Blaise et al. 2008; Lee et al. 2009; Pradhan et al. 2012; Miller et al. 2012). Federici et al. (2007) reported that the gills of *Oncorhynchus mykiss* showed sensitivity toward TiO_2 nanoparticles. Furthermore, TiO_2 nanoparticles were reported to inhibit the growth of *Desmodesmus subspicatus* at higher concentrations (Hund-Rinke and Simon 2006). However, the toxicity of metallic nano-sized particles in an aquatic ecosystem is debatable (Sharma 2009) as their physicochemical properties are dissimilar from their ionic and bulk form (Christian et al. 2008). Moreover, soluble ions released from metallic nanoparticles appear be the main cause of ecotoxicity (Aruoja et al. 2009). Green plants are also affected by metallic nanoparticles as these particles enter into the plant by various means such as stomata, cuts or wounds, and through the roots. Zn and ZnO nanoparticles negatively affect the growth of *Raphanus sativus* (radish), *Brassica napus* (rape), and *Lolium perenne* (ryegrass) (Lin and Xing 2007). Similarly, Yang and Watts (2005) reported the toxicity of alumina nanoparticles in *Brassica oleracea* (cabbage), *Daucus carota* (carrot), *Zea mays* (corn), *Cucumis sativus* (cucumber), and *Glycine max* (soybean). Metallic nanoparticles are known to induce effects on human health, as they cause gastroduodenal corrosive injury (Liu et al. 2004), a cytotoxic effect on glomerular and

Table 3.1 Toxic effect of metallic nanoparticles on organisms

Nanoparticles	Organisms	Effects	References
TiO_2 nanoparticles	*Daphnia magna*	Bioaccumulation may interfere with food intake and ultimately affect growth and reproduction	Zhu et al. (2010)
TiO_2 nanoparticles	*Daphnia magna*	Caused mortality	Lovern and Klaper (2006)
ZnO and TiO_2 nanoparticles	*Escherichia coli*	Induced oxidative stress and DNA damage leading to reduced viability of *E. coli*	Kumar et al. (2011a)
ZnO and TiO_2 nanoparticles	*Salmonella typhimurium*	Both ZnO and TiO_2 nanoparticles were significantly internalized in the *S. typhimurium* cells in a concentration-dependent manner and these nanoparticles exhibited weak mutagenic potential	Kumar et al. (2011b)
Ag nanoparticles	Bacteria	Ag nanoparticles inhibited soil-denitrifying bacteria	Throbäck et al. (2007)
Ag nanoparticles	Bacteria	Inhibited the nitrifying organisms	Choi et al. (2008)
TiO_2 and ZnO nanoparticles	Bacteria	TiO_2 and ZnO nanoparticles reduced both microbial biomass, bacterial diversity, and composition	Ge et al. (2011)
CuO, NiO, ZnO, and Sb_2O_3 nanoparticles	*Escherichia coli*, *Bacillus subtilis*, and *Streptococcus aureus*	Toxic to microorganisms: toxicity order was CuO > ZnO > NiO > Sb_2O_3 nanoparticles	Baek and An (2011)
CeO_2 nanoparticles	*Synechocystis* PCC6803 and *Escherichia coli*	*E. coli* was sensitive to the 'direct' effects of nanoparticles, whereas *Synechocystis* was protected by extracellular polymeric substances, preventing direct cellular contacts	Thill et al. (2006)
Ag nanoparticles	Zebrafish	Induced oxidative stress and apoptosis	Choi et al. (2010)
Au nanoparticles	Phytoplanktonic alga (*Scenedesmus subspicatus*) and a benthic bivalve (*Corbicula fluminea*)	Mortality was 20% at lowest contamination condition; the highest reached 50% in algae. Au nanoparticles were adsorbed by the algal cell wall, leading to progressive intracellular and wall disturbances. In bivalves these nanoparticles bioaccumulated and penetrated into the gills and digestive epithelia to cause oxidative stress	Renault et al. (2008)
Ag nanoparticles	*Mytilus edulis*	Au nanoparticles accumulated in digestive gland causing oxidative stress	Tedesco et al. (2010)

tubular renal cells (Pujalté et al. 2011), and toxic effects on the pulmonary system (Moos et al. 2010). Karlsson et al. (2009) assessed the effect of metallic nanoparticles on human cell lines and reported that nanoparticles are perhaps more toxic than their bulk forms.

3.4.1 Uptake of Metallic Nanoparticles

The detailed mechanisms of biological uptake of metallic nanoparticles are not well known. However, it has been hypothesized that uptake of nanoparticles in animal bodies takes place through the gut (Baun et al. 2008) by various mechanisms such as diffusion through cell membranes, via endocytosis and adhesion (Geiser et al. 2005; Kim et al. 2007). In other biotic components (plants, algae, fungi) the cell wall restricted the entry of nanoparticles as it acts as a barrier. However, small-sized nanoparticles enter the cell via the pores in the cell wall (Zemke-White et al. 2000), and further interaction of these small-sized nanoparticles with the cell wall might increase the pore size, resulting in the internalization of large-sized nanoparticles (Navarro et al. 2008). After passing through the cell wall, endocytosis takes place (Navarro et al. 2008), and inside the cell nanoparticles bind with various cellular structures, thereby inhibiting cellular activity or damaging the cell organelles. Plants interact more frequently with nanoparticles by the presence of stomata, cuts or wounds, and the large surface area of leaf and roots (Navarro et al. 2008).

3.4.2 Mode of Action of Nanoparticles

Internalized metallic nanoparticles inside the cell interfere with several biological mechanisms, such as causing disruption of the membrane potential, and destabilization and oxidation of protein, and affect the stability of nucleic acid, stimulate the production of free radical species called reactive oxygen species (ROS), disrupting energy flow and releasing toxic compounds (Klaine et al. 2008). Gold nanoparticles have been reported to puncture the cell membrane (Tsao et al. 1999) and alter the cell shape and enzymatic activity (Liu et al. 2004).

Metallic nanoparticles also generate oxidative stress in biological systems by the production of ROS. ROS disrupt the influx and efflux of electrons and ions, disrupt membrane permeability, and reduce glutathione content inside the cell (Limbach et al. 2007). ROS increase the permeability of cell membrane by oxidization of double bonds of fatty acids. It has also been reported that TiO_2 nanoparticles have photocatalyst properties (Khus et al. 2006) and, with exposure to ultraviolet radiation (Zhao et al. 2007) produce ROS thereby causing DNA damage. Photosensitive silver nanoparticles have been shown to break the double-stranded DNA upon exposure of to UV light (Badireddy et al. 2007). It has also been reported that CeO_2 nanoparticles cause oxidation of membrane-bound complexes of respiratory electron transport chain (Thill et al. 2006). Moreover, quantum dots cause oxidative destruction (Hardman 2006), and heavy metals or metal ions released from quantum dots are toxic to the living cells (Klaine et al. 2008). Silver ions discharged from the metallic nanoparticles interact with functional thiol groups (derived from the cysteine residues) of enzymes (Matsumura et al. 2003) and inhibit the respiratory enzymes (Kim et al. 2007).

3.5 Ecotoxicology Assessment and Possible Strategies

According to the U. S. Environmental Protection Agency (EPA), "risk is a measure of the probability that cause damage to life, health, property, and/or the environment." Before assessing the biotic hazard, it is desirable to assess the physical and chemical properties of nanoparticles. The various techniques for analysis and characterization of metallic nanoparticles include membrane filtration (Akthakul et al. 2005; Howell et al. 2006), size-exclusion chromatography (Wang et al. 2006), and photon correlation spectroscopy, used to determine the size and sometimes the shape of metallic nanoparticles (Chrastina and Schnitzer 2010). Additionally, transmission electron microscopy (TEM) (Jose-Yacaman et al. 2001; Chrastina and Schnitzer 2010), the scanning electron microscope (SEM) (Rai et al. 2009), and atomic force microscopy (AFM) (Viguie et al. 2007) are used to gather evidence about the configuration, arrangement, charges, and force of nanoparticles. The complex arrangement of crystal metallic nanoparticles can be resolved by X-ray diffraction (XRD) and energy-dispersive X-ray (EDX) techniques (Rai et al. 2009). However, it is difficult to analyze the properties of nanoparticles because their concentration in the environment is below the detection limit and test samples also carry unwanted nanoparticles (Lead and Wilkinson 2006). After the analysis and characterization of metallic nanoparticles, standard toxicity tests on organisms are used to assess the impacts of nanoparticles. For aquatic threat evaluation, an algal growth inhibition assay is commonly used, and *Pseudokirchneriella subcapitata* is an ideal organism for envisaging lethal threats to primary producers (Aruoja et al. 2009). Toxicological effects of metallic nanoparticles on diverse algal species, such as *Chlamydomonas reinhardtii* (Navarro et al. 2008; Chen et al. 2012; Melegari et al. 2013), *Desmodesmus subspicatus* (Hartmann et al. 2010), *Dunaliella tertiolecta*, and *Chlorella vulgaris* (Oukarroum et al. 2012), have also been studied. Also, *Vibrio fischeri*, a naturally luminescent bacterium, is widely used for ecotoxicological studies, and the bacterial luminescence inhibition assay is economical and easy to perform (Mortimer et al. 2008). There are various aspects for understanding the risk assessment of nanoparticles, such as dose of nanoparticles, exposure time, and endpoint measurement. Furthermore, short- and long-term laboratory experiments and development of a coordinated approach are still needed for assessing the toxicity.

3.6 Conclusions

Because of the wide application of metallic nanoparticles, their unintentional release and exposure pose a serious hazard to organisms and their environments. Only a few areas have been covered for the assessment and testing of the hazardous effects of metallic nanoparticles. Therefore, there is a requirement for data on the long-term effects of metallic nanoparticles, in vivo interactions of metallic nanoparticles, applied methods, databases of well-established toxicity tests, and the establishment of testing guidelines to enhance the transparency and comparability of obtained data.

Acknowledgments The authors thank the Head of the Botany Department, University of Allahabad, Allahabad, for providing the necessary facilities, and are also grateful to UGC and CSIR for providing financial support to Ifra Zoomi, Pragya Srivastava, Dheeraj Pandey, and Ovaid Akhtar.

References

Abboud Y, Saffaj T, Chagraoui A, El Bouari A, Brouzi K, Tanane O, Ihssane B (2014) Biosynthesis, characterization and antimicrobial activity of copper oxide nanoparticles (CONPs) produced using brown alga extract (*Bifurcariabifurcata*). Appl Nanosci 4(5):571–576

Abdeen S, Geo S, Sukanya S, Praseetha PK, Dhanya RP (2014) Biosynthesis of silver nanoparticles from Actinomycetes for therapeutic applications. Int J Nano Dimens 5(2):155–162

Ahmad A, Mukherjee P, Senapati S, Mandal D, Khan MI, Kumar R, Sastry M (2003) Extracellular biosynthesis of silver nanoparticles using the fungus *Fusariumoxysporum*. Colloids Surf B Bioint 28(4):313–318

Akthakul A, Hochbaum AI, Stellacci F, Mayes AM (2005) Size fractionation of metal nanoparticles by membrane filtration. Adv Mater 17:532

Ankamwar B, Chaudhary M, Sastry M (2005) Gold nanotriangles biologically synthesized using tamarind leaf extract and potential application in vapor sensing. Synth React Inorg Metal-Org Nano-Metal Chem 35(1):19–26

Arif N, Yadav V, Singh S, Tripathi DK, Dubey NK, Chauhan DK, Giorgetti L (2018) Interaction of copper oxide nanoparticles with plants: uptake, accumulation, and toxicity. In: Nanomaterials in plants, algae, and microorganisms. Academic Press, London, pp 297–310

Aruoja V, Dubourguier HC, Kasemets K, Kahru A (2009) Toxicity of nanoparticles of CuO, ZnO and TiO2 to microalgae *Pseudokirchneriellasubcapitata*. Sci Total Environ 407(4):1461–1468

Auffan M, Rose J, Bottero JY, Lowry GV, Jolivet JP, Wiesner MR (2009) Towards a definition of inorganic nanoparticles from an environmental, health and safety perspective. Nat Nanotechnol 4(10):634

Badireddy AR, Hotze EM, Chellam S, Alvarez P, Wiesner MR (2007) Inactivation of bacteriophages via photosensitization of fullerol nanoparticles. Environ Sci Technol 41:6627–6632

Baek YW, An YJ (2011) Microbial toxicity of metal oxide nanoparticles (CuO, NiO, ZnO, and Sb2O3) to *Escherichia coli*, *Bacillus subtilis*, and *Streptococcus aureus*. Sci Total Environ 409(8):1603–1608

Baun A, Hartmann NB, Grieger K, Kusk KO (2008) Ecotoxicity of engineered nanoparticles to aquatic invertebrates: a brief review and recommendations for future toxicity testing. Ecotoxicology 17(5):387–395

Blaise C, Gagné F, Ferard JF, Eullaffroy P (2008) Ecotoxicity of selected nano-materials to aquatic organisms. Environ Toxicol Int J 5:591–598

Bosetti M, Masse A, Tobin E, Cannas M (2002) Silver coated materials for external fixation devices: in vitro biocompatibility and genotoxicity. Biomaterials 23(3):887–892

Carnes CL, Klabunde KJ (2003) The catalytic methanol synthesis over nanoparticle metal oxide catalysts. J Mol Catal A Chem 194(1–2):227–236

Chan NY, Zhao M, Wang N, Au K, Wang J, Chan LW, Dai J (2013) Palladium nanoparticle enhanced giant photoresponse at LaAlO3/SrTiO3 two-dimensional electron gas heterostructures. ACS Nano 7(10):8673–8679

Chen L, Zhou L, Liu Y, Deng S, Wu H, Wang G (2012) Toxicological effects of nanometer titanium dioxide (nano-TiO2) on *Chlamydomonasreinhardtii*. Ecotox Environ Safe 84:155–162

Cheng N, Tian J, Liu Q, Ge C, Qusti AH, Asiri AM, Al-Youbi AO, Sun X (2013) Au-nanoparticle-loaded graphitic carbon nitride nanosheets: green photocatalytic synthesis and application toward the degradation of organic pollutants. ACS Appl Mater Inter 5(15):6815–6819

Cherchi C, Chernenko T, Diem M, Gu AZ (2011) Impact of nano titanium dioxide exposure on cellular structure of *Anabaena variabilis* and evidence of internalization. Environ Toxicol Chem 30(4):861–869

Cho M, Chung H, Choi W, Yoon J (2005) Different inactivation behavior of MS-2 phage and Escherichia coli in TiO2 photocatalytic disinfection. Appl Environ Microbiol 71(1):270–275

Choi O, Deng KK, Kim NJ, Ross L Jr, Surampalli RY, Hu Z (2008) The inhibitory effects of silver nanoparticles, silver ions, and silver chloride colloids on microbial growth. Water Res 42(12):3066–3074

Choi JE, Kim S, Ahn JH, Youn P, Kang JS, Park K, Yi J, Ryu DY (2010) Induction of oxidative stress and apoptosis by silver nanoparticles in the liver of adult zebrafish. AquatToxicol 100(2):151–159

Choi H, Ko SJ, Choi Y, Joo P, Kim T, Lee BR, Jung JW, Choi HJ, Cha M, Jeong JR, Hwang IW (2013) Versatile surface plasmon resonance of carbon-dot-supported silver nanoparticles in polymer optoelectronic devices. Nat Photon 7(9):732

Choopun S, Tubtimtae A, Santhaveesuk T, Nilphai S, Wongrat E, Hongsith N (2009) Zinc oxide nanostructures for applications as ethanol sensors and dye-sensitized solar cells. Appl Surf Sci 256(4):998–1002

Chrastina A, Schnitzer JE (2010) Iodine-125 radiolabeling of silver nanoparticles for in vivo SPECT imaging. Int J Nanomedicine 5:653

Christian P, Von der Kammer F, Baalousha M, Hofmann T (2008) Nanoparticles: structure, properties, preparation and behaviour in environmental media. Ecotoxicology 17(5):326–343

Djurišić AB, Ng AM, Chen XY (2010) ZnO nanostructures for optoelectronics: material properties and device applications. Prog Quantum Electron 34(4):191–259

Durán N, Marcato PD, De Souza GI, Alves OL, Esposito E (2007) Antibacterial effect of silver nanoparticles produced by fungal process on textile fabrics and their effluent treatment. J Biomed Nanotechnol 3(2):203–208

El-Sayed IH, Huang X, El-Sayed MA (2005) Surface plasmon resonance scattering and absorption of anti-EGFR antibody conjugated gold nanoparticles in cancer diagnostics: applications in oral cancer. Nano Lett 5(5):829–834

Espitia PJ, Soares ND, dos Reis Coimbra JS, de Andrade NJ, Cruz RS, Medeiros EA (2012) Zinc oxide nanoparticles: synthesis, antimicrobial activity and food packaging applications. Food Bioprocess Tech 5(5):1447–1464

Federici G, Shaw BJ, Handy RD (2007) Toxicity of titanium dioxide nanoparticles to rainbow trout (*Oncorhynchusmykiss*): gill injury, oxidative stress, and other physiological effects. Aquat Toxicol 84(4):415–430

Ge Y, Schimel JP, Holden PA (2011) Evidence for negative effects of TiO2 and ZnO nanoparticles on soil bacterial communities. Environ Sci Technol 4:1659–1664

Geiser M, Rothen-Rutishauser B, Kapp N, Schürch S, Kreyling W, Schulz H, Semmler M, ImHof V, Heyder J, Gehr P (2005) Ultrafine particles cross cellular membranes by nonphagocytic mechanisms in lungs and in cultured cells. Environ Health Perspect 113(11):1555

Golinska P, Wypij M, Ingle AP, Gupta I, Dahm H, Rai M (2014) Biogenic synthesis of metal nanoparticles from actinomycetes: biomedical applications and cytotoxicity. Appl Microbiol Biotechnol 98(19):8083–8097

Gulson B, McCall MJ, Bowman DM, Pinheiro T (2015) A review of critical factors for assessing the dermal absorption of metal oxide nanoparticles from sunscreens applied to humans, and a research strategy to address current deficiencies. Arch Toxicol 89(11):1909–1930

Guo JZ, Cui H, Zhou W, Wang W (2008) Ag nanoparticle-catalyzed chemiluminescent reaction between luminol and hydrogen peroxide. J Photochem Photobiol A Chem 193(2–3):89–96

Gupta N, Singh HP, Sharma RK (2010) Single-pot synthesis: plant mediated gold nanoparticles catalyzed reduction of methylene blue in presence of stannous chloride. Colloids Surf A Physicochem Eng Asp 367(1–3):102–107

Hardman R (2006) Atoxicologic review of quantum dots: toxicity depends on physicochemical and environmental factors. Environ Health Perspect 114(2):165

Hartmann NB, Von der Kammer F, Hofmann T, Baalousha M, Ottofuelling S, Baun A (2010) Algal testing of titanium dioxide nanoparticles—testing considerations, inhibitory effects and modification of cadmium bioavailability. Toxicology 269(2–3):190–197

Hoch LB, Mack EJ, Hydutsky BW, Hershman JM, Skluzacek JM, Mallouk TE (2008) Carbothermal synthesis of carbon-supported nanoscale zero-valent iron particles for the remediation of hexavalent chromium. Environ Sci Technol 42(7):2600–2605

Hola K, Markova Z, Zoppellaro G, Tucek J, Zboril R (2015) Tailored functionalization of iron oxide nanoparticles for MRI, drug delivery, magnetic separation and immobilization of biosubstances. Biotechnol Adv 33(6):1162–1176

Howell KA, Achterberg EP, Tappin AD, Worsfold PJ (2006) Colloidal metals in the tamar estuary and their influence on metal fractionation by membrane filtration. Environ Chem 3:199–207

Hund-Rinke K, Simon M (2006) Ecotoxic effect of photocatalytic active nanoparticles (TiO2) on algae and daphnids. Environ Sci Pollut Res 13(4):225–232

Hvolbæk B, Janssens TV, Clausen BS, Falsig H, Christensen CH, Nørskov JK (2007) Catalytic activity of Au nanoparticles. Nano Today 2(4):14–18

Jose-Yacaman M, Marin-Almazo M, Ascencio JA (2001) High resolution TEM studies on palladium nanoparticles. J Mol Catal A 173:61–74

Kamat PV, Meisel D (2003) Nanoscience opportunities in environmental remediation. C R Chim 6(8–10):999–1007

Karlsson HL, Gustafsson J, Cronholm P, Möller L (2009) Size-dependent toxicity of metal oxide particles—a comparison between nano-and micrometer size. Toxicol Lett 188(2):112–118

Khatami M, Pourseyedi S, Khatami M, Hamidi H, Zaeifi M, Soltani L (2015) Synthesis of silver nanoparticles using seed exudates of Sinapisarvensis as a novel bioresource, and evaluation of their antifungal activity. Bioresour Bioprocess 2(1):19

Khus M, Gernjak W, Ibanez PF, Rodriguez SM, Galvez JB, Icli S (2006) A comparative study of supported TiO_2 as photocatalyst in water decontamination at solar pilot plant scale. J Sol Energy 128:331–337

Kim JS, Kuk E, Yu KN, Kim JH, Park SJ, Lee HJ, Kim SH, Park YK, Park YH, Hwang CY, Kim YK, Lee YS, Jeong DH, Cho MH (2007) Antimicrobial effects of silver nanoparticles. Nanomed Nanotechnol Biol Med 3:95–101

Klaine SJ, Alvarez PJ, Batley GE, Fernandes TF, Handy RD, Lyon DY, Mahendra S, McLaughlin MJ, Lead JR (2008) Nanomaterials in the environment: behavior, fate, bioavailability, and effects. Environ Toxicol Chem 27(9):1825–1851

Köhler JM, Abahmane L, Wagner J, Albert J, Mayer G (2008) Preparation of metal nanoparticles with varied composition for catalytical applications in microreactors. Chem Eng Sci 63(20):5048–5055

Koul A, Kumar A, Singh VK, Tripathi DK, Mallubhotla S (2018) Exploring plant-mediated copper, iron, titanium, and cerium oxide nanoparticles and their impacts. In: Nanomaterials in plants, algae, and microorganisms. Academic Press, London, pp 175–194

Kumar A, Pandey AK, Singh SS, Shanker R, Dhawan A (2011a) Cellular uptake and mutagenic potential of metal oxide nanoparticles in bacterial cells. Chemosphere 83(8):1124–1132

Kumar A, Pandey AK, Singh SS, Shanker R, Dhawan A (2011b) Engineered ZnO and TiO2 nanoparticles induce oxidative stress and DNA damage leading to reduced viability of *Escherichia coli*. Free Radic Biol Med 51(10):1872–1881

Kumar B, Smita K, Cumbal L, Debut A, Pathak RN (2014) Sonochemical synthesis of silver nanoparticles using starch: a comparison. Bioinorg Chem Appl 2014:1–8

Kundu S, Ghosh SK, Mandal M, Pal T (2002) Silver and gold nanocluster catalyzed reduction of methylene blue by arsine in micellar medium. Bull Mater Sci 25(6):577–579

Lead JR, Wilkinson KJ (2006) Aquatic colloids and nanoparticles: current knowledge and future trends. Environ Chem 3(3):159–171

Lee SW, Kim SM, Choi J (2009) Genotoxicity and ecotoxicity assays using the freshwater crustacean Daphnia magna and the larva of the aquatic midge *Chironomusriparius* to screen the ecological risks of nanoparticle exposure. Environ Toxicol Pharmacol 28(1):86–91

Li XQ, Elliott DW, Zhang WX (2006) Zero-valent iron nanoparticles for abatement of environmental pollutants: materials and engineering aspects. Crit Rev Solid State Mat Sci 31(4):111–122

Li S, Shen Y, Xie A, Yu X, Qiu L, Zhang L, Zhang Q (2007) Green synthesis of silver nanoparticles using *Capsicum annuum* L. extract. Green Chem 9(8):852–858

Li H, Wang W, Gong Z, Yu Y, Chen H, Xia J (2015) Shape-controlled synthesis of nickel phosphide nanocrystals and their application as hydrogen evolution reaction catalyst. J Phys Chem Solids 80:22–25

Limbach LK, Wick P, Manser P, Grass RN, Bruinink A, Stark WJ (2007) Exposure of engineered nanoparticles to human lung epithelial cells: influence of chemical composition and catalytic activity on oxidative stress. Environ Sci Technol 41(11):4158–4163

Lin D, Xing B (2007) Phytotoxicity of nanoparticles: inhibition of seed germination and root growth. Environ Pollut 150(2):243–250

Liu S, Dai Z, Chen H, Ju H (2004) Immobilization of hemoglobin on zirconium dioxide nanoparticles for preparation of a novel hydrogen peroxide biosensor. Biosens Bioelectron 19(9):963–969

Lovern SB, Klaper R (2006) *Daphnia magna* mortality when exposed to titanium dioxide and fullerene (C60) nanoparticles. Environ Toxicol Chem Int J 25(4):1132–1137

Mak SY, Chen DH (2004) Fast adsorption of methylene blue on polyacrylic acid-bound iron oxide magnetic nanoparticles. Dyes Pigments 61(1):93–98

Mann S (ed) (1996) Biomimetic materials chemistry. VCH, New York, pp 1–40

Matsumura Y, Yoshikata K, Kunisaki S, Tsuchido T (2003) Mode of bactericidal action of silver zeolite and its comparison with that of silver nitrate. Appl Environ Microbiol 69:4278–4281

Melegari SP, Perreault F, Costa RH, Popovic R, Matias WG (2013) Evaluation of toxicity and oxidative stress induced by copper oxide nanoparticles in the green alga *Chlamydomonasreinhardtii.* Aquat Toxicol 142:431–440

Miller RJ, Bennett S, Keller AA, Pease S, Lenihan HS (2012) TiO2 nanoparticles are phototoxic to marine phytoplankton. PLoS One 7(1):e30321

Moos PJ, Chung K, Woessner D, Honeggar M, Cutler NS, Veranth JM (2010) ZnO particulate matter requires cell contact for toxicity in human colon cancer cells. Chem Res Toxicol 23(4):733–739

Mornet S, Vasseur S, Grasse F, Veverka P, Goglio G, Demourgues A, Portier J, Pollert E, Duguet E (2006) Magnetic nanoparticle design for medical applications. Prog Solid State Chem 34(2–4):237–247

Mortimer M, Kasemets K, Heinlaan M, Kurvet I, Kahru A (2008) High throughput kinetic *Vibrio fischeri* bioluminescence inhibition assay for study of toxic effects of nanoparticles. Toxicol In Vitro 22(5):1412–1417

Mukherjee P, Ahmad A, Mandal D, Senapati S, Sainkar SR, Khan MI, Parishcha R, Ajaykumar PV, Alam M, Kumar R, Sastry M (2001) Fungus-mediated synthesis of silver nanoparticles and their immobilization in the mycelial matrix: a novel biological approach to nanoparticle synthesis. Nano Lett 1(10):515–519

Mukherjee P, Bhattacharya R, Bone N, Lee YK, Patra CR, Wang S, Lu L, Secreto C, Banerjee PC, Yaszemski MJ, Kay NE (2007) Potential therapeutic application of gold nanoparticles in B-chronic lymphocytic leukemia (BCLL): enhancing apoptosis. J Nanobiotechnol 5(1):4

Nasrollahzadeh M, Sajadi SM, Maham M (2015) Green synthesis of palladium nanoparticles using Hippophaerhamnoides Linn leaf extract and their catalytic activity for the Suzuki–Miyaura coupling in water. J Mol Catal A Chem 396:297–303

Navarro E, Piccapietra F, Wagner B, Marconi F, Kaegi R, Odzak N, Sigg L, Behra R (2008) Toxicity of silver nanoparticles to *Chlamydomonasreinhardtii.* Environ Sci Technol 42(23):8959–8964

Oukarroum A, Bras S, Perreault F, Popovic R (2012) Inhibitory effects of silver nanoparticles in two green algae, *Chlorella vulgaris* and *Dunaliellatertiolecta.* Ecotoxicol Environ Safe 78:80–85

Padmavathy N, Vijayaraghavan R (2008) Enhanced bioactivity of ZnO nanoparticles—an antimicrobial study. Sci Technol Adv Mat 9(3):035004

Pantidos N, Horsfall LE (2014) Biological synthesis of metallic nanoparticles by bacteria, fungi and plants. J Nanomed Nanotechnol 5(5):1

Pradhan A, Seena S, Pascoal C, Cássio F (2012) Copper oxide nanoparticles can induce toxicity to the freshwater shredder *Allogamusligonifer*. Chemosphere 89(9):1142–1150

Pujalté I, Passagne I, Brouillaud B, Tréguer M, Durand E, Ohayon-Courtès C, L'Azou B (2011) Cytotoxicity and oxidative stress induced by different metallic nanoparticles on human kidney cells. Part Fibre Toxicol 8(1):1

Rai M, Yadav A, Gade A (2009) Silver nanoparticles as a new generation of antimicrobials. Biotechnol Adv 27(1):76–83

Rai M, Deshmukh SD, Ingle AP, Gupta IR, Galdiero M, Galdiero S (2016) Metal nanoparticles: the protective nanoshield against virus infection. Crit Rev Microbiol 42(1):46–56

Raman CD, Kanmani S (2016) Textile dye degradation using nano zero valent iron: a review. J Environ Manag 177:341–355

Renault S, Baudrimont M, Mesmer-Dudons N, Gonzalez P, Mornet S, Brisson A (2008) Impacts of gold nanoparticle exposure on two freshwater species: a phytoplanktonic alga (*Scenedesmussubspicatus*) and a benthic bivalve (*Corbiculafluminea*). Gold Bull 41(2):116–126

Saifuddin N, Wong CW, Yasumira AA (2009) Rapid biosynthesis of silver nanoparticles using culture supernatant of bacteria with microwave irradiation. J Chem 6(1):61–70

Sau TK, Rogach AL, Jäckel F, Klar TA, Feldmann J (2010) Properties and applications of colloidal nonspherical noble metal nanoparticles. Adv Mat 22(16):1805–1825

Shahverdi AR, Minaeian S, Shahverdi HR, Jamalifar H, Nohi AA (2007) Rapid synthesis of silver nanoparticles using culture supernatants of *Enterobacteria*: a novel biological approach. Process Biochem 42(5):919–923

Sharma VK (2009) Aggregation and toxicity of titanium dioxide nanoparticles in aquatic environment—a review. J Environ Sci Health A 44(14):1485–1495

Shweta, Vishwakarma K, Sharma S, Narayan RP, Srivastava P, Khan AS, Dubey NK, Tripathi DK, Chauhan DK (2017) Plants and carbon nanotubes (CNTs) interface: present status and future prospects. In: Nanotechnology. Springer, Singapore, pp 317–340

Shweta, Tripathi DK, Chauhan DK, Peralta-Videa JR (2018) Availability and risk assessment of nanoparticles in living systems: a virtue or a peril? In: Nanomaterials in plants, algae, and microorganisms. Academic Press, London, pp 1–31

Simkiss K, Wilbur KM (1989) Biomineralization. Academic Press, New York

Singaravelu G, Arockiamary JS, Kumar VG, Govindaraju K (2007) A novel extracellular synthesis of monodisperse gold nanoparticles using marine alga, *Sargassumwightii* Greville. Colloids Surf B Biointer 57(1):97–101

Singh M, Singh S, Prasad S, Gambhir IS (2008) Nanotechnology in medicine and antibacterial effect of silver nanoparticles. Dig J Nanomater Bios 3(3):115–122

Singh S, Tripathi DK, Dubey NK, Chauhan DK (2016) Effects of nano-materials on seed germination and seedling growth: striking the slight balance between the concepts and controversies. Mater Focus 5(3):195–201

Singh J, Vishwakarma K, Ramawat N, Rai P, Singh VK, Mishra RK, Kumar V, Tripathi DK, Sharma S (2019) Nanomaterials and microbes' interactions: a contemporary overview. 3 Biotech 9(3):68

Tedesco S, Doyle H, Blasco J, Redmond G, Sheehan D (2010) Oxidative stress and toxicity of gold nanoparticles in *Mytilusedulis*. Aquat Toxicol 100(2):178–186

Thanh NT, Rosenzweig Z (2002) Development of an aggregation-based immunoassay for anti-protein a using gold nanoparticles. Anal Chem 74(7):1624–1628

Thill A, Zeyons O, Spalla O, Chauvat F, Rose J, Auffan M, Flank AM (2006) Cytotoxicity of CeO2 nanoparticles for Escherichia coli. Physico-chemical insight of the cytotoxicity mechanism. Environ Scitechnol 40(19):6151–6156

Throbäck IN, Johansson M, Rosenquist M, Pell M, Hansson M, Hallin S (2007) Silver (Ag^{+}) reduces denitrification and induces enrichment of novel nirK genotypes in soil. FEMS Microbiol Lett 270(2):189–194

Tomczak MM, Slocik JM, Stone MO, Naik RR (2007) Bio-based approaches to inorganic material synthesis. Biochem Soc Trans 35:512–515

Tripathi DK, Ahmad P, Sharma S, Chauhan DK, Dubey NK (eds) (2017) Nanomaterials in plants, algae, and microorganisms: concepts and controversies, vol vol. 1. Academic Press, London

Tsao N, Kanakamma PP, Luh TY, Chou CK, Lei HY (1999) Inhibition of Escherichia coli-induced meningitis by carboxyfullerene. Antimicrob Agents Chemother 43:2273–2277

Viguie JR, Sukmanowski J, Nolting B, Royer FX (2007) Study of agglomeration of alumina nanoparticles by atomic force microscopy (AFM) and photon correlation spectroscopy (PCS). Colloids Surf A Physicochem Eng Asp 302:269–275

Vishwakarma K, Upadhyay N, Kumar N, Tripathi DK, Chauhan DK, Sharma S, Sahi S (2018) Potential applications and avenues of nanotechnology in sustainable agriculture. In: Nanomaterials in plants, algae, and microorganisms. Academic Press, London, pp 473–500

Wang MF, Dykstra TE, Lou XD, Salvador MR, Scholes GD, Winnik MA (2006) Colloidal CdSenanocrystalspassivated by a dye-labeled multidentate polymer: quantitative analysis by size-exclusion chromatography. Angew Chem Int Ed 45:2221–2224

Wiley B, Sun Y, Mayers B, Xia Y (2005) Shape-controlled synthesis of metal nanostructures: the case of silver. Chem Eur J 11(2):454–463

Wise JP Sr, Goodale BC, Wise SS, Craig GA, Pongan AF, Walter RB, Thompson WD, Ng AK, Aboueissa AM, Mitani H, Spalding MJ (2010) Silver nanospheres are cytotoxic and genotoxic to fish cells. Aquat Toxicol 97(1):34–41

Xia Y, Xiong Y, Lim B, Skrabalak SE (2009) Shape-controlled synthesis of metal nanocrystals: simple chemistry meets complex physics? Angew Chem Int Ed 48(1):60–103

Yadav A, Kon K, Kratosova G, Duran N, Ingle AP, Rai M (2015) Fungi as an efficient mycosystem for the synthesis of metal nanoparticles: progress and key aspects of research. Biotechnol Lett 37(11):2099–2120

Yang L, Watts DJ (2005) Particle surface characteristics may play an important role in phytotoxicity of alumina nanoparticles. Toxicol Lett 158(2):122–132

Zemke-White WL, Clements KD, Harris PJ (2000) Acid lysis of macroalgae by marine herbivorous fishes: effects of acid pH on cell wall porosity. J Exp Mar Bio Ecol 245:57–68

Zhao XU, Liz W, Chen Y, Ahi LY, Zhu YF (2007) Solid-phase photocatalytic degradation of polyethylene plastic under UV and solar light irradiation. J Mol Catal A Chem 268:101–106

Zhao X, Liu W, Cai Z, Han B, Qian T, Zhao D (2016) An overview of preparation and applications of stabilized zero-valent iron nanoparticles for soil and groundwater remediation. Water Res 100:245–266

Zhu X, Chang Y, Chen Y (2010) Toxicity and bioaccumulation of TiO2 nanoparticle aggregates in *Daphnia magna*. Chemosphere 78(3):209–215

Chapter 4
Tripartite Interaction Among Nanoparticles, Symbiotic Microbes, and Plants: Current Scenario and Future Perspectives

Ovaid Akhtar, Ifra Zoomi, Dheeraj Pandey, Harbans Kaur Kehri, and Raghvendra Pratap Narayan

Abbreviations

AM	Arbuscular mycorrhiza
CNT	Carbon nanotubes
DOC	Dissolved organic carbon
GRSPs	Glomalin-related soil proteins
MWCNTs	Multi-walled carbon nanotubes
NP	Nanoparticles
ROS	Reactive oxygen species

4.1 Introduction

Nanoparticles (NPs) are defined as the particles with a size of 1–100 nm in diameter (Auffan et al. 2009). Because of their high surface-to-volume ratio, they are very reactive. They can pass even through the cell membrane. These fascinating characteristics make them unique and hence, overwhelming researches have been carried out to explore their possible roles in biotechnology and agriculture production (Singh et al. 2016; Mishra et al. 2017; Shweta et al. 2017, 2018; Arif et al. 2018;

O. Akhtar
Department of Botany, Kamla Nehru Institute of Physical and Social Sciences, Sultanpur, India

I. Zoomi · D. Pandey · H. K. Kehri
Sadasivan Mycopathology Laboratory, Department of Botany, University of Allahabad, Allahabad, India

R. P. Narayan (✉)
Netaji Subhash Chandra Bose Government Girls P.G. College, Lucknow, India

I. Bhushan et al. (eds.), *Nanomaterials and Environmental Biotechnology*, Nanotechnology in the Life Sciences,
https://doi.org/10.1007/978-3-030-34544-0_4

Vishwakarma et al. 2018; Kehri et al. 2019; Rastogi et al. 2019a, b; Singh et al. 2019). NPs have promising application in the field of nanofertilizer, nanopesticide, nanoherbicide, nanosensor and as a delivery system for the sustained release of agrochemicals (Mishra et al. 2017; Koul et al. 2018; Singh et al. 2018; Tripathi et al. 2018; Vishwakarma et al. 2018). It was of thought that NPs could be used for boosting the agricultural economic growth in the near future (Sabourin and Ayande 2015). But, unregulated use of NPs has threw the ecosystems at the brink of risk caused by toxicity of these NPs (Hong et al. 2014). Although NPs have gained the rapid attention of plant scientists during the recent few years, their fates in relation to sustainable agriculture are still unexplored.

Despite of achievements in the field of NPs in agronomy, their role is still in a topsy-turvy stage (Yang et al. 2017). Different types of plants behave NPs differently. Also, same plant behaves different NPs differently. The response of NPs on plants as well as microbes depends upon the nature and concentration of NPs, nature of coating material, nature of growth media, mode of application, part of the plants, and even stage of plant development (Yang et al. 2017). As most of the NPs are made up of heavy metals, their elevated concentration in the biological systems could be detrimental to both aboveground and belowground flora as well.

Yang et al. (2017) have concluded the interaction of NPs with plants and microbes in their review. They have discussed pros and cons of NPs in details. But unfortunately, no conclusion could be established about the effect of NPs on symbioses in plants. In this chapter, extensive literature mining has been carried out to collect all the reports on interactions where effects of NPs have been studied on plants in relation to the symbiotic bacteria and/or fungi. Although there are very limited reports on tripartite interaction, it has been tried to find out the nature of interaction. Also, the possible mechanism behind these interactions is elucidated with the help of artwork. Eventually, this chapter puts forward the future perspectives and direction of research to study the behavior of NPs with reference to plant-microbe symbioses in greater depth.

4.2 Nanoparticles Versus Plant Growth

First stage of NP and plant interaction is the uptake of NPs inside the plant cell. Usually, plants can accumulate NPs of 40–50 nm dimension (Sabo-Attwood et al. 2012; Taylor et al. 2014). Besides the size, morphology, chemical nature, and coating properties play crucial roles in the absorption of NPs in plants (Raliya et al. 2016).

After absorption of NPs through roots, they follow either apoplastic pathway (extracellular spaces and xylem vessels) or symplastic pathway (across the living cell and through plasmodesmata). On foliar application, most of NPs are absorbed through the stomata (Pérez-de-Luque 2017). After absorption NPs are transported to the sink and internalized into the cell by pore formation, carrier proteins, and plasmodesmata or through the ion channels (Pérez-de-Luque 2017).

Fe_3O_4, TiO_2, and carbon NPs were reported to inhibit the seed germination rate, root elongation, and germination index in cucumber plant (Mushtaq 2011). Cu NPs were found inhibitory on emerging root length, and Ag NPs decreased the plant biomass and transpiration (Stampoulis et al. 2009). ZnO NPs caused cytotoxic (lipid peroxidation) and genotoxic effects (decreasing mitotic index, increasing micronuclei and chromosomal aberration index) in *Allium cepa* (Kumari et al. 2011). CuO NPs have DNA-damaging effect in *Raphanus sativus*, *Lolium perenne*, and *Lolium rigidum* (Atha et al. 2012).

On the contrary to above findings, carbon nanotubes (CNTs) were reported to increase the seed germination and growth in tomato (Khodakovskaya et al. 2009). TiO_2 NPs improved the energy utilization and conversion efficiency in D1/D2/Cyt b559 complex in spinach (Su et al. 2009). Stampoulis et al. (2009) reported no negative effects of multi-walled carbon nanotubes (MWCNTs) and Ag, Cu, ZnO, and Si NPs on seed germination in *Cucurbita pepo*. Lee et al. (2010) reported no negative or positive effect of Al_2O_3 NPs on *Arabidopsis thaliana*. TiO_2 NPs were absorbed by the *Triticum aestivum*, *Brassica napus*, and *Arabidopsis thaliana* but did not affect their germination and root elongation (Larue et al. 2011).

Despite enormous researches carried out on NP-plant interaction, no clear conclusion has been made till date. NPs are reported to play positive, negative, as well as neutral roles on plant growth. Various fates of NPs in plants and their interaction have been reviewed in details by Pérez-de-Luque (2017). Accordingly, NPs alter the physiology, biochemistry, and genetics of plants. The interaction depends upon various factors, e.g., nature and concentration of NPs, nature of coating material, nature of growth media, mode of application, part of the plants, and even stage of plant development (Yang et al. 2017).

4.3 Nanoparticles Versus Soil Microorganisms

Plants are immobile organisms. They are dependent upon root system for water and nutrient absorption. In the vicinity of root systems, millions of microbes including plant growth-promoting bacteria (PGPR) dwell. They affect the root systems through their activity and themselves get affected by the exudates of roots (Philippot et al. 2013; Zoomi et al. 2017). More often, after continuous absorption, nutrient and water adjacent to the root systems get depleted. Because of the limited growth of roots, they cannot extract their need from soil from farther distance. In such conditions it is arbuscular mycorrhizal (AM) fungi and other microbes play key roles in the survival of plants (Zoomi et al. 2017; Kehri et al. 2018). NPs in the soil also interact with these microbes, and this interaction can control the growth and survival of aboveground plants. Therefore, the consequences of NPs on belowground biodiversity are of prime importance to understand.

The research study carried out on the effect of NPs on soil microbes is still in the beginning phase. NPs are reported as a good stimulator of soil microbes, but their harmful effect cannot be neglected (Rajput et al. 2018). Some research studies, however, report NPs as soil microbe neutral. The harmful aspect of NPs on soil microorganisms has been reviewed by Rajput et al. (2018) in details. The interaction of NPs with soil microbes depends upon particle configuration and coating (McKee and Filser 2016). McKee and Filser (2016) summarized the morphological nature of bacteria with reference to the NPs. In most of the reports, they have found gram-negative bacteria as less sensitive than gram-positive bacteria. However, any general conclusion could not be drawn. Further, toxicity may be related to specific nanoscale properties of the NPs.

McKee and Filser (2016) concluded that Ag- and Cu-based NPs exhibit antimicrobial activities. Suresh et al. (2013) described the toxic effects of Ag, Al_2O_3, TiO_2, CeO_2, CuO, CdSe, CdTe, FePt, and ZnO NPs. 1.0 g/l citrate-coated Ag NPs (c-Ag NPs) inhibited the growth of *E. coli* by 90 ± 5% (Doody et al. 2014). Simonin et al. (2016) reported very strong negative impact of TiO_2 NPs on nitrification enzyme activities and the abundances of ammonia-oxidizing microorganism just after 90 days of exposure to even the lowest, realistic concentration. High concentration of TiO_2 (Du et al. 2011; Ge et al. 2011; Simonin et al. 2015) and ZnO (Du et al. 2011; Ge et al. 2011) reported to inhibit microbial respiration and enzyme activity in the soil. CuO and Fe_2O_3 NPs were reported as potentially harmful to soil environment (Frenk et al. 2013). Besides the negative effects, NPs were also reported to pose positive effects on soil microorganisms. 1.0 g/l citrate-coated Ag NPs (c-Ag NPs) enhanced the growth of *Bacillus subtilis* by 127 ± 23% (Doody et al. 2014). Also, Si, Fe, Au, Pd, Ag_2S, and Pt NPs exhibit no or little effect on soil microbes (Suresh et al. 2013).

4.4 Nanoparticles Versus Symbioses

4.4.1 ZnO Nanoparticle Versus Symbioses

AM fungi reduced the Zn accumulation in plants (Wang et al. 2016; Li et al. 2015). Under low phosphorus condition, AM fungi reduced translocation, uptake, and accumulation of Zn (Jing et al. 2016; Wang et al. 2014, 2018) in plant. AM fungi-mediated reduction in the soil pH is responsible for decreased Zn bioavailability (Jing et al. 2016). Further, under low phosphorus conditions, AM fungi reduce the translocation and accumulation of Zn released from Zn NP (Jing et al. 2016). AM fungi secrete a group of glycoproteins called glomalin-related soil proteins (GRSPs) in the soil. GRSPs bind ZnO NP in the soil (Ghasemi Siani et al. 2017), making them immobilized. The efficiency of AM fungi is further increased under organic phosphorus (Wang et al. 2018) and low phosphorus conditions (Jing et al. 2016). Under low phosphorus condition, the efficiency of AM fungi is at maximum. High level of phosphorus in the soil inhibits the AM colonization, as plants get adequate phosphorus and the mutualism starts shifting toward parasitism. Under $ZnSO_4$

treatment the efficiency of AM fungi becomes more responsive toward the ZnO NP (Li et al. 2015). It is because AM fungi perceive heavy metal toxicity more quickly. Other mechanism of amelioration of ZnO NP toxicity in plants is the AM fungi-mediated increase in the accumulation of P, N, K, Fe, and Cu (Wang et al. 2014, 2016, 2018; Jing et al. 2016). Decrease in ZnO NP toxicity was confirmed by reduced ROS production in AM fungi-colonized plants (Wang et al. 2016).

ZnO NPs enhance the growth performance in plants inhabited by non-AM symbiotic microbes (Medina-Velo et al. 2017; Singhal et al. 2017; Bandyopadhyay et al. 2015). ZnO nanorods stimulate the number of fungal pellets, spore size, early sporulation, and thick hyphae in *Piriformospora indica*. This caused increased crop productivity (Singhal et al. 2017). ZnO NPs increased the growth performance in leguminous plants by accumulation of essential elements – sulfur and magnesium – assisted by higher nodule formation in the roots.

Contrary to above findings, ZnO NPs were found bactericidal to *Sinorhizobium meliloti* (Bandyopadhyay et al. 2012). However, it is less toxic than $ZnCl_2$ for *Medicago sativa* L. – *Sinorhizobium meliloti* association (Bandyopadhyay et al. 2015). Ultra-high-resolution scanning transmission electron microscopy (STEM) revealed that ZnO NPs were accumulated on the bacterial cell wall and internalized into the periplasmic space. ZnO NPs also altered the polysaccharide structures of extracellular polymeric substances on bacterial cell wall (Bandyopadhyay et al. 2012). ZnO NPs also reduced the root shoot biomass (Bandyopadhyay et al. 2015).

4.4.2 Ag Nanoparticle Versus Symbioses

Ag NPs over 0.01 mg/kg enhanced the growth and ecological behavior and also decreased the antioxidants' activity in AM fungi-associated plants by decreasing the Ag content in plants (Feng et al. 2013). They through X-ray microcomputed tomography found that AM fungi decreased the Ag content in plants.

Ag NPs in low concentration (50 ppm) improved growth parameters in leguminous plant (Pallavi et al. 2016). They have found that it increased the total bacteria, nitrogen fixer, and phosphate solubilizer count. Also, Ag NPs increase root nodulation in the roots of legumes. However, higher concentration (75 ppm) of Ag NPs was found toxic to nitrogen fixers and other bacteria (Pallavi et al. 2016).

4.4.3 CeO_2 Nanoparticle Versus Symbioses

CeO_2 NPs were found as bacteriostatic to the symbiotic N_2-fixer *Sinorhizobium meliloti*. It was accumulated on the surface of bacterial cell. CeO_2 NPs altered the protein and polysaccharide structures of extracellular polymeric substances on bacterial cell wall (Bandyopadhyay et al. 2012).

4.4.4 Fe_3O_4 Nanoparticle Versus Symbioses

Fe_3O_4 NPs in higher concentration reduced the biomass production in mycorrhizal clover (Feng et al. 2013). It also decreases the bacterial abundance and dissolved organic content (DOC) in maize plant rhizosphere (Cao et al. 2016) and shifted the community composition. At high concentration Fe_3O_4 NPs reduced the root mycorrhizal colonization, soil GRSP content, and alkaline phosphatase activity. This also decreased the availability of P nutrition in plants (Cao et al. 2017).

AM fungi reduced the negative effect of Fe_3O_4 NPs on the plant-microbe interaction (Cao et al. 2016). They have found that AM fungi enhanced the growth of plant and organic matter release from root (Cao et al. 2016).

4.5 Conclusions

On the account of controversial results obtained on the interaction of NPs with plants, soil microbes, and symbionts, it is hard to draw any straightforward mechanism of interaction. However, several common behaviors could be attributed to this interaction. NPs in higher concentration cause adverse effects on plants as well as their microbial symbionts including AM fungi. Higher concentration of NPs reduces AM colonization, soil GRSP content, and alkaline phosphatase activity (Fig. 4.1a). This decreases the availability of P nutrition in plants (Fig. 4.1a). NPs in higher concentration are bacteriostatic and sometimes toxic to nitrogen fixers and other symbiotic bacteria.

In symbiotically associated plants, the effect of NPs toxicity is significantly lower as compared to the plants without symbionts (Fig. 4.1b). AM fungal and bacterial symbionts reduce the NP toxicity to the plants by the mechanism yet to be much understood. AM fungi have more than one mechanism through which they dilute the NP toxicity in host plants (Fig. 4.1b). These include (1) immobilization and, hence, reduced accumulation of heavy metallic NP by secretion of GRSPs, (2) reduction in NP accumulation by altering the soil pH and production of more DOC, (3) accumulation of more nutrient resulting in more biomass accumulation, and (4) reduction in reactive oxygen species in plants (Fig. 4.1b).

Besides the negative roles, regulated dose of NPs has positive roles in plants, especially those inhabited by endophytic microbes and/or nitrogen fixers. The interaction among non-AM symbionts, NPs, and plants is also poorly understood. However, based on researches carried out, several conclusions could be drawn. Accordingly, NPs stimulate the plant growth by (1) increased accumulation of essential elements – sulfur and magnesium – assisted by higher nodule formation in the roots; (2) stimulation in the number of fungal pellets, spore size, early sporulation, and thick hyphae in endophytes; and (3) increasing the total bacteria, nitrogen fixer, and phosphate solubilizer count.

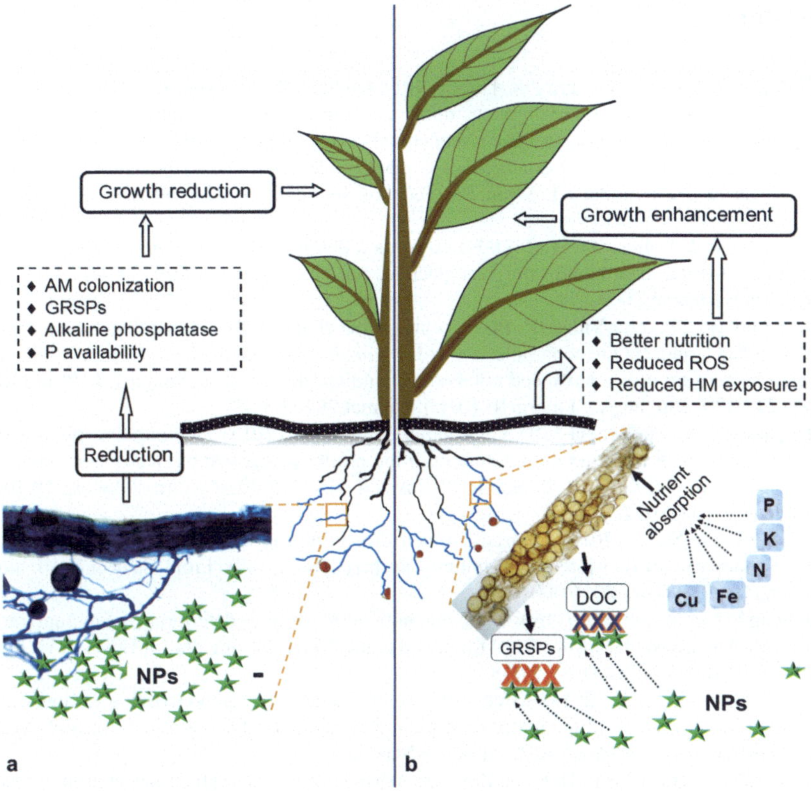

Fig. 4.1 Mechanism of tripartite interaction among NPs, AM fungi, and plants. (**a**) High concentration of NPs in soil. (**b**) Low concentration of NPs in soil

4.6 Future Perspectives

Use of engineered NPs in soil which is increasing day by day may pose detrimental effects on belowground biodiversity. Our understating on the interaction of NPs with belowground biodiversity is at its preliminary phase. For sustainable use of NPs in agriculture production, there is a need of extensive research to explore the effects of various NPs and their optimum dose, coating material, carrier, mode, time, and frequency of application on different symbiotic microbes with reference to the host plant. *Extensive research is also needed to trace and study the movement and localization of NPs in symbiotic microbes. Also, consequences of NPs internalization on gene expression and cellular functions should be studied.*

References

Arif N, Yadav V, Singh S, Tripathi DK, Dubey NK, Chauhan DK, Giorgetti L (2018) Interaction of copper oxide nanoparticles with plants: uptake, accumulation, and toxicity. In: Nanomaterials in plants, algae, and microorganisms. Academic Press, London, pp 297–310

Atha DH, Wang H, Petersen EJ et al (2012) Copper oxide nanoparticle mediated DNA damage in terrestrial plant models. Environ Sci Technol 46(3):1819–1827. https://doi.org/10.1021/es202660k

Auffan M, Rose J, Bottero JY et al (2009) Towards a definition of inorganic nanoparticles from an environmental, health and safety perspective. Nat Nanotechnol 4(10):634–641. https://doi.org/10.1038/nnano.2009.242

Bandyopadhyay S, Peralta-Videa JR, Plascencia-Villa G et al (2012) Comparative toxicity assessment of CeO_2 and ZnO nanoparticles towards *Sinorhizobium meliloti*, a symbiotic alfalfa associated bacterium: use of advanced microscopic and spectroscopic techniques. J Hazard Mater 241–242:379–386. https://doi.org/10.1016/j.jhazmat.2012.09.056

Bandyopadhyay S, Plascencia-Villa G, Mukherjee A et al (2015) Comparative phytotoxicity of ZnO NPs, bulk ZnO, and ionic zinc onto the alfalfa plants symbiotically associated with *Sinorhizobium meliloti* in soil. Sci Total Environ 515–516:60–69. https://doi.org/10.1016/j.scitotenv.2015.02.014

Cao J, Feng Y, Lin X et al (2016) Arbuscular mycorrhizal fungi alleviate the negative effects of iron oxide nanoparticles on bacterial community in rhizospheric soils. Front Environ 4:10. https://doi.org/10.3389/fenvs.2016.00010

Cao J, Feng Y, Lin X et al (2017) Iron oxide magnetic nanoparticles deteriorate the mutual interaction between arbuscular mycorrhizal fungi and plant. J Soils Sediments 17(3):841–851. https://doi.org/10.1007/s11368-016-1561-8

Doody M, Bais H, Jin Y (2014) Effects of citrate-coated silver nanoparticles on interactions between soil bacteria and the major crop plant Zea mays. In: EGU general assembly conference abstracts, Vienna, Austria, 27 April – 2 May, 2014

Du W, Sun Y, Ji R et al (2011) TiO_2 and ZnO nanoparticles negatively affect wheat growth and soil enzyme activities in agricultural soil. J Environ Monit 13(4):822–828. https://doi.org/10.1039/c0em00611d

Feng Y, Cui X, He S et al (2013) The role of metal nanoparticles in influencing arbuscular mycorrhizal fungi effects on plant growth. Environ Sci Technol 47(16):9496–9504. https://doi.org/10.1021/es402109n

Frenk S, Ben-Moshe T, Dror I et al (2013) Effect of metal oxide nanoparticles on microbial community structure and function in two different soil types. PLoS One 8(12):e84441. https://doi.org/10.1371/journal.pone.0084441

Ge Y, Schimel JP, Holden PA (2011) Evidence for negative effects of TiO_2 and ZnO nanoparticles on soil bacterial communities. Environ Sci Technol 45(4):1659–1664. https://doi.org/10.1021/es103040t

Ghasemi Siani N, Fallah S, Pokhrel LR et al (2017) Natural amelioration of Zinc oxide nanoparticle toxicity in fenugreek (*Trigonella foenum-graecum*) by arbuscular mycorrhizal (*Glomus intraradices*) secretion of glomalin. Plant Physiol Biochem 112:227–238. https://doi.org/10.1016/j.plaphy.2017.01.001

Hong J, Peralta-Videa JR et al (2014) Evidence of translocation and physiological impacts of foliar applied CeO_2 nanoparticles on cucumber (*Cucumis sativus*) plants. Environ Sci Technol 48(8):4376–4385. https://doi.org/10.1021/es404931g

Jing XX, Su ZZ, Xing HE et al (2016) Biological effects of ZnO nanoparticles as influenced by arbuscular mycorrhizal inoculation and phosphorus fertilization. Huan Jing Ke Xue 37(8):3208–3215. https://doi.org/10.13277/j.hjkx.2016.08.049

Kehri HK, Akhtar O, Zoomi I et al (2018) Arbuscular Mycorrhizal Fungi: Taxonomy and its Systematics. Int J Life Sci Res 6(4):58–71

Kehri HK, Zoomi I, Singh U et al (2019) Review on biogenic synthesis of nanoparticles and their therapeutic applications: List of novel eco-friendly approaches. J Nanosci Tech 5(4):810–816

Khodakovskaya M, Dervishi E, Mahmood M et al (2009) Carbon nanotubes are able to penetrate plant seed coat and dramatically affect seed germination and plant growth. ACS Nano 3(10):3221–3227. https://doi.org/10.1021/nn900887m

Koul A, Kumar A, Singh VK, Tripathi DK, Mallubhotla S (2018) Exploring plant-mediated copper, iron, titanium, and cerium oxide nanoparticles and their impacts. In: Nanomaterials in plants, algae, and microorganisms. Academic Press, London, pp 175–194

Kumari M, Khan SS, Pakrashi S et al (2011) Cytogenetic and genotoxic effects of zinc oxide nanoparticles on root cells of *Allium cepa*. J Hazard Mater 190(1–3):613–621. https://doi.org/10.1016/j.jhazmat.2011.03.095

Larue C, Khodja H, Herlin-Boime N et al (2011) Investigation of titanium dioxide nanoparticles toxicity and uptake by plants. J Phys Conf Ser 304:012057. https://doi.org/10.1088/1742-6596/304/1/012057

Lee CW, Mahendra S, Zodrow K et al (2010) Developmental phytotoxicity of metal oxide nanoparticles to *Arabidopsis thaliana*. Environ Toxicol Chem 29(3):669–675. https://doi.org/10.1002/etc.58

Li S, Liu X, Wang F et al (2015) Effects of ZnO nanoparticles, $ZnSO_4$ and arbuscular mycorrhizal fungus on the growth of maize. Huan Jing Ke Xue 36(12):4615–4622. http://www.ncbi.nlm.nih.gov/pubmed/27012001

McKee MS, Filser J (2016) Impacts of metal-based engineered nanomaterials on soil communities. Environ Sci Nano 3(3):506–533. https://doi.org/10.1039/C6EN00007J

Medina-Velo IA, Barrios AC, Zuverza-Mena N et al (2017) Comparison of the effects of commercial coated and uncoated ZnO nanomaterials and Zn compounds in kidney bean (*Phaseolus vulgaris*) plants. J Hazard Mater 332:214–222. https://doi.org/10.1016/j.jhazmat.2017.03.008

Mishra S, Keswani C, Abhilash PC et al (2017) Integrated approach of agri-nanotechnology: challenges and future trends. Front Plant Sci 8:471. https://doi.org/10.3389/fpls.2017.00471

Mushtaq YK (2011) Effect of nanoscale Fe_3O_4, TiO_2 and carbon particles on cucumber seed germination. J Environ Sci Health A 46(14):1732–1735. https://doi.org/10.1080/10934529.2011.633403

Pallavi, Mehta CM, Srivastava R et al (2016) Impact assessment of silver nanoparticles on plant growth and soil bacterial diversity. 3 Biotech 6(2):254. https://doi.org/10.1007/s13205-016-0567-7

Pérez-de-Luque A (2017) Interaction of nanomaterials with plants: what do we need for real applications in agriculture? Front Environ Sci 5:12. https://doi.org/10.3389/fenvs.2017.00012

Philippot L, Raaijmakers JM, Lemanceau P et al (2013) Going back to the roots: the microbial ecology of the rhizosphere. Nat Rev Microbiol 11(11):789–799. https://doi.org/10.1038/nrmicro3109

Rajput VD, Minkina T, Sushkova S et al (2018) Effect of nanoparticles on crops and soil microbial communities. J Soils Sediments 18(6):2179–2187. https://doi.org/10.1007/s11368-017-1793-2

Raliya R, Franke C, Chavalmane S et al (2016) Quantitative understanding of nanoparticle uptake in watermelon plants. Front Plant Sci 7:1288. https://doi.org/10.3389/fpls.2016.01288

Rastogi A, Tripathi DK, Yadav S, Chauhan DK, Živčák M, Ghorbanpour M, El-Sheery NI, Brestic M (2019a) Application of silicon nanoparticles in agriculture. 3 Biotech 9(3):90

Rastogi A, Zivcak M, Tripathi DK, Yadav S, Kalaji HM, Brestic M (2019b) Phytotoxic effect of silver nanoparticles in Triticum aestivum: improper regulation of photosystem I activity as the reason for oxidative damage in the chloroplast. Photosynthetica 57(1):209–216

Sabo-Attwood T, Unrine JM, Stone JW et al (2012) Uptake, distribution and toxicity of gold nanoparticles in tobacco (*Nicotiana xanthi*) seedlings. Nanotoxicology 6(4):353–360. https://doi.org/10.3109/17435390.2011.579631

Sabourin V, Ayande A (2015) Commercial opportunities and market demand for nanotechnologies in agribusiness sector. J Technol Manag Innov 10(1):40–51. https://doi.org/10.4067/S0718-27242015000100004

Shweta, Vishwakarma K, Sharma S, Narayan RP, Srivastava P, Khan AS, Dubey NK, Tripathi DK, Chauhan DK (2017) Plants and carbon nanotubes (CNTs) interface: present status and future prospects. In: Nanotechnology. Springer Singapore, Singapore, pp 317–340

Shweta, Tripathi DK, Chauhan DK, Peralta-Videa JR (2018) Availability and risk assessment of nanoparticles in living systems: a virtue or a peril? In: Nanomaterials in plants, algae, and microorganisms. Academic Press, London, pp 1–31

Simonin M, Guyonnet JP, Martins JMF et al (2015) Influence of soil properties on the toxicity of TiO_2 nanoparticles on carbon mineralization and bacterial abundance. J Hazard Mater 283:529–535. https://doi.org/10.1016/j.jhazmat.2014.10.004

Simonin M, Richaume A, Guyonnet JP et al (2016) Titanium dioxide nanoparticles strongly impact soil microbial function by affecting archaeal nitrifiers. Sci Rep 6(1):33643. https://doi.org/10.1038/srep33643

Singh S, Tripathi DK, Dubey NK, Chauhan DK (2016) Effects of nano-materials on seed germination and seedling growth: striking the slight balance between the concepts and controversies. Mater Focus 5(3):195–201

Singh A, Singh NB, Hussain I et al (2018) Plant-nanoparticle interaction: an approach to improve agricultural practices and plant productivity. Int J Pharm Sci Invent 4(8):25–40

Singh J, Vishwakarma K, Ramawat N, Rai P, Singh VK, Mishra RK, Kumar V, Tripathi DK, Sharma S (2019) Nanomaterials and microbes' interactions: a contemporary overview. 3 Biotech 9(3):68

Singhal U, Khanuja M, Prasad R et al (2017) Impact of synergistic association of ZnO-nanorods and symbiotic fungus *Piriformospora indica* DSM 11827 on *Brassica oleracea* var. botrytis (Broccoli). Front Microbiol 8:1909. https://doi.org/10.3389/fmicb.2017.01909

Stampoulis D, Sinha SK, White JC (2009) Assay-dependent phytotoxicity of nanoparticles to plants. Environ Sci Technol 43(24):9473–9479. https://doi.org/10.1021/es901695c

Su M, Liu H, Liu C et al (2009) Promotion of nano-anatase TiO_2 on the spectral responses and photochemical activities of D1/D2/Cyt b559 complex of spinach. Spectrochim Acta A Mol Biomol Spectrosc 72(5):1112–1116. https://doi.org/10.1016/j.saa.2009.01.010

Suresh AK, Pelletier DA, Doktycz MJ (2013) Relating nanomaterial properties and microbial toxicity. Nanoscale 5(2):463–474. https://doi.org/10.1039/C2NR32447D

Taylor AF, Rylott EL, Anderson CWN et al (2014) Investigating the toxicity, uptake, nanoparticle formation and genetic response of plants to gold. PLoS One 9(4):e93793. https://doi.org/10.1371/journal.pone.0093793

Tripathi DK, Ahmad P, Sharma S, Chauhan DK, Dubey NK (eds) (2018) Nanomaterials in plants, algae, and microorganisms: concepts and controversies, vol 1. Academic Press, London

Vishwakarma K, Upadhyay N, Kumar N, Tripathi DK, Chauhan DK, Sharma S, Sahi S (2018) Potential applications and avenues of nanotechnology in sustainable agriculture. In: Nanomaterials in plants, algae, and microorganisms. Academic Press, London, pp 473–500

Wang WZ, Wang FY, Li S et al (2014) Arbuscular mycorrhizal symbiosis influences the biological effects of nano-ZnO on maize. Huan Jing Ke Xue 35(8):3135–3141. http://www.ncbi.nlm.nih.gov/pubmed/25338390

Wang F, Liu X, Shi Z et al (2016) Arbuscular mycorrhizae alleviate negative effects of zinc oxide nanoparticle and zinc accumulation in maize plants – a soil microcosm experiment. Chemosphere 147:88–97. https://doi.org/10.1016/j.chemosphere.2015.12.076

Wang F, Jing X, Adams CA et al (2018) Decreased ZnO nanoparticle phytotoxicity to maize by arbuscular mycorrhizal fungus and organic phosphorus. Environ Sci Pollut Res 25(24):23736–23747. https://doi.org/10.1007/s11356-018-2452-x

Yang J, Cao W, Rui Y (2017) Interactions between nanoparticles and plants: phytotoxicity and defense mechanisms. J Plant Interact 12(1):158–169. https://doi.org/10.1080/17429145.2017.1310944

Zoomi I, Narayan RP, Akhtar O et al (2017) Role of plant growth promoting rhizobacteria in reclamation of wasteland. In: Microbial biotechnology. Springer Singapore, Singapore, pp 61–80. https://doi.org/10.1007/978-981-10-6847-8_3

Chapter 5
Effect of Nanoparticles on Plant Growth and Physiology and on Soil Microbes

Muhammad Nafees, Shafaqat Ali, Muhammad Rizwan, Asma Aziz, Muhammad Adrees, Syed Makhdoom Hussain, Qasim Ali, and Muhammad Junaid

5.1 Introduction

The use of nanotechnology for potential benefits in agriculture is enormous and has been increasing day by day (Shapira and Youtie 2015; Resham et al. 2015; Nath 2015). Novel applications of nanotechnology have been developed in biotechnology and agriculture (Siddiqui et al. 2015; Singh et al. 2016, 2019; Shweta et al. 2017, 2018; Arif et al. 2018; Vishwakarma et al. 2018) to manage food productivity (Kumari et al. 2014). Nanoparticles (NPs) are very tiny particles, defined as the 10^{-9} part of 1 m (1 m^{-9}) (Huang et al. 2015). NP efficiency relies on their surface area, size, composition, shape, and above all the effective concentration at which they work efficiently (Khodakovskaya et al. 2012; Ranjan et al. 2014;

M. Nafees
State Key Laboratory of Pollution Control and Resource Reuse, School of the Environment, Nanjing University, Nanjing, China

S. Ali (✉)
Department of Environmental Sciences and Engineering, Government College University, Faisalabad, Pakistan

Department of Biological Sciences and Technology, China Medical University (CMU), Taichung, Taiwan
e-mail: shafaqataligill@gcuf.edu.pk

M. Rizwan · M. Adrees
Department of Environmental Sciences and Engineering, Government College University, Faisalabad, Pakistan

A. Aziz · S. M. Hussain
Department of Zoology, Government College University, Faisalabad, Pakistan

Q. Ali
Department of Botany, Government College University, Faisalabad, Pakistan

M. Junaid
Key Laboratory for Heavy Metal Pollution Control and Reutilization, School of Environment and Energy, Peking University Shenzhen Graduate School, Shenzhen, China

I. Bhushan et al. (eds.), *Nanomaterials and Environmental Biotechnology*, Nanotechnology in the Life Sciences,
https://doi.org/10.1007/978-3-030-34544-0_5

Dasgupta et al. 2016; Jain et al. 2016; Maddineni et al. 2015). Nanotechnology provides a very large variety of techniques and devices to formulate NPs, detect biotic and abiotic stress in plants, and provide genetic manipulation that allows more precise plant breeding (Perez-de-Luque and Hermosin 2013; Fraceto et al. 2016). Fertilizers are very important in the growth, development, and metabolism of plants (Giraldo et al. 2014), but at most concentrations applied fertilizers are not available to plants because of leaching, runoff, and degradation. Thus, it is very important to control or minimize chemical fertilizer loss. With their unique properties, NPs encapsulate nutrients, which, released as required, control the discharge of chemical fertilizers for plant growth (Derosa et al. 2010; Nair et al. 2010; Shweta et al. 2018). Several studies have shown that particular low doses of NPs enhance plant physiology (Zheng et al. 2005; Klaine et al. 2008). NPs can enter plant cells through the stomata of leaves and roots to transport nutrients, DNA, and chemicals (Galbraith 2007; Torney et al. 2007). Nanomaterials can break down the plasma membrane, inducing pore formation to enter into the plant cells (Wong et al. 2016) and reach the cytosol (Serag et al. 2011). These NPs enhance chlorophyll activity, water uptake, and specific microbial communities in the soil (Fig. 5.1).

With unique physicochemical properties, NPs can enhance the biochemical processes of plants (Giraldo et al. 2014). The application of carbon nanotubes (CNTs) to activate the growth and physiology of different plants has been well documented;

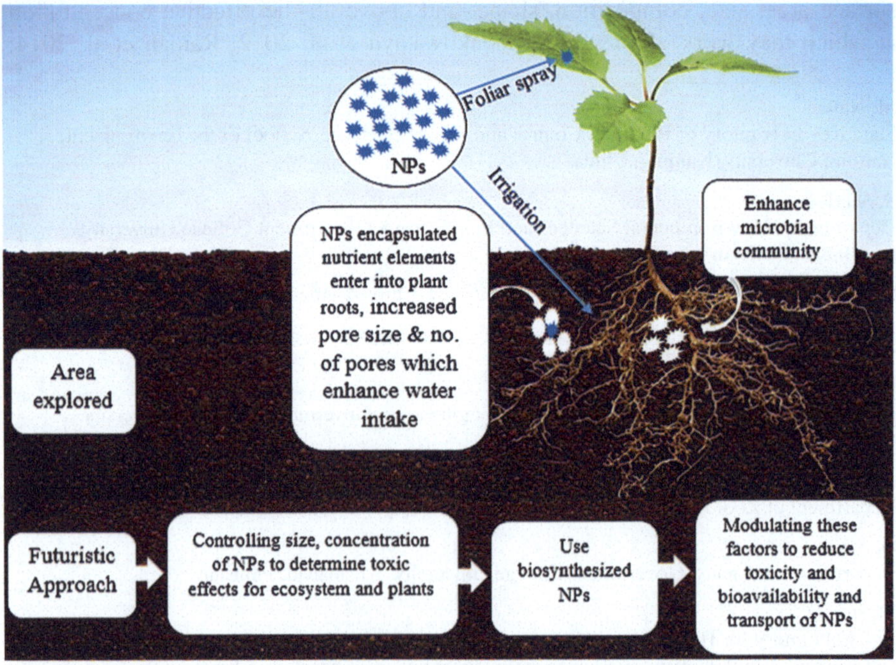

Fig. 5.1 Nanoparticle spray or irrigation and the effects on plant growth and the soil microbial community

for example, the root growth of ryegrass, onion, and cucumber was increased by CNTs (Lin and Xing 2007; Canas et al. 2008; Shweta et al. 2017). NPs have some toxic effects on plants and other living organisms, but also increase the growth, physiology, and photosynthesis of plants. This review discusses the impact of nanoparticles on plants and microbial communities.

5.2 Effect of Nanoparticles on Plants

The impact of nanoparticles on plants depends upon the plant species and the NP variety (Table 5.1) (Nair 2016; Servin and White 2016; Singh et al. 2016; Vishwakarma et al. 2018; Tripathi et al. 2017; Rastogi et al. 2019). Minerals such as nitrogen and phosphorus act as growth factors, regulating plant growth and also increasing crop productivity. Phosphorus fertilizer increases the availability of phosphorus in the soil and increases the uptake of phosphorus from the root surfaces. In phosphorus-solubilizing enzymes in which Zn is a cofactor, phosphatase and phytase enzyme activity was increased by 84–108%. ZnO NPs also enhanced root length, root volume, and the chlorophyll and protein content of the leaves in mung bean plants. ZnO NPs also maintained soil health by influencing the soil microbial community (Raliya et al. 2016).

Germination of cucumber seed was enhanced by exposure to various concentrations of ZnO NPs (de la Rosa et al. 2013). ZnO NPs not only were absorbed by *Vigna radiata* and *Cicer arietinum* roots but also improved the length and biomass of the roots and shoots of these species (Mahajan et al. 2011). This NP also enhanced somatic embryogenesis by shoot regeneration, induced the synthesis of proline, and increased tolerance against stress by increasing the activity of different enzymes (Helaly et al. 2014). Gold (Au) NPs enhanced the seed germination of *Brassica juncea*, *Boswellia ovalifoliolata*, and *Gloriosa superba* (Arora et al. 2012; Gopinath et al. 2014). The Au NPs increased the number of leaves, leaf area, and length of the plant and its chlorophyll and carbohydrate content, which increased growth, development, and crop yield (Arora et al. 2012; Gopinath et al. 2014). The Au NPs demonstrated importance in seed germination, in antioxidants, and altered the expression of micro-RNAs that regulate morphological, physiological, and metabolic processes in plants (Kumar et al. 2013).

The effects of CeO_2 were collectively found on seed germination, vegetative parts, the cotyledon, floral parts, and ripening of fruits. The rate of seed germination (97%) was high in a 10 mg/l concentration of CeO_2. No negative effect on germination and no significant effect on production of chlorophyll was seen with any concentration of CeO_2 NPs on tomato plants, although there was a significant difference in the growth of the vegetative parts of the tomato plant; faster growth was found at 10 mg/l CeO_2 NPs. The number of floral buds was slightly higher in the control and the 10 mg/l concentration of CeO_2 NPs, and 67% of buds were converted into the flower. Fruit size, production, and ripening were enhanced by increasing concentrations of CeO_2 NPs; large, heavy fruits were found at 10 mg/l (Wang et al. 2012a).

Table 5.1 Effect of nanoparticles on plant growth/physiology/tolerance against stress

Nanoparticles	Plant	Impact on plant parts/process	References
Al_2O_3	*Lemna minor*	Increased root length, photosynthetic activity, biomass accumulation	Juhel et al. (2011)
TiO_2	*Triticum aestivum*	Increased root length	Larue et al. (2012)
CeO_2, ZnO	*Zea mays*	Reduced yield	Zhao et al. (2012)
CuO	*Brassica napus*	Increased plant growth	Rahmani et al. (2016)
$FeCl_3$	*Lepidium sativum* *Sinapis alba* *Sorghum saccharatum*	Seed germination, seedling length, biomass	Libralato et al. (2016)
Ag NO_3	Lentil seed	Seed germination/elongation of root and shoot	Hojjat and Hojjat (2016)
Fe_2O_3	Soybean	Increased root length, regulated the enzyme	Alidoust and Isoda (2013)
Cu, Zn	Wheat seedling	Increased RWC and stabilized photosynthetic pigments	Taran et al. (2017)
$Ca_3(PO_4)_2$	Rice	Increased growth, micro-fertilizer and promoter of growth	Upadhyaya et al. (2017)
Fe_3O_4, TiO_2	Soya bean	Enhanced plant growth, crop yield, effect on leaf carbon and phosphorus	Burke David et al. (2015)
CeO_2	Soya bean	Stimulated plant growth, rubisco carboxylase activity, relative water content	Cao et al. (2017)
ZnO	Chickpea	Effect on root, accumulation of biomass in seedlings, lowered ROS, promoted antioxidant activity	Burmana et al. (2013)
Ag	Wheat	Increased shoot fresh and dry weight, enhanced salt tolerance ability of crop	Mohamed et al. (2017)
SiO_2	*Zea mays* L., *Phaseolus vulgaris* L., *Hyssopus officinalis* L., *Nigella sativa* L., *Amaranthus retroflexus* L., *Taraxacum officinale* F. H. Wigg	Seed germination, root and shoot length, fresh weight (except *Hyssopus officinalis* L.) and dry weight, photosynthetic pigments, total protein and total amino acids (except *Hyssopus officinalis* L.) significantly increased at 400 mg l^{-1}; these parameters were decreased in weeds, and total carbohydrates decreased in all plants except *A. retroflexus*	Sharifi-Rad et al. (2016)

(continued)

Table 5.1 (continued)

Nanoparticles	Plant	Impact on plant parts/process	References
Ag	Wheat (*Triticum astivium* var. UP2338), cowpea (*Vigna sinensis* var. Pusa Komal), brassica (*Brassica juncea* var. Pusa jai Kisan), oat	Wheat was unaffected by Ag NPs, but overall growth of cowpea and *Brassica* plants was influenced	Pallavi et al. (2016)
TiO_2	*Arabidopsis thaliana* (L.) Heynh, corn, cabbage, lettuce, oat, *Brassica napus* L. Cucumber, fennel, onion, tomato Parsley (*Petroselinum crispum* Mill.), red clover, soybean, spinach, wheat	Enhanced germination, root elongation and seedling growth	Szymanska et al. (2016), Andersen et al. (2016), Mahmoodzadeh et al. (2013), Servin et al. (2012), Feizi et al. (2013, 2012), Haghighi and Teixeira da Silva (2014), Dehkourdi and Mosavi (2013), Gogos et al. (2016), Rezaei et al. (2015), Zheng et al. (2005), Mahmoodzadeh and Aghili (2014).
TiO_2	Chickpea (*Cicer arietinum* L.), tomato, wheat, Flax (*Linum usitatissium* L.)	Enhanced tolerance against cold in chickpea, heat in tomato, drought in wheat and flax	Mohammadi et al. (2013, 2014), Qi et al. (2013), Jaberzadeh et al. (2013), Aghdam et al. (2016)
TiO_2	Tomato, oilseed rape, *Arabidopsis*, spinach, basil (*Ocimum basilicum* L.)	Increased chlorophyll contents of tomato and oil seed rape, promoted activity of rubisco and net photosynthesis in *Arabidopsis*, spinach, tomato, and basil (*Ocimum basilicum* L.)	Raliya et al. (2015a), Li et al. (2015), Ze et al. (2011), Lei et al. (2008), Kiapour et al. (2015)
TiO_2	Barley, corn, mung bean, snail clover, tomato, wheat	Enhanced crop yield and biomass	Moaveni and Kheiri (2011), Morteza et al. (2013), Raliya et al. (2015b), Rafique et al. (2015)

Clement et al. (2013) determined the effect of TiO_2 NPs on algae, rotifers, and plants. High concentrations of TiO_2 NPs have antimicrobial activity and also promoted the growth of roots. The collective effect of SiO_2 NPs on germination of seeds, elongation of roots and shoots, and water content of *Zea mays* L. was determined. SiO_2 NP uptake by plants from a hydroponic environment and increased growth of seed and elongation of roots was high as compared to control. Seed germination was increased at 400 mg/l but decreased at 2000 and 4000 mg/l concentrations of SiO_2 NPs. SiO_2 NPs increased root length but decreased shoot length of plants at concentrations from 0 to 4000 mg/l. SiO_2 NPs showed a negative correlation between NP concentration and relative water content (RWC) in plants. The RWC was decreased as the concentration of SiO_2 increased from 0 to 4000 mg/l. It was observed that SiO_2 NPs had a significant effect on photosynthetic pigments (chlorophyll *a*, *b*, and carotenoids), which increased at 400–4000 mg/l NP concentration in *Z. mays*. High photosynthetic content was found at 400 mg/l SiO_2 NPs (Rad et al. 2014).

5.2.1 Effects of NPs on Photosynthesis

Photosynthesis is the key mechanism that transforms light energy into chemical energy. Rubisco is an enzyme used in carbon fixation during light reactions. SiO_2 NP increased the photosynthesis rate by increasing the activity of carbonic anhydrase and the formation of photosynthetic pigments (Xie et al. 2012; Siddiqui and Al-Whaibi 2014). Carbon anhydrase acts as a supplier for CO_2 to the rubisco enzyme, which enhances photosynthesis (Siddiqui et al. 2012). TiO_2 has photocatalytic properties that not only increase the efficiency of light absorbance but also increase the conversion of light energy into chemical energy. TiO_2 also improved fixation of CO_2, prevented the plant from aging, and ultimately enhanced the photosynthesis process (Hong et al. 2005; Yang et al. 2006).

TiO_2 NPs increased CO_2 fixation by increasing the activity of rubisco and ultimately improving plant growth. TiO_2 NPs enhanced the net rate of photosynthesis, water conduction, and plant transpiration (Ma et al. 2008; Qi et al. 2013). ZnO NPs showed a positive effect on the growth of cotton (*Gossypium hirsutum* L.). The growth (130.6%) and biomass (131%) of cotton were significantly enhanced by ZnO NPs.

ZnO NPs increased the level of chlorophyll *a*, *b*, and carotenoids (141.6%, 134.7%, 138.6%, respectively) and increased soluble protein (179.4%) but reduced malondialdehyde (MDA) level in plant leaves. Various enzymatic activities of catalase, superoxide dismutase (264.2%), and peroxidase (182.8%) were also increased and improved the growth of cotton plants (Venkatachalam et al. 2016).

5.3 Effect of Nanoparticles on the Soil Microbial Community

Soil microbes have a significant role in soil health, plant growth, productivity, and biological and chemical reactions within soil and plants (Table 5.2) (Falkowski et al. 2008; Schimel and Schaeffer 2012; Philippot et al. 2013; Vacheron et al. 2013; Singh et al. 2019). NPs enter into the soil through several ways including human

Table 5.2 Effect of nanoparticles on the soil microbial community

Nanoparticles	Impact on soil microbial community/processes	References
Fe_3O_4, TiO_2	Changed the soil microbial community, influenced the colonies of nitrifying bacteria associated with roots	Burke David et al. (2015)
CuO	Influenced the composition and activity of the bacterial community, decreased the oxidative potential of the soil	Schlich and Hund-Rinke (2015)
ZnO	Ammonification, dehydrogenase, and hydrolase activity	Shen et al. (2015)
TiO_2	Influenced carbon mineralization, pH of soil, organic matter; identified soil type and moisture	Simonin et al. (2015)
CeO_2, Fe_3O_4, SnO	No effect on microbial biomass C and N	VittoriAntisari et al. (2013)
Ag	Different impact on ion release shape and function of the natural soil microbes	Zhai et al. (2016)
TiO_2, ZnO	Altered soil microbes, enhanced the degradation of organic pollutants	Ge et al. (2012)
Ag	Influenced soil microbial diversity and functional bacterial diversity	Pallavi et al. (2016)
Ag	Increased biomass of *Aspergillus niger and Penicillium chrysogenum* Enhanced soil extract and inhibited antifungal activity of Ag	Pietrzak and Gutarowska (2015)
Ag	Affected functional diversity of soil microbial community and associated ecosystem processes	Zhai et al. (2016)
CuO, Fe_3O_4	Increased toxicity toward microbial community	Frenk et al. (2013)
SiO_2, Pd, Au, Cu	Increased number of microbial colonies in soil, enhanced metabolic rate of soil community	Shah and Belozerova 2009

activity, sewage, and industrial waste. NPs of silica, palladium, gold, and copper have beneficial effects on soil microbes and seed germination of lettuce (Shah and Belozerova 2009). Biological and physicochemical properties determined their health and increased soil productivity. Biosolids have been used as organic fertilizers for decades; silver and titanium NPs were detected above the threshold level and adversely affected soil microbiota (Kim et al. 2010; Rottman et al. 2012; Wang et al. 2012a, b). Zinc oxide and copper NPs did not show harmful effects on soil microbes although silver and titanium NPs showed an adverse effect on the microbial biomass richness (Cardoso et al. 2013; Shah et al. 2014).

Asadishad et al. (2017) investigated the efficacy of gold nanoparticles coated with citrate (50 nm) and polyvinylpyrrolidone (PVP) (5, 50, and 100 nm) on soil enzymatic activity and soil microbes. They noted that a low concentration of Au NPs (0.1 mg/kg) reduced the size of PVP. Au NPs stimulate soil enzymatic activity; the Au NP size and soil enzymatic activity showed no correlation at a high dose (100 mg/kg). Citrate-coated Au NPs significantly increased soil enzymatic activity as compared to PVP-coated Au NPs at 50 nm size of both particles. Biomass of the important soil bacteria Actinobacteria and Proteobacteria was increased by the addition of citrate-coated Au NPs.

5.4 Impact of Carbon Nanotubes on Plants

Carbon nanotubes are allotropic forms of carbon nanoparticles, open or closed nano-structure cylindrical tubes that are single-walled carbon nanotubes (SWCNTs) or multi-walled carbon nanotube (MWCNTs). These layers are composed of rolled sheets of graphene. These nanotubes vary from 100 nm to some centimeters in length; the outer diameter of SWCNTs varies from 0.8 to 2 nm and that of MWCNTs from 5 to 20 nm (De Volder et al. 2013). CNTs were shown to act as growth regulators for plants (Khot et al. 2012). It was also noted that different sizes and composition of CNTs affect different plant growth parameters (Table 5.3). The stress-related gene of the tomato seed was regulated by MWCNTs that enhance seed germination and growth (Khodakovskaya et al. 2009).

CNTs are involved in major cellular processes of plants such as up- or downregulation of gene expression. MWCNTs induced the expression of a gene that codes for water channels and increased the water intake ability of root cells. CNTs are very small in diameter, so they can easily pass through the pores of the cell wall and also can increase the cell-wall pores. CNTs induced pores in the cell wall that enhanced water uptake, which regulates the activity of starch hydrolase enzymes and increases seed germination (Santos et al. 2013; Vithanage et al. 2017). These CNTs also act as a slow-release fertilizer that promotes plant growth (Wu 2013).

MWCNTs are also frequently used in hydroponic culture; CNTs (2000 mg/l) increase the root length of ryegrass (Lin and Xing 2007). Canas et al. (2008) showed that CNTs enhanced the physiology of six crops: cucumber, carrot, onion, tomato, cabbage, and lettuce. Plants were treated with uncoated (0, 104, 315, or 1750 mg/l) or coated (0, 160, 900, or 5000 mg/l) CNTs for 48 h. The uncoated CNTs significantly boosted root length of onion and cucumber more than the coated CNTs, with an inverse proportion between time and root elongation in these hydroponic crops. More effective results were seen on the first day as compared to the second day. It was hypothesized that CNTs may have an obligatory effect on the root length of plants by obstructing the relationship between roots and microbes, altering vital biological and chemical reactions. CNTs not only were absorbed by the plant but accumulated in the epidermal tissue of wheat roots (Wild and Jones 2009). Citrate-coated CNTs enhanced the growth and physiology of plants by increasing water uptake capability and also the uptake of nutrients and minerals, which directly affected the photosynthesis of the plants. CNTs increased plant length and also increased the number of leaves, which enhanced plant photosynthetic activity (Tripathi et al. 2011).

MWCNTs regulated the gene expression of the aquaporin gene (*NtPIPI*), and of two water channel genes (*CycB* and *NtLRX*), which increased cell permeability for water absorption and also helped in formation of the cell wall and regulation of mitosis (Khodakovskaya et al. 2012). MWCNTs also had a significant effect on root

Table 5.3 Effect of carbon nanoparticles and nanotubes on plant growth processes

Plant name	CNPs/CNT	Impact on plant parts, growth/process	References
Lycopersicon esculentum	CNTs	Seed germination and growth	Anjum et al. (2014)
Medicago sativa, Triticum aestivum	CNTs	Root elongation	Miralles et al. (2012)
Allium cepa, Cucumis sativus	SWCNTs	Root elongation	Canas et al. 2008
Hordeum vulgare L., *Glycine max, Zea mays*	MWCNTs	Growth (leaf, root and shoot)/germination	Lahiani et al. (2013)
Wheat	MWCNTs	Root growth and yield	Wang et al. (2012a)
Lycopersicon esculentum	MWCNTs	Increased uptake of water and nutrients	Tiwari et al. (2013)
Zea mays	MWCNTs	Increased nutrient transport and yield	Tiwari et al. (2014)
Mustard plant *(Brassica juncea)*	MWCNTs	Increased seed germination, root elongation	Mondal et al. (2011)
Tomato	MWCNTs	Increased plant growth (flower and fruit) and yield	Khodakovskaya et al. (2013), Alimohammadi et al. (2011)
Wheat, maize, peanut, garlic	CNTs	Increase in root and shoot length	Rao and Srivastava (2014)
Red spinach, lettuce, rice, cucumber, chili, lady finger (okra), soybean	CNTs	Increased growth, root and shoot length	Begum et al. (2014)
Corn	CNTs	Increased growth, root and shoot length, biomass	De La Torre-Roche et al. (2013)
Hyoscyamus niger	SWCNTs	Enhanced plant performance, antioxidant activity, and biosynthesis of protein	Hatami et al. 2017
Zucchini	SWCNTs, MWCNTs	No significant change in seed germination	Stampoulis et al. (2009)
Solanum lycopersicum	CNPs	Seed coat permeability	Ratnikova et al. 2015
Buckypaper	CNTs (SWCNTs, MWCNTs)	Increased permeability (pore size)	Shen et al. (2017)
Broccoli	CNTs	Positive effect on growth, enhanced CO_2 assimilation	Martinez-Ballesta et al. (2016)
Arabidopsis thaliana	MWCNTs	Effect on efficiency of photosynthesis and physiological mechanism	Voleti (2015)

length of wheat seedlings, and on germination and growth of soya bean, corn, and barley (Wang et al. 2012a, b; Lahiani et al. 2013). The root length of wheat seedlings increased 32% with MWCNTs at 40–160 μg/l for 3 to 7 days (Wang et al. 2012a, b). CNTs impacted early plant growth by germination of seed, expression of genes, cell culturing, and physiological processes such as photosynthesis and antioxidant activities (Canas et al. 2008).

SWCNTs enhanced photosynthetic activity threefold as compared to normal photosynthesis, and increased the rate of electron transport because SWCNTs combine with the chloroplast and enable the leaf to enhance the rate of electron transport by a photo-absorption mechanism (Giraldo et al. 2014). The germination ability of seed might be enhanced by increasing concentrations of MWCNTs. The highest seed germination rate was noted at 60 μg/ml CNTs; increasing CNT concentrations increased plant growth and also enhanced the yield of cotton per plant. The highest yield of cotton was found at 100 μg/ml CNTs (Sawant 2016): there was a linear correlation between seed germination and CNT concentration. It was observed that the length of plants (62 ± 5.58cm), boll' number/ plant (5.8 ± 0.64) and size of boll (3.41 ± 0.27cm) and yield of cotton (3.4 ± 0.37/hectare) was found highest at 120 μg/ml, 80 μg/ml, 60 μg/ml, 100 μg/ml of CNTs respectively (Sawant 2016).

Various studies have shown that SWCNTs and MWCNTs positively affect germination and growth of tomato, rice, common gram, and tobacco by increasing their water uptake ability, which improves germination processes (Khodakovskaya et al. 2009; Nair 2016). The toxic levels of Ag, ZnO, and Al_2O_3 induced oxidative stress and produced reactive oxygen and nitrogen species, which reduced plant growth (Zhao et al. 2012; Thwala et al. 2013; Hossain et al. 2015; Xia et al. 2015). Oxidative species reduced rubisco activity and decreased the photo-protective activity of photosystem II (Jiang et al. 2017). The defensive system of plant consists of nonenzymatic antioxidants, which include thiols, glutathione, phenolics, ascorbate and enzymatic CAT, SOD, APX, GR, GPX, and GST (Singh et al. 2015). Oxidative stresses were caused by NPs that decreased photosynthetic rate, ultimately inhibiting plant growth (Da Costa and Sharma 2016; Li et al. 2016).

Chegini et al. (2017) observed that physiological parameters were affected by MWCNTs, drought conditions, and their interactions in *Salvia mirzayanii.* The leaf water content and chlorophyll index showed significant alterations under drought conditions. The various levels of MWCNTs affected electrolyte leakage index and caused a significant difference in phenolic compounds under the interactions of the experimental treatments. Phenolic content was significantly influenced at MWCNT 50 and 200 mg/l, to 25% of field capacity (FC), respectively. The concentration of MWCNTs (50 mg/l) in moderate drought condition changed the physiological traits and antioxidant activity of *S. mirzayanii.*

Barbinta-Patrascu et al. (2017) reported an effect of carbon nanotubes coated with chlorophyll *a* and laden biomimetic membrane. The multilamellar lipid vesicles increased antioxidant (85%) activity and antibacterial activity against *Staphylococcus aureus*, and the highest antioxidant ability was found in hybrid CNTs that originated through the multilamellar lipid vesicles (TP3). They were widely dispersed and increased the reaction sites for removal of ROS by increasing their surface area. The TP3 sample showed the highest antibacterial activity resulting from good dispersion because a large surface area was provided to destroy bacterial contamination. The SWCNTs react directly with bacterial cells and physically break down their cell membrane by puncture, causing the death of the bacterial cells (*S. aureus*) (Bai et al. 2011; Smith and Rodrigues 2015).

5.4.1 Effect of CNTs on Photosynthesis Mechanism

Sunlight is the most available source of energy, which is conserved in many ways in an ecosystem. One of the most efficient methods for the conservation of sunlight is photosynthesis. For this purpose, the higher green plants, algae, and bacteria contain special pigments that use water and CO_2 to form organic molecules. These photosynthetic organisms contain the photo-elements chlorophyll *a*, *b*, *d*, and *f*, and a series of electron carrier redox reactions (Blankenship et al. 2011). The thylakoid membrane of plastids acts as a photo-current producer in the presence of potassium ferrocyanide. The cell surface (1 cm^2) produced maximum electric power, 24 mW, at 625 nm of red light. The thylakoid membrane immobilized with MWCNTs acts as an anode with MWCNTs as a cathode, which produced the maximum current density, 38 mA/cm^2. The maximum electric power produced at this current density is 5.3 mW/cm^2 (Calkins et al. 2013). The effect of CNTs on chlorophyll *f* and *d* was more than that on chlorophyll *a* and *b*: it enhanced the absorption ability of far-red and infrared light (700–750 nm) and also enhanced the ability of photo-convertors (Voloshin et al. 2015). The CNTs were synthetic NPs that penetrate into the biological matrix and have multifunctional properties such as water uptake and conduction for electricity in biological systems. MWCNTs were most electro-conductive in BY-2 tobacco cells as compared to balsam fir wood at high temperature (Di Giacomo et al. 2013; Leslie et al. 2014).

It was investigated whether CNTs had a positive effect on photosystem I of cyanobacteria by enhancing the ability of conversion of light into current. The MWCNTs were non-encroaching because a carboxylate pyrene derivative formed the fixed covalent structure of photosystem I (PS I). The PS I was ascribed as the transporter of photo-current to the electrode (MWCNTs) (Ciornii et al. 2017). MWCNTs have a combined effect on thylakoid, the multi-protein complexes PS I and II, and photo-electrochemical properties. SWCNTs enhanced immobilization of the reaction center of the bacterium *Rhodobacter (Rb.) sphaeroides* (sp.) and also enhanced the photo-electrochemical activity (Ham et al. 2010; Calkins et al. 2013). MWCNTs significantly enhanced direct transfer of electron in the thylakoid of spinach and of the cyanobacterium *Nostoc* sp. (Sekar et al. 2014). CNTs enhanced the expression of *Arabidopsis* aquaporin in tobacco plant and enhanced photosynthetic activity by production of the photo-electric current. It was observed that CNTs activate gene and protein expression of aquaporin in tobacco cells (Khodakovskaya et al. 2012).

5.5 Effect of CNTs on Soil Microbial Community

The soil contains different microorganisms that form the biota of the soil as the main source of nutrients which are significant in plant growth (Table 5.4). Microorganisms have a key role in recycling of nutrients by decomposition of organic matter (Simonet and Valcarcel 2009; Dinesh et al. 2012). Some microorganisms associate with plant roots; the soil microbial community normally consists of gram-positive bacteria, gram-negative bacteria, and fungi (Luongo and Zhang 2010; Santos et al. 2013). The major challenge in the agriculture sector is the conservation of biodiversity and protection of the biomass of these soil microbes. CNTs can change a microbial community by increasing or decreasing the toxins present in organic compounds (Dinesh et al. 2012). Limited literature is available on the impact of CNTs on soil microbial communities. It has been also reported that CNTs had no significant effect on soil microbes. So, there is a need to thoroughly explore CNT impacts on soil microbes.

Mukherjee et al. (2016) reported that low and high concentrations of CNTs have no adverse effect on soil microbiota. High (10–10,000 mg/kg) and low (10–1,000 mg/ml) concentrations of CNTs were used to investigate effects on soil microbial community and enzymatic activity, but it was found that CNTs had no visible effect on soil microbes and enzymatic activity, although these high and low CNT concentrations reduced selected species of bacteria. These specific concentrations increased the amount of polycyclic aromatic hydrocarbon (PAH)-degrading bacteria. Similarly, when red clover was treated with MWCNTs, the activity of symbiotic microorganisms as nitrogen fixers was slightly increased at 3000 mg/kg MWCNTs (Moll et al. 2016).

Table 5.4 Effect of carbon nanoparticles (CNP) and carbon nanotubes (CT) on the soil microbial community

CNPs/ CNT	Impact on soil microbial community/processes	References
MWCNTs	Enhanced activity of anaerobic ammonium oxidation bacteria, high carbohydrate and protein	Wang et al. (2013)
SWCNTs	Strong antimicrobial activity	Kang et al. (2007)
SWCNTs	Effect on both gram-positive and gram-negative bacteria	Jin et al. (2014)
MWCNTs	Effect on soil enzyme activity, soil microbial biomass	Chung et al. (2015)
MWCNTs	Conditionally affect soil microbial community	Kerfahi et al. (2015)
CNTs	Effects on composition of soil microbes	Khodakovskaya et al. (2013)
CNTs	Affect growth of gram-negative bacteria	Cordeiro et al. (2014)
SWCNTs	Effects on antimicrobial activity of surface bacteria	Jackson et al. (2013)
CNTs	Toxic effect on microbes	Petersen et al. (2014)

5.6 Future Possibilities

Nanoparticles have great potential to promote plant growth and development by increasing nutrient uptake, improving water uptake efficiency, and enhancing photosynthetic activity. However, there is a need to improve NP use in agriculture by developing target-specific NPs to enhance plant growth, physiological parameters, and the soil microbial community. There is an urgent need to utilize NPs having great potential to enhance photosynthesis mechanism because minimal attention is being given to this area of research. Biosynthesized NPs should be used: by controlling their size and concentration we can determine the mechanism of toxicity in plants. Modulating these factors, we can reduce transportation, toxicity, and bioavailability to the ecosystem. There is further need to explore the function of NPs beneath plant roots.

References

Aghdam MTB, Mohammadi H, Ghorbanpour M (2016) Effects of nanoparticulate anatase titanium dioxide on physiological and biochemical performance of *Linumusitatissimum* (Linaceae) under well-watered and drought stress conditions. Braz J Bot 39:139–146. https://doi.org/10.1007/s40415-015-0227-x

Alidoust D, Isoda A (2013) Effect of Fe_2O_3 NPs on photosynthetic characteristic of soybean (Glycine max (L.) Merr.): foliar spray versus soil amendment. Acta Physiol Plant 35:3365–3375. https://doi.org/10.1007/s11738-013-1369-8

Alimohammadi M, Xu Y, Wang D, Biris AS, Khodakovskaya MV (2011) Physiological responses induced in tomato plants by a two-component nanostructural system composed of carbon nanotubes conjugated with quantum dots and its in vivo multimodal detection. Nanotechnology 22(29):295101

Andersen CP, King G, Plocher M, Storm M, Pokhrel LR, Johnson MG et al (2016) Germination and early plant development of ten plant species exposed to titanium dioxide and cerium oxide NPs. Environ Toxicol Chem 35:2223–2229. https://doi.org/10.1002/etc.3374

Anjum NA, Singh N, Singh MK, Sayeed I, Duarte AC, Pereira E, Ahmad I (2014) Single-bilayer graphene oxide sheet impacts and underlying potential mechanism assessment in germinating faba bean (*Viciafaba*L.). Sci Total Environ 472:834–841

Arif N, Yadav V, Singh S, Tripathi DK, Dubey NK, Chauhan DK, Giorgetti L (2018) Interaction of copper oxide nanoparticles with plants: uptake, accumulation, and toxicity. In: Nanomaterials in plants, algae, and microorganisms. Academic Press, London, pp 297–310

Arora S, Sharma P, Kumar S, Nayan R, Khanna PK, Zaidi MGH (2012) Gold-nanoparticle inducedenhancement in growth and seed yield of *Brassica juncea*. Plant Growth Regul 66:303–310

Asadishad B, Chaha S, Cianciarelli V, Zhou K, Tufenkji N (2017) Effect of gold NPs on extracellular nutrient-cycling enzyme activity and bacterial community in soil slurries: role of nanoparticle size and surface coating. Environ Sci Nano 4:907–918

Bai Y, Park IS, Lee SJ, Bae TS, Watari F, Uo M, Lee MH (2011) Aqueous dispersion of surfactant-modified multiwalled carbon nanotubes and their application as an antibacterial agent. Carbon 49:3663–3671. https://doi.org/10.1128/JCM.01091-11

Barbinta-Patrascu ME, Badea N, Ungureanu C, Pirvu C, Iftimie V, Antohe S (2017) Photophysical studies on biocomposites based on carbon nanotubes and chlorophyll-loaded biomimetic membranes. Romanian Rep Phys 69:604

Begum P, Ikhtiari R, Fugetsu B (2014) Potential impact of multi-walled carbon nanotubes exposure to the seedling stage of selected plant species. Nano 4:203–221

Blankenship RE, Tiede DM, Barber J, Brudvig GW, Fleming G, Ghirardi MR, Gunner M, Junge W, Kramer DM, Melis A, Moore TA, Moser CC, Nocera DG, Nozik AJ, Ort DR, Parson WW, Prince RC, Sayr RT (2011) Comparing photosynthetic and photovoltaic efficiencies and recognizing the potential for improvement. Science 332:805–809

Burke David J, Pietrasiak N, Situ SF, Abenojar EC, Porche M, Kraj P, Lakliang Y, Samia ACS (2015) Iron oxide and titanium dioxide nanoparticle effects on plant performance and root associated microbes. Int J Mol Sci 16:23630–23650. https://doi.org/10.3390/ijms161023630

Burmana U, Sainib M, Kumar P (2013) Effect of zinc oxide NPs on growth and antioxidant system of chickpea seedlings. Toxicol Environ Chem 95:605–612. https://doi.org/10.1080/02772248.2013.803796

Calkins JO, Umasankar Y, O'Neill H, Ramasamy RP (2013) High photo-electrochemical activity of thylakoid-carbon nanotube composites for photosynthetic energy conversion. Energy Environ Sci 6:1891–1900

Canas JE, Long M, Nations S, Vadan R, Dai L, Luo M et al (2008) Effects of functionalized and nonfunctionalized single-walled carbon nanotubes on root elongation of select crop species. Environ Toxicol Chem 27:1922–1931. https://doi.org/10.1897/08-117.1

Cao Y, Li S, Wang Z, Chang F, Kong J, Gai J, Zhao T (2017) Identification of major quantitative trait loci for seed oil content in soybeans by combining linkage and genome-wide association mapping. Front Plant Sci 8: 1222.

Cardoso EJ, Vasconcellos RL, Marina YH, Santos CA, Alves PR et al (2013) Soil health: looking for suitable indicators. What should be considered to assess the effects of use and management on soil health? Sci Agric 70:274–289. Link: https://goo.gl/IgzmGx

Chegini E, Ghorbanpour M, Hatami M, Taghizadeh M (2017) Effect of multi-walled carbon nanotubes on physiological traits, phenolic contents and antioxidant capacity of *Salvia mirzayanii* Rech f & Esfand under drought stress. J Med Plants 139(2):191–207

Chung H, Kim MJ, Ko K, Kim JH, Kwon H-A, Hong I et al (2015) Effects of graphene oxides on soil enzyme activity and microbial biomass. Sci Total Environ 514:307–313

Ciornii D, Feifel SC, Hejazi M, Kolsch A, Lokstein H, Zouni A, Phys FL, Solidi SA (2017) Construction of photobiocathodes using multi-walled carbon nanotubes and photosystem I. Phys Status Solidi 214:1700017. https://doi.org/10.1002/pssa.201700017

Clement L, Hurel C, Marmier N (2013) Toxicity of TiO2 NPs to cladocerans, algae, rotifers and plants – effects of size and crystalline structure. Chemosphere 90:1083–1090

Cordeiro LF, Marques BF, Kist LW, Bogo MR, Lopez G, Pagano G et al (2014) Toxicity of fullerene and nanosilver nanomaterials against bacteria associated to the body surface of the estuarine worm *Laeonereisacuta* (Polychaeta. Nereididae). Mar Environ Res 99:52–59. https://doi.org/10.1016/j.marenvres.2014.05.011

Da Costa MVJ, Sharma PK (2016) Effect of copper oxide NPs on growth, morphology, photosynthesis, and antioxidant response in *Oryzasativa*. Photosynthetica 54:110–119. https://doi.org/10.1007/s11099-015-0167-5

Dasgupta N, Shivendu R, Bhavapriya R, Venkatraman M, Chidambaram R, Avadhani GS, Ashutosh K (2016) Thermal co-reduction approach to vary size of silver nanoparticle: itsmicrobial and cellular toxicology. Environ Sci Pollut Res 23:4149–4163. https://doi.org/10.1007/s11356-015-4570-z

de la Rosa G, Lopez-Moreno ML, de Haro D, Botez CE, Peralta-Videa JR, Gardea-Torresdey JL (2013) Effects of ZnO NPs in alfalfa, tomato, and cucumber at the germination stage: root development and X-ray absorption spectroscopy studies. Pure Appl Chem 85:2161–2174

De La Torre-Roche R, Hawthorne J, Deng Y, Xing B, Cai W, Newman LA et al (2013) Multiwalled carbon nanotubes and c60 fullerenes differentially impact the accumulation of weathered pesticides in four agricultural plants. Environ Sci Technol 47:12539–12547. https://doi.org/10.1021/es4034809

De Volder MF, Tawfick SH, Baughman RH, Hart AJ (2013) Carbon nanotubes: present and future commercial applications. Science 339:535–539. https://doi.org/10.1126/science.1222453

Dehkourdi EH, Mosavi M (2013) Effect of anatase NPs (TiO_2) on parsley seed germination (Petroselinum crispum) in vitro. Biol Trace Elem Res 155:283–286. https://doi.org/10.1007/s12011-013-9788-3

Derosa MC, Monreal C, Schnitzer M, Walsh R, Sultan Y (2010) Nanotechnology in fertilizers. Nat Nanotechnol 5:91. https://doi.org/10.1038/nnano.2010.2

Di Giacomo R et al (2013) Candida albicans /MWCNTs: a stable conductive bio nanocomposite and its temperature sensing properties. IEEE Trans Nano Technol 12:111–114

Dinesh R, Anandaraj M, Srinivasan V, Hamza S (2012) Engineered NPs in the soil and their potential implications to microbial activity. Geoderma 173–174:19–27. https://doi.org/10.1016/j.geoderma.2011.12.018

Falkowski PG, Fenchel T, Delong EF (2008) The microbial engines that drive earth's biogeochemical cycles. Science 320:1034–1039

Feizi H, Rezvani Moghaddam P, Shahtahmassebi N, Fotovat A (2012) Impact of bulk and nanosized titanium dioxide (TiO2) on wheat seed germination and seedling growth. Biol Trace Elem Res 146:101–106. https://doi.org/10.1007/s12011-011-9222-7

Feizi H, Kamali M, Jafari L, Rezvani Moghaddam P (2013) Phytotoxicity and stimulatory impacts of nanosized and bulk titanium dioxide on fennel (*Foeniculum vulgare* Mill). Chemosphere 91:506–511. https://doi.org/10.1016/j.chemosphere.2012.12.012

Fraceto LF, Grillo R, de Medeiros GA, Scognamiglio V, Rea G, Bartolucci C (2016) Nanotechnology in agriculture: which innovation potential does it have? Front Environ Sci 4:20. https://doi.org/10.3389/fenvs.2016.00020

Frenk S, Ben-Moshe T, Dror I, Berkowitz B, Minz D (2013) Effect of metal oxide nanoparticles on microbial community structure and function in two different soil types. PLoS One 8:84441. https://doi.org/10.1371/journal.pone.0084441

Galbraith DW (2007) Nanobiotechnology: silica breaks through in plants. Nat Nanotechnol 2:272–273

Ge Y, Schimel PJ, Holdena AP (2012) Identification of soil bacteria susceptible to TiO2 and ZnO NPs. Appl Environ Microbiol 78:6749–6758

Giraldo PJ, Landry PM, Faltermeier MS, McNicholas PT, Iverson MN, Boghossian AA, Reuel FN, Hilmer JA, Sen F, Brew AJ, Michael S, Strano SM (2014) Plant nanobionics approach to augment photosynthesis and biochemical sensing. Nat Mater 13:400–408. https://doi.org/10.1038/nmat3890

Gogos A, Moll J, Klingenfuss F, van der Heijden M, Irin F, Green MJ et al (2016) Vertical transport and plant uptake of NPs in a soil mesocosm experiment. J Nanobiotechnol 14:40. https://doi.org/10.1186/s12951-016-0191-z

Gopinath K, Gowri S, Karthika V, Arumugam A (2014) Green synthesis of gold NPs from fruit extract of *Terminalia arjuna*, for the enhanced seed germination activity of *Gloriosa superba*. J Nanostruct Chem 4:1–11

Haghighi M, Teixeira da Silva JA (2014) The effect of N-TiO_2 on tomato, onion, and radish seed germination. J Crop Sci Biotechnol 17:221–227. https://doi.org/10.1007/s12892-014-0056-7

Ham M-H, Choi JH, Boghossian AA, Jeng ES, Graff RA, Heller DA, Chang AC, Mattis A, Bayburt TH, Grinkova YV, Zeiger AS, Van Vliet KJ, Hobbie EK, Sligar SG, Wraight CA, Strano MS (2010) Photo-electrochemical complexes for solar energy conversion that chemically and autonomously regenerate. Nat Chem 2:929–936

Hatami M, Hadian J, Ghorbanpour M (2017) Mechanisms underlying toxicity and stimulatory role of singlewalled carbon nanotubes in *Hyoscyamusniger* during drought stress simulated by polyethylene glycol. J Hazard Mater 324:Part B. https://doi.org/10.1016/j.jhazmat.2016.10.064

Helaly MN, El-Metwally MA, El-Hoseiny H, Omar SA, El-Sheery NI (2014) Effect of NPs on biological contamination of in vitro cultures and organogenic regeneration of banana. Aust J Crop Sci 8:612–624

Hong F, Zhou J, Liu C, Yang F, Wu C, Zheng L, Yang P (2005) Effect of nano-TiO2 on photochemical reaction of chloroplasts of spinach. Biol Trace Elem Res 105(1–3):269–279

Hojjat SS, Hojjat H (2016) Effects of silver nanoparticle exposure on germination of Lentil (Lens culinarisMedik.). IJFAS 5:248–252

Hossain Z, Mustafa G, Komatsu S (2015) Plant responses to nanoparticle stress. Int J Mol Sci 16:26644–26653. https://doi.org/10.3390/ijms161125980

Huang S, Wang L, Liu L, Hou Y, Li L (2015) Nanotechnology in agriculture, livestock, and aquaculture in China: a review. Agron Sustain Dev 35:369–400. https://doi.org/10.1007/s13593-014-0274-x

Jaberzadeh A, Moaveni P, Moghadam HRT, Zahedi H (2013) Influence of bulk and NPs titanium foliar application on some agronomic traits, seed gluten and starch contents of wheat subjected to water deficit stress. Not Bot Horti Agrobot Cluj Napoca 41:201–207

Jackson P, Jacobsen NR, Baun A, Birkedal R, Kuhnel D, Jensen KA et al (2013) Bioaccumulation and ecotoxicity of carbon nanotubes. Chem Cent J 7:154. https://doi.org/10.1186/1752-153X-7-154

Jain A, Shivendu R, Nandita D, Cidambaram R (2016) Nanomaterials in food and agriculture: an overview on their safety concerns and regulatory issues. Crit Rev Food Sci Nutr 58:297–317. https://doi.org/10.1080/10408398.2016.1160363

Jiang HS, Yin LY, Ren NN, Zhao ST, Li Z, Zhi Y et al (2017) Silver NPs induced reactive oxygen species via photosynthetic energy transport imbalance in an aquatic plant. Nanotoxicology 11:157–167. https://doi.org/10.1080/17435390.2017.1278802

Jin L, Son Y, DeForest JL, Kang YJ, Kim W, Chung H (2014) Single-walled carbon nanotubes alter soil microbial community composition. Sci Total Environ 466:533–538

Juhel G, Batisse E, Hugues Q, Daly D, Frank NA, van Pelt M, Halloran J, Marcel A, Jansen K (2011) Alumina NPs enhance growth of *Lemnamino*. Aquat Toxicol 105:328–336

Kang S, Pinault M, Pfefferle LD, Elimelech M (2007) Single-walled carbon nanotubes exhibit strong antimicrobial activity. Langmuir 23:8670–8673

Kerfahi D, Tripathi BM, Singh D, Kim H, Lee S, Lee J et al (2015) Effects of functionalized and raw multiwalled carbon nanotubes on soil bacterial community composition. PLoS One 10:0123042. https://doi.org/10.1371/journal.pone.0123042

Khodakovskaya MV, Dervishi E, Mahmood M, Xu Y, Li Z, Watanabe F et al (2009) Carbon nanotubes are able to penetrate plant seed coat and dramatically affect seed germination and plant growth. ACS Nano 3:3221–3227. https://doi.org/10.1021/nn900887m

Khodakovskaya MV, De Silva K, Biris AS, Dervishi E, Villagarcia H (2012) Carbon nanotubes induce growth enhancement of tobacco cells. ACS Nano 6:2128–2135. https://doi.org/10.1021/nn204643g

Khodakovskaya MV, Kim BS, Kim JN, Alimohammadi M, Dervishi E, Mustafa T, Cernigla CE (2013) Carbon nanotubes as plant growth regulators: effects on tomato growth, reproductive system, and soil microbial community. Small Nano Micro 9:115–123. https://doi.org/10.1002/smll.201201225

Khot LR, Sankaran S, Maja JM, Ehsani R, Schuster EW (2012) Applications of nanomaterials in agricultural production and crop protection: a review. Crop Protect 35:64–70. https://doi.org/10.1016/j.cropro.2012.01.007

Kiapour H, Moaveni P, Habibi D, Sani B (2015) Evaluation of the application of gibbrellic acid and titanium dioxide NPs under drought stress on some traits of basil (OcimumbasilicumL.). Int J Agron Agric Res 6:138–150

Kim B, Park CS, Murayama M, Hochella MF (2010) Discovery and characterization of silver sulfide NPs in final sewage sludge products. Environ Sci Technol 44:7509–7514. Link: https://goo.gl/OS5yfM

Klaine SJ, Alvarez PJ, Batley GE, Fernandes TF, Handy RD, Lyon DY, Mahendra S, McLaughlin MJ, Lead JR (2008) Nanomaterials in the environment: behavior, fate, bioavailability, and effects. Environ Toxicol Chem 27:1825

Kumar V, Guleria P, Kumar V, Yadav SK (2013) Gold nanoparticle exposure induces growth and yield enhancement in *Arabidopsis thaliana*. Sci Total Environ 461:462–468

Kumari A, Singla R, Guliani A, Yadav SK (2014) Nano encapsulation for drug delivery. EXCLI J 13:265–286

Lahiani MH, Dervishi E, Chen J, Nima Z, Gaume A, Biris AS et al (2013) Impact of carbon nanotube exposure to seeds of valuable crops. ACS Appl Mater Interfaces 5:7965–7973. https://doi.org/10.1021/am402052x

Larue C, Laurette J, Herlin-Boime N, Khodja H, Barbara Fayard B (2012) Accumulation, translocation and impact of TiO2 NPs in wheat (Triticum aestivum spp.): influence of diameter and crystal phase. Sci Total Environ 431:197–208

Lei Z, Mingy S, Xiao W, Chao L, Chunxiang Q, Liang C et al (2008) Antioxidant stress is promoted by nano-anatase in spinach chloroplasts under UV-B radiation. Biol. Trace Elem Res 121:69–79. https://doi.org/10.1007/s12011-007-8028-0

Leslie DC et al (2014) A bioinspired omni-phobic surface coating on medical devices prevents thrombosis and biofouling. Nat Biotechnol 32:1134–1140

Li J, Naeem MS, Wang X, Liu L, Chen C, Ma N et al (2015) Nano-TiO2 is not phytotoxic as revealed by the oilseed rape growth and photosynthetic apparatus ultra-structural response. PLoS One 10:0143885. https://doi.org/10.1371/journal.pone.0143885

Li M, Ahammed GJ, Li C, Bao X, Yu J, Huang C et al (2016) Brassino-steroid ameliorates zinc oxide NPs-induced oxidative stress by improving antioxidant potential and redox homeostasis in tomato seedling. Front Plant Sci 7:615. https://doi.org/10.3389/fpls.2016.00615

Libralato G, Costa Devoti A, Zanella M, Sabbioni E, Micetic I, Manodori L, Pigozzo A, Manenti S, Groppi F, Volpi GA (2016) Phytotoxicity of ionic, micro- and nano-sized iron in three plant species. Ecotoxicol Environ Safe 123:81–88

Lin D, Xing B (2007) Phytotoxicity of NPs: inhibition of seed germination and root growth. Environ Pollut 150:243–250. https://doi.org/10.1016/j.envpol.2007.01.016

Luongo LA, Zhang XJ (2010) Toxicity of carbon nanotubes to the activated sludge process. J Hazard Mater 178:356–362

Ma L, Liu C, Qu C, Yin S, Liu J, Gao F, Hong F (2008) Rubisco activase mRNA expression in spinach: modulation by nanoanatase treatment. Biol Trace Elem Res 122(2):168–178

Maddineni SB, Badal KM, Shivendu R, Nandita D (2015) Diastase assisted green synthesis of size-controllable gold NPs. RSC Adv 5:26727–26733. https://doi.org/10.1039/C5RA03117F

Mahajan P, Dhoke SK, Khanna AS (2011) Effect of nano-ZnO particle suspension on growth of mung (*Vigna radiata*) and gram (*Cicer arietinum*) seedlings using plant agar method. J Nanotechnol 2011:1–7. https://doi.org/10.1155/2011/696535

Mahmoodzadeh H, Aghili R (2014) Effect on germination and early growth characteristics in wheat plants (Triticum aestivumL.) seeds exposed to TiO2 NPs. J Chem Health Risks 4:467–472

Mahmoodzadeh H, Nabavi M, Kashefi H (2013) Effect of nanoscale titanium dioxide particles on the germination and growth of canola (Brassica napus). J Ornamental Hort Plants 3:25–32

Martinez-Ballesta MC, Zapata L, Chalbi N, Carvaja M (2016) Multiwalled carbon nanotubes enter broccoli cells enhancing growth and water uptake of plants exposed to salinity. J Nanobiotechnol 14:42. https://doi.org/10.1186/s12951-016-0199-4

Miralles P, Johnson E, Church TL, Harris AT (2012) Multiwalled carbon nanotubes in alfalfa and wheat: toxicology and uptake. J Rl Soc Interface 9(77):3514–3527

Moaveni P, Kheiri T (2011) TiO2 nano particles affected on maize (Zea mays L). In: 2nd international conference on agricultural and animal science, vol 22. IACSIT Press, Singapore, pp 160–163

Mohamed AKSH, Qayyum MF, Abdel-Hadi AM, Rehman RA, Ali S, Rizwan M (2017) Interactive effect of salinity and silver NPs on photosynthetic and biochemical parameters of wheat. Arch Agron Soil Sci 63:1736. https://doi.org/10.1080/03650340.2017.1300256

Mohammadi R, Maali-Amiri R, Abbasi A (2013) Effect of TiO2 NPs on chickpea response to cold stress. Biol Trace Elem Res 152:152403–152410. https://doi.org/10.1007/s12011-013-9631-x

Mohammadi R, Maali-Amiri R, Mantri NL (2014) Effect of TiO2 NPs on oxidative damage and antioxidant defense systems in chickpea seedlings during cold stress. Russian J Plant Physiol 61:768–775. https://doi.org/10.1134/S1021443714050124

Moll J, Gogos A, Bucheli DT, Widmer F, van der Heijden AGM (2016) Effect of NPs on red clover and its symbiotic microorganisms. J Nanobiotechnol 14:36. https://doi.org/10.1186/s12951-016-0188-7

Mondal A, Basu R, Das S, Nandy P (2011) Beneficial role of carbon nanotubes on mustard plant growth: an agricultural prospect. J Nanopart Res 13:4519–4528. https://doi.org/10.1007/s11051-011-0406-z

Morteza E, Moaveni P, Farahani HA, Kiyani M (2013) Study of photosynthetic pigments changes of maize (Zea mays L.) under nano TiO2 spraying at various growth stages. Springer Plus 2:247. https://doi.org/10.1186/2193-1801-2-247

Mukherjee A, Majumdar S, Servin AD, Pagano L, Dhankher OP, White JC (2016) Carbon nanomaterials in agriculture: a critical review. Front Plant Sci 7:172. https://doi.org/10.3389/fpls.2016.00172

Nair R (2016) Effects of NPs on plant growth and development. Plant Nanotechnol:95–118. Link: https://goo.gl/6ZOTj5

Nair R, Varghese SH, Nair BG, Maekawa T, Yoshida Y, Kumar DS (2010) Nanoparticulate material delivery to plants. Plant Sci 179:154–163

Nath D (2015) Safer nano formulation for the next decade. In: Basiuk VA, Basiuk EV (eds) Green processes for nanotechnology. Springer, Cham, pp 327–352. https://doi.org/10.1007/978-3-319-15461-9_12

Pallavi MCM, Srivastava R, Arora S, Sharma AK (2016) Impact assessment of silver NPs on plant growth and soil bacterial diversity. Biotech 6:254. https://doi.org/10.1007/s13205-016-0567-7

Perez-de-Luque A, Hermosin C (2013) Nanotechnology and its use in agriculture. In: Bagchi D, Bagchi M, Moriyama H, Shahidi F (eds) Bio-nanotechnology: a revolution in food, biomedical and health sciences. Blackwell Publishing Ltd, Oxford, pp 383–398

Petersen EJ, Henry TB, Zhao J, Maccuspie RI, Kirschling TL, Dobrovolskaia MA et al (2014) Identification and avoidance of potential artifacts and misinterpretations in nanomaterial ecotoxicity measurements. Environ Sci Technol 48:4226–4246. https://doi.org/10.1021/es4052999

Philippot L, Raaijmakers JM, Lemanceau P, van der Putten WH (2013) Going back to the roots: the microbial ecology of the rhizosphere. Nat Rev Microbiol 11:789–799

Pietrzak K, Gutarowska B (2015) Influence of the silver nanoparticles on microbial community indifferent environments. Acta Biochimica Polonica 62:72–724. https://doi.org/10.18388/abp.2015_1118

Qi M, Liu Y, Li T (2013) Nano-TiO2 improve the photosynthesis of tomato leaves under mild heat stress. Biol Trace Element Res 156:323–328. https://doi.org/10.1007/s12011-013-9833-2

Rad JS, Karimi J, Mohsenzadeh S, Rad MS, Moradgholi J (2014) Evaluating SiO_2 NPs effects on developmental characteristic and photosynthetic pigment contents of *Zea mays* L. Bull Env Pharmacol Life Sci 3:194–201

Rafique R, Arshad M, Khokhar M, Qazi I, Hamza A, Virk N (2015) Growth response of wheat to titania NPs application. NUST J Eng Sci 7:42–46

Rahmani F, Peymani A, Daneshvand E, Biparva P (2016) Impact of zinc oxide and copper oxide NPs on physiological and molecular processes in Brassica napus L. Indian J Plant Physiol 21:122–128. https://goo.gl/anwCLO

Raliya R, Biswas P, Tarafdar JC (2015a) TiO2 nanoparticle biosynthesis and its physiological effect on mung bean (Vigna radiata L.). Biotechnol Rep 5:22–26. https://doi.org/10.1016/j.btre.2014.10.009

Raliya R, Nair R, Chavalmane S, Wang WN, Biswas P (2015b) Mechanistic evaluation of translocation and physiological impact of titanium dioxide and zinc oxide NPs on the tomato (*Solanum lycopersicum*L.) plant. Metallomics 7:1584–1594. https://doi.org/10.1039/C5MT00168D

Raliya R, Tarafdar CJ, Pratim Biswas P (2016) Enhancing the mobilization of native phosphorous in mung bean rhizosphere using ZnO NPs synthesized by soil fungi. J Agric Food Chem 64:3111–3118. https://doi.org/10.1021/acs.jafc.5b05224

Ranjan S, Nandita D, Arkadyuti RC, Melvin SS, Chidambaram R, Rishi S, Ashutosh K (2014) Nanoscience and nanotechnologies in food industries: opportunities and research trends. J Nanopart Res 16:2464. https://doi.org/10.1007/s11051-014-2464-5

Rao DP, Srivastava A (2014) Enhancement of seed germination and plant growth of wheat, maize, peanut, and garlic using multiwalled carbon nanotubes. Euro Chemical Bull 3:502–504

Rastogi A, Tripathi DK, Yadav S, Chauhan DK, Živčák M, Ghorbanpour M, El-Sheery NI, Brestic M (2019) Application of silicon nanoparticles in agriculture. 3 Biotech 9(3):90

Ratnikova TA, Podila R, Rao AM, Taylor AG (2015) Tomato seed coat permeability to selected carbon nanomaterials and enhancement of germination and seedling growth. Sci World J 2015:419215. https://doi.org/10.1155/2015/419215

Resham S, Khalid M, Gul Kazi A (2015) Nano biotechnology in agricultural development. In: Barh D et al (eds) Plant Omics: the omics of plant science. Springer, New Delhi, pp 683–698. https://doi.org/10.1007/978-81-322-2172-2_24

Rezaei F, Moaveni P, Mozafari H (2015) Effect of different concentrations and time of nano TiO2 spraying on quantitative and qualitative yield of soybean (Glycine max L.) at Shahr-e-Qods, Iran. Biol Forum 7:957–964

Rottman J, Shadman F, Sierra-Alvarez R (2012) Interactions of inorganic oxide NPs with sewage biosolids. Water Sci Technol 66:1821–1827. Link: https://goo.gl/RzB8U2

Santos SMA, Dinis AM, Rodrigues DMF, Peixoto F, Videira RA, Jurado AS (2013) Studies on the toxicity of an aqueous suspension of C60 NPs using a bacterium (gen. Bacillus) and an aquatic plant (Lemnagibba) as in vitro model systems. Aqua Toxicol 142–143:347–354. https://doi.org/10.1016/j.aquatox.2013.09.001

Sawant D (2016) Effect of carbon nanotubes on seed germination, growth and yield of hybrid Bt cotton Var. 7383 BG II.

Schimel JP, Schaeffer SM (2012) Microbial control over carbon cycling in soil. Front Microbiol 3:348. https://doi.org/10.3389/fmicb.2012.00348

Schlich K, Hund-Rinke K (2015) Influence of soil properties on the effect of silver nanomaterials on microbial activity in five soils. Environ Pollut 196:321–330. https://doi.org/10.1016/j.envpol.2014.10.021

Sekar N, Umasankar Y, Ramasamy R (2014) Photocurrent generation by immobilized cyanobacteria via direct electron transport in photo-bioelectrochemical cells. Phys Chem Chem Phys 16:7862–7871

Serag MF, Kaji N, Gaillard C, Okamoto Y, Terasaka K, Jabasini M et al (2011) Trafficking and subcellular localization of multi walled carbon nanotubes in plant cells. ACS Nano 5:493–499. https://doi.org/10.1021/nn102344t

Servin AD, White JC (2016) Nanotechnology in agriculture: next steps for understanding engineered nanoparticle exposure and risk. Nano Impact 1:9–12. Link: https://goo.gl/oBZC8s

Servin AD, Castillo-Michel H, Hernandez-Viezcas JA, Diaz BC, PeraltaVidea JR, Gardea-Torresdey JL (2012) Synchrotron micro-XRF and micro-XANES confirmation of the uptake and translocation of TiO_2 NPs in cucumber (*Cucumis sativus*) plants. Environ Sci Technol 46:7637–7643. https://doi.org/10.1021/es300955b

Shah V, Belozerova I (2009) Influence of metal nanoparticles on the soil microbial community and germination of lettuce seeds. Water Air Soil Pollut 197:143–148. https://doi.org/10.1007/s11270-008-9797-6

Shah V, Jones J, Dickman J, Greenman S (2014) Response of soil bacterial community to metal NPs in biosolids. J Haz Mater 274:399–403. Link: https://goo.gl/ewuXud

Shapira P, Youtie J (2015) The economic contributions of nanotechnology to green and sustainable growth. In: Basiuk VA, Basiuk EV (eds) Green processes for nanotechnology. Springer, Cham, pp 409–434. https://doi.org/10.1007/978-3-319-15461-9_15

Sharifi-Rad J, Sharifi-Rad M, Teixeira da Silva JA (2016) Morphological, physiological and biochemical responses of crops (*Zea mays* L., *Phaseolus vulgaris* L.), medicinal plants (*Hyssopus officinalis* L., *Nigella sativa* L.), and weeds (*Amaranthus retroflexus*L., *Taraxacumofficinale*F. H. Wigg) exposed to SiO2 NPs. J Agr Sci Tech 18:1027–1040

Shen Z, Chen Z, Hou Z, Li T, Lu X (2015) Ecotoxicological effect of zinc oxide NPs on soil microorganisms. Front Environ Sci Eng 9:912–918. https://doi.org/10.1007/s11783-015-0789-7

Shen Z, Röding M, Kröger M, Li Y (2017) Carbon nanotube length governs the viscoelasticity and permeability of bucky paper. Polymers 9:115. https://doi.org/10.3390/polym9040115

Shweta, Vishwakarma K, Sharma S, Narayan RP, Srivastava P, Khan AS, Dubey NK, Tripathi DK, Chauhan DK (2017) Plants and carbon nanotubes (CNTs) interface: present status and future prospects. In: Nanotechnology. Springer, Singapore, pp 317–340

Shweta, Tripathi DK, Chauhan DK, Peralta-Videa JR (2018) Availability and risk assessment of nanoparticles in living systems: a virtue or a peril? In: Nanomaterials in plants, algae, and microorganisms. Academic Press, London, pp 1–31

Siddiqui MH, Al-Whaibi MH (2014) Role of nano-SiO_2 in germination of tomato (*Lycopersicum esculentum*seeds Mill.). Saudi Biol Sci 21:13–17

Siddiqui MH, Mohammad F, Khan MMA, Al-Whaibi MH (2012) Cumulative effect of nitrogen and sulphur on *Brassica juncea*L. genotypes under NaCl stress. Protoplasma 249:139–153

Siddiqui MH, Al-Whaibi MH, Mohammad F (2015) Nanotechnology and plant sciences NPs and their impact on plants. Springer, Cham. https://doi.org/10.1007/978-3-319-14502-0

Simonet BM, Valcarcel M (2009) Monitoring NPs in the environment. Anal Bioanal Chem 393:17–21. https://doi.org/10.1007/s00216-008-2484-z

Simonin M, Guyonnet JP, Martins JM, Ginot M, Richaume A (2015) Influence of soil properties on the toxicity of TiO_2 NPs on carbon mineralization and bacterial abundance. J Hazard Mater 283:529–535. https://doi.org/10.1016/j.jhazmat.2014.10.004

Singh VP, Singh S, Kumar J, Prasad SM (2015) Investigating the roles of ascorbate-glutathione cycle and thiol metabolism in arsenate tolerance in ridged luffa seedlings. Protoplasma 252:1217–1229

Singh S, Tripathi DK, Dubey NK, Chauhan DK (2016) Effects of nano-materials on seed germination and seedling growth: striking the slight balance between the concepts and controversies. Mater Focus 5(3):195–201

Singh J, Vishwakarma K, Ramawat N, Rai P, Singh VK, Mishra RK, Kumar V, Tripathi DK, Sharma S (2019) Nanomaterials and microbes' interactions: a contemporary overview. 3 Biotech 9(3):68

Smith SC, Rodrigues DF (2015) Carbon-based nanomaterials for removal of chemical and biological contaminants from water: a review of mechanisms and applications. Carbon 91:122–143

Stampoulis D, Sinha SK, White JC (2009) Assay dependent phytotoxicity of nanoparticles to plants. Environ Sci Technol 43:9473–9479. https://doi.org/10.1021/es901695c

Szymanska R, Kolodziej K, Slesak I, Zimak-Piekarczyk P, Orzechowska A, Gabruk M et al (2016) Titanium dioxide NPs (100-1000 mg/l) can affect vitamin E response in *Arabidopsis thaliana*. Environ Pollut 213:957–965. https://doi.org/10.1016/j.envpol.2016.03.026

Taran N, Storozhenko V, Svietlova N, Batsmanova L, Shvartau V, Kovalenko M (2017) Effect of zinc and copper NPs on drought resistance of wheat seedlings. Nanoscale Resea Lett 12:60

Thwala M, Musee N, Sikhwivhilu L, Wepener V (2013) The oxidative toxicity of Ag and ZnO NPs towards the aquatic plant *Spirodelapunctat a*and the role of testing media parameters. Environ Sci Process Impacts 15:1830–1843. https://doi.org/10.1039/c3em00235g

Tiwari DK, Dasgupta–Schubert N, Villaseñor LM, Tripathi D, Villegas J (2013) Interaction of carbon nanotubes with mineral nutrients for the promotion of growth of tomato seedlings. Nano Stud 7:87–96

Tiwari DK, Dasgupta-Schubert N, Villaseñor-Cendejas LM, Villegas J, Carreto-Montoya L, Borjas-García SE (2014) Interfacing carbon nanotubes (CNT) with plants: enhancement of growth, water and ionic nutrient uptake in maize (*Zea Mays*) and implications for nanoagriculture. Appl Nanosci 4:577–591

Torney F, Trewyn BG, Lin VS-Y, Wang K (2007) Mesoporous silica NPs deliver DNA and chemicals into plants. Nat Nanotechnol 2:295–300

Tripathi S, Sonkar SK, Sarkar S (2011) Growth stimulation of gram (*Cicer arietinum*) plant by water soluble carbon nanotubes. Nanoscale 3:1176–1181. https://doi.org/10.1039/c0nr00722f

Tripathi DK, Ahmad P, Sharma S, Chauhan DK, Dubey NK (eds) (2017) Nanomaterials in plants, algae, and microorganisms: concepts and controversies, vol 1. Academic Press, London

Upadhyaya H, Begum L, Dey B, Nath PK, Panda SK (2017) Impact of calcium phosphateNPs on rice plant. J Plant Sci Phytopathol 1:001–0010

Vacheron J et al (2013) Plant growth-promoting rhizobacteria and root system functioning. Front Plant Sci 4:356. https://doi.org/10.3389/fpls.2013.00356

Venkatachalam P, Priyanka N, Manikandan K, Ganeshbabu I, Indiraarulselvi P, Geetha N, Muralikrishna K, Bhattacharya RC, Tiwari M, Sharma N, Sahi SV (2016) Enhanced plant growth promoting role of phycomolecules coated zinc oxide NPs with P supplementation in cotton (Gossypium hirsutum L.). Plant Physiol Biochem PII S0981-9428(16):30354–0. https://doi.org/10.1016/j.plaphy.2016.09.004

Vishwakarma K, Upadhyay N, Kumar N, Tripathi DK, Chauhan DK, Sharma S, Sahi S (2018) Potential applications and avenues of nanotechnology in sustainable agriculture. In: Nanomaterials in plants, algae, and microorganisms. Academic Press, London, pp 473–500

Vithanage M, Seneviratne M, Ahmad M, Sarkar B, Sik Ok Y (2017) Contrasting effects of engineered carbon nanotubeson plants: a review. Environ Geochem Health 39:1421. https://doi.org/10.1007/s10653-017-9957-y

VittoriAntisari L, Carbone S, Gatti A, Vianello G, Nannipieri P (2013) Toxicity of metal oxide (CeO_2, Fe_3O_4, SnO_2) engineered NPs on soil microbial biomass and their distribution in soil. Soil Biol Biochem 60:87–94. https://doi.org/10.1016/j.soilbio.2013.01.016

Voleti, R (2015) Effects of low concentrations of carbon nanotubes on growth and gas exchange in Arabidopsis Thaliana. MSU Graduate Theses 2933. http://bearworks.missouristate.edu/theses/2933

Voloshin RA, Kreslavski VD, Zharmukhamedov SK, Bedbenov VS, Ramakrishna S, Allakhverdiev SI (2015) Photo-electrochemical cells based on photosynthetic systems: a review. Biofuel Resea J 6:227–235

Wang Q, Ma X, Zhang W, Peia H, Chen Y (2012a) The impact of cerium oxide NPs on tomato (*Solanum lycopersicum* L.) and its implications for food safety. Metallomics 4:1105–1112

Wang X, Han H, Liu X, Gu X, Chen K, Lu D (2012b) Multi-walled carbon nanotubes can enhance root elongation of wheat (*Triticum aestivum*) plants. J Nanopart Res 14:841. https://doi.org/10.1007/s11051-012-0841-5

Wang D, Wang G, Zhang G, Xu X, Yang F (2013) Using graphene oxide to enhance the activity of anammox bacteria for nitrogen removal. Bioresour Technol 131:527–530

Wild E, Jones KC (2009) Novel method for the direct visualization of in vivo nanomaterials and chemical interactions in plants. Environ Sci Technol 43:5290–5294. https://doi.org/10.1021/es900065h

Wong MH, Misra RP, Giraldo JP, Kwak SY, Son Y, Landry MP et al (2016) Lipid exchange envelope penetration (LEEP) of NPs for plant engineering: a universal localization mechanism. Nano Lett 16:1161–1172. https://doi.org/10.1021/acs.nanolett.5b04467

Wu M-Y (2013) Effects of incorporation of nano-carbon into slow released fertilizer on rice yield and nitrogen loss in surface water of paddy soil. In Intelligent System Design and Engineering Applications (ISDEA). Third International Conference IEEE, pp. 676–681. https://doi.org/10.1109/ISDEA.2012.161

Xia B, Chen B, Sun X, Qu K, Ma F, Du M (2015) Interaction of TiO2 NPs with the marine microalga *Nitzschiaclosterium*: growth inhibition, oxidative stress and internalization. Sci Total Environ 508:525–533. https://doi.org/10.1016/j.scitotenv.2014.11.066

Xie Y, Li B, Zhang Q, Zhang C (2012) Effects of nano-silicon dioxide on photosynthetic fluorescence characteristics of *Indocalamus barbatus* McClure. J Nanjing Forest Univ (Natural Science Edition) 2:59–63

Yang F, Hong F, You W, Liu C, Gao F, Wu C, Yang P (2006) Influence of nano-anatase TiO_2 on the nitrogen metabolism of growing spinach. Biol Trace Elem Res 110(2):179–190

Ze Y, Liu C, Wang L, Hong M, Hong F (2011) The regulation of TiO2 NPs on the expression of light-harvesting complex II and photosynthesis of chloroplasts of Arabidopsis thaliana. Biol Trace Elem Res 143:1131–1141. https://doi.org/10.1007/s12011-010-8901-0

Zhai Y, Hunting RE, Wouters M, Willie Peijnenburg WJGM, Vijver GM (2016) Silver nanoparticles, ions, and shape governing soil microbial functional diversity: nano shapes micro. Front Microbiol 7:1–9. https://doi.org/10.3389/fmicb.2016.01123

Zhao L, Peng B, Hernandez-Viezcas JA, Rico C, Sun Y, Peralta-Videa JR et al (2012) Stress response and tolerance of *Zea mays* to CeO_2 NPs: cross talk among H_2O_2, heat shock protein, and lipid peroxidation. ACS Nano 6:9615–9622. https://doi.org/10.1021/nn302975u

Zheng L, Hong F, Lu S, Liu C (2005) Effect of nano-TiO2 on strength of naturally aged seeds and growth of spinach. Bio Trace Element Res 104:83–91. https://doi.org/10.1385/BTER:104:1:083

Chapter 6
Recent Trends and Advancement Toward Phyto-mediated Fabrication of Noble Metallic Nanomaterials: Focus on Silver, Gold, Platinum, and Palladium

Satarudra Prakash Singh, Mohammad Israil Ansari, Brijesh Pandey, Janmejai Kumar Srivastava, Thakur Prasad Yadav, Humaira Rani, Ashna Parveen, Jyoti Mala, and Akhilesh Kumar Singh

6.1 Introduction

Materials having sizes smaller than 100 nm are not unusual on our planet because such materials are naturally produced as a result of photochemical volcanic activity as well as combustion including food cooking and automobile exhausts. However, Professor Norio Taniguchi initially introduced the multidisciplinary science in the year 1974 that covers research as well as technology from chemistry and physics together with biology usually called nanotechnology (Haleemkhan et al. 2015). Nanotechnology has emerged as one of the important areas of contemporary science that deals with methodology, fabrication, as well as tailoring of materials' size within 1–100 nm range. In such range, the entire properties of individual atoms or molecules along with their corresponding bulk counterparts undergo drastic transformation that gives rise to a wide range of interesting novel and astonishing physicochemical properties based on their morphology, size, as well as size distribution

S. P. Singh · B. Pandey · A. K. Singh (✉)
Department of Biotech and Genome, School of Life Sciences, Mahatma Gandhi Central University, Motihari, India
e-mail: akhileshsingh@mgcub.ac.in

M. I. Ansari
Department of Botany, University of Lucknow, Lucknow, India

J. K. Srivastava · H. Rani · A. Parveen
Amity Institute of Biotechnology, Amity University Uttar Pradesh, Lucknow Campus, Lucknow, India

T. P. Yadav
Department of Physics, Institute of Science, Banaras Hindu University, Varanasi, India

J. Mala
D. P. C. S. S. (Miller) Senior Secondary School, Bihar, Patna, India

I. Bhushan et al. (eds.), *Nanomaterials and Environmental Biotechnology*, Nanotechnology in the Life Sciences,
https://doi.org/10.1007/978-3-030-34544-0_6

(Baker et al. 2013; Ahmed et al. 2016). This is attributable to fact that when the materials' size reaches to nanoscale, the % atoms existing at the surface of material turn out to be substantial, where the greater surface area of nanomaterials (NMs) results, the role played by small bulk of material (Akhtar et al. 2013). The NMs depict exceptional features owing to their high surface energy, large fraction of surface atoms, diminished imperfections, as well as spatial confinement (Bai et al. 2009). Apart from this, NMs are advantageous compared to corresponding bulk counterparts because of their surface-enhanced Raman scattering, surface plasmon resonance, surface-enhanced Rayleigh scattering, and surface plasmon light scattering properties. Owing to such possessions, NMs could be exploited as building blocks toward many optoelectronic, electronic, chemical sensing, and biological applications (Wong and Schwaneberg 2003; Ramanavicius et al. 2005; Singh et al. 2018a, b; Pandey et al. 2018).

A noteworthy field of research in nanotechnology is the fabrication of NMs having distinct shapes as well as sizes and chemical structures including regulated disparities. Since the past two decades, the formation of noble metallic NMs (silver, gold, platinum, and palladium) has been gaining substantial interest owing to a wide range of potential uses like shampoos, detergents, cosmetic products, toothpaste, etc. (Kim and Song 2010). Table 6.1 summarizes the potential uses of silver, gold, platinum, and palladium as noble metallic NMs. Such valuable metallic NMs are produced as a result of either top to bottom or bottom to up techniques. Figure 6.1 summarizes the top to bottom and bottom to up approaches for the fabrication of metallic NMs. In top to bottom strategy, production of metal NMs is carried out as a reduction of appropriate bulk compounds into nano-size by employing different physical/chemical approaches (Mittal et al. 2013). In this method,

Table 6.1 Promising applications of gold, platinum, silver, and palladium NMs

Noble metallic NMs	Promising applications	References
Gold	Medicine, disease diagnostic, drug delivery systems, and antibacterial activity	Bhumkar et al. (2007); Bhattacharya and Murkherjee (2008); Torres-Chavolla et al. (2010); Puvanakrishnan et al. (2012); Bindhu and Umadevi (2014); Liu et al. (2017)
Platinum	Catalysts as well as in various biomedical uses in amalgamation with other NMs in alloy, core shell and bimetallic nanostructure, and antibacterial activity	Hrapovic et al. (2004); Narayanan and El-Sayed (2004); Bhattacharya and Murkherjee (2008); Cheong et al. (2010); Lin et al. (2011); Sharma (2017); Rokade et al. (2017)
Silver	Sensor technology, antimicrobial activity, biological leveling, and various other biomedical uses	Gomez-Romero (2001); Qiu et al. (2004); AshaRani et al. (2009); Li et al. (2011); Pollini et al. (2011); Patil et al. (2012); Bindhu and Umadevi (2014); Ashokkumar et al. (2015)
Palladium	Catalysis, electrocatalysis, sensing, as well as plasmonic wave guiding and antibacterial activity	Gopidas et al. (2003); Chen et al. (2007); Chen et al. (2010); Coccia et al. (2012); West et al. (2010); Rokade et al. (2017)

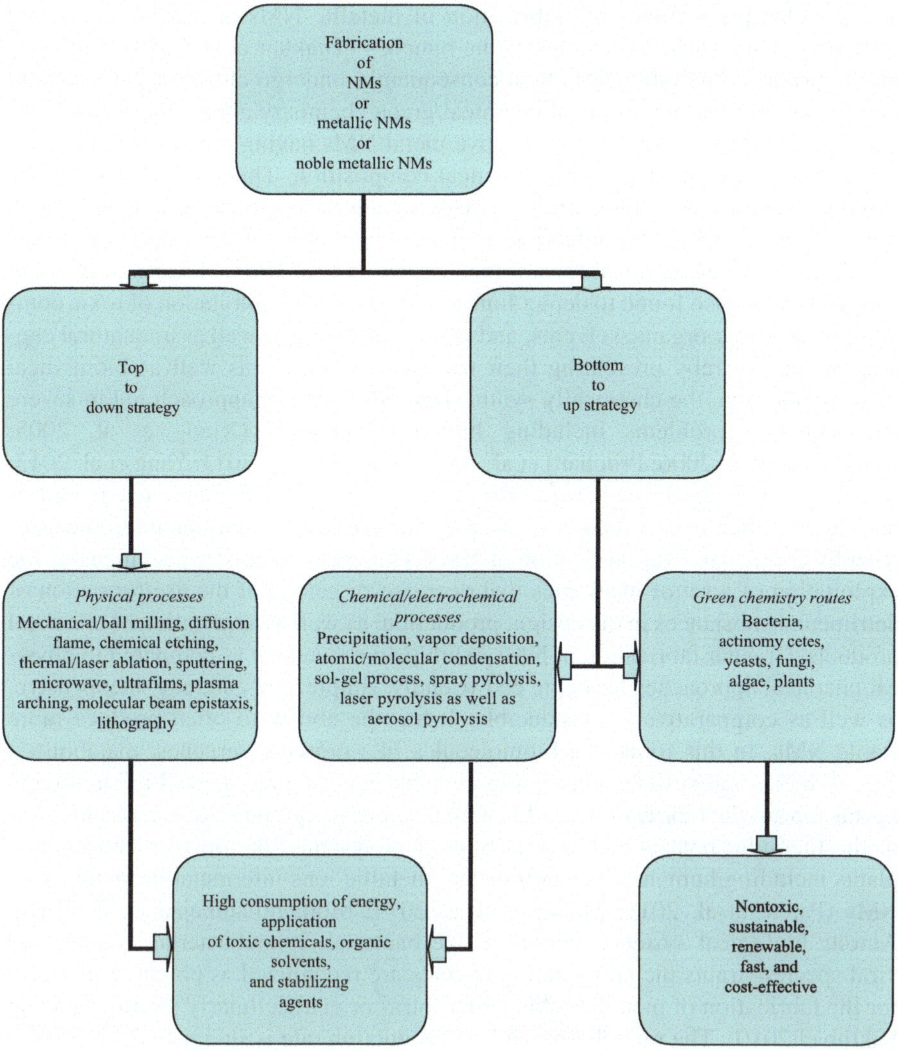

Fig. 6.1 Various strategies employed for the fabrication of metallic NMs including noble metallic NMs

outwardly precise apparatuses are employed so as to cut and mill for shaping the substances to anticipated order. Many physical fabrication approaches such as lithography, pyrolysis, thermolysis, etc., are included in this group (Li et al. 1999; Tsai et al. 2004; Shubin et al. 2012). The main limitation associated with top to down method is the inadequacy of the surface assembly of metal NMs that exhibit a substantial impact upon their physical as well as surface chemistry characteristics (Treguer et al. 1998). Additional drawback involves requirement of mammoth energy for maintaining the high magnitude of pressure and temperature during such fabrication process (Li et al. 1999; Treguer et al. 1998). However, the bottom to top technique involves the fabrication of metallic NMs primarily because of gathering of the atoms/molecules or the bunches (Thakkar et al. 2010). Such primarily produced metallic NMs then consequently undergo clustering, which lead to ultimate material by means of chemical/green chemistry approaches. This technique facilitates to achieve cost-effective metal NMs having lower surface imperfection with more homogeneous chemical composition. This method is normally exploited toward wet fabrication processes such as chemical, electrochemical, sonochemical, and polyol reduction. Remarkably, these wet fabrication processes are shown to be economical toward bulk production of NMs. Nevertheless, these approaches are also found to depict limitations like wide exploitation of toxic compounds, nonpolar organic solvents, and many stabilizing as well as nonnatural capping agents, thereby preventing their uses in the clinical as well as biomedical field. In addition, the chemically synthesized NMs by this approach led to severe environmental problems including biological systems (Xiong et al. 2005; Nemamcha et al. 2006; Pritchard et al. 2010; Gangula et al. 2011; Yang et al. 2013; Mizukoshi et al. 2010; Devi et al. 2016). Considering these, there is a need to introduce green chemistry strategy as simple, comparatively reproducible, and eco-friendly alternative over fabrication of NMs. The green chemistry encompasses the exploitation of a set of ideologies that decrease or get rid of the use/formation of detrimental substances in the design, production, as well as application of chemical products. Greener fabrication of NMs exhibits improvement as compared to physical/chemical approaches for being eco-friendly, simple, eye-catching, inexpensive, as well as comparatively reproducible. It has the ability to often produce more stable NMs. In this route, the biomolecules like proteins, terpenes, metabolites, etc., of biological systems play a role as reducing, capping, as well as stabilizing agents toward the fabrication of NMs with discrete shape and size. A critical review of the literature reveals that several biological systems like microorganisms and plants including human cells can translate metallic ions into metal or metal oxide NMs (Patete et al. 2011; Makarov et al. 2014; Mohammadinejad et al. 2016). Among biological systems, microbes like bacterial, cyanobacterial, fungal, and viral species/strains including actinomycetes are recognized as potential platform for the fabrication of metallic NMs either intra- or extracellularly (Narayanan and Sakthivel 2010). The microbe-based NM production rate is sluggish and results in merely restricted number of sizes as well as shapes over plant-based substances. In this context, binary function of phytochemicals as reducing and stabilizing agents (inexpensive), easy accessibility, the economical cultivation, simple process with requirement of normal pressure as well as temperature (economical), lack of complex as well as multistage processes such as isolation of microbes, culturing, maintenance, etc., biocompatibility of plant extracts (appropriate toward medicinal application), lesser or no contamination (environmental friendly along with safe toward human therapeutic application), and faster fabrication of NMs together with the ability to easily scale up bulk production of NMs make plant systems as eye-catching route for the NM fabrication together with metal/metal oxide over other biological systems (Shankar et al. 2004a; Iravani 2011; Dhillon et al. 2012; Dauthal and Mukhopadhyay 2016; Mohammadinejad et al. 2016). Overall, the potential of plant system as bio-factory for the production of NMs is not only an underexplored but also a profitable strategy (Harris and Bali 2008; Thakkar et al. 2010; Iravani 2011). Furthermore, plant system is found to be rich in a wide variety of phytochemicals/biomolecules. Nevertheless, their potential is yet to be completely exploited under full control for fabricating NMs. Interestingly, it is feasible for the fabrication of metallic NMs at commercial level with the help of plant tissue culture methodologies coupled with optimizing the downstream processing (Jha et al. 2009). This chapter provides an insight toward the recent trends and advances of phyto-mediated synthesis of noble metallic NMs (silver, gold, platinum, and palladium).

6.2 An Overview on Phyto-mediated Fabrication of Metallic NMs/Noble Metallic NMs

Bio-fabrication of metallic NMs by employing living plants (intracellular), plant extracts (extracellular), as well as phytochemicals has received substantial interest as suitable alternative over conventional physical as well as chemical approaches (Iravani 2011). Figure 6.2 depicts the different strategies for the plant-based fabrication of metallic NMs including noble metallic NMs. The significance and advantage of plant system-based fabrication of NMs enhances severalfold when the extracts of cost-effective agricultural wastes are exploited toward the production of metallic NMs. Nowadays, the investigators are exploiting the individual phytoconstituents like polyphenols, proteins organic acids, etc. toward the production of metallic NMs so as to achieve more control on size as well as morphology of NMs (Basha et al. 2010; Tamuly et al. 2014).

A critical review of literature revealed little reports on the intracellular production of metallic NMs, which are not only tedious but also involve high-priced processing stages for the recovery of the NMs from the biomass. As a result, merely *Medicago sativa*, *Brassica juncea*, *Sesbania drummondii*, *Chilopsis linearis*, *Triticum aestivum*, *Avena sativa*, as well as *Festuca rubra* are reported to date toward the fabrication of different metallic NMs via intracellular strategy (Gardea-Torresdey et al. 2002, 2003, 2005; Armendariz et al. 2004a, b; Sharma et al. 2007; Harris and Bali 2008; Marchiol et al. 2014). The foremost limitation of intracellular approach involves the dissimilarities in reducing as well as stabilizing the ability of phytochemicals/biomolecules present in various portions of plant system, which leads to the production of extremely polydisperse metallic NMs of varied morphologies. Also, the tedious recovery of metallic NMs from plant biomass along with little commercial feasibility of plant biomass-mediated fabrication of metallic NMs has diminished the potential of this route.

In contrast, the extracellular fabrication of metallic NMs turns out to be an attractive strategy owing to effortless downstream processing as well as scaling up. The extracellular approach of metallic NM fabrication employing plant extracts is more common than intracellular approach because of its potential in various commercial

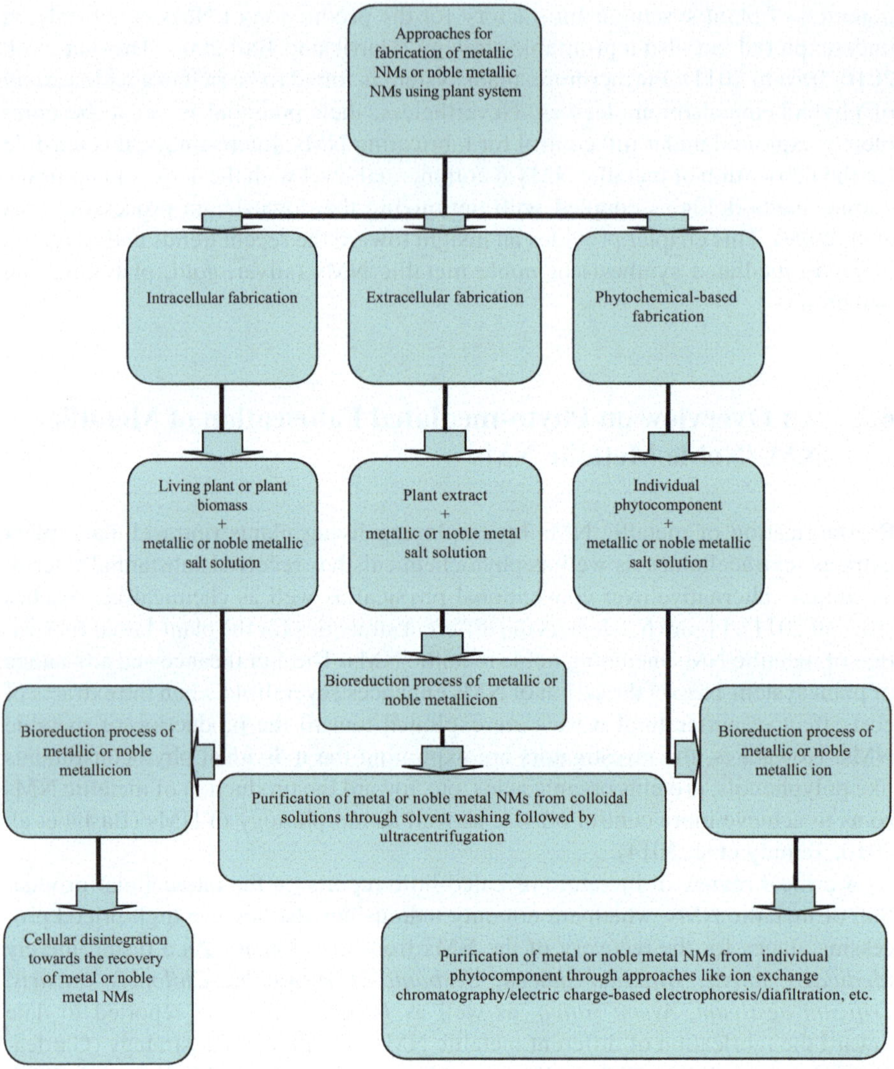

Fig. 6.2 Various plant system-based strategies toward the production of metallic NMs/noble metallic NMs

uses. Usually, the exploitation of plant extract is found to assist the formation of NMs with highly energetic favorable spherical shape as well as more reactive fcc structure. Interestingly, the spherical shape and desired growth together with (111) plane facilitate the reactivity of NMs toward many industrial uses. This fact advocates that plant extract-based reduction is a regulated equilibrium process, which controls the formation of NMs up to particular shape as well as structure, thereby revealing the significance of this route. Nevertheless, it is extremely challenging to reproduce metallic NMs having monodispersity with definite surface morphology because of definitely varying phytochemical components along with structure of plants from diverse origins (Ahmed et al. 2014). Besides, there is often more possibility for the presence of contaminations with plant biomasses/agricultural wastes. Thus, it is extremely problematic for the synthesis of metallic NMs having desired size as well as morphology. However, such desired metallic NMs could be fabricated merely as a result of exploitation of particular phytochemical/biomolecule derived from specific portion of plant biomass/living plants. This must be followed when the focus is to fabricate metallic NMs with high monodispersity as well as desired surface morphology. Several investigators have found the biosynthesis of metallic NMs using plant extracts with their promising uses in various fields (Niraimathi et al. 2013; Seralathan et al. 2014).

The fabrication of metallic NMs using plant extract is linked with the issue of varied phytochemical composition (Ahmed et al. 2014). As a result, it is extremely challenging to forecast the precise mechanism of metallic NM production employing plant extracts. Nevertheless, there are few investigations which forecasted the function of polyphenols as well as flavonoids toward the formation and stabilization of metallic NMs (Basha et al. 2010; Tamuly et al. 2014). Considering this, it is essential to recognize the exact phytoconstituents accountable toward production of NM. Therefore, investigators are concentrating on the isolation of specific antioxidant phytoconstituents for formation of metallic NMs. To date, merely limited studies have been carried out concerning phytoconstituents that are straightforwardly exploited toward the production of metallic NMs (Kasthuri et al. 2009a, b; Basha et al. 2010; Safaepour et al. 2009; Singh et al. 2010; Govindaraju et al. 2011; Leonard et al. 2011; Ahmed et al. 2014; Durai et al. 2014; Iram et al. 2014; Tamuly et al. 2014).

6.3 Recent Fabrication Trends of Silver, Gold, Platinum, and Palladium NMs Using Plant System

A perusal of literature exhibited that plant system can be effectively employed toward the fast as well as extracellular fabrication of noble metallic NMs (Shankar et al. 2004a, b; Kasthuri et al. 2009a; Kim and Song 2010). For instance, Shankar et al. (2003) conducted investigations on the fabrication of gold and silver NMs using geranium leaves, where it was concluded that such noble metallic NMs can be produced through the exploitation of plant extracts with rates analogous to chemical approaches. Shankar et al. (2004b) also successfully carried out the phyto-fabrication of triangular gold nanoprisms as a result of reaction between lemongrass leaf extract and aqueous $AuCl_4^-$ ions. Furthermore, the leaf extracts of many plant systems like *Aloe vera*, geranium, lemongrass, neem, and tamarind were not only exploited for the reduction of gold ions into gold NMs but also silver ions to silver NMs (Chandran et al. 2006; Shankar et al. 2003; Ankamwar et al. 2005a, b). In the same way, fabrication of platinum as well as palladium NMs was also carried out by exploiting the extracts of various portions of different plant species (Nadagouda and Varma 2008; Lin et al. 2011; Coccia et al. 2012; Soundarrajan et al. 2012). Table 6.2 provides the recent fabrication trends of gold, platinum, silver, and palladium NMs using different plant systems.

Table 6.2 Phyto-mediated synthesis of silver, gold, platinum, and palladium NMs

Plant system	Nature of noble metallic NMs	Size (nm)	Morphology/ Shape	References
Tamarindus indica	Gold	20–40	Triangular	Ankamwar et al. (2005a)
Aloe vera	Gold and silver	Not reported	Spherical and triangular	Chandran et al. (2006)
Cinnamomum camphora	Gold and silver	55–80	Triangular, spherical (gold) and quasi-spherical (silver)	Huang et al. (2007)
Azadirachta indica	Gold	5.5–7.5	Crystalline	Ramezani et al. (2008)
Eucalyptus camaldulensis	Gold	5.5–7.5	Crystalline	Ramezani et al. (2008)
Pelargonium roseum	Gold	5.5–7.5	Crystalline	Ramezani et al. (2008)
Cinnamomum zeylanicum	Palladium	15–20	Crystalline	Sathishkumar et al. (2009a)
Curcuma longa	Palladium	10–15	Spherical	Sathishkumar et al. (2009b)
Gardenia jasminoides	Palladium	3–5	Not reported	Jia et al. (2009)
Chenopodium album	Gold and silver	10–30	Quasi-spherical	Dwivedi and Gopal (2010)
Musa paradisiaca	Palladium	50	Crystalline irregular	Bankar et al. (2010)
Ocimum sanctum	Silver	5–10	Spherical	Ahmad et al. (2010)
Mangifera indica	Gold	17–20	Spherical	Philip (2010)
Diospyros kaki	Platinum	2–12	Crystalline	Song et al. (2010)
Nyctanthes arbor-tristis	Gold	19.8	Spherical, triangular, and hexagonal	Das et al. (2011)
Memecylon edule	Silver	20–50	Triangular, circular and hexagonal	Elavazhagan and Arunachalam (2011)
Cuminum cyminum	Gold	1–10	Spherical	Krishnamurthy et al. (2011)
Murraya koenigii	Gold	20	Spherical and triangular	Philip et al. (2011)
Vitex negundo	Silver	5 and 10–30	Spherical and fcc	Zargar et al. (2011)
Beta vulgaris	Gold	Not reported	Spherical, rod-shaped, and nanowires	Castro et al. (2011)
Cymbopogon citratus	Silver	32	Not reported	Masurkar et al. (2011)
Sphearanthus amaranthoids	Gold	39	Spherical	Nellore et al. (2012)
Pinus resinosa	Platinum	6–8	Irregular	Coccia et al. (2012)
Pinus resinosa	Palladium	16–20	Spherical	Coccia et al. (2012)

(continued)

Table 6.2 (continued)

Plant system	Nature of noble metallic NMs	Size (nm)	Morphology/ Shape	References
Trigonella foneumgraecum	Gold	15–25	Spherical	Aromal and Philip (2012)
Glycine max	Palladium	15	Spherical	Petla et al. (2012)
Thevetia peruviana	Silver	10–30	Spherical	Rupiasih et al. (2013)
Rosa hybrid	Gold	10	Spherical, triangular, and hexagonal	Singh et al. (2013)
Alternanthera dentata	Silver	50–100	Spherical	Kumar et al. (2014)
Angelica, Hypericum, and Hamamelis	Gold	4–8	Spherical, ovals, and polyhedral	Pasca et al. (2014)
Acorus calamus	Silver	31.83	Spherical	Nakkala et al. (2014)
Podophyllum hexandrum L.	Gold	5–35	Spherical and triangle	Jeyaraj et al. (2014)
Gardenia jasminoides	Palladium	3–5	Not reported	Khan et al. (2014)
Pistacia atlantica	Silver	10–50	Spherical	Sadeghi et al. (2015)
Punica granatum	Platinum	16–23	Spherical	Dauthal and Mukhopadhyay (2015)
Ficus carica	Silver	13	Not reported	Ulug et al. (2015)
Stachys lavandulifolia	Gold	34–80	Spherical and triangle	Khademi-Azandehi and Moghaddam (2015)
Diospyros paniculata	Silver	14–28	Spherical	Rao et al. (2016)
Ocimum sanctum	Gold	1–50, 10–300, 50–300, and >200	Circular disks with rough edges, spherical and large, anisotropic platelets with defined edges	Lee et al. (2016)
Parkia roxburghii	Gold and silver	5–25	Spherical and quasi-spherical	Paul et al. (2016)
Prunus × *yedoensis*	Platinum	10–50	Spherical and oval	Velmurugan et al. (2016)
Piper longum	Gold	20–200	Spherical	Nakkala et al. (2016)
Dioscorea alata	Silver	10–20	Polycrystalline and face centered cubic crystalline	Pugazhendhi et al. (2016)
Pimpinella tirupatiensis	Palladium	12.25	Spherical	Narasaiah et al. (2017)
Rhus chinensis	Gold	20–40	Oval and spherical	Patil et al. (2017)
Gloriosa superba	Platinum	0.8–3	Spherical	Rokade et al. (2018)

(continued)

Table 6.2 (continued)

Plant system	Nature of noble metallic NMs	Size (nm)	Morphology/ Shape	References
Gloriosa superba	Palladium	5–8	Spherical	Rokade et al. (2018)
Saudi's Dates	Platinum	1.3–2.6	Spherical	Al-Radadi (2019)
Cannabis sativa	Gold	12–18	Quasi-spherical, spherical with few triangular, rods, and hexagonal	Singh et al. (2018)
Cannabis sativa	Silver	20–40	Spherical	Singh et al. (2018)
Caesalpinia ferrea (Tul.)	Silver	30–50	Spheroidal	Soares et al. (2018)
Phoenix sylvestris L.	Silver	43	Not reported	Qidwai et al. (2018)

6.4 General Mechanism of Silver, Gold, Platinum, and Palladium NM Fabrication in Plant System

The general mechanism of fabrication of metallic NMs like silver, gold, platinum, and palladium using plant system/plant extract as detailed in Iravani (2011) consists of three steps: (a) activation step, (b) growth step, and (c) termination step. In the activation step, reduction of metallic ions as well as nucleation of the reduced metallic atoms takes place. In the subsequent step (growth stage), the small adjacent NMs spontaneously combine into larger size particles with enhancement in thermodynamic stability of NMs. Remarkably, the termination step is accountable for defining the ultimate shape of the NMs. Figure 6.3 summarized the general mechanism of fabrication of metallic NMs including silver, gold, platinum, and palladium. As the time period of the growth step enhances, NMs undergo aggregation that results into the formation of nanotubes, nanoprisms, and nanohexahedrons including various other arbitrarily shaped NMs.

6.5 Key Factors/Parameters for Optimal Fabrication of Silver, Gold, Platinum, and Palladium

The following important factors/conditions must be considered for optimal synthesis including fabrication of extremely stable metallic NMs like silver, gold, platinum, and palladium using plant system (Iravani 2011):

(a) *Selection of the best plant system*: Among plant system, investigators have emphasized the key inherent features of the candidates like enzyme activities as well as biochemical routes. For instance, plant systems having massive

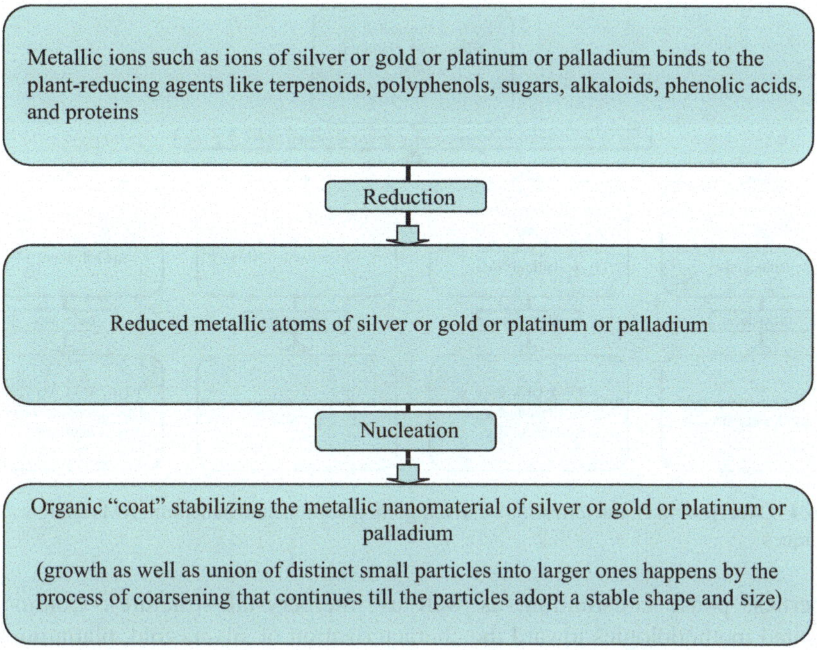

Fig. 6.3 Mechanism of fabrication of metallic NMs (like silver, gold, platinum, and palladium) using plant system/plant extract

potential for accumulation of heavy metal as well as detoxification are the best organisms for the fabrication of metallic NMs including silver, gold, platinum, and palladium.

(b) *Optimal reaction conditions:* In an attempt to exploit the plant systems toward the fabrication of metallic NMs such as silver, gold, platinum, and palladium at industrial scale, the yield and the production rate are the key issues that need to be focused. Further, there is a need to carry out optimization of the bioreduction environments of the reaction mixture. The metal salt concentration, the plant extract quantity, pH, incubation period of reaction mixture, temperature, and buffer strength need to be regulated and optimized. Optimization of these important parameters might regulate morphologies as well as other properties of metallic NMs including silver, gold, platinum, and palladium.

6.6 Characterization of Metallic NMs (Silver, Gold, Platinum, and Palladium)

Characterization of fabricated silver, gold, platinum, and palladium is essential as it allows to comprehend as well as to confirm the effectiveness of fabrication approach. Even though the technique is exploited toward fabrication, the NMs are usually characterized for different physicochemical properties (size, shape, surface

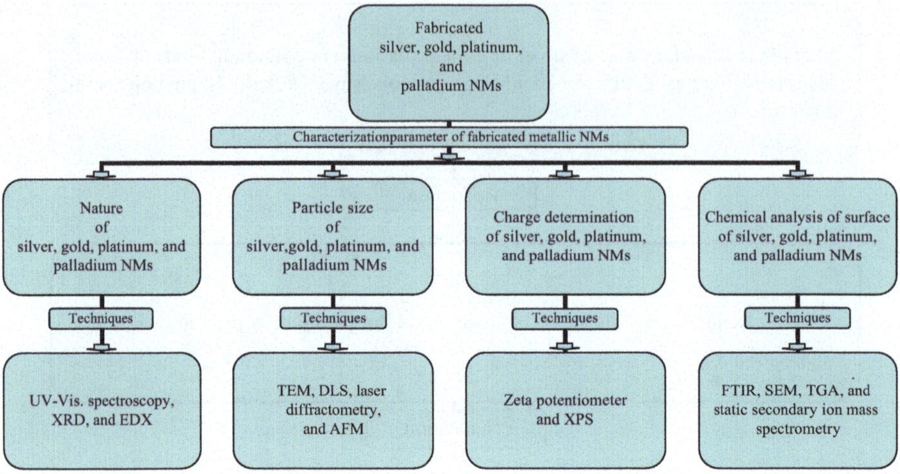

Fig. 6.4 Overview on characterization of silver, gold, platinum, and palladium NMs using various techniques

properties, phase constitution, as well as microcrystal structure). Commonly exploited methodologies toward the characterization of silver, gold, platinum, and palladium NMs are UV-Vis spectroscopy, atomic force microscopy (AFM), infrared spectroscopy, scanning electron microscopy (SEM), X-ray photoelectron spectroscopy (XPS), transmission electron microscopy (TEM), dynamic light scattering (DLS), powder X-ray diffraction (XRD), energy-dispersive X-ray spectroscopy (EDX), Fourier-transform infrared spectroscopy (FTIR), extended X-ray absorption fine structure (EXAFS), thermogravimetric analysis (TGA), nuclear magnetic resonance (NMR), and nanoparticle tracking analysis (NTA). Figure 6.4 summarizes the various techniques used for the characterization of silver, gold, platinum, and palladium NMs.

6.7 Conclusions and Future Perspective

It is concluded that the conventional physical and chemical approaches toward NM fabrication are associated with at least one or more disadvantages. In view of the benefits as well as disadvantages of conventional NM fabrication approaches, one should select an ideal approach that is eco-friendly, economical, and effortlessly scalable and controllable with monodisperse fabrication with minimum steps along with least energy prerequisite. In this context, biogenic fabrication of NMs including metallic NMs emerged as the utmost cost-effective as well as eco-friendly approach over conventional physical and chemical approaches. Among different biogenic fabrication, microorganisms can be exploited for silver, gold, platinum, as well as palladium NM synthesis, but the rate of production is slow, and merely a limited number of sizes and shapes are attainable over plant system. The plant system is rich with a wide variety of phytochemicals/biomolecules, which act as reducing, capping, and stabilizing agents during fabrication of silver, gold, platinum, as well as palladium NMs. The potential of plant system as bio-factory for the production of metallic NMs including silver, gold, platinum, as well as palladium is under exploration as beneficial strategy. But their potential is yet to be completely exploited under full control for fabricating NMs together with noble metals. Furthermore, it is feasible for the fabrication of metallic NMs at commercial level with the help of in vitro culture techniques along with optimization of downstream processing. A detailed understanding of biochemical mechanisms associated with the phyto-mediated NM fabrication is a prerequisite so as to enable this strategy to be economically competitive with the traditional approaches. Nevertheless, it is menacing because of the complexity as well as huge variability among various plant systems. A critical review of literature revealed various reports with a proposed mechanism of phyto-mediated NM fabrication, where nearly all are just reasonable hypotheses lacking substantial investigational backing. As the hypotheses also differ with different plant species, therefore comprehensive assessment is needed so as to comprehend the real mechanism of NM fabrication of a specific plant system, which still remains a key problem. Achieving high reproducibility, maximizing homogeneity, as well as scaling up of NMs production are important bottlenecks toward metallic NM synthesis employing plant systems. Further experimentations are also needed in order to appraise the long-term stability of phyto-fabricated NMs including silver, gold, platinum, as well as palladium.

References

Ahmad N, Sharma S, Alam MK, Singh VN, Shamsi SF, Mehta BR, Fatma A (2010) Rapid synthesis of silver nanoparticles using dried medicinal plant of basil. Colloids Surf B Biointerfaces 81:81–86

Ahmed KBA, Subramaniam S, Veerappan G, Hari N, Sivasubramanian A, Veerappan A (2014) β-Sitosterol-d-glucopyranoside isolated from *Desmostachya bipinnata* mediates photoinduced rapid green synthesis of silver nanoparticles. RSC Adv 4:59130–59136

Ahmed S, Ahmad M, Swami BL, Ikram S (2016) A review on plants extract mediated synthesis of silver nanoparticles for antimicrobial applications: a green expertise. J Adv Res 7:17–28

Akhtar MS, Panwar J, Yun YS (2013) Biogenic synthesis of metallic nanoparticles by plant extracts. ACS Sustain Chem Eng 1:591–602

Al-Radadi NS (2019) Green synthesis of platinum nanoparticles using Saudi's Dates extract and their usage on the cancer cell treatment. Arab J Chem 12(3):330–349. https://doi.org/10.1016/j.arabjc.2018.05.008

Ankamwar B, Chaudhary M, Sastry M (2005a) Gold nanotriangles biologically synthesized using tamarind leaf extract and potential application in vapor sensing. Synth React Inorg, Met-Org, Nano-Met Chem 35:19–26

Ankamwar B, Damle C, Ahmad A, Satry M (2005b) Biosynthesis of gold and silver nanoparticles using *Emblica officinalis* fruit extract, their phase transfer and transmetallation in an organic solution. J Nanosci Nanotechnol 5:1665–1671

Armendariz V, Herrera I, Peralta-Videa JR, Jose-Yacaman M, Troiani H, Santiago P, Gardea-Torresdey JL (2004a) Size controlled gold nanoparticle formation by *Avena sativa* biomass: use of plants in nanobiotechnology. J Nanopart Res 6:377–382

Armendariz V, Jose-Yacaman M, Moller AD, Peralta-Videa JR, Troiani H, Herrera I, Gardea-Torresdey JL (2004b) HRTEM characterization of gold nanoparticles produced by wheat biomass. Rev Mex Fis 50:7–11
Aromal SA, Philip D (2012) Green synthesis of gold nanoparticles using *Trigonella foenumgraecum* and its size-dependent catalytic activity. Spectrochim Acta A Mol Biomol Spectrosc 97:1–5
AshaRani PV, Low Kah Mun G, Hande MP, Valiyaveettil S (2009) Cytotoxicity and genotoxicity of silver nanoparticles in human cells. ACS Nano 3:279–290
Ashokkumar S, Ravi S, Kathiravan V, Velmurugan S (2015) Synthesis of silver nanoparticles using A. indicum leaf extract and their antibacterial activity. Spectrochim Acta A Mol Biomol Spectrosc 134:34–39
Bai X, Gao Y, Liu HG, Zheng L (2009) Synthesis of amphiphilic ionic liquids terminated gold nanorods and their superior catalytic activity for the reduction of nitro compounds. J Phys Chem C 113:17730–17736
Baker S, Rakshith D, Kavitha KS, Santosh P, Kavitha HU, Rao Y, Satish S (2013) Plants: emerging as nanofactories towards facile route in synthesis of nanoparticles. Bioimpacts 3:111–117
Bankar A, Joshi B, Kumar AR, Zinjarde S (2010) Banana peeled extract mediated noval route for the synthesis of palladium nanoparticles. Mater Lett 64:1951–1953
Basha SK, Govindaraju K, Manikandan R, Ahn JS, Bae EY, Singaravelu G (2010) Phytochemical mediated gold nanoparticles and their PTP 1B inhibitory activity. Colloids Surf B Biointerfaces 75:405–409
Bhattacharya R, Murkherjee P (2008) Biological properties of "naked" metal nanoparticles. Adv Drug Deliv Rev 60:128–1306
Bhumkar DR, Joshi HM, Sastry M, Pokharkar VB (2007) Chitosan reduced gold nanoparticles as novel carriers for transmucosal delivery of insulin. Pharm Res 24:1415–1426
Bindhu MR, Umadevi M (2014) Silver and gold nanoparticles for sensor and antibacterial applications. Spectrochim Acta A Mol Biomol Spectrosc 128:37–45
Castro L, Blazquez ML, Munoz JA, Gonzalez F, Garcia-Balboa C, Ballester A (2011) Biosynthesis of gold nanowires using sugar beet pulp. Process Biochem 46:1076–1082
Chandran SP, Chaudhary M, Pasricha R, Ahmad A, Sastry M (2006) Synthesis of gold nanotriangles and silver nanoparticles using *Aloe vera* plant extract. Biotechnol Prog 22:577–583
Chen ZH, Jie JS, Luo LB, Wang H, Lee CS, Lee ST (2007) Applications of silicon nanowires functionalized with palladium nanoparticles in hydrogen sensors. Nanotechnology 18:345502
Chen H, Wei G, Ispas A, Hickey SG, Eychmuller A (2010) Synthesis of palladium nanoparticles and their applications for surface enhanced Raman scattering and electrocatalysis. J Phys Chem C 114:21976–21981
Cheong S, Watt JD, Tilley RD (2010) Shape control of platinum and palladium nanoparticles for catalysis. Nanoscale 2:2045–2053
Coccia F, Tonucci L, Bosco D, Bressan M, d'Alessandro N (2012) One pot synthesis of lignin-stabilized platinum and palladium nanoparticles and their catalytic behaviours in oxidation and reduction reactions. Green Chem 14:1073–1078
Das RK, Gogoi N, Bora U (2011) Green synthesis of gold nanoparticles using *Nyctanthes arbortristis* flower extract. Bioprocess Biosyst Eng 34:615–619
Dauthal P, Mukhopadhyay M (2015) Biofabrication, characterization, and possible bio-reduction mechanism of platinum nanoparticles mediated by agro-industrial waste and their catalytic activity. J Ind Eng Chem 22:185–191
Dauthal P, Mukhopadhyay M (2016) Noble metal nanoparticles: plant-mediated synthesis, mechanistic aspects of synthesis, and applications. *Ind Eng Chem Res* 55:9557–9577
Devi TB, Ahmaruzzaman M, Begum S (2016) A rapid, facile and green synthesis of Ag@AgCl nanoparticles for the effective reduction of 2,4-dinitrophenyl hydrazine. New J Chem 40:1497–1506
Dhillon GS, Brar SK, Kaur S, Verma M (2012) Green approach for nanoparticle biosynthesis by fungi: current trends and applications. Crit Rev Biotechnol 32:49–73

Durai P, Gajendran ACB, Ramar M, Pappu S, Kasivelu G, Thirunavukkarasu A (2014) Synthesis and characterization of silver nanoparticles using crystal compound of sodium para-hydroxybenzoate tetrahydrate isolated from *Vitex negundo*. L leaves and its apoptotic effect on human colon cancer cell lines. Eur J Med Chem 84:90–99

Dwivedi AD, Gopal K (2010) Biosynthesis of gold and silver nanoparticles using *Chenopodium album* leaf extract. Colloids Surf A Physicochem Eng Asp 369:27–33

Elavazhagan T, Arunachalam KD (2011) *Memecylon edule* leaf extract mediated green synthesis of silver and gold nanoparticles. Int J Nanomedicine 6:1265–1278

Gangula A, Podila R, Ramakrishna M, Karanam L, Janardhana C, Rao AM (2011) Catalytic reduction of 4-nitrophenol using biogenic gold and silver nanoparticles derived from *Breynia rhamnoides*. Langmuir 27:15268–15274

Gardea-Torresdey JL, Parsons JG, Gomez E, Peralta-Videa J, Troiani HE, Santiago P, José-Yacamán M (2002) Formation and growth of Au nanoparticles inside live alfalfa plants. Nano Lett 2:397–401

Gardea-Torresdey JL, Gomez E, Peralta-Videa JR, Parsons JG, Troiani H, Jose-Yacaman M (2003) Alfalfa sprouts: a natural source for the synthesis of silver nanoparticles. Langmuir 19:1357–1361

Gardea-Torresdey JL, Rodriguez E, Parsons JG, Peralta-Videa JR, Meitzner G, Cruz-Jimenez G (2005) Use of ICP and XAS to determine the enhancement of gold phytoextraction by *Chilopsis linearis* using thiocyanate as a complexing agent. Anal Bioanal Chem 382:347–352

Gomez-Romero P (2001) Hybrid organic-inorganic materials: In search of synergic activity. Adv Mater 13:163–174

Gopidas KR, Whitesell JK, Fox MA (2003) Synthesis, characterization and catalytic activity of a palladium nanoparticle cored dendrimier. Nano Lett 3:1757–1760

Govindaraju K, Kiruthiga V, Manikandan R, Ashokkumar T, Singaravelu G (2011) β-Glucosidase assisted biosynthesis of gold nanoparticles: a green chemistry approach. Mater Lett 65:256–259

Haleemkhan AA, Naseem, Vidya Vardhini B (2015) Synthesis of nanoparticles from plant extracts. Int J Mod Chem Appl Sci 2:195–203

Harris AT, Bali R (2008) On the formation and extent of uptake of silver nanoparticles by live plants. J Nanopart Res 10:691–695

Hrapovic S, Liu Y, Male KB, Luong JHT (2004) Electrochemical biosensing plateform using platinum nanoparticles and carbon nanotubes. Anal Chem 76:1083–1088

Huang J, Li Q, Sun D, Lu Y, Su Y, Yang X, Wang H, Wang Y, Shao W, Hong NJ, Chen C (2007) Biosynthesis of silver and gold nanoparticles by novel sundried *Cinnamomum camphora* leaf. Nanotechnology 18:105104–105115

Iram F, Iqbal MS, Athar MM, Saeed MZ, Yasmeen A, Ahmad R (2014) Glucoxylan-mediated green synthesis of gold and silver nanoparticles and their phyto-toxicity study. Carbohydr Polym 104:29–33

Iravani S (2011) Green synthesis of metal nanoparticles using plants. Green Chem 13:2638–2650

Jeyaraj M, Arun R, Sathishkumar G, MubarakAli D, Rajesh M, Sivanandhan G, Kapildev G, Manickavasagam M, Thajuddin N, Ganapathi A (2014) An evidence on G2/M arrest, DNA damage and caspase mediated apoptotic effect of biosynthesized gold nanoparticles on human cervical carcinoma cells (HeLa). Mater Res Bull 52:15–24

Jha AK, Prasad K, Kulkarni AR (2009) Plant system: nature's nanofactory. Colloids Surf B Biointerfaces 73:219–223

Jia L, Zhang Q, Li Q, Song H (2009) The biosynthesis of palladium nanoparticles by antioxidants in *Gardenia jasminoides* Ellis: long lifetime nanocatalysts for *p*-nitrotoluene hydrogenation. Nanotechnology 20:385601–385607

Kasthuri J, Kathiravan K, Rajendiran N (2009a) Phyllanthin-assisted biosynthesis of silver and gold nanoparticles: a novel biological approach. J Nanopart Res 11:1075–1085

Kasthuri J, Veerapandian S, Rajendiran N (2009b) Biological synthesis of silver and gold nanoparticles using Apiin as reducing agent. Colloids Surf B Biointerfaces 68:55–60

Khademi-Azandehi P, Moghaddam J (2015) Green synthesis, characterization and physiological stability of gold nanoparticles from *Stachys lavandulifolia* Vahl extract. Particuology 19:22–26

Khan M, Khan M, Kuniyil M, Adil SF, Al-Warthan A, Alkhathlan HZ, Tremel W, Tahir MN, Siddiqui MRH (2014) Biogenic synthesis of palladium nanoparticles using Pulicaria glutinosa extract and their catalytic activity towards the Suzuki coupling reaction. Int J Inorg Organomet Bioinorg Chem 43:9026–9031

Kim BS, Song JY (2010) Biological synthesis of gold and silver nanoparticles using plant leaf extracts and antimicrobial applications. In: Hou CT, Shaw JF (eds) Biocatalysis and biomolecular engineering. John Wiley & Sons, Inc, Hoboken, pp 447–457

Krishnamurthy S, Sathishkumar M, Lee SY, Bae MA, Yun YS (2011) Biosynthesis of Au nanoparticles using cumin seed powder extract. J Nanosci Nanotechnol 11:1811–1814

Kumar DA, Palanichamy V, Roopan SM (2014) Green synthesis of silver nanoparticles using *Alternanthera dentata* leaf extract at room temperature and their antimicrobial activity. Spectrochim Acta A Mol Biomol Spectrosc 127:168–171

Lee SY, Krishnamurthy S, Cho CW, Yun YS (2016) Biosynthesis of gold nanoparticles using *Ocimum sanctum* extracts by solvents with different polarity. ACS Sustain Chem Eng 4:2651–2659

Leonard K, Ahmmad B, Okamura H, Kurawaki J (2011) In situ green synthesis of biocompatible ginseng capped gold nanoparticles with remarkable stability. Colloids Surf B Biointerfaces 82:391–396

Li Y, Duan X, Qian Y, Li Y, Liao H (1999) Nanocrystalline silver particles: synthesis, agglomeration, and sputtering induced by electron beam. J Colloid Interface Sci 209:347–349

Li WR, Xie XB, Shi QS, Duan SS, Ouyang YS, Chen YB (2011) Antibacterial effect of silver nanoparticles on *Staphylococcus aureus*. Biometals 24:135–141

Lin X, Wu M, Wu B, Kuga S, Endo T, Huang Y (2011) Platinum nanoparticles using wood nanomaterials: eco-friendly synthesis, shape control and catalytic activity for p-nitrophenol reduction. Green Chem 13:283–287

Liu H, Lian T, Liu Y, Hong Y, Sun D, Li Q (2017) Plant-mediated synthesis of Au nanoparticles: separation and identification of active biomolecule in the water extract of Cacumen Platycladi. Ind Eng Chem Res 56:5262–5270

Makarov VV, Love AJ, Sinitsyna OV, Makarova SS, Yaminsky IV, Taliansky ME, Kalinina NO (2014) "Green" nanotechnologies: synthesis of metal nanoparticles using plants. Acta Nat 6:35–44

Marchiol L, Mattiello A, Pošcic F, Giordano C, Musetti R (2014) *In vivo* synthesis of nanomaterials in plants: location of silver nanoparticles and plant metabolism. Nanoscale Res Lett 9:101–111

Masurkar SA, Chaudhari PR, Shidore VB, Kamble SP (2011) Rapid biosynthesis of silver nanoparticles using *Cymbopogan Citratus* (Lemongrass) and its antimicrobial activity. Nanomicro Lett 3:189–194

Mittal AK, Chisti Y, Banerjee UC (2013) Synthesis of metallic nanoparticles using plant extracts. Biotechnol Adv 31:346–356

Mizukoshi Y, Sato K, Konno TJ, Masahashi N (2010) Dependence of photocatalytic activities upon the structures of Au/Pd bimetallic nanoparticles immobilized on TiO surface. Appl Catal B 94:248–253

Mohammadinejad R, Karimi S, Iravani S, Varma RS (2016) Plant-derived nanostructures: types and applications. Green Chem 18:20–52

Nadagouda MN, Varma RS (2008) Green synthesis of silver and palladium nanoparticles at room temperature using coffee and tea extract. Green Chem 10:859–862

Nakkala JR, Mata R, Kumar Gupta A, Rani Sadras S (2014) Biological activities of green silver nanoparticles synthesized with *Acorous calamus* rhizome extract. Eur J Med Chem 85:784–794

Nakkala JR, Mata R, Sadras SR (2016) The antioxidant and catalytic activities of green synthesized gold nanoparticles from *Piper longum* fruit extract. Process Saf Environ Prot 100:288–294

Narasaiah P, Mandal BK, Sarada NC (2017) Green synthesis of Pd NPs from Pimpinella tirupatiensis plant extract and their application in photocatalytic activity dye degradation. IOP Conf Ser Mater Sci Eng 263:022013

Narayanan R, El-Sayed MA (2004) Shape dependent catalytic activity of platinum nanoparticles in colloidal solution. Nano Lett 4:1343–1348

Narayanan KB, Sakthivel N (2010) Biological synthesis of metal nanoparticles by microbes. Adv Colloid Interf Sci 156:1–13

Nellore J, Pauline PC, Amarnath K (2012) Biogenic synthesis of *Spheararanthus amaranthoids* towards the efficient production of the biocompatible gold nanoparticles. Dig J Nanomater Biostruct 7:123–133

Nemamcha A, Rehspringer JL, Khatmi D (2006) Synthesis of palladium nanoparticles by sonochemical reduction of palladium (II) nitrate in aqueous solution. J Phys Chem B 110:383–387

Niraimathi KL, Sudha V, Lavanya R, Brindha P (2013) Biosynthesis of silver nanoparticles using *Alternanthera sessilis* (Linn.) extract and their antimicrobial, antioxidant activities. Colloids Surf B Biointerfaces 102:288–291

Pandey B, Singh AK, Singh SP (2018) Nanoparticles mediated gene knock out through miRNA replacement: recent progress and challenges. In: Mishra RK, Thomas S, Mohapatra S, Dasgupta N, Ranjan S (eds) Applications of targeted nano-drugs and delivery systems. Elsevier, Amsterdam

Pasca RD, Mocanu A, Cobzac SC, Petean I, Horovitz O, Cotisel MT (2014) Biogenic syntheses of gold nanoparticles using plant extracts. Part Sci Technol 32:131

Patete JM, Peng X, Koenigsmann C, Xu Y, Karnb B, Wong SS (2011) Viable methodologies for the synthesis of high-quality nanostructures. Green Chem 13:482–519

Patil RS, Kokate MR, Kolekar SS (2012) Bioinspired synthesis of highly stabilized silver nanoparticles using *Ocimum tenuiflorum* leaf extract and their antibacterial activity. Spectrochim Acta A Mol Biomol Spectrosc 91:234–238

Patil MP, Ngabire D, Thi HHP, Kim M, Kim G (2017) Eco-friendly synthesis of gold nanoparticles and evaluation of their cytotoxic activity on cancer cells. J Clust Sci 28:119–132

Paul B, Bhuyan B, Purkayastha DD, Dhar SS (2016) Photocatalytic and antibacterial activities of gold and silver nanoparticles synthesized using biomass of *Parkia roxburghii* leaf. J Photochem Photobiol B Biol 154:1–7

Petla RK, Vivekanandhan S, Misra M, Mohanty AK, Satyanarayana N (2012) Soybean (Glycine max) leaf extract based green synthesis of palladium nanoparticles. J Biomater Nanobiotechnol 3:14–19

Philip D (2010) Rapid green synthesis of spherical gold nanoparticles using *Mangifera indica* leaf. Spectrochim Acta A Mol Biomol Spectrosc 77:807–810

Philip D, Unni C, Aromal SA, Vidhu VK (2011) *Murraya koenigii* leaf-assisted rapid green synthesis of silver and gold nanoparticles. Spectrochim Acta A Mol Biomol Spectrosc 78:899–904

Pollini M, Paladini F, Catalno M, Taurino A, Licciulli A, Maffezzoli A, Sannio A (2011) Antibacterial coatings on haemodialysis catheters by photochemical deposition of silver nanoparticles. J Mater Sci Mater Med 22:2005–2012

Pritchard J, Kesavan L, Piccinini M, He Q, Tiruvalam R, Dimitratos N, Lopez-Sanchez JA, Carley AF, Edwards JK, Kiely CJ, Hutchings GJ (2010) Direct synthesis of hydrogen peroxide and benzyl alcohol oxidation using Au-Pd catalysts prepared by sol immobilization. Langmuir 26:16568–16577

Pugazhendhi S, Sathya P, Palanisamy PK, Gopalakrishnan R (2016) Synthesis of silver nanoparticles through green approach using *Dioscorea alata* and their characterization on antibacterial activities and optical limiting behavior. J Photochem Photobiol B Biol 159:155–160

Puvanakrishnan P, Park J, Chatterjee D, Krishnan S, Tunnel JW (2012) In vivo tumor targeting of gold nanoparticles: effect of particle type and dosing strategy. Int J Nanomedicine 7:1251–1258

Qidwai A, Kumar R, Dikshit A (2018) Green synthesis of silver nanoparticles by seed of *Phoenix sylvestris* L. and their role in the management of cosmetics embarrassment. Green Chem Lett Rev 11:176–188. https://doi.org/10.1080/17518253.2018.1445301

Qiu H, Rieger B, Gilbert R, Jerome R, Jerome C (2004) PLA-coated gold nanoparticles for leveling of PLA biocarriers. Chem Mater 16:850–856

Ramanavicius A, Kausaite A, Ramanaviciene A (2005) Biofuel cell based on direct bioelectrocatalysis. Biosens Bioelectron 20:1962–1967

Ramezani N, Ehsanfar Z, Shamsa F, Amin G, Shahverdi HR, Esfahani HRM, Shamsaie A, Bazaz RD, Shahverdi AR (2008) Screening of medicinal plant methanol extracts for the synthesis of gold nanoparticles by their reducing potential. Z Naturforsch 63:903–908

Rao NH, Lakshmidevi N, Pammi SVN, Kollu P, Ganapaty S, Lakshmi P (2016) Green synthesis of silver nanoparticles using methanolic root extracts of *Diospyros paniculata* and their antimicrobial activities. Mater Sci Eng C 62:553–557

Rokade SS, Joshi KA, Mahajan K, Tomar G, Dubal DS, Parihar VS, Kitture R, Bellare J, Ghosh S (2017) Novel anticancer platinum and palladium nanoparticles from Barleria prionitis. Glob J Nanomed 2:555600. https://doi.org/10.19080/GJN.2017.02.555600

Rokade SS, Joshi KA, Mahajan K, Patil S, Tomar G, Dubal DS, Parihar VS, Kitture R, Bellare JR, Ghosh S (2018) Gloriosa superba mediated synthesis of platinum and palladium nanoparticles for induction of apoptosis in breast cancer. Bioinorg Chem Appl 2018:4924186, 1–9

Rupiasih NN, Aher A, Gosavi S, Vidyasagar PB (2013) Green synthesis of silver nanoparticles using latex extract of *Thevetia peruviana*: a novel approach towards poisonous plant utilization. J Phys Conf Ser 423:1–8

Sadeghi B, Rostami A, Momeni SS (2015) Facile green synthesis of silver nanoparticles using seed aqueous extract of *Pistacia atlantica* and its antibacterial activity. Spectrochim Acta A Mol Biomol Spectrosc 134:326–332

Safaepour M, Shahverdi AR, Shahverdi HR, Khorramizadeh MR, Gohari AR (2009) Green synthesis of small silver nanoparticles using geraniol and its cytotoxicity against Fibrosarcoma-Wehi 164. Avicenna J Med Biotechnol 1:111–115

Sathishkumar M, Sneha K, Kwak IS, Mao J, Tripathy SJ, Yun YS (2009a) Phyto-crystallization of palladium through reduction process using *Cinnamom zeylanicum* bark extract. J Hazard Mater 171:400–404

Sathishkumar M, Sneha K, Yun YS (2009b) Palladium nanocrystals synthesis using Curcuma longa tuber extract. Int J Mater Sci 4:11–17

Seralathan J, Stevenson P, Subramaniam S, Raghavan R, Pemaiah B, Sivasubramanian A, Veerappan A (2014) Spectroscopy investigation on chemo-catalytic, free radical scavenging and bactericidal properties of biogenic silver nanoparticles synthesized using *Salicornia brachiata* aqueous extract. Spectrochim Acta A Mol Biomol Spectrosc 118:349–355

Shankar SS, Ahmad A, Pasricha R, Sastry M (2003) Bioreduction of chloroaurate ions by geranium leaves and its endophytic fungus yields gold nanoparticles of different shapes. J Mater Chem 13:1822–1826

Shankar SS, Rai A, Ahmad A, Sastry M (2004a) Rapid synthesis of Au, Ag, and bimetallic Au core-Ag shell nanoparticles using neem (*Azadirachta indica*) leaf broth. J Colloid Interface Sci 275:496–502

Shankar SS, Rai A, Ankamwar B, Singh A, Ahmad A, Sastry M (2004b) Biological synthesis of triangular gold nanoprisms. Nat Mater 3:482–488

Sharma KD (2017) Antibacterial activity of biogenic platinum nanoparticles: an invitro study. Int J Curr Microbiol App Sci 6:801–808

Sharma NC, Sahi SV, Nath S, Parsons JG, Gardea-Torresdey JL, Pal T (2007) Synthesis of plant-mediated gold nanoparticles and catalytic role of biomatrix-embedded nanomaterials. Environ Sci Technol 41:5137–5142

Shubin Y, Plyusnin P, Sharafutdinov M (2012) In situ synchrotron study of Au-Pd nanoporous alloy formation by single-source precursor thermolysis. Nanotechnology 23:405302–405307

Singh AK, Talat M, Singh DP, Srivastava ON (2010) Biosynthesis of gold and silver nanoparticles by natural precursor clove and their functionalization with amine group. J Nanopart Res 12:1667–1675

Singh A, Sharma MM, Batra A (2013) Synthesis of gold nanoparticles using chick pea leaf extract using green chemistry. J Optoelectron Biomed Mater 5:27–32

Singh P, Pandit S, Garnæs J, Tunjic S, Mokkapati VR, Sultan A, Thygesen A, Mackevica A, Mateiu RV, Daugaard AE, Baun A, Mijakovic I (2018) Green synthesis of gold and silver nanoparticles from *Cannabis sativa* (industrial hemp) and their capacity for biofilm inhibition. Int J Nanomedicine 13:3571–3591

Singh AK, Pal P, Gupta V, Yadav TP, Gupta V, Singh SP (2018a) Green synthesis, characterization and antimicrobial activity of zinc oxide quantum dots using *Eclipta alba*. Mater Chem Phys 203:40–48

Singh AK, Yadav TP, Pandey B, Gupta V, Singh SP (2018b) Engineering nanomaterials for smart drug release: recent advances and challenges. In: Mishra RK, Thomas S, Mohapatra S, Dasgupta N, Ranjan S (eds) Applications of targeted nano-drugs and delivery systems. Elsevier, Amsterdam

Soares MRPS, Corrêa RO, Stroppa PHF, Marques FC, Andrade GFS, Corrêa CC, Brandão MAF, Raposo NRB (2018) Biosynthesis of silver nanoparticles using *Caesalpinia ferrea* (Tul.) Martius extract: physicochemical characterization, antifungal activity and cytotoxicity. PeerJ 6:e4361. https://doi.org/10.7717/peerj.4361

Song JY, Kwon EY, Kim BS (2010) Biological synthesis of platinum nanoparticles using *Diopyros kaki* leaf extract. Bioprocess Biosyst Eng 33:159–164

Soundarrajan C, Sankari A, Dhandapani P, Maruthamuthu S, Ravichandran S, Sozhan G, Palaniswamy N (2012) Rapid biological synthesis of platinum nanoparticles using *Ocimum sanctum* for water electrolysis applications. Bioprocess Biosyst Eng 35:827–833

Tamuly C, Hazarika M, Bordoloi M, Bhattacharyya PK, Kar R (2014) Biosynthesis of Ag nanoparticles using pedicellamide and its photocatalytic activity: an eco-friendly approach. Spectrochim Acta A Mol Biomol Spectrosc 132:687–691

Thakkar KN, Mhatre SS, Parikh RY (2010) Biological synthesis of metallic nanoparticles. Nanomedicine: NBM 6:257–262

Torres-Chavolla E, Ranasinghe RJ, Alocilja EC (2010) Characterization and functionalization of biogenic gold nanoparticles for biosensing enhancement. IEEE Trans Nanobiotechnol 9:533–538

Treguer M, Cointet CD, Remita H, Khatouri J, Mostafavi M, Amblard J, Belloni J (1998) Dose rate effects on radiolytic synthesis of gold-silver bimetallic clusters in solution. J Phys Chem B 102:4310–4321

Tsai SC, Song YL, Tsai CS, Yang CC, Chiu WY, Lin HM (2004) Ultrasonic spray pyrolysis for nanoparticles synthesis. J Mater Sci 39:3647–3657

Ulug B, Haluk Turkdemir M, Cicek A, Mete A (2015) Role of irradiation in the green synthesis of silver nanoparticles mediated by fig (*Ficus carica*) leaf extract. Spectrochim Acta A Mol Biomol Spectrosc 135:153–161

Velmurugan P, Shim J, Kim K, Oh BT (2016) Prunus × yedoensis tree gum mediated synthesis of platinum nanoparticles with antifungal activity against phytopathogens. Mater Lett 174:61–65

West PR, Ishii S, Naik GV, Emani NK, Shalaev VM, Boltasseva A (2010) Searching for better plasmonic materials. Laser Photon Rev 4:795–808

Wong TS, Schwaneberg U (2003) Protein engineering in bioelectrocatalysis. Curr Opin Biotechnol 14:590–596

Xiong Y, Chen J, Wiley B, Xia Y (2005) Understanding the role of oxidative etching in the polyol synthesis of Pd nanoparticles with uniform shape and size. J Am Chem Soc 127:7332–7333

Yang F, Cheng K, Wu T, Zhang Y, Yin J, Wang G, Cao D (2013) Au-Pd nanoparticles supported on carbon fiber cloth as the electrocatalyst for H_2O_2 electroreduction in acid medium. J Power Sources 233:252–258

Zargar M, Hamid AA, Bakar FA, Shamsudin MN, Shameli K, Jahanshiri F (2011) Green synthesis and antibacterial effect of silver nanoparticles using *Vitex negundo* L. Molecules 16:6667–6676

Chapter 7
Development of Environmental Biosensors for Detection, Monitoring, and Assessment

Shagun Gupta and Vipan Kakkar

7.1 Introduction

Whatever that can be around us comprises our environment. Environment is everything that boosts our ability to live on the planet earth including the chemical, physical, and various other natural forces. It makes up our surroundings along with all the nonliving and living things like the water covering most of the earth's surface, the air we breathe, the soil, the animals and plants near us, the insects, and much more. Living things such as human beings constantly interact with each other and with their environment and try to adapt themselves as per its conditions in order to survive, forming an ecosystem – a functional unit of environment (Health & Science 2011). Since several years, scientists have been carefully analyzing this interaction and examining the different ways in which the humans affect their environment. Human beings are playing with natural environment to make their own lives' comfortable. As we know, every coin has two sides, and a rose has also a thorn: the same is true with modernization. There is no doubt that modern techniques play a significant role in making humans' lives easy, but they are also associated with major drawbacks that affect our environment badly. Scientists have observed increasing cases of air and water pollution, acid rains, deforestation, change in climate such as global warming, and many other problems which are dangerous not only to the earth but also to ourselves (Fact Monster 2017). Figure 7.1 shows the effect of human activities on the environment. There have also been substantial occurrences of various natural calamities like cloud bursting, earthquakes, tsunamis, etc. in the past few years. So, proper monitoring of environment is the

S. Gupta · V. Kakkar (✉)
School of Electronics & Communication Engineering, Shri Mata Vaishno Devi University, Katra, India
e-mail: vipan.kakar@smvdu.ac.in

I. Bhushan et al. (eds.), *Nanomaterials and Environmental Biotechnology*, Nanotechnology in the Life Sciences,
https://doi.org/10.1007/978-3-030-34544-0_7

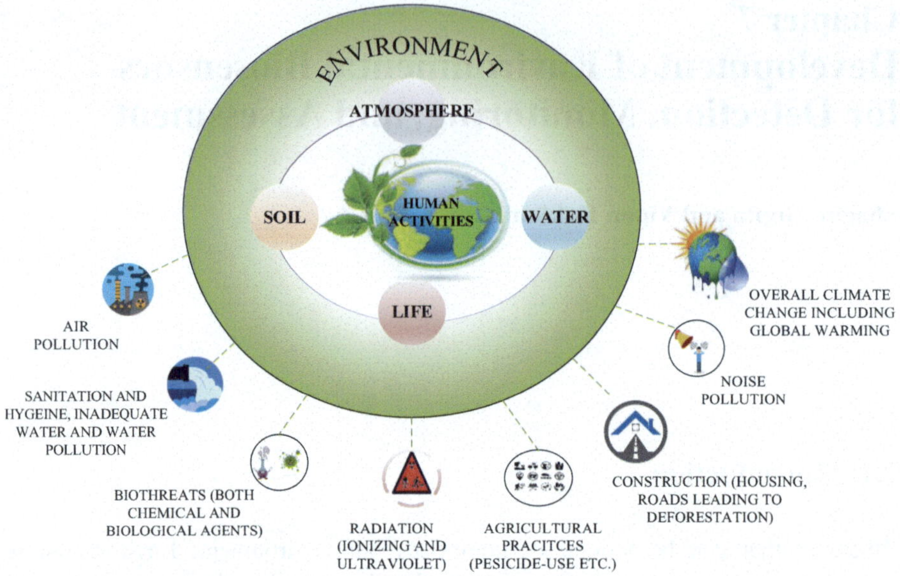

Fig. 7.1 Impact of human activities on environment

need of the hour so as to protect ourselves and the whole planet in general from such life-threatening situations and also to prepare ourselves well in advance in order to fight and survive such situations (Fact Monster 2017; Health & Science 2011).

Environmental monitoring basically refers to the ways by which we can interpret the overall condition of our planet indicating how healthy it is. Global management of environment firstly requires the thorough investigation of various pollutants that are contaminating the air, water, food, soil, etc., which ultimately pose greater ecological threats, and then to find and adopt ways to eliminate these pollutants completely in the environment (Koedrith et al. 2015). Air pollutants originate from various sources such as power plants, refineries, laboratory and industrial processes, vehicle emissions, etc. Sulfur and nitrogen dioxide, volatile organic compounds, carbon monoxide, particulate matter, etc. are some of the major air contaminants. Water and soil pollutants are generally categorized as either radioactive elements (such as tritium), or microbiological (such as coliform) or inorganic compounds (like arsenic), or synthetic organic elements (like pesticides, herbicides, weedicides, etc.). Some volatile organic compounds like benzene also contribute in increasing water and soil pollution. Nowadays, food safety and biothreats are other major concerns gaining much attention (Singh 2017). Figure 7.2 specifies the various ways needed to be adopted in order to deal with these several environmental contaminants.

Strict regulations have been imposed by the US Environmental Protection Agency (US EPA) on the concentrations of various air and water contaminants present in the environment. However, traditional detection and monitoring techniques (like different chromatographic techniques (Justino et al. 2017; Lang et al. 2016;

Fig. 7.2 Different ways to reduce the impact of environmental pollutants on health

Hassani et al. 2017) are complex, time-intensive, and costly and require skilled labor and proper laboratory facilities with existing limitations in analytical and sampling methods (Ho et al. 2005; Looney and Falta 2000; Koedrith et al. 2015). Moreover, these conventional techniques also lack the efficiency when used for in situ measurements essential in emergency situations such as in the cases of acute poisoning, epidemics, or accidental release of biothreats or pesticides (Arduini et al. 2013; Guo et al. 2017; Zhang et al. 2014). Thus, in order to overcome the magnification of various environmental issues, there is urgent need of highly sensitive, rapid, cost-effective, miniaturized, portable, and easy-to-operate techniques and devices that can detect and monitor various contaminants responsible for the destruction of ecosystem (Zhang et al. 2014; Guo et al. 2017; Arduini et al. 2013). Advancements in nanotechnology and biosensing techniques have made it possible to design such devices that can achieve all the aforesaid objectives. The analytical performance such as limit of detection, sensitivity, and specificity, especially for in situ applications, has been improved to a greater extent with the development of smart and rapid biosensors employing various nanocomposites or nanomaterials to detect distinct environmental pollutants (Maduraiveeran and Jin 2017; Zhao et al. 2013).

The subsequent sections of this chapter describe the various available biosensing techniques and explore their potential for the monitoring and detection of environmental pollutants.

7.2 Biosensing Techniques

The term "biosensor" is short form of "biological sensor." A biosensor is an analytical device that is made up of a biological component with a physicochemical detector. The biological element (e.g., an enzyme, cell receptors, organelles, a nucleic acid, an antibody, etc.) interacts with the analyte being tested producing a biological response, and then the transducer converts this biological response mostly into an electrical signal (Robertson 2019) which can be more easily measured and quantified. The detector element or the transducer works in either one of the following physicochemical ways: optical, piezoelectric, electrochemical, electrostatic, or electromechanical. Depending on their specific application, biosensors are also known as optrodes, immunosensors, resonant mirrors, biochips, chemical canaries, biocomputers, and glucometers (Robertson 2019; Monošík et al. 2012).

7.2.1 Biosensor System

Every biosensor comprises a bio-recognition component, bio-transducer component, and electronic system which includes a signal amplifier, processor, and display. A typical biosensor system with its working principle is shown in Fig. 7.3.

7.2.2 Classification of Biosensors

As shown in Fig. 7.3, biosensors can be designed to detect and identify different bio-recognition components using different transduction principles. So, they can be classified based on these two criteria as shown in Fig. 7.4 (Monošík et al. 2012). The brief description of each biosensor type is given below:

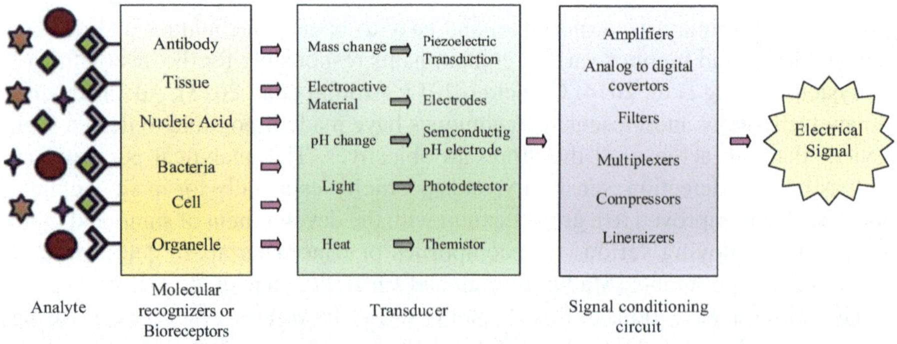

Fig. 7.3 Working principle of biosensors. (Vargas-Bernal et al. 2012)

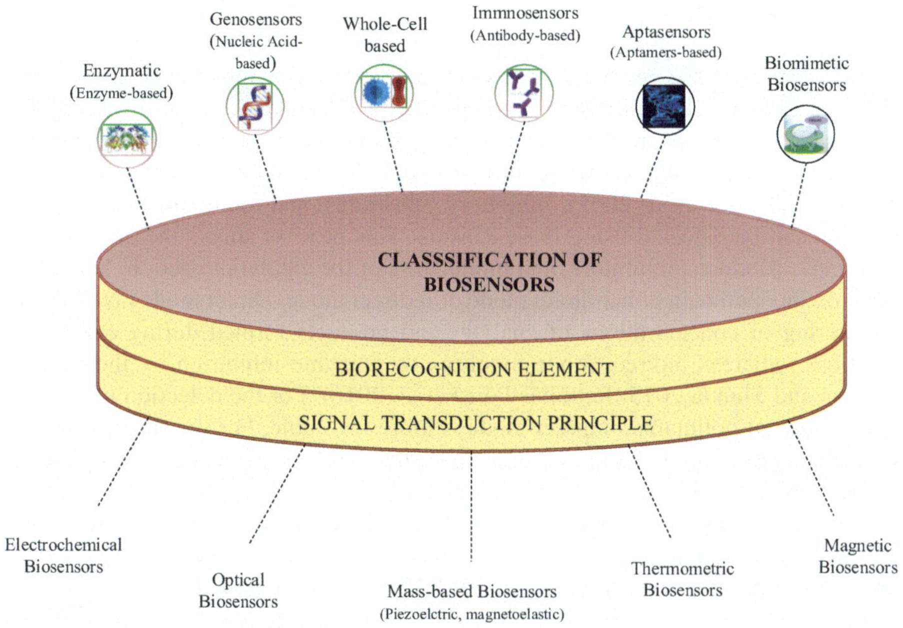

Fig. 7.4 Classification of biosensors

7.2.2.1 On the Basis of Bio-recognition Element

Biosensors can be used to detect either natural biological materials (such as microorganisms, antibodies, nucleic acids, enzymes, cell receptors, tissues, organelles, other natural products, etc.), or biologically derived materials (like aptamers, various engineered proteins including DARPins (designed ankyrin repeat proteins), bones, biomass, recombinant antibodies, etc.), or biomimics (such as synthetic receptors, combinatorial ligands, biomimetic catalysts, imprinted polymers, etc.) (Hashemi Goradel et al. 2018; Justino et al. 2017; Lu et al. 2017). Based on these recognition elements, biosensors are classified as:

Immunosensors

Immunosensors make use of the specific binding feature of an antibody and antigen (Salvador et al. 2007). In these types of biosensors, antibodies are used as recognition receptors. Antibodies are organic compounds that have the capability to control the grafting process and physicochemical superficial properties so as to improve the detection specificity and sensitivity (Vargas-Bernal et al. 2012). Such biosensors are highly selective in detecting various analytes owing to the availability of a wide range of affinities. Further, monoclonal antibody technology (Peterson 2005) if used with immunosensors can result in better screening of antibodies with defined biological and chemical properties at affordable costs.

Enzymatic Biosensors

These biosensors use enzymes as recognition elements owing to their easy availability, varied functionality, and high specificity (Hashemi Goradel et al. 2018). Enzymatic biosensors are catalytic biosensors based on the principle of the conversion of a non-detectable substrate into an electrochemically or optically detectable product (Salvador et al. 2007). Inhibitors, substrates, and modulators of catalytic reaction can be detected using these sensors. This process allows the detection of substrates, products, inhibitors, and modulators of the catalytic reaction. Such biosensors can operate in either direct mode or indirect mode. Direct mode involves the monitoring of concentrations of analytes and products formed during enzymatic reactions, whereas indirect mode monitors the enzyme inhibition by the analyte (Nigam and Shukla 2015; Rebollar-Pérez et al. 2016). For the detection of various environmental pollutants, indirect mode is more favorable. For the successful biosensor design, immobilization of enzyme on the solid surface is one of the crucial steps. Immobilized enzymatic biosensors are more advantageous in comparison with the free enzyme-based sensors as they are rapid, require minimal amounts of sample, and are more sensitive as interference by the differential mode is less in these biosensors (Vargas-Bernal et al. 2012; Nigam and Shukla 2015).

Whole-Cell Based Biosensors

Whole cells (fungi, yeast, bacteria, or eukaryotic cells) can also be used as bio-recognition element as they are better sources of enzymatic activities than isolated enzymes (Salvador et al. 2007; Gutiérrez et al. 2015). Whole-cell-based biosensors, like yeast biosensors, microbial biosensors, etc. utilize the immobilization techniques in order to immobilize the cells on transducer electrode. Such sensors are useful especially when detection of noncommercially available or expensive enzymes is needed (Ahmed et al. 2014; Hashemi Goradel et al. 2018).

Genosensors

In these types of biosensors, bio-recognition component is a class of small biomolecules or biopolymers known as nucleic acids. Nucleic acids include DNA and RNA which are very large and complex forms of organic molecules and contain the genetic information and code of an organism. These are composed of nucleotides and are essential to all known forms of life (Antonio Blanco 2017). The working of DNA-based biosensors relies on the principle of hybridization of complementary strands of DNA in a highly specific manner. Selection of probe and immobilization plays a significant role in the construction of a genosensor (Hashemi Goradel et al. 2018). Genosensors have a wide range of applications in various fields like in medical diagnostics for the detection of genetic diseases, in forensic medicine, in environmental monitoring to control food and water contamination, for the detection of various pathogens, and so on (Valle 2011; Orozco et al. 2016).

Aptasensors

Aptasensors are also DNA-based sensors in which selected DNA plays the role of a highly specific receptor of chemical or biological species (Hashemi Goradel et al. 2018). Aptasensors use aptamer, an artificial nucleic acid ligand, as bio-recognition agent. Aptamers can be generated against proteins, amino acids, drugs, and other molecules (O'sullivan 2002). The utilization of aptamers in biosensors for the detection of biomolecules has many advantages when compared with the use of antibodies due to the chemical stability, cost-effectiveness, and smaller size of aptamers (Lim et al. 2010).

Genosensors and aptasensors are the affinity-based DNA biosensors in which a DNA probe is used to capture an analyte or target. The other category of DNA biosensors based on the catalytic capability of DNA is termed as catalytic-based DNA biosensors. DNAzymes and aptazymes fall under this category of DNA-based biosensors (Hashemi Goradel et al. 2018). DNAzymes are not found in nature and are basically catalytically active DNA molecules that are obtained and isolated via in vitro selection (Zhou et al. 2017). Aptazymes are ligand-activated self-cleaving ribozymes containing integrated aptamer domains and generated by combining DNAzymes, used for chemically modifying nucleic acids, and aptamers, required for binding a broad range of molecules (Zhou et al. 2017; Tang et al. 2017; Hashemi Goradel et al. 2018).

Biomimetic Biosensors

In these biosensors, non-biological receptors are used as bio-recognition element as these receptors have the capability of mimicking the behavior of natural biomolecules. Biomimetic biosensors can be used to analyze the sense of taste, an indispensable perception of human sensory organ. Sense of taste refers to the five basic tastes including bitterness, sourness, sweetness, umami, and saltiness (Lu et al. 2017). These biosensors also find applications in medical diagnostics, in monitoring environment and food safety, etc. (Kozitsina et al. 2018; Huffman et al. 2017).

7.2.2.2 On the Basis of the Transduction Principle

The transducer is an imperative part of a biosensor since it is responsible for the conversion of a biological or biochemical response into an analyzable and measurable electrical signal. Selection of a transducer for a particular application is crucial because appropriate transducers have a dramatic role in providing the optimum detection with high sensitivity and specificity (Salvador et al. 2007). Different transduction mechanisms are available for designing a biosensor, and according to these signal transduction principles, biosensors can be classified in the following ways:

Electrochemical Biosensors

The electrochemical transducers are one of the most developed biosensors available in the literature (Rebollar-Pérez et al. 2016). An electrochemical biosensor exploits the various physicochemical properties of electroactive substances and detects these properties in order to realize the bio-recognition so as to provide a measurable signal such as electrical voltage, current, impedance, or superficial charge (Vargas-Bernal et al. 2012). Depending on the measured electrical signal and mode of electrical measurement, electrochemical biosensors can be subdivided into different categories such as potentiometric (measurement of voltage), amperometric (measurement of current), chronoamperometric, conductometric, impedimetric (resistive, capacitative), coulometric, and field-effect transistors (Vargas-Bernal et al. 2012; Salvador et al. 2007; Hashemi Goradel et al. 2018; Semantic Scholar 2013). In chronoamperometric transduction technique, the working electrode potential is stepped which produces the current as a function of time by using faradaic processes. This technique is useful for analyzing the kinetics of chemical reactions, for monitoring the adsorption and diffusion processes, and for measuring protein and protease activity (El Harrad and Amine 2017). Conductometric technique depends on the current produced by ions in solution, while impedimetric transduction refers to the measurement of impedance (voltage as a function of the current). The performance of both these biosensing methods is highly influenced by a sensing layer placed between the two electrodes (Salvador et al. 2007). Field-Effect Transistor (FET) based Biosensor mechanism is based on the measurement of current which is varied due to the potentiometric effect at a gate electrode (Guiseppi-Elie and Lingerfelt 2005). Coulometric biosensing principle is based on the exhaustive electrolysis of analyte by which we can determine the unknown analyte concentration in a solution. This is done by converting the analyte completely from one oxidation state to another (Chemicool Dictionary 2017; David Harvey 2019).

Optical Biosensors

Optical biosensors rely on the principle of detecting the alterations in various optical properties such as adsorption, reflectance, refractive index, scattering of light, absorbance, luminescence (bioluminescence, chemiluminescence), fluorescence, etc. (Mishra et al. 2017). Optical techniques can be used for the quantification of an optical spectrum and for interpreting the changes in a mechanical measure and are required for overall biological or chemical analysis (Fortin 2009). The detection of analyte is usually done by employing electromagnetic radiations, like ultraviolet (UV), infrared (IR), or other radiations (Hashemi Goradel et al. 2018).

Piezoelectric Biosensors

Piezoelectricity is an electrical effect caused by the variations in the material strain (Fortin 2009). Piezoelectric biosensors are sensitive to changes in density, mass, or viscosity of samples in the vicinity of its active surface (Hashemi Goradel et al. 2018). Acoustic wave biosensors employ the piezoelectric transduction mechanism for the detection of biomolecules. Such sensors are based on an oscillating crystal resonating at the fundamental frequency. The bio-recognition element needed to interact selectively with the target analyte is coated as a layer on the crystal element, and when the binding of analyte on sensing surface takes place, the mass of crystal changes which eventually changes the resonance frequency (Monošík et al. 2012). Such biosensors are easy to operate and offer label-free on-line analysis at affordable costs. Cantilever biosensors are the other emerging class of mass-sensitive biosensors that can achieve lower limit of detections when compared with conventional methods (Salvador et al. 2007).

Thermometric Biosensors

Quantification of the release of thermal energy or the evolution and absorption of heat during a biochemical reaction is the fundamental principle of thermometric biosensors (Monošík et al. 2012; Hashemi Goradel et al. 2018). In these biosensors, immobilized enzyme or cell is directly attached to a thermistor, and the resistance of transducer changes as a result of the temperature variations in reaction medium (Salvador et al. 2007; Monošík et al. 2012).

Magnetic Biosensors

Magnetic biosensors utilize magnetic particles (such as paramagnetic or superparamagnetic particles or crystals) for detecting biological interactions (Nabaei et al. 2018). The size of these particles ranges from micro- to nanometers, and their surfaces are modified and functionalized in order to recognize and identify specific biomolecules by measuring the variations in magnetic properties (magneto-optical properties, resistance, etc.) or by analyzing magnetically induced effects such as coil inductance alterations, Hall effect, giant magnetoresistance (GMR), etc. (Monošík et al. 2012; Li and Kosel 2013; Rife et al. 2003).

7.3 Environmental Biosensors

Environmental monitoring is the need of the hour as it is required to protect the whole ecosystem from various toxic pathogens and pollutants that are present and can be released into different media including water, soil, food, and air. These environmental contaminants are associated with various diseases and thus are harmful to human health. Such pollutants in environment can be categorized into two groups: organic and inorganic compounds. Organic pollutants include pesticides, bisphenol A, surfactants, polycyclic aromatic hydrocarbons, hormones, linear alkylbenzene sulfonates, polychlorinated biphenyls, phenols, dioxins, antibiotics, alkanes, and toxins. Inorganic phosphates, metals, and nitrate are the types of inorganic pollutants. In recent years, various biosensors including genosensors, aptasensors, immunosensors, and enzymatic biosensors employing different transduction mechanisms (Monošík et al. 2012) as mentioned in the previous section have been reported for the monitoring and detection of several environmental pollutants (Rebollar-Pérez et al. 2016). Some of these pollutants are explained in this section, and the different biosensor detection systems developed to detect these pollutants are also discussed. Table 7.1 provides a summary of recently developed environmental biosensors based on different principles with their detection limits.

7.3.1 Pesticides

Pesticides are one among the most important environmental pollutants and are significantly present in the environment due to exhaustive utilization in agricultural practices in order to increase the yield and crop productivity (Justino et al. 2017; Hashemi Goradel et al. 2018). Pesticides can be classified on the basis of their chemical structures into organophosphate (parathion), synthetic pyrethroids, organochlorine (atrazine), carbamate, and inorganic pesticides (Verma and Bhardwaj 2015). These pesticides are dangerous to health as they can cause respiratory and immunological problems, nervous system disorders, and even cancers due to their carcinogenic nature (Sassolas et al. 2012). For highly sensitive detection of these pesticides without the requirement of sample pretreatment, various simple and miniaturized in situ biosensors have been developed (Justino et al. 2017; Vogrinc et al. 2015; Kim et al. 2015), and few of them are summarized in Table 7.1.

7.3.2 Pathogens

Pathogens are present in environmental matrices, especially in water compartments (Koedrith et al. 2015). These pathogens (like *Escherichia coli* or *E. coli*) are dangerous to health and can cause several diseases. Some biosensors have been recently proposed for the environmental monitoring of these pathogens and are described in Table 7.1 (Justino et al. 2017; Liong et al. 2013; Vidic et al. 2017; Hashemi Goradel et al. 2018).

Table 7.1 Recently developed biosensors for the detection of various environmental pollutants

Pollutant category	Target/analyte	Transduction principle	Bio-recognition element	Limit of detection	References
Pesticides	Methyl parathion	Electrochemical (impedimetric)	Enzyme (hydrolase [16], acetylcholinesterase (AChE) [49])	0.1 ng mL^{-1} 0.42 pg mL^{-1}	Zhao et al. (2013), Peng et al. (2017)
		Electrochemical	Enzyme (AChE)	5 fg mL^{-1}	Deng et al. (2016)
		Electrochemical (amperometric)		1.5–1.8 ng mL^{-1}	Nunes et al. (2014)
		Optical	*Sphingomonas* sp. cells	0.01 ppm	Mishra et al. (2017)
	Atrazine	Electrochemical (impedimetric)	Aptamers	2.2 pg mL^{-1}	Madianos et al. (2018)
		Electrochemical (amperometric)	Phage/antibody (monoclonal) complex	0.2 pg mL^{-1}	González-Techera et al. (2015)
		Electrochemical (FET)	Antibodies (monoclonal)	0.01 ng mL^{-1}	Belkhamssa et al. (2016b)
		Electrochemical (voltametric)		0.016 ng mL^{-1}	Liu et al. (2014)
	Carbofuran	Electrochemical (amperometric)	Enzyme (AChE)	0.5 pM, 136 pM	Yang et al. (2013), Li et al. (2017)
		Electrochemical (voltametric)		3.6 nM	Jeyapragasam and Saraswathi (2014)

(continued)

Table 7.1 (continued)

Pollutant category	Target/analyte	Transduction principle	Bio-recognition element	Limit of detection	References
Pathogens	*Escherichia coli* (*E. coli*)	Electrochemical (capacitive)	Polymerizable form of histidine	70 CFU mL^{-1} (CFU-colony forming unit)	Idil et al. (2017)
		Piezoelectric (QCM (quartz crystal microbalance))		1.54×10^6 CFU mL^{-1}	Yilmaz et al. (2015)
		Optical (SPR (surface plasmon resonance))		3.72×10^5 CFU mL^{-1}	
		Optical (electrochemiluminescence)		8 CFU mL^{-1}	Chen et al. (2017)
	Legionella pneumophila	Electrochemical (amperometric)	Antibodies (polyclonal)	10^4 CFU mL^{-1}	Martín et al. (2015)
		Optical (SPR)		10^3 CFU mL^{-1}, 10 CFU mL^{-1}	Meneghello et al. (2017), Manera et al. (2013)
		Optical (SPR)	Nucleic acids	10^4 CFU mL^{-1}	Yilmaz et al. (2015)
	Bacillus subtilis	Electrochemical (amperometric)	Antibodies (polyclonal)	10^2 CFU mL^{-1}	Yoo et al. (2017)
Potentially toxic elements	Mercury ion (Hg^{2+})	Optical (fluorescence)	DNA	17.6 nM	Wu et al. (2016)
		Electrochemical (voltametric)	Nucleic acids	3 fM	Shi et al. (2017)
		Optical (evanescent-wave optical fiber)		1.2 nM	Long et al. (2013)
		Optical (SERS – surface enhancement Raman spectrum)		0.84 pM	Yang et al. (2017)
	Zinc, cadmium	Optical	Microorganism (recombinant *E. coli* DH5)	–	Kim et al. (2015)
	Lead (Pb^{2+})	Optical (fluorescence)	Aptamers	61 nM	Chen et al. (2018)
		Optical (fluorescence)	DNAzymes	5 nM, 16.7 nM	Niu et al. (2018), Ravikumar et al. (2017)

Toxins	Saxitoxin	Optical (interferometry)	Aptamers	0.5 ng mL^{-1}	Gao et al. (2017)
		Electrochemical (voltametric)	Cardiomyocyte cells	0.35 ng mL^{-1}	Wang et al. (2015)
	Domoic acid	Electrochemical (FET)	Antibodies (monoclonal)	10 ng mL^{-1}	Marques et al. (2017)
		Optical (SPR)	Antibodies	1.66 $ngmL^{-1}$, 0.1 ng mL^{-1}	McNamee et al. (2013), Colas et al. (2016)
	Brevetoxin-2	Electrochemical (voltametric)	Cardiomyocyte cells	1.55 $ngmL^{-1}$	Wang et al. (2015)
		Electrochemical (impedimetric)	Aptamers	106 $pgmL^{-1}$	Eissa et al. (2015)
Endocrine-disrupting chemicals	Bisphenol A	Optical (fluorescence)	Aptamers	0.1 ng mL^{-1}, 0.23 ng mL^{-1}	Ragavan et al. (2013), He et al. (2017)
		Optical (evanescent-wave optical fiber)		0.45 ng mL^{-1}	Yildirim et al. (2014)
	Nonylphenol	Electrochemical (FET)	Antibodies (monoclonal)	5 ng mL^{-1}	Belkhamssa et al. (2016a)
	17 β-Estradiol	Photo-electrochemical	Aptamers	33 fM	Fan et al. (2014)
		Electrochemical (capacitive)	Antibodies	1 pg mL^{-1}	Singh et al. (2017)
		Electrochemical (voltametric)	Antibodies	2.25 mL^{-}	Dai and Liu (2017)

7.3.3 Potentially Toxic Elements or Heavy Metals

Owing to industrialization, the amount of heavy metals (like zinc, mercury, copper, etc.) in the environment has increased. Over the time, these metals accumulate in the environment due to their nondegradable nature, and almost all these metals have toxic effects and produce reactive oxygen species (ROS) (Gutiérrez et al. 2015). Water pollution caused by heavy metals and their respective ions can result in several human health hazards. So, the rapid and highly sensitive on-field analysis of heavy metals at affordable costs is a primary concern worldwide. Different portable biosensors developed (Long et al. 2013; Kim et al. 2015; Hashemi Goradel et al. 2018; Justino et al. 2017; Vogrinc et al. 2015) for the detection of few of these metals are described in Table 7.1.

7.3.4 Toxins

Eutrophication of aquatic systems results in the algal blooms of cyanobacteria which produces harmful toxins such as microcystins and brevetoxins, causing the water pollution (Justino et al. 2017; Vogrinc et al. 2015). So, in order to prevent the life-threatening situations, the early detection of such toxins is much needed. Cost-effective and reliable biosensor systems have been developed (Marques et al. 2017; McNamee et al. 2013; Wang et al. 2015) in the recent years for the sensitive detection of various toxins as specified in Table 7.1.

7.3.5 Endocrine-Disrupting Chemicals (EDCs)

EDCs (such as bisphenol A, 4-nonylphenol, etc.) are present in our environment, food, and consumer and personal care products and can interfere with our metabolism, hormone biosynthesis, etc. (Diamanti-Kandarakis et al. 2009). They have been suspected to be associated with the increased breast cancer incidences, immune function variations, altered reproductive functions in females and males, abnormal growth patterns, and neurodevelopmental delays in children (World Health Organization 2012). Different biosensors (Justino et al. 2017; Hashemi Goradel et al. 2018; Vogrinc et al. 2015; Kim et al. 2015) developed for the detection of these deadly EDCs are summarized in Table 7.1.

7.3.6 *Other Environmental Compounds*

Apart from the abovementioned environmental contaminants, pollutants like harmful algal blooms, pharmaceutical wastes, etc. are also present in the environment. Electrochemical genosensors have been developed for the detection of Algal RNA (Orozco et al. 2016; McPartlin et al. 2017). Pharmaceutical wastes contaminate the water due to the improper excretion of beta-blockers, analgesics/anti-inflammatories, antihypertensive drugs, antibiotics, psychiatric drugs, and so on. Therefore, highly sensitive and specific detection of these compounds is essential in order to prevent life-threating effects on living organisms including humans. For pharmaceutical detections, many enzyme-based biosensors based on the use of tyrosinases, laccases, and peroxidases have been developed (Rebollar-Pérez et al. 2016).

7.4 Summary

In this chapter, different biosensing techniques and their applications in the detection of various environmental pollutants have been summarized. Development of specific, rapid, sensitive, and cost-effective biosensors is crucial for the protection of the ecosystem from life-threatening situations and for its survival in fluctuating and harsh environmental conditions. Conclusively, for the overall environmental management, we need to find and adopt certain ways and techniques that can reduce the side effects of modernization and industrialization. The combined use of such corrective measure techniques and biosensors will help in improving the environment quality and will protect the mankind from deadly situations, eventually making the planet earth a healthy place to live in.

References

Ahmed A, Rushworth JV, Hirst NA, Millner PA (2014) Biosensors for whole-cell bacterial detection. Clin Microbiol Rev 27:631–646

Antonio Blanco GB (2017) Nucleic acids. In: Antonio Blanco GB (ed) Medical biochemistry. Academic Press, London

Arduini F, Guidone S, Amine A, Palleschi G, Moscone D (2013) Acetylcholinesterase biosensor based on self-assembled monolayer-modified gold-screen printed electrodes for organophosphorus insecticide detection. Sensors Actuators B Chem 179:201–208

Belkhamssa N, Da Costa JP, Justino CI, Santos PS, Cardoso S, Duarte AC, Rocha-Santos T, Ksibi M (2016a) Development of an electrochemical biosensor for alkylphenol detection. Talanta 158:30–34

Belkhamssa N, Justino CI, Santos PS, Cardoso S, Lopes I, Duarte AC, Rocha-Santos T, Ksibi M (2016b) Label-free disposable immunosensor for detection of atrazine. Talanta 146:430–434

Chemicool Dictionary (2017). What is a coulometry? [Online]. Available: https://www.chemicool.com/definition/coulometry.html

Chen S, Chen X, Zhang L, Gao J, Ma Q (2017) Electrochemiluminescence detection of Escherichia coli O157: H7 based on a novel polydopamine surface imprinted polymer biosensor. ACS Appl Mater Interfaces 9:5430–5436
Chen Y, Li H, Gao T, Zhang T, Xu L, Wang B, Wang J, Pei R (2018) Selection of DNA aptamers for the development of light-up biosensor to detect Pb (II). Sensors Actuators B Chem 254:214–221
Colas F, Crassous MP, Laurent S, Litaker RW, Rinnert E, Le Gall E, Lunven M, Delauney L, Compère C (2016) A surface plasmon resonance system for the underwater detection of domoic acid. Limnol Oceanogr Methods 14:456–465
David Harvey (2019). Coulometric methods. [Online]. Available: https://chem.libretexts.org/LibreTexts/Northeastern/11%3A_Electrochemical_Methods/11.3%3A_Coulometric_Methods
Dai Y, Liu C (2017) Detection of 17 β-estradiol in environmental samples and for health care using a single-use, cost-effective biosensor based on differential pulse voltammetry (DPV). Biosensors 7:15
Deng Y, Liu K, Liu Y, Dong H, Li S (2016) An novel acetylcholinesterase biosensor based on nano-porous pseudo carbon paste electrode modified with gold nanoparticles for detection of methyl parathion. J Nanosci Nanotechnol 16:9460–9467
Diamanti-Kandarakis E, Bourguignon J-P, Giudice LC, Hauser R, Prins GS, Soto AM, Zoeller RT, Gore AC (2009) Endocrine-disrupting chemicals: an Endocrine Society scientific statement. Endocr Rev 30:293–342
Eissa S, Siaj M, Zourob M (2015) Aptamer-based competitive electrochemical biosensor for brevetoxin-2. Biosens Bioelectron 69:148–154
El Harrad L, Amine A (2017) Chronoamperometric biosensor for protease activity assay and inhibitor screening. Electroanalysis 29:2395–2400
Fact Monster (2017). What is the environment? [Online]. Available: https://www.factmonster.com/science/environment/what-environment
Fan L, Zhao G, Shi H, Liu M, Wang Y, Ke H (2014) A femtomolar level and highly selective 17β-estradiol photoelectrochemical aptasensor applied in environmental water samples analysis. Environ Sci Technol 48:5754–5761
Fortin J (2009) Transduction principles. Functional thin films and nanostructures for sensors. Springer, Boston
Gao S, Zheng X, Wu J (2017) A biolayer interferometry-based competitive biosensor for rapid and sensitive detection of saxitoxin. Sensors Actuators B Chem 246:169–174
González-Techera A, Zon MA, Molina PG, Fernández H, González-Sapienza G, Arévalo FJ (2015) Development of a highly sensitive noncompetitive electrochemical immunosensor for the detection of atrazine by phage anti-immunocomplex assay. Biosens Bioelectron 64:650–656
Guiseppi-Elie A, Lingerfelt L (2005) Impedimetric detection of DNA hybridization: towards near-patient DNA diagnostics. Immobilisation of DNA on Chips I. Springer, Berlin/Heidelberg
Guo L, Li Z, Chen H, Wu Y, Chen L, Song Z, Lin T (2017) Colorimetric biosensor for the assay of paraoxon in environmental water samples based on the iodine-starch color reaction. Anal Chim Acta 967:59–63
Gutiérrez JC, Amaro F, Martín-González A (2015) Heavy metal whole-cell biosensors using eukaryotic microorganisms: an updated critical review. Front Microbiol 6:48
Harvey D. Coulometric methods. [Online]. Available: https://chem.libretexts.org/LibreTexts/Northeastern/11%3A_Electrochemical_Methods/11.3%3A_Coulometric_Methods
Hashemi Goradel N, Mirzaei H, Sahebkar A, Poursadeghiyan M, Masoudifar A, Malekshahi ZV, Negahdari B (2018) Biosensors for the detection of environmental and urban pollutions. J Cell Biochem 119:207–212
Hassani S, Momtaz S, Vakhshiteh F, Maghsoudi AS, Ganjali MR, Norouzi P, Abdollahi M (2017) Biosensors and their applications in detection of organophosphorus pesticides in the environment. Arch Toxicol 91:109–130
He M-Q, Wang K, Wang J, Yu Y-L, He R-H (2017) A sensitive aptasensor based on molybdenum carbide nanotubes and label-free aptamer for detection of bisphenol a. Anal Bioanal Chem 409:1797–1803

Health & Science O (2011) Environment is everything around us [Online]. Available: https://www.groundreport.com/environment-is-everything-around-us/
Ho C, Robinson A, Miller D, Davis M (2005) Overview of sensors and needs for environmental monitoring. Sensors 5:4–37
Huffman B, Mazrouei R, Bevelheimer J, Shavezipur M (2017) Three-dimensional biomimetic biosensors for food safety applications. ASME 2017 International Design Engineering Technical Conferences and Computers and Information in Engineering Conference. American Society of Mechanical Engineers, V004T09A002-V004T09A002
Idil N, Hedström M, Denizli A, Mattiasson B (2017) Whole cell based microcontact imprinted capacitive biosensor for the detection of Escherichia coli. Biosens Bioelectron 87:807–815
Jeyapragasam T, Saraswathi R (2014) Electrochemical biosensing of carbofuran based on acetylcholinesterase immobilized onto iron oxide–chitosan nanocomposite. Sensors Actuators B Chem 191:681–687
Justino C, Duarte A, Rocha-Santos T (2017) Recent progress in biosensors for environmental monitoring: a review. Sensors 17:2918
Kim M, Lim JW, Kim HJ, Lee SK, Lee SJ, Kim T (2015) Chemostat-like microfluidic platform for highly sensitive detection of heavy metal ions using microbial biosensors. Biosens Bioelectron 65:257–264
Koedrith P, Thasiphu T, Weon J-I, Boonprasert R, Tuitemwong K, Tuitemwong P (2015) Recent trends in rapid environmental monitoring of pathogens and toxicants: potential of nanoparticle-based biosensor and applications. Sci World J 2015:510982
Kozitsina A, Svalova T, Malysheva N, Okhokhonin A, Vidrevich M, Brainina K (2018) Sensors based on bio and biomimetic receptors in medical diagnostic, environment, and food analysis. Biosensors 8:35
Lang Q, Han L, Hou C, Wang F, Liu A (2016) A sensitive acetylcholinesterase biosensor based on gold nanorods modified electrode for detection of organophosphate pesticide. Talanta 156:34–41
Li F, Kosel J (2013) A magnetic biosensor system for detection of E. coli. IEEE Trans Magn 49:3492–3495
Li Z, Qu S, Cui L, Zhang S (2017) Detection of carbofuran pesticide in seawater by using an enzyme biosensor. J Coast Res 80:1–5
Lim Y, Kouzani A, Duan W (2010) Aptasensors: a review. J Biomed Nanotechnol 6:93–105
Liong M, Hoang AN, Chung J, Gural N, Ford CB, Min C, Shah RR, Ahmad R, Fernandez-Suarez M, Fortune SM (2013) Magnetic barcode assay for genetic detection of pathogens. Nat Commun 4:1752
Liu X, Li W-J, Li L, Yang Y, Mao L-G, Peng Z (2014) A label-free electrochemical immunosensor based on gold nanoparticles for direct detection of atrazine. Sensors Actuators B Chem 191:408–414
Long F, Zhu A, Shi H, Wang H, Liu J (2013) Rapid on-site/in-situ detection of heavy metal ions in environmental water using a structure-switching DNA optical biosensor. Sci Rep 3:2308
Looney BB, Falta RW (2000) Vadose zone science and technology solutions. Battelle Press, Columbus
Lu L, Hu X, Zhu Z (2017) Biomimetic sensors and biosensors for qualitative and quantitative analyses of five basic tastes. TrAC Trends Anal Chem 87:58–70
Madianos L, Tsekenis G, Skotadis E, Patsiouras L, Tsoukalas D (2018) A highly sensitive impedimetric aptasensor for the selective detection of acetamiprid and atrazine based on microwires formed by platinum nanoparticles. Biosens Bioelectron 101:268–274
Maduraiveeran G, Jin W (2017) Nanomaterials based electrochemical sensor and biosensor platforms for environmental applications. Trends Environ Anal Chem 13:10–23
Manera MG, Montagna G, Cimaglia F, Chiesa M, Poltronieri P, Santino A, Rella R (2013) SPR based immunosensor for detection of Legionella pneumophila in water samples. Opt Commun 294:420–426

Marques I, Pinto Da Costa J, Justino C, Santos P, Duarte K, Freitas A, Cardoso S, Duarte A, Rocha-Santos T (2017) Carbon nanotube field effect transistor biosensor for the detection of toxins in seawater. Int J Environ Anal Chem 97:597–605

Martín M, Salazar P, Jiménez C, Lecuona M, Ramos MJ, Ode J, Alcoba J, Roche R, Villalonga R, Campuzano S (2015) Rapid Legionella pneumophila determination based on a disposable core–shell Fe3O4@ poly (dopamine) magnetic nanoparticles immunoplatform. Anal Chim Acta 887:51–58

McNamee SE, Elliott CT, Delahaut P, Campbell K (2013) Multiplex biotoxin surface plasmon resonance method for marine biotoxins in algal and seawater samples. Environ Sci Pollut Res 20:6794–6807

McPartlin DA, Loftus JH, Crawley AS, Silke J, Murphy CS, O'Kennedy RJ (2017) Biosensors for the monitoring of harmful algal blooms. Curr Opin Biotechnol 45:164–169

Meneghello A, Sonato A, Ruffato G, Zacco G, Romanato F (2017) A novel high sensitive surface plasmon resonance Legionella pneumophila sensing platform. Sensors Actuators B Chem 250:351–355

Mishra A, Kumar J, Melo JS (2017) An optical microplate biosensor for the detection of methyl parathion pesticide using a biohybrid of Sphingomonas sp. cells-silica nanoparticles. Biosens Bioelectron 87:332–338

Monošík R, Streďanský M, Šturdík E (2012) Biosensors-classification, characterization and new trends. Acta Chim Slov 5:109–120

Nabaei V, Chandrawati R, Heidari H (2018) Magnetic biosensors: modelling and simulation. Biosens Bioelectron 103:69–86

Nigam VK, Shukla P (2015) Enzyme based biosensors for detection of environmental pollutants-a review. J Microbiol Biotechnol 25:1773–1781

Niu X, Zhong Y, Chen R, Wang F, Liu Y, Luo D (2018) A "turn-on" fluorescence sensor for Pb2+ detection based on graphene quantum dots and gold nanoparticles. Sensors Actuators B Chem 255:1577–1581

Nunes G, Lins J, Silva F, Araujo L, Silva F, Mendonça C, Badea M, Hayat A, Marty J-L (2014) Design of a macroalgae amperometric biosensor; application to the rapid monitoring of organophosphate insecticides in an agroecosystem. Chemosphere 111:623–630

O'sullivan CK (2002) Aptasensors–the future of biosensing? Anal Bioanal Chem 372:44–48

Orozco J, Villa E, Manes C-L, Medlin LK, Guillebault D (2016) Electrochemical RNA genosensors for toxic algal species: enhancing selectivity and sensitivity. Talanta 161:560–566

Peng L, Dong S, Wei W, Yuan X, Huang T (2017) Synthesis of reticulated hollow spheres structure NiCo2S4 and its application in organophosphate pesticides biosensor. Biosens Bioelectron 92:563–569

Peterson NC (2005) Advances in monoclonal antibody technology: genetic engineering of mice, cells, and immunoglobulins. ILAR J 46:314–319

Ragavan K, Selvakumar L, Thakur M (2013) Functionalized aptamers as nano-bioprobes for ultrasensitive detection of bisphenol-A. Chem Commun 49:5960–5962

Ravikumar A, Panneerselvam P, Radhakrishnan K, Morad N, Anuradha C, Sivanesan S (2017) DNAzyme based amplified biosensor on ultrasensitive fluorescence detection of Pb (II) ions from aqueous system. J Fluoresc 27:2101–2109

Rebollar-Pérez G, Campos-Terán J, Ornelas-Soto N, Méndez-Albores A, Torres E (2016) Biosensors based on oxidative enzymes for detection of environmental pollutants. Biocatalysis 1:118–129

Rife J, Miller M, Sheehan P, Tamanaha C, Tondra M, Whitman L (2003) Design and performance of GMR sensors for the detection of magnetic microbeads in biosensors. Sensors Actuators A Phys 107:209–218

Robertson S. (2019). What are biosensors? [Online]. Available: http://www.news-medical.net/health/What-are-Biosensors.aspx

Salvador J-P, Adrian J, Galve R, Pinacho DG, Kreuzer M, Sanchez-Baeza F, Marco M-P (2007) Application of bioassays/biosensors for the analysis of pharmaceuticals in environmental samples. Compr Anal Chem 50:279–334

Sassolas A, Prieto-Simón B, Marty J-L (2012) Biosensors for pesticide detection: new trends. Am J Anal Chem 3:210

Semantic Scholar (2013) In-situ label free impedimetric nanowire based DNA biosensor. [ONLINE]. Available: https://www.semanticscholar.org/paper/IN-SITU-LABEL-FREE-IMPEDIMETRIC-NANOWIRE-BASEDDNA/552e54d906466fb5070d33ca9c9f6ab07f68f7e6#paper-header

Shi L, Wang Y, Chu Z, Yin Y, Jiang D, Luo J, Ding S, Jin W (2017) A highly sensitive and reusable electrochemical mercury biosensor based on tunable vertical single-walled carbon nanotubes and a target recycling strategy. J Mater Chem B 5:1073–1080

Singh V (2017) Environmantal Health. [ONLINE]. Available: https://www.slideshare.net/VineetaSingh33/environmental-health-70767665

Singh AC, Bacher G, Bhand S (2017) A label free immunosensor for ultrasensitive detection of 17β-estradiol in water. Electrochim Acta 232:30–37

Tang W, Hu JH, Liu DR (2017) Aptazyme-embedded guide RNAs enable ligand-responsive genome editing and transcriptional activation. Nat Commun 8:15939

Valle MD (2011). Genosensors, sensors made with DNA. Available: https://www.uab.cat/PDF/PDF_1309759280308_en.pdf

Vargas-Bernal R, Rodríguez-Miranda E, Herrera-Pérez G (2012) Evolution and expectations of enzymatic biosensors for pesticides. Pesticides-advances in chemical and botanical pesticides. IntechOpen. Available from: https://doi.org/10.5772/46227

Verma N, Bhardwaj A (2015) Biosensor technology for pesticides—a review. Appl Biochem Biotechnol 175:3093–3119

Vidic J, Manzano M, Chang C-M, Jaffrezic-Renault N (2017) Advanced biosensors for detection of pathogens related to livestock and poultry. Vet Res 48:11

Vogrinc D, Vodovnik M, Marinšek-Logar R (2015) Microbial biosensors for environmental monitoring. Acta Agric Slov 106:67–75

Wang Q, Fang J, Cao D, Li H, Su K, Hu N, Wang P (2015) An improved functional assay for rapid detection of marine toxins, saxitoxin and brevetoxin using a portable cardiomyocyte-based potential biosensor. Biosens Bioelectron 72:10–17

World Health Organization (2012). Endocrine Disrupting Chemicals (EDCs) [Online]. Available: https://www.who.int/ceh/risks/cehemerging2/en/

Wu LL, Wang Z, Zhao SN, Meng X, Song XZ, Feng J, Song SY, Zhang HJ (2016) A metal–organic framework/DNA hybrid system as a novel fluorescent biosensor for mercury (II) ion detection. Chem Eur J 22:477–480

Yang L, Wang G, Liu Y, Wang M (2013) Development of a biosensor based on immobilization of acetylcholinesterase on NiO nanoparticles–carboxylic graphene–nafion modified electrode for detection of pesticides. Talanta 113:135–141

Yang X, He Y, Wang X, Yuan R (2017) A SERS biosensor with magnetic substrate CoFe2O4@ Ag for sensitive detection of Hg2+. Appl Surf Sci 416:581–586

Yildirim N, Long F, He M, Shi H-C, Gu AZ (2014) A portable optic fiber aptasensor for sensitive, specific and rapid detection of bisphenol-A in water samples. Environ Sci: Processes Impacts 16:1379–1386

Yilmaz E, Majidi D, Ozgur E, Denizli A (2015) Whole cell imprinting based Escherichia coli sensors: a study for SPR and QCM. Sensors Actuators B Chem 209:714–721

Yoo M-S, Shin M, Kim Y, Jang M, Choi Y-E, Park SJ, Choi J, Lee J, Park C (2017) Development of electrochemical biosensor for detection of pathogenic microorganism in Asian dust events. Chemosphere 175:269–274

Zhang W, Asiri AM, Liu D, Du D, Lin Y (2014) Nanomaterial-based biosensors for environmental and biological monitoring of organophosphorus pesticides and nerve agents. TrAC Trends Anal Chem 54:1–10

Zhao Y, Zhang W, Lin Y, Du D (2013) The vital function of Fe 3 O 4@ Au nanocomposites for hydrolase biosensor design and its application in detection of methyl parathion. Nanoscale 5:1121–1126

Zhou W, Ding J, Liu J (2017) Theranostic dnazymes. Theranostics 7:1010

Chapter 8
Nano-Based Materials and Their Synthesis

Shalini Chaudhary, Atin Kumar Pathak, and Shamshad Ahmad

8.1 Introduction

In recent years, a new promising research area – nanotechnology – has a great deal of interest in various fields. Nanoparticles (NPs) are particulates characterized as scatterings of strong particles with not less than one measurement of 10–1000 nm in size. The surface region is the main vital element of NPs to volume proportion, permitting corporation with different particles. However, it specifically implies the precise manipulation of molecules and atoms in order to design and control of properties of the nanomaterial. Recently, in a couple of years, nanotechnology has attracted an extraordinary interest due to its potential effects on various logical ranges, for example, pharmaceutical industries, space and hardware businesses, etc. Innovation of nanoparticles allows little estimated materials and structures in the scope of couple of nanometers to fewer than 100 nm. Nanoparticles with the same synthetic organization show extensively changed substance, organic and physical properties, in view of their high surface-to-volume extent. Recently these particles have been used in different fields of drug transport, hyperthermia of tumors, antimicrobial activities etc. (Fig. 8.1). A key domain of examination in nanotechnology discusses about the combination of nanometer ranging particles of diverse sizes, shapes, and monodispersity. Therefore, development of environmental-friendly and cost-effective synthesis of nanoparticles is a crucial task. There are different

S. Chaudhary (✉)
School for Environmental and Sustainable Development, Central University of Gujarat, Gandhinagar, India

A. K. Pathak
Department of Energy Management, Sri Mata Visnao Devi University, Katra, India

S. Ahmad
Department of Environmental Science, Babasaheb Bhimrao Ambedkar University, Lucknow, India

I. Bhushan et al. (eds.), *Nanomaterials and Environmental Biotechnology*, Nanotechnology in the Life Sciences,
https://doi.org/10.1007/978-3-030-34544-0_8

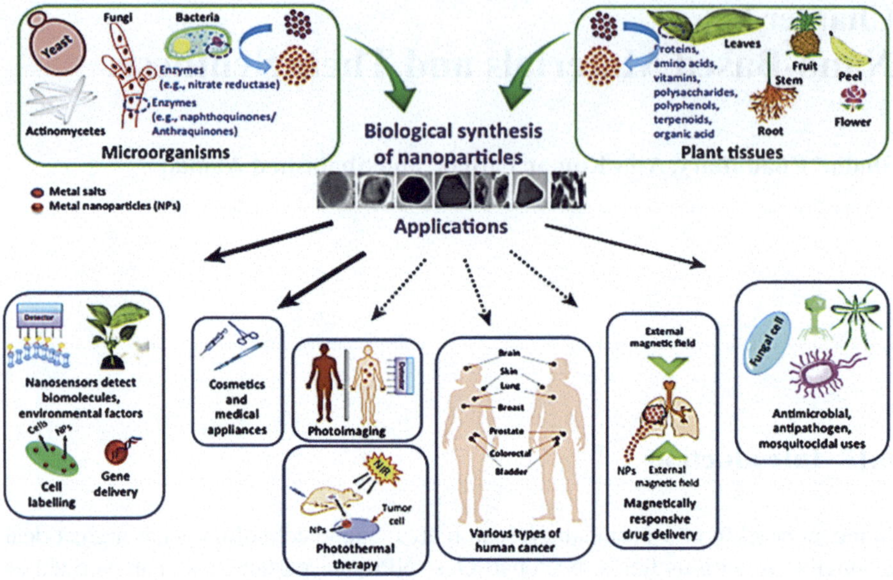

Fig. 8.1 Application of nanoparticle in various systems

microorganisms possessing the ability to synthesize nanoparticles, having the ability to be modified and exploited to fulfill the purpose. The metallic nanoparticles are becoming increasingly important due to their potential application in many fields. Thus using nanotechnology may extend service life, limit the potential for environmental damage, and reduce failure rates. The present chapter focuses on providing an overview and a discussion of metallic nanoparticle synthesis by various biological and non-biological ways. Therefore, a green, nontoxic way of synthesizing metallic nanoparticles is needed in order to allow them to be used in a wider range of industries. This could potentially be achieved by using biological methods. This review will focus on how material science and biology can work together to create a green way of synthesizing metal nanoparticles for a wide range of uses.

8.2 Green Synthesis of MNPs (Biological/Bioreduction)

Much literature has been reported till date on the biological synthesis of MNPs using microbes, i.e., bacteria and fungi, and plants, because of their antioxidant/reducing properties typically responsible for reduction of metallic compounds to their respective MNPs (Fig. 8.2). Since the early 19th century, plant extracts have been known to have the ability to reduce MNPs, but the mechanism involved was not well understood. Recently, in medical science, MNPs synthesized using green technology have attracted a great deal of interest. Magnetic nanoparticles synthesized using green technologies over chemical-based methods have several advantages, i.e., more eco-friendly, easily scaled up, cost-effective, and comparatively require

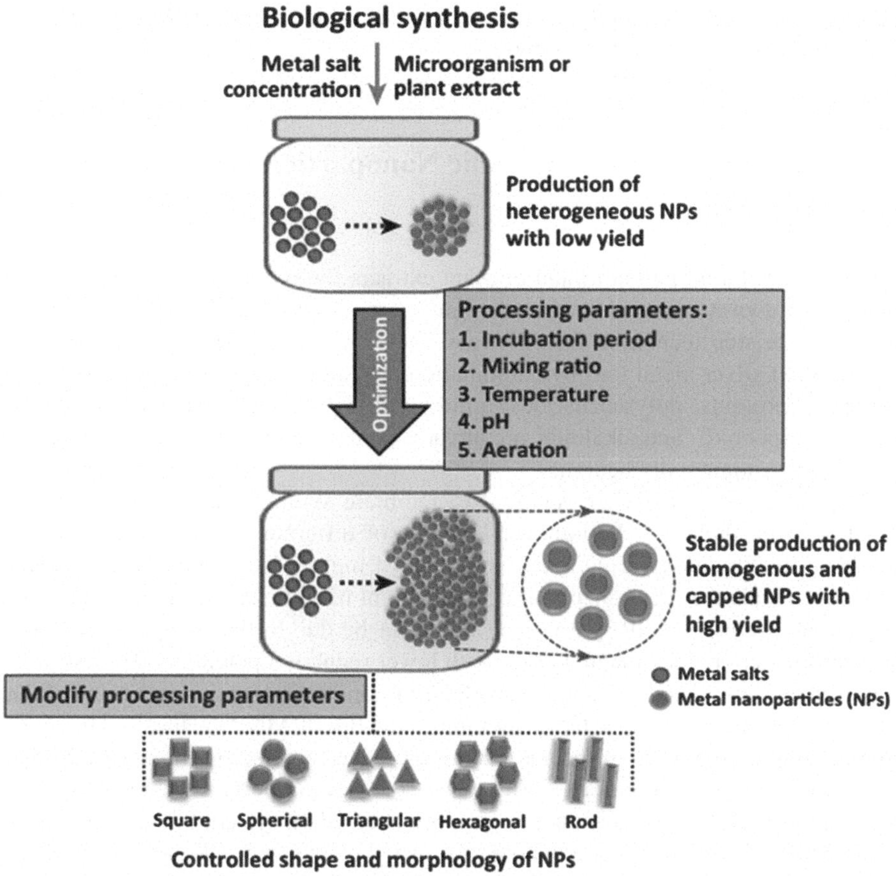

Fig. 8.2 Parameters for producing biological nanoparticles

less energy, pressure, temperature, and other toxic chemicals. In the synthesis of MNPs, plant extracts may act as both stabilizing and reducing agents which directly influence the characteristics of nanoparticles (NPs). Different plant extracts contain different phytoconstituents like catechins, flavonoids, polyphenols, enzymes, alkaloids, vitamins, and functional groups, i.e., plant pigments, polysaccharide, tannins, which are responsible for bioreduction process of metal salts to MNPs. Bioreduction process typically involves simple mixing of aqueous extract with aqueous metallic salt solution, and reaction is carried out at room temperature. The biological synthesis of MNPs involves microorganisms and is considered as bottom-up approach. In this process nanoparticle formation occurs due to oxidation of metal ions occurs in presence of biomolecules i.e. sugars, enzymes, proteins secreted by microorganisms. However, the complete mechanism for synthesis of metallic nanoparticles is not well explored due to the fact that a different microorganism interacts differently with metal ions. Thus the formation of nanoparticles by biochemical processing, environmental factors, and its interaction ultimately determine the formation, size, and morphology of the nanoparticle.

8.3 Green Synthesis of Metallic Nanoparticles Using Plant Extracts

The use of different parts of plant or plant extracts for synthesis of silver nanoparticles has drawn attention, due to its economical, rapid, eco-friendly, non-pathogenic, and a single step technique for the biosynthesis processes. The stabilization and reduction of silver metal ions by combination of biomolecules of plant extract, i.e., saponins, proteins, polysaccharides, amino acids, phenolics, enzymes, vitamins, tannins, terpenoids, and alkaloids, are already established to have medicinal values and are environmentally benign.

Various plant extracts are reported to facilitate synthesis of nanoparticles, as mentioned in Table 8.1. Various MNP types (of different metals) were achieved using GS methodology. This is an indication of high reducing potentials of plant phytochemicals that neutralize uni- or multivalent metallic cations (Mn^{+}) into neutral atoms (M^{o}) for MNP synthesis. This might be due to the failure of the plant biomolecules to reduce metal cations with lower reduction potentials. Thus screening of plant biomolecules having high electron-donating capacity can be one of the efficient options for making GS a common modality of MNP synthesis. The innumerable angiospermic plant species are being used as remedies for diseases and also as a part of our diet. Therefore, during plant selection, important technical aspects, such as availability, socioeconomic importance, edibility, ethnobotanical background apart from the reducing nature of selected plant species. Randomized selection procedure might result in MNP synthesis, but biocompatibility issue may put it in a nonprogressive research interest. Angiospermic plant species, such as *Centella asiatica, Azadirachta indica, Camellia sinensis, and Aloe vera*, are the frontline examples of plant species that have been highly explored for their medicinal values and also have clinical relevancy (Table 8.1).

8.4 Nanoparticle Synthesis Using Microorganisms

Microorganisms are the main nanofactories that hold massive, eco-friendly, and cost-effective tools, which avoid toxic, harsh chemicals and decrease high energy demand required during physiochemical synthesis. Microorganisms have the capability to accumulate and detoxify heavy metal potential due to several enzymatic action and reaction, which reduces metallic salts to metallic nanoparticles (MNPs) with a narrow range of size distribution with less polydispersity. The mechanism and experimental methods of synthesizing nanoparticles in microorganisms. Over the past decades, microorganisms, such as bacteria, fungi, and yeasts, have been used

Table 8.1 Plant species used for green synthesis of different metallic nanoparticles (MNPs)

Plant species	Plant material	Type of MNPs	Mechanism/causative agent	Size of MNPs (nm)	References
Azadirachta indica	Kernel	Au, Ag	Azadirachtin	50–100	Shukla et al. (2012)
Camellia sinensis	Leaves	Au	Catechins	15–42	Nune et al. (2009)
Jatropha curcas	Latex	Pb	Curcacycline A and B	10–12	Joglekar et al. (2011)
Geranium	Leaves	Ag	Terpenoids (not specified)	16–40	Shankar et al. (2003a)
Nelumbo nucifera	Leaves	Ag	Not mentioned	25–80	Santhoshkumar et al. (2011)
Lemongrass plant extract	Leaves	Au	Sugar derivative molecules	200–500	Shankar et al. (2005)
Avena sativa	Stems	Au	Not specified	5–85	Armendariz et al. (2004)
Aloe vera	Leaves	Ag	Not specified	70	Medda et al. (2015)
Cinnamomum camphora	Leaves	Ag	Polyol compounds	64.8	Huang et al. (2007)
Syzygium aromaticum	Flower buds	Cu	Eugenol	5–40	Subhankari and Nayak (2013)
Euphorbia esula	Leaves	Cu	Flavonoids and phenolic acids	20–110	Nasrollahzadeh et al. (2014)
Camellia sinensis	Leaves	Fe_2O_3	Polyphenols	5–15	Hoag et al. (2009)
Eucalyptus	Leaves	Fe_2O_3	Epicatechin and quercetin glucuronide	20–80	Wang et al. (2014)
Aloe barbadensis miller	Leaves	ZnO	Phenolic compounds, terpenoids	25–40	Sangeetha et al. (2011)
Nephelium lappaceum L.	Peels	MgO	Not mentioned	100	Suresh et al. (2014)
Nephelium lappaceum L.	Peels	NiO	Nickel-ellagate complex formation	50	Yuvakkumar et al. (2014)
Clitoria ternatea	Whole plant	MgO	Bioactive compounds	50–400	Sushma et al. (2016)
Cintella asiatica	Leaves	Au	Phenolic compounds	9.3 and 11.4	Das et al. (2010)
Camellia sinensis	Leaves	Pt	Pure tea polyphenol	30–60	Alshatwi et al. (2015)
Asparagus racemosus	Tuber cortex	Pd	Bioactive compounds	1–6	Raut et al. (2013)

to synthesized extra- and intracellular MNPs given in Table 8.2. A biological procedure for MNP synthesis has been informed using bacterial biomass, supernatant, and derived constituents. Among the several methodologies, extracellular synthesis of MNPs has received much attention because it removes the downstream processing process which is required for the recovery of MNPs in intracellular methodologies. Moreover, metal-resistant genes, proteins, enzymes, cofactors, and organic substance

Table 8.2 Microbial synthesis of different metallic nanoparticles (MNPs)

Bacteria	Nanoparticle	Size	Morphology	References
Bacterial				
Aeromonas sp.	Ag	6.4	Spherical	Rai et al. (2009)
Bacillus megaterium	Au	1.9	Spherical	Sanpo et al. (2013)
Escherichia coli (DH5?)	Au, Ag	10–50	Spherical, triangular	Mahanty et al. (2013)
Klebsiella (aerogenes, pneumoniae)	CdS, Ag	~52	Spherical	Shahverdi et al. (2007)
Nocardiopsis sp. MBRC-1	Ag	45	Spherical	Manivasagan et al. (2013)
Shewanella	Au, Fe_3O_4			Konishi et al. (2006)
Thermoanaerobacter ethanolicusTOR-39	Co, Cr Ni and Fe_3O_4	5–25	Octahedral	Rai et al. (2008)
Pseudomonas stutzeri AG259	Ag, Ag_2S	20–50	Nanocrystal	Joerger et al. (2000)
Lactobacillus sp.	Au, Ag, Au-Ag	50	Hexagonal/counter	Nair and Pradeep (2002)
Desulfovibrio desulfuricans	Pd	10–15		Yong et al. (2002)
Corynebacterium sp. SH09	Ag	0.3–30		Zhang et al. (2005)
Lactobacillus sp.	Ti		Spherical	Prasad et al. (2007)
Shewanella oneidensis	Fe_3O_4	40–50	Rectangular, hexagonal	Perez-Gonzalez et al. (2010)
Yeast cells	Fe_3O_4		Wormhole-like	Zhou et al. (2009)
Saccharomyces cerevisiae	Sb_2O_3	2–10	Spherical	Jha et al. (2009)
Lactobacillus sp.	$BaTiO_3$	20–80	Tetragonal	Jha and Prasad (2010)
Fusarium oxysporum	TiO_2	6–13	Spherical	Bansal et al. (2005)
Fusarium oxysporum	ZrO_2	3–11	Spherical	Bansal et al. (2004)
Fungus				
Fusarium oxysporum	Au	20–40	Spherical, triangular	Mukherjee et al. (2002)
F. oxysporum	Zr	3–11	Quasi-spherical	Bansal et al. (2004)
F. oxysporum	Au-Ag	8–14		Senapati et al. (2005)
F. oxysporum	Si	5–15	Quasi-spherical	Bansal et al. (2005)
F. oxysporum	Pt	10–50	Triangle, hexagons, square, rectangles	Riddin et al. (2006)

(continued)

Table 8.2 (continued)

Bacteria	Nanoparticle	Size	Morphology	References
F. oxysporum	$BaTiO_3$	4		Bansal et al. (2006)
V. luteoalbum	Au	10	Spherical	Gericke and Pinches (2006)
Aspergillus flavus	Ag	8.9		Vigneshwaran et al. (2007)
Coriolus versicolor	Ag	25–75	Spherical	Sanghi and Verma (2009)
Fusarium oxysporum	$PbCO_3$, $CdCO_3$	120–200	Spherical	Sanyal et al. (2005)
Fusarium oxysporum	$SrCO_3$	10–50	Needle like	Rautaray et al. (2004)
Brevibacterium casei	PHB	100–125	–	Pandian et al. (2009)
Yeasts	$Zn_3(PO_4)_2$	10–80	Rectangular	Yan et al. (2009)
Fusarium oxysporum	CdSe	9–15	Spherical	Kumar et al. (2007)
Aspergillus fumigatus	ZnO	1.2–6.8	Spherical and hexagonal	Raliya and Tarafdar (2013)
Aspergillus oryzae	$FeCl_3$	10–24.6	Spherical	Raliya (2013)
Aspergillus tubingensis	$Ca_3P_2O_8$	28.2	Spherical	Tarafdar et al. (2012)
Rhizopus oryzae	Au	10	Nanocrystalline	Das et al. (2009)
Aureobasidium pullulans	Au	29 ± 6	Spherical	Zhang et al. (2011)
Colletotrichum sp.	Au	20–40	Decahedral and icosahedral	Shankar et al. (2003b)
Helminthosporium solani	Au	2–70	Polydispersed	Kumar et al. (2008)
Neurospora crassa	Au	32	Spherical	Castro-Longoria et al. (2011)
Penicillium brevicompactum	Au	10–50	Spherical	Mishra et al. (2011)
Verticillium luteoalbum	Au	<10	Spheres and rods	Gericke and Pinches (2006)
Cylindrocladium floridanum	Au	19.5	Spherical	Narayanan and Sakthivel (2013)
Coriolis versicolor	Au	20–100	Spherical and ellipsoidal	Sanghi and Verma (2010)
Verticillium sp.	Ag	25	Spherical	Mukherjee et al. (2001)
Aspergillus fumigatus	Ag	5–25	Mostly spherical, triangular	Bhainsa and Souza (2006)
Pleurotus sajorcaju	Ag	30.5	Spherical	Vigneshwaran and Kathe (2007)

(continued)

Table 8.2 (continued)

Bacteria	Nanoparticle	Size	Morphology	References
Aspergillus flavus	Ag	8.92	Spherical	Vigneshwaran et al. (2007)
Trichoderma asperellum	Ag	13–18	Nanocrystalline	Mukherjee et al. (2008)
Penicillium fellutanum	Ag	5–25	Mostly spherical	Kathiresan et al. (2009)
Penicillium strain J3	Ag	10–100	Mostly spherical	Maliszewska et al. (2009)
Cladosporium cladosporioides	Ag	10–100	Mostly spherical	Balaji et al. (2009)
Phoma glomerata	Ag	60–80	Spherical	Birla et al. (2009)
Coriolus versicolor	Ag	25–75	Spherical	Sanghi and Verma (2009)
Trichoderma viride	Ag	5–40	Spherical, rodlike	Fayaz et al. (2009)
Amylomyces rouxii KSU-09	Ag	5–27	Spherical	Musarrat et al. (2010)
Aspergillus flavus NJP08	Ag	17	Spherical	Jain et al. (2011)
Aspergillus terreus CZR-1	Ag	2.5	Spherical	Raliya and Tarafdar (2012)
Fusarium oxysporum	Au-Ag	8–14	Quasi-spherical	Senapati et al. (2005)
Verticillium sp.	Fe_3O_4	100–400	Cubo-octahedral, quasi-spherical	Bharde et al. (2006)
Aspergillus flavus	TiO_2	62–74	Spherical	Rajakumar et al. (2012)
Aspergillus flavus TFR7	TiO_2	12–15	Extracellular	Raliya et al. (2015)
Fusarium spp.	Zn	100–200	Irregular, spherical	Velmurugan et al. (2010)
Aspergillus versicolor	Hg	20.5	Alteration	Das et al. (2008)
Algae				
Shewanella algae	Au	9.6	Spherical	Sau and Murphy (2004)
Sargassum muticum	Zn	30–57	Spherical	Azizi et al. (2014)
Chlorococcum humicola	Ag	16		Jena et al. (2013)
Plectonemaboryanum	Pt	<300	Spherical	Lengke et al. (2006)
Sargassum bovinum	Pd	5–10		Momeni and Nabipour (2015)
Phormidium tenue	Cd	5.1		MubarakAli et al. (2012)
Phormidium valderianum	Ag	15		Parial et al. (2012)
Chlorella vulgaris	Ag	44		Xie et al. (2007)
Chlorella pyrenoidusa	Au	25–30		Oza et al. (2012)

play significant roles by working as reducing agents. Moreover, these provide natural topping to synthesize MNPs, thus avoiding the accumulation with and increasing stability of MNPs. In recent research, bacteria, including *Pseudomonas deceptionensis* (Jo et al. 2015), *Weissella oryzae* (Singh et al. 2015), *Bacillus methylotrophicus* (Wang et al. 2015), *Brevibacterium frigoritolerans* (Singh et al. 2016), and *Bhargavaea indica* (Singh et al. 2015), have been explored for synthesis of Ag and Au NPs. Similar potential for producing NPs has been showed by using several *Bacillus* and other species, including *B. licheniformis, B. amyloliquefaciens, Rhodobacter sphaeroides*, *Listeria* monocytogenes, *B. subtilis*, and *Streptomyces anulatus*. Various genera of microorganisms have been reported for metal nanoparticle synthesis, including *Bacillus, Pseudomonas, Klebsiella, Escherichia, Enterobacter, Aeromonas, Corynebacterium, Lactobacillus, Pseudomonas, Weissella, Rhodobacter, Rhodococcus, Brevibacterium, Streptomyces, Trichoderma, Desulfovibrio, Sargassum, Shewanella, Plectonemaboryanum, Rhodopseudomonas, Pyrobaculum*, etc. These investigations suggest that the main mechanism of the synthesis of nanoparticles using bacteria depends on enzymes, for instance, the nitrate reductase enzyme was found to be responsible for silver nanoparticle synthesis in *B. licheniformis* (Singh et al. 2016).

Rather than using bacteria, mycosynthesis is a straightforward approach for achieving stable and easy biological nanoparticle synthesis. Most fungi containing important metabolites with higher bioaccumulation ability and simple downstream processing are easy to culture for the efficient, low-cost production of nanoparticles. Moreover, compared with bacteria, fungi have higher tolerances to, and uptake competences for, metals, particularly in terms of the high wall-binding capability of metal salts with fungal biomass for the high-yield production of nanoparticles. Three possible mechanisms have been proposed to explain the mycosynthesis of metal nanoparticles: nitrate reductase action; electron shuttle quinones; or both. Fungal enzymes, such as the reductase enzymes from *Penicillium* species and *Fusarium oxysporum*, nitrate reductase, and /-NADPH-dependent reductases, were found to have a significant role in nanoparticle synthesis, similar to the mechanism found in bacteria. The synthesis of nanoparticles using actinomycetes has not been well explored, even though actinomycete-mediated nanoparticles have good monodispersity and stability and significant biocidal activities against various pathogens. The synthesis of silver, copper, and zinc nanoparticles using *Streptomyces* sp. has demonstrated that the reductase enzyme from *Streptomyces* sp. has a vital role in the reduction of metal salts. Similar to other microorganisms, yeasts have also been widely investigated for the extracellular synthesis of the nanoparticles on a large scale, with straightforward downstream processing. Furthermore, virus-mediated synthesis of nanoparticles is also possible. Viruses can be used to synthesize nanowires with functional components that are assembled for various applications, such as battery electrodes, photovoltaic devices, and super capacitors. Nanoparticles are slow with low productivity, and the recovery of nanoparticles requires downstream processing. Furthermore, problems related to microorganism-based synthesis for nanoparticles also include the complex steps, such as microbial sampling, isolation, culturing, and maintenance.

8.5 Conclusion

The green method for nanoparticle synthesis, which is rapidly replacing traditional chemical syntheses, is of great interest because of eco-friendliness, economic views, feasibility, and a wide range of applications in several areas such as nano-medicine and catalysis medicine. Recently, various types of biological units which serve a dual role as both the reducing and stabilizing agents have been used in the synthesis of bioactive nanoparticles.

References

Alshatwi AA, Athinarayanan J, Subbarayan PV (2015) Green synthesis of platinum nanoparticles that induce cell death and G2/M-phase cell cycle arrest in human cervical cancer cells. J Mater Sci Mater Med 26(1):1–9

Armendariz V et al (2004) Size controlled gold nanoparticle formation by Avena sativa biomass use of plants in nanobiotechnology. J Nanopart Res 6(4):377–382

Azizi S et al (2014) Green biosynthesis and characterization of zinc oxide nanoparticles using brown marine macroalga *Sargassum muticum* aqueous extract. Mater Lett 116:275–277

Balaji DS, Basavaraja S, Deshpande R, Mahesh DB, Prabhakar BK, Venkataraman A (2009) Extracellular biosynthesis of functionalized silver nanoparticles by strains of *Cladosporium cladosporioides* fungus. Colloids Surf B Biointerfaces 68:88–92

Bansal V, Rautaray D, Ahmad A, Sastry M (2004) Biosynthesis of zirconia nanoparticles using the fungus *Fusarium oxysporum*. J Mater Chem 14(22):3303–3305

Bansal V, Rautaray D, Bharde A et al (2005) Fungus-mediated biosynthesis of silica and titania particles. J Mater Chem 15(26):2583–2589

Bansal V et al (2006) Room-temperature biosynthesis of ferro-electric barium titanate nanoparticles. J Am Chem Soc 128(36):11958–11963

Bhainsa KCD, Souza SF (2006) Extracellular biosynthesis of silver nanoparticles using the fungus *Aspergillus fumigatus*. Colloids Surf B Biointerfaces 47:160–164

Bharde A, Rautaray D, Bansal V, Ahmad A, Sarkar I, Yusuf SM, Sanyal M, Sastry M (2006) Extracellular biosynthesis of magnetite using fungi. Small 2:135–141

Birla SS, Tiwari VV, Gade AK, Ingle AP, Yadav AP, Rai MK (2009) Fabrication of silver nanoparticles by *Phomaglomerata* and its combined effect against *Escherichia coli*, *Pseudomonas aeruginosa* and *Staphylococcus aureus*. Lett Appl Microbiol 48:173–179

Castro-Longoria E, Vilchis-Nestor AR, Avalos-Borja M (2011) Biosynthesis of silver, gold and bimetallic nanoparticles using the filamentous fungus *Neurospora crassa*. Colloids Surf B Biointerfaces 83:42–48

Das S, Das A, Guha A (2008) Adsorption behavior of mercury on functionalized Aspergillus versicolor mycelia: atomic force microscopic study. Langmuir 25:360–366

Das SK, Das AR, Guha AK (2009) Gold nanoparticles: microbial synthesis and application in water hygiene management. Langmuir 25:8192–8199

Das RK, Borthakur BB, Bora U (2010) Green synthesis of gold nanoparticles using ethanolic leaf extract of Centella asiatica. Mater Lett 64(13):1445–1447

Fayaz AM, Balaji K, Girilal M, Kalaichelvan PT, Venkatesan R (2009) Mycobased synthesis of silver nanoparticles and their incorporation into sodium alginate films for vegetable and fruit preservation. J Agric Food Chem 57:6246–6252

Gericke M, Pinches A (2006) Microbial production of gold nanoparticles. Gold Bull 39(1):22–28

Hoag GE et al (2009) Degradation of bromothymol blue by 'greener' nano-scale zero-valent iron synthesized using tea polyphenols. J Mater Chem 19(45):8671–8677

Huang J et al (2007) Biosynthesis of silver and gold nanoparticles by novel sundried *Cinnamomum camphora* leaf. Nanotechnology 18(10):105104

Jain N, Bhargava A, Majumdar S, Tarafdar JC, Panwar J (2011) Extracellular biosynthesis and characterization of silver nanoparticles using *Aspergillus flavus* NJP08: a mechanism perspective. Nanoscale 3:635–641

Jena J et al (2013) Biosynthesis and characterization of silver nanoparticles using microalga *Chlorococcum humicola* and its antibacterial activity. Int J Nanomater Biostruct 3:1–8

Jha AK, Prasad K (2010) Ferroelectric BaTiO3 nanoparticles biosynthesis and characterization. Colloids Surf B Biointerfaces 75(1):330–334

Jha AK, Prasad K, Prasad K (2009) A green low-cost biosynthesis of Sb_2O_3 nanoparticles. Biochem Eng J 43(3):303–306

Jo JH, Singh P, Kim YJ, Wang C, Mathiyalagan R (2015) Pseudomonas deceptionensis DC5-mediated synthesis of extracellular silver nanoparticles. Artif Cells Nanomed Biotechnol 44: 1576–1581

Joerger R, Klaus T, Granqvist C (2000) Biologically produced silver-carbon composite materials for optically functional thin film coatings. Adv Mater 12(6):407–409

Joglekar S et al (2011) Novel route for rapid biosynthesis of lead nanoparticles using aqueous extract of Jatropha curcas L. latex. Mater Lett 65(19):3170–3172

Kathiresan K, Manivannan S, Nabeel M, Dhivya B (2009) Studies on silver nanoparticles synthesized by a marine fungus, *Penicillium fellutanum* isolated from coastal mangrove sediment. Colloids Surf B Biointerfaces 71:133–137

Konishi Y, sukiyama T, Ohno K et al (2006) Intracellular recovery of gold by microbial reduction of AuCl-4ions using the anaerobic bacterium Shewanella algae. Hydrometallurgy 81(1):24–29

Kumar SA, Ansary AA, Abroad A, Khan MI (2007) Extracellular biosynthesis of CdSe quantum dots by the fungus, *Fusarium oxysporum*. J Biomed Nanotechnol 3(2):190–194

Kumar SA, Peter YA, Nadeau JL (2008) Facile biosynthesis, separation and conjugation of gold nanoparticles to doxorubicin. Nanotechnology 19:495101. https://doi.org/10.1088/0957-4484/19/49/495101

Lengke MF, Fleet ME, Southam G (2006) Synthesis of platinum nanoparticles by reaction of filamentous cyanobacteria with platinum (IV)-chloride complex. Langmuir 22(17):7318–7323

Mahanty A, Bosu R, Panda P et al (2013) Microwave assisted rapid combinatorial synthesis of silver nanoparticles using *E. coli* culture supernatant. Inter J Pharma Bio Sci 4(2):1030–1035

Maliszewska I, Szewczyk K, Waszak K (2009) Biological synthesis of silver nanoparticles. J Phys Conf Ser 146. https://doi.org/10.1088/1742-6596/146/1/012025

Manivasagan P, Venkatesan J, Senthilkumar K et al (2013) Biosynthesis, antimicrobial and cytotoxic effect of silver nanoparticles using a novel *Nocardiopsis* sp. MBRC-1. Biomed Res Int 2013:287638-9

Medda S et al (2015) Biosynthesis of silver nanoparticles from Aloe vera leaf extract and antifungal activity against *Rhizopus* sp. and *Aspergillus* sp. Appl Nanosci 5(7):875–880

Mishra A, Tripathy S, Wahab R, Jeong SH, Hwang I, Yang YB, Kim YS, Shin HS, Yun SI (2011) Microbial synthesis of gold nanoparticles using the fungus *Penicillium brevicompactum* and their cytotoxic effects against mouse mayo blast cancer C2C12 cells. Appl Microbiol Biotechnol 92:617–630

Momeni S, Nabipour I (2015) A simple green synthesis of palladium nanoparticles with sargassum alga and their electrocatalytic activities towards hydrogen peroxide. Appl Biochem Biotechnol 176:1–13

MubarakAli D et al (2012) Synthesis and characterization of CdS nanoparticles using C-phycoerythrin from the marine. Mater Lett 74:8–11

Mukherjee P, Ahmad A, Mandal D, Senapati S, Sainkar SR, Khan MI, Parishcha R, Ajaykumar PV, Alam M, Kumar R et al (2001) Fungus-mediated synthesis of silver nanoparticles and their immobilization in the mycelial matrix: a novel biological approach to nanoparticle synthesis. Nano Lett 1:515–519

Mukherjee P et al (2002) Extracellular synthesis of gold nanoparticles by the fungus *Fusarium oxysporum*. Chembiochem 3(5):461–463

Mukherjee P, Roy M, Mandal BP, Dey GK, Mukherjee PK, Ghatak J, Tyagi AK, Kale SP (2008) Green synthesis of highly stabilized nanocrystalline silver particles by a non-pathogenic and agriculturally important fungus *T. asperellum*. Nanotechnology 19:1–7

Musarrat J, Dwivedi S, Singh BR, Al-Khedhairy AA, Azam A, Naqvi A (2010) Production of antimicrobial silver nanoparticles in water extracts of the fungus *Amylomycesrouxii* strain KSU-09. Bioresour Technol 101:8772–8776

Nair B, Pradeep T (2002) Coalescence of nanoclusters and formation of submicron crystallites assisted by Lactobacillus strains. Cryst Growth Des 2(4):293–298

Narayanan K, Sakthivel N (2013) Mycocrystallization of gold ions by the fungus *Cylindrocladium floridanum*. World J Microbiol Biotechnol 29:2207–2211

Nasrollahzadeh M, Sajadi SM, Khalaj M (2014) Green synthesis of copper nanoparticles using aqueous extract of the leaves of *Euphorbia esula* L and their catalytic activity for ligand-free Ullmann-coupling reaction and reduction of 4-nitrophenol. RSC Adv 4(88):47313–47318

Nune SK et al (2009) Green nanotechnology from tea: phytochemicals in tea as building blocks for production of bio-compatible gold nanoparticles. J Mater Chem 19(19):2912–2920

Oza G et al (2012) Facile biosynthesis of gold nanoparticles exploiting optimum pH and temperature of fresh water algae *Chlorella pyrenoidusa*. Adv Appl Sci Res 3(3):1405–1412

Pandian SRK, Deepak V, Kalishwaralal K, Muniyandi J, Rameshkumar N, Gurunathan S (2009) Synthesis of PHB nanoparticles from optimized medium utilizing dairy industrial waste using *Brevibacteriumcasei* SRKP2: a green chemistry approach. Colloids Surf B 74(1):266–273

Parial D et al (2012) Screening of different algae for green synthesis of gold nanoparticles. Eur J Phycol 47(1):22–29

Perez-Gonzalez T, Jimenez-Lopez C, Neal AL et al (2010) Magnetite biomineralization induced by *Shewanella oneidensis*. Geochim Cosmochim Acta 74(3):967–979

Prasad K, Jha AK, Kulkarni A (2007) Lactobacillus assisted synthesis of titanium nanoparticles. Nanoscale Res Lett 2(5):248–250

Rai M, Yadav A, Gade A (2008) CRC 675—current trends in phytosynthesis of metal nanoparticles. Crit Rev Biotechnol 28(4):277–284

Rai M, Yadav A, Gade A (2009) Silver nanoparticles as a new generation of antimicrobials. Biotechnol Adv 27:76–83

Rajakumar G, Rahuman A, Roopan SM, Khanna VG, Elango G, Kamaraj C, Zahir AA, Velayutham K (2012) Fungus-mediated biosynthesis and characterization of TiO2 nanoparticles and their activity against pathogenic bacteria. Spectrochim Acta A Mol Biomol Spectrosc 91:23–29

Raliya R (2013) Rapid, low-cost, and ecofriendly approach for iron nanoparticle synthesis using *Aspergillus oryzae*TFR9. J Nanopart. https://doi.org/10.1155/2013/141274

Raliya R, Tarafdar JC (2012) Novel approach for silver nanoparticle synthesis using *Aspergillusterreus*CZR-1: mechanism perspective. J Bionanosci 6:12–16

Raliya R, Tarafdar JC (2013) ZnO nanoparticle biosynthesis and its effect on phosphorous-mobilizing enzyme secretion and gum contents in Clusterbean (*Cyamopsis tetragonoloba* L.). Agirc Res 2:48–57

Raliya R, Biswas P, Tarafdar JC (2015) TiO_2 nanoparticle biosynthesis and its physiological effect on mung bean (*Vigna radiata* L.). Biotechnol Rep 5:22–26

Raut RW et al (2013) Rapid biosynthesis of platinum and palladium metal nanoparticles using root extract of *Asparagus racemosus* Linn. Adv Mater Lett 4(8):650–654

Rautaray D, Sanyal A, Adyanthaya SD, Ahmad A, Sastry M (2004) Biological synthesis of strontium carbonate crystals using the fungus *Fusarium oxysporum*. Langmuir 20(16):6827–6833

Riddin T, Gericke M, Whiteley C (2006) Analysis of the inter-and extracellular formation of platinum nanoparticles by *Fusarium oxysporum* f. sp. *lycopersici* using response surface methodology. Nanotechnology 17(14):3482

Sangeetha G, Rajeshwari S, Venckatesh R (2011) Green synthesis of zinc oxide nanoparticles by aloe barbadensis miller leaf extract: structure and optical properties. Mater Res Bull 46(12):2560–2566

Sanghi R, Verma P (2009) Biomimetic synthesis and characterisation of protein capped silver nanoparticles. Bioresour Technol 100(1):501–504

Sanghi R, Verma P (2010) pH dependent fungal proteins in the green synthesis of gold nanoparticles. Adv Mater Lett 1:193–199

Sanpo N, Wen C, Berndt CC et al (2013) Antibacterial properties of spinel ferrite nanoparticles. In: Mendez A (ed) Microbial pathogens and strategies for combating them: science, technology and education. Formatex Research Centre, Badajoz, pp 239–250

Santhoshkumar T et al (2011) Synthesis of silver nanoparticles using Nelumbo nucifera leaf extract and its larvicidal activity against malaria and filariasis vectors. Parasitol Res 108(3):693–702

Sanyal A, Rautaray D, Bansal V, Ahmad A, Sastry M (2005) Heavy-metal remediation by a fungus as a means of production of lead and cadmium carbonate crystals. Langmuir 21(16):7220–7224

Sau TK, Murphy CJ (2004) Room temperature, high-yield synthesis of multiple shapes of gold nanoparticles in aqueous solution. J Am Chem Soc 126(28):8648–8649

Senapati S, Ahmad A, Khan MI, Sastry M, Kumar R (2005) Extracellular biosynthesis of bimetallic Au-Ag alloy nanoparticles. Small 1(5):517–520

Shahverdi AR, Minaeian S, Jamalifar H et al (2007) Rapid synthesis of silver nanoparticles using culture supernatants of Enterobacteria: a novel biological approach. Process Biochem 42:919–923

Shankar SS, Ahmad A, Sastry M (2003a) Geranium leaf assisted biosynthesis of silver nanoparticles. Biotechnol Prog 19(6):1627–1631

Shankar SS, Ahmad A, Pasricha R, Sastry M (2003b) Bioreduction of chloroaurate ions by geranium leaves and its endophytic fungus yields gold nanoparticles of different shapes. J Mater Chem 13:1822–1826

Shankar SS et al (2005) Controlling the optical properties of lemongrass extract synthesized gold nanotriangles and potential application in infrared-absorbing optical coatings. Chem Mater 17(3):566–572

Shukla VK et al (2012) Green synthesis of nanosilver as a sensor for detection of hydrogen peroxide in water. J Hazard Mater 213:161–166

Singh P, Kim Y-J, Wang C, Mathiyalagan R, Yang DC (2015) Artificial cells. Nanomed Biotechnol 44(6):1569–1575

Singh P, Kim Y-J, Zhang D, Yang D-C (2016) Biological synthesis of nanoparticles from plants and microorganisms. Trends Biotechnol 34:7

Subhankari I, Nayak P (2013) Synthesis of copper nanoparticles using *Syzygium aromaticum* (Cloves) aqueous extract by using green chemistry. World J Nano Sci Technol 2(1):14–17

Suresh J, Yuvakkumar R, Sundrarajan M, Hong SI (2014) Green synthesis of magnesium oxide nanoparticles. In Advanced Materials Research (Vol. 952, pp. 141–144). Trans Tech Publications

Sushma NJ et al (2016) Facile approach to synthesize magnesium oxide nanoparticles by using Clitoria ternatea—characterization and in vitro antioxidant studies. Appl Nanosci 6(3):437–444

Tarafdar JC, Raliya R, Rathore I (2012) Microbial synthesis of phosphorous nanoparticle from tricalcium phosphate using *Aspergillus tubingensis* TFR-5. J Bionanosci 6:84–89

Velmurugan P, Shim J, You Y, Choi S, Kamala-Kannan S, Lee KJ, Kim HJ, Oh BT (2010) Removal of zinc by live, dead, and dried biomass of *Fusarium* spp. isolated from the abandoned-metal mine in South Korea and its perspective of producing nanocrystals. J Hazard Mater 182:317–324

Vigneshwaran N, Kathe A (2007) Silver-protein (core-shell) nanoparticle production using spent mushroom substrate. Langmuir 23:7113–7117

Vigneshwaran N, Ashtaputre NM, Varadarajan PV, Nachane RP, Paralikar KM, Balasubramanya RH (2007) Biological synthesis of silver nanoparticles using the fungus *Aspergillus flavus*. Mater Lett 61(6):1413–1418

Wang T et al (2014) Green synthesis of Fe nanoparticles using eucalyptus leaf extracts for treatment of eutrophic wastewater. Sci Total Environ 466:210–213

Wang C, Kim YJ, Singh P, Mathiyalagan R, Jin Y et al (2015) Green synthesis of silver nanoparticles by Bacillus methylotrophicus, and their antimicrobial activity. Artif Cells Nanomed Biotechnol 44:1127–1132

Xie J et al (2007) Silver nanoplates: from biological to biomimetic synthesis. ACS Nano 1(5):429–439

Yan S, He W, Sun C et al (2009) The biomimetic synthesis of zinc phosphate nanoparticles. Dyes Pigments 80(2):254–258

Yong P et al (2002) Bioreduction and biocrystallization of palladium by *Desulfovibrio desulfuricans* NCIMB 8307. Biotechnol Bioeng 80(4):369–379

Yuvakkumar R et al (2014) Rambutan *(Nephelium lappaceum* L.) peel extract assisted biomimetic synthesis of nickel oxide nanocrystals. Mater Lett 128:170–174

Zhang H et al (2005) Biosorption and bioreduction of diamine silver complex by Corynebacterium. J Chem Technol Biotechnol 80(3):285–290

Zhang X, He X, Wang K, Yang X (2011) Different active biomolecules involved in biosynthesis of gold nanoparticles by three fungus species. J Biomed Nanotechnol 7:245–254

Zhou W, He W, Zhong S et al (2009) Biosynthesis and magnetic properties of mesoporous Fe3O4 composites. J Magn Magn Mater 321(8):1025–1028

Chapter 9
Nano-based Composites and Their Synthesis

Shalok Bharti, Lalit Thakur, Sanjeev Anand, and Varun Dutta

9.1 Introduction

In today's world the type of material used in a device plays a significant role. The conventional materials like metals, polymers and ceramics have their limitations. Therefore the use of composites is introduced in various applications. In composites two or more materials are combined to increase the property of the material. This approach is further increased by introducing one of the components in the composite at nanoscale level (Hussain et al. 2006). Such a composite is called nanocomposite. Nanocomposites are multiphase solid materials in which one of its parts has dimensions in nanometre series (i.e. 10^{9} m) (Roy et al. 1986). Because of the increase in the area of surface and the effect of quantum at such a small scale, it's mechanical, electrical, chemical and other properties get improved. Nanocomposites are different from the conventional composite materials. In conventional composite materials, reinforcements are added to the base in order to increase its properties, e.g. SiC when added to the aluminium alloy, its hardness and tensile strength increase (Gupta et al. 2018). But nanocomposites are different from the conventional composites due to the fact that it contains exceptionally high aspect ratio. It is known as the material of the present era because of the property and design uniqueness which is not found in the traditional form of composites. There are still many areas in nanocomposites which are to be understood (Schmidt et al. 2002); on the other hand, it is to be noted that the first research was carried out in 1992

S. Bharti · V. Dutta (✉)
School of Mechanical Engineering, SMVD University, Katra, J&K, India

L. Thakur
Department of Mechanical Engineering, NIT Kurukshetra, Haryana, India

S. Anand
School of Energy Management, SMVD University, Katra, J&K, India

I. Bhushan et al. (eds.), *Nanomaterials and Environmental Biotechnology*, Nanotechnology in the Life Sciences,
https://doi.org/10.1007/978-3-030-34544-0_9

(Gleiter 1992). For understanding the structure-property relationship, the area/volume ratio of reinforcement materials which is present in the synthesis of nanocomposites is very crucial. There found the opportunity to use the spinning CNTs into the products of composites (Dalton et al. 2003) which opens the way for the use of CNTs in the nanocomposites. In these days nanocomposites seem to provide the new technology to the industry along with its eco-friendly nature (Choa et al. 2003). Nanocomposites can be differentiated into three categories according to the material matrix as (Henrique et al. 2009):

1. Polymer matrix nanocomposites (PMNC)
2. Ceramic matrix nanocomposites (CMNC)
3. Metal matrix nanocomposites (MMNC)

They are appropriately in Table 9.1 (Henrique et al. 2009).

It must be kept in mind that the properties of the composites based upon the polymer are yet to be reached. The properties include electrical, chemical and others. Various recent reports are present which show recent development in the characterization, properties, preparation and crystallization behaviour of the matrix and layered nanocomposites (Ray and Bousmina 2005; Pandey et al. 2005). Ceramics has high thermal and chemical stability along with good wear resistance. But they are brittle in nature which limits its commercial use in the industry. Due to this reason, ceramic matrix nanocomposites come to existence which help to enhance its properties. A significant work on ceramic matrix nanocomposites (CMNC), primarily on the Al_2O_3/SiC system, is done by Niihara. The use of high-strength nanofibre in ceramic matrices helps the preparation of advanced nanocomposites which results in high toughness and superior failure properties as compared to the rapid failure of ceramic materials (She et al. 2000). Similarly metal matrix nanocomposites (MMNC) are the materials which contain the alloy matrix or ductile metal in which the nanosized reinforcement is entrenched. These materials increase the properties, e.g. ductility and toughness, and help to compensate the limitation of each other. Hence metal matrix nanocomposites are appropriate for the manufacturing of materials with more service temperature abilities and more strength in shear/compression practices. MMNC show potential for their use in areas such as aerospace and automobile industries (Tjong and Wang 2004). Due to the easy process of manufacturing polymer materials and due to its properties, they are widely used in the industry. However, they have the limitations of having low modulus and strength as compared to metals or ceramics.

Table 9.1 Different types of nanocomposites with examples (Henrique et al. 2009)

Category	Examples
Polymer	Thermoplastic/layered silicates/thermoset polymer, TiO_2/polyester, CNT/polymer, layered double hydroxides/polymer
Metal	Co/Cr, $2O_3$, Fe/MgO, Ni/Al^2O_3, Al/CNT, Mg/CNT Fe-Cr/Al
Ceramic	Al_2O_3/CNT, Al_2O_3/TiO_2, $2O_3/SiO_2$, Al_2O_3/TiO_2, Al_2O_3/SiC, SiO_2/Ni, Al

The addition of reinforcement to polymer resins results in great improvement of its biodegradability. There are incredible rewards for the use of nanocomposites, and hence worldwide attention is given to them. In this paper the synthesis and overview of the three different types of nanocomposites are reviewed. Various developments and mechanisms related to them are also discussed along with their potential uses in present and future perspectives.

9.2 Synthesis of Nanocomposites

Although the three types of nanocomposites are of nanosize, the preparation techniques for all of them are similar to that of microcomposites. These techniques are the same to those of CNT-reinforced composites. The techniques for the synthesis of the three types of nanocomposites are discussed below.

9.2.1 Ceramic Matrix Nanocomposites (CMNC)

The ceramic matrix nanocomposites are the nanocomposites in which the main elements consist of ceramic (a chemical compound made from nitrides, silicides, oxides, etc.) and have metal as the secondary element. There are various techniques from which the ceramic matrix nanocomposites can be synthesized. The most common technique is the "powder process" technique.

Several techniques which are useful for the synthesis of various nanocomposites by powder process (Palmero 2015) are shown in Fig. 9.1 (Ravichandran et al. 2018).

There are many different methods that have been given for the manufacturing of ceramic matrix nanocomposites (Thostenson et al. 2001; Stearns et al. 1992; Balázsi et al. 2005). The most common methodologies, as applied for microcomposites, are spray pyrolysis; polymer precursor route; vapour techniques (PVD and CVD); conventional powder method; and chemical methods. They include the colloidal and precipitation approaches; sol-gel process; and the template synthesis.

Table 9.2 gives the list of the systems manufactured by certain techniques, whereas Table 9.3 presents their limitations and advantages.

The flexible control of the chemical and structural properties of the final oxide materials is affected by various factors that affect the sol-gel process like timing, precursor, solvent, water/metal ratio, pH, etc. (Ennas et al. 1998).

9.2.1.1 Synthesis of Ceramic CNT Nanocomposites

There are various different methods which are available for the synthesis of CNT-reinforced nanocomposites (Thostenson et al. 2001; Ning et al. 2003; Balázsi et al. 2005). Some of them are given in Table 9.4.

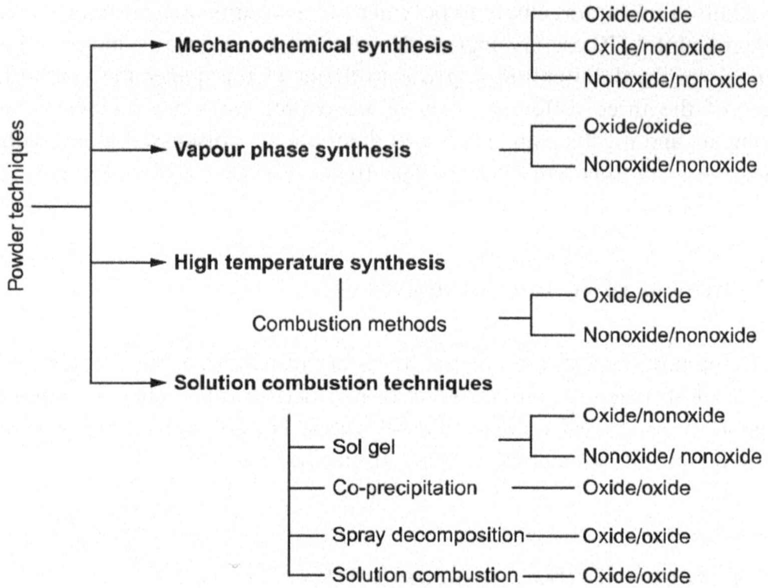

Fig. 9.1 Different powder process methods for the manufacturing of ceramic nanocomposites

9.2.2 Metal Matrix Nanocomposites

When the hard ceramics or soft reinforcement contains the hard metallic matrix, then the resultant nanocomposite is known as the metal matrix nanocomposite (Shiva Shankar and Basavarajappa 2015). There are various benefits of metal matrix nanocomposites as compared to monolithic metals such as the low thermal expansion coefficient, improved wear resistance and greater modulus and specific strength (Henrique et al. 2009; Kori et al. 2014). Some of the common approaches in the preparation of metal matrix nanocomposites include liquid metal infiltration; vapour techniques (CVD, PVD); chemical methods; spray pyrolysis; rapid solidification; and electrodeposition. The chemical method consists of the sol-gel and colloidal processes (Baker et al. 2004; Schadler et al. 1998).

Table 9.5 shows the list of these methods, whereas Table 9.6 shows their benefits and drawbacks.

The techniques which are used for the preparation of metal matrix composites can be divided into mainly five types. These types are mainly based upon the temperature which is used for the preparation of the metal matrix during the process (Sealy 2004). These types can be named as:

1. Solid-phase processes
2. Deposition techniques
3. Liquid-phase processes
4. In situ processes
5. Two-phase (solid-liquid) processes

Table 9.2 Techniques for the synthesis of ceramic nanocomposites (Henrique et al. 2009)

Method	System	Procedure	Reference
Powder process	Al_2O_3/SiC	(i) Choose the correct raw material. The raw materials are generally powders having small size, high purity and uniformity (ii) In aqueous or organic media, mix the material by attrition milling or wet-milling method (iii) After mixing, dry them with the help of lamps or oven or with the help of freeze drying (iv) After that the solid material must be merged by means of hot pressing or gas pressure or injection moulding or slip casting or sintering or pressure filtration method	Stearns et al. (1992)
Polymer precursor process	SiN/SiC, Al_2O_3/SiC	Silicon polymeric predecessors are merged in matrix material. After that the pyrolysis of the mixture should be done in the microwave oven which will ultimately produce the reinforced particle	Sternitzke (1997), Riedel et al. (1995)
Sol-gel process	TiO_2/Fe_2O_3, ZnO/Co, La_2O_3/TiO_2, Al_2O_3/ SiC, TiO_2/Al_2O_3, Al_2O_3/SiO_2, SiO_2/Ni, $Al_2O_3/SiO_2/ZrO_2$, TiO_2/Fe_2, TiO_5, $NdAlO_3/Al_2O_3$	(i) Perform the polycondensation reaction and hydrolysis of the organic/inorganic molecule whose parent molecule dissolved in the organic media (ii) Due to the reactions in them, there will be formation of three-dimensional polymers which will consist of metal and oxygen bonds (iii) The obtained material is then lead to the drying stage. This will convert it to a solid material, and consolidation process can be obtained by thermal treatment further	Mathur et al. (2002)

1. *Solid-Phase Processes*

In it the elemental powders are blended together to form the particular reinforced metal matrix composite. The main method consists of powder metallurgy. The main steps in this technique consist of:

- Preparing the powder
- Blending the powder
- Squashing
- Sintering
- Sizing
- Machining

Table 9.3 Benefits and drawbacks of ceramic nanocomposite synthesis techniques (Henrique et al. 2009)

Technique	Benefits	Drawbacks	Reference
Polymer precursor process	In this technique there is the probability for synthesising the particles that are finer in nature and have better reinforcement dispersion	There is the dispersion of the ultra-fine particles and accumulation in the product which result into the phase-segregated and inhomogeneous material	Sternitzke (1997), Riedel et al. (1995)
Powder process	It is simple to synthesize	High temperature, poor phase dispersion, low formation rate, secondary phases are formed in the product, accumulation	Stearns et al. (1992)
Sol-gel process	Production of polymers which consist of metal-oxygen bonds has three dimensions. Products are highly pure and simple; chemical homogeneity is high; in it the processing time is low; matrices may be single or multiple and versatile in nature; this technique is exclusively applicable for those composite materials that have liquid or viscous fluids in it which is obtained from the alkoxides	In comparison with other mixing methods, it has lower amount of voids and has more shrinkage	Mathur et al. (2002)

Table 9.4 Techniques for synthesising the ceramic CNT nanocomposites (Henrique et al. 2009)

Method	System	Procedure	Reference
Catalytic decomposition	Al_2O_3/CNT	In it the Al_2O_3 powder is saturated with iron catalysts by using acetylene over it	Ning et al. (2003)
Solvothermal process	Fe_3O_4/CNT	With the help of ultrasonic treatment, carbon nanotubes are dispersed in the ethylenediamine (EDA); iron (III)-urea complex is then added; at 200 °C the Teflon-lined autoclaves are heated for about 50 hours and then cooled at room temperature	Jiang and Gao (2003)
Hot pressing	SiO_2/CNT; SiC/CNT	(i) Firstly the glass powders of CNTs and SiO_2 are dispersed into the ethanol (ii) Ultrasonic treatment and stirring process are then applied (iv) The obtained product is then taken for drying process by hot pressing in atmosphere having pure N_2 (i) Nanoparticles of SiC and CNTs are then mixed with each other	Thostenson et al. (2001), Ning et al. (2003)
Spray pyrolysis or CVD	Al_2O_3/CNT	In the porous walls, anodization of the growth of carbon nanotubes is performed to produce the alumina matrix. From substrate to matrix surface, carbon nanotubes develop into the hexagonal straight pores	Xia et al. (2004a, b), Kamalakaran et al. (2003)

Table 9.5 Techniques for the processing of metal matrix nanocomposite (Henrique et al. 2009)

Process	System	Procedure	Reference
Liquid infiltration	Pb/Fe; Nb/Cu; Al-C60; Pb/Cu; W/Cu; Nb/Fe	(I) Firstly the matrix metal material and the reinforcement particles are mixed together (II) The thermal treatment is carried away which helps the reinforcement material to be surrounded by the melted metal matrix. This is done with the help of liquid infiltration (III) In order to get rid of the internal porosity and support the merging of the metal matrix and reinforcement, further heat treatment is carried away below the melting temperature of the metal matrix	Choa et al. (2003), Yoon et al. (2002), Khalid et al. (2003)
RSP with ultrasonic	Al/SiC	To improve the properties like wettability and blending of the reinforcement and the metal matrix, this method is used	Li et al. (2004)
PVD/CVD	Cu/Pb, Al/Mo, Cu/W	CVD: In it the material's vapours are obtained with the help of chemical reactions. After which the consolidation process is carried away PVD: (I) Vapour phase of various components is formed by evaporation/sputtering (II) In order to support the condensation of the metal nanoparticles, the vapour phase is carried out by super saturation in the inert atmosphere (III) In the inert atmosphere, merging of the nanocomposites is done with the help of thermal treatment	Choy (2003), Holtz and Provenzano (1994), Chow et al. (1990)
Spray pyrolysis	W/Cu; Fe/MgO	(I) Firstly the inorganic material is dissolved into the relevant solvent in order to have the liquid source (II) With the help of the ultrasonic atomizer, a mist is generated in the form of liquid (III) The mist is then carried towards the chamber which is already heated with the help of carrier gas. (IV) The generated vapours are then vaporized. These vapours are then trapped with the help of filters. It is done in order to assure the decomposition which will further help to form the oxides of the respective materials. (V) Some of the metal oxides are reduced so that particular metallic materials are produced respectively	Choa et al. (2003)

(continued)

Table 9.5 (continued)

Process	System	Procedure	Reference
Rapid solidification process (RSP)	Al/X/Zr (where X = Cu, Ni, Si); Iron (Fe) alloy; Al/Pb	(I) In his technique, firstly the metal components are melted with each other (II) The melting is done in such a way that the homogeneity is ensured. It is done by ensuring the melting temperature more than the miscibility gap critical line (III) With the processing like melt spinning, the melts are then taken to rapid solidification	Bhattacharya and Chattopadhyay (2004), Branagan and Tang (2002)
Chemical processes (sol-gel, colloidal)	Au/Fe/Au, Ag/Au, Fe/ SiO_2	For Fe/Au consisting of nanocomposites: (I) Firstly iron shell is produced (II) Secondly shell is prepared, and after that the second gold coating is done which is followed by drying the powders (III) The final material is then obtained by pressing the mixture Colloidal method: (I) In order to form the metal particles, the inorganic salts are removed from the solution chemically (II) Dry materials are then undergoes consolidation (III) To promote the preparation of the metal component and to encourage the reduction of the selective oxides from the solution, the thermal treatment and drying of the solid are carried away in atmosphere such as h_2, i.e. reducing atmosphere Sol-gel process: (I) With the help of mesoporous silica comprising of 0.1 chloroauric acid (aq.) and 0.6 M sodium borohydride (aq.), two micelle solutions are prepared (II) The mixing of the solution is carried away until the whole gold is removed from it	Cushing et al. (2004), West et al. (2003) Carpenter et al. (2000)
High-energy ball milling	Cu/Al_2O_3	Powders are milled with each other till we get nanocomposites	Ying and Zhang (2000)

Table 9.6 Benefits and drawbacks of the production techniques of metal matrix nanocomposites (Henrique et al. 2009)

Technique	Benefits	Drawbacks	Reference
Liquid infiltration	Different types of material with better wear resistance and better stiffness can be mould in various shapes and near net shapes Solidification is rapid The contact time between the reinforcement and matrix is short By this method production at industrial level as well as lab level can be done	Reinforcement can be separated High temperature can be used Through the processing stage, the undesired formation of various products can be seen	Khalid et al. (2003), Yoon et al. (2002), Choa et al. (2003)
RSP with ultrasonic	In this technique the distribution without the accumulation of even fine particles is good		Baker et al. (2004)
PVD/CVD	Films are uniformly thick Reproducibility is better This technique has the ability to form pure and highly dense materials In it the adhesion is seen at the rate when deposition is at high rate	Price Comparative complexity Development of various parameters	Choy (2003), Holtz and Provenzano (1994)
Spray pyrolysis	In it the homogeneous, ultra-fine and spherical powder can be prepared with quality, multicomponent system and reproductive size	In order to produce the uniformed particles in nanosize at large quantity, high cost is involved	Choa et al. (2003)
RSP (rapid solidification process)	Efficient and simple	It encourages separation of particles In it nanocomposites can be prepared which are having metal-metal nature In it the fine particles are distributed in the non-homogeneous way	Bhattacharya and Chattopadhyay (2004), Branagan and Tang (2002)
Chemical processes (sol-gel, colloidal)	It is flexible in nature Treating temperature is low Severe stoichiometry control Chemical homogeneity is high It is a simple process Products are highly pure	Permeability is really high Wear resistance is low Porosity control is difficult Bonding is weak	Cushing et al. (2004), Carpenter et al. (2000), West et al. (2003)
High-energy ball milling	Distribution is uniform Mixing is homogeneous		Ying and Zhang (2000)

2. *Deposition Technique*

Some of the important deposition methods are PVD, CVD, electroplating, spray deposition and immersion plating. In this technique, the matrix material which is required to form the composites is coated upon the individual fibres in a tow. After this it is observed that the structural shape or consolidated composite plate is formed by the action of successive diffusion.

3. *Liquid-Phase Processes*

With the help of various property methods, the particular ceramics are merged in the metal matrix. After that the material is then mixed with each other which is then followed by casting of the resultant composite. The eligibility for the selection of the reinforcement material consists of:

(i) Temperature of the melting
(ii) Thermal stability
(iii) Shape and size of the reinforcement particles
(iv) Matrix alloy
(v) Cost
(vi) Tensile strength
(vii) Modulus of elasticity

4. *In Situ Processes*

With the help of in situ process, the reinforced phase is prepared. From the particular alloy at starting stage, the production of composite material is performed.

5. *Two-Phase Processes*

In this technique the matrix and the ceramics are mixed at the location in the phase diagram where the matrix consists of both the liquid and the solid phases. In this technique, the major method includes the rheocasting/compocasting and spray deposition.

9.2.2.1 Synthesis of CNT-Reinforced Metal Matrix Composites

There are various different techniques by which the carbon nanotube-reinforced metal matrix composites can be prepared (Laurent et al. 1998; Chen et al. 2000; Yang and Schaller 2004). Some of the techniques are shown in Table 9.7.

Table 9.7 Techniques for the preparation of metal-CNT nanocomposites (Henrique et al. 2009)

Technique	System	Procedure	Reference
Nanoscale dispersion	Al/CNT	With the help of ethylene propylene and natural rubber, the precursor of MWCNT is prepared which is 10–50 mm long and 13 mm in diameter It is then mixed with aluminium powder With the help of compression moulding, it is then rolled into sheets at 353 K temperature The procurer is then placed over the 28-micrometre grain-sized aluminium plate It is then heated in N_2 atmosphere for 1 hour at 1073 K temperature It is then finally cooled	Noguchi et al. (2004)
Hot pressing	Al/CNT	With the help of grinding, powders are mixed for 30 minutes Under the pressure of 25 MPa, hot pressing is done at 793 K	Xu et al. (1999)
PM/ infiltration	Mg-Al_2O_3f-CNT	(I) With the help of acid and alcohol, the mg powders are then mechanically mixed with multiwalled carbon nanotube which is 1% in volume After that the sinterization process is applied under 25 MPa pressure at 550 °C (II) The Al_2O_3 fibres which are 40–100 mm in length and 25% in volume are then covered under pressure with the multiwalled carbon nanotubes which are then used for the infiltration of the molten mg	Yang and Schaller (2004)
Electroless coating	Sn_2Sb/CNT, SnSb 0.5/CNT, Sn/CNT	In the occurrence of CNTs, the $SbCl_3$ and $SnCl_2$ precursors are reduced by potassium borohydride	Chen et al. (2003), Chen et al. (2002)
Electroless coating	Co-CNT	(I) The electrolysis plating bath is used which consists of cobalt precursor, activated CNT, a buffer, reducing agent $CoSO_4.7H_2O$ and the complexing agent. It results when the CNTs are coated with CO (II) The thermal treatment is done with 200 torr pressure in 10% H_2/N_2 at 873 K temperature	Chen et al. (2000)

9.2.3 Polymer Nanocomposites

Polymers have many applications in many fields. The polymers are known for their ductility, capability of being easily processed, less weight, etc.

But in comparison with metals or ceramics, polymers show several limitations such as low stability, less strength, poor mechanical properties, etc.

But these limitations can be overcome with the help of polymer nanocomposites. With the help of fillers made of nanofibres or nanoparticles, polymer nanocomposites emerge as the material which is more efficient and is widely used in areas such as solar cells, aerospace, sports material, supercapacitors, etc.

The improvement in the properties of polymer nanocomposites is due to the nanostructures having larger aspect ratio. But this is not wholly dependent upon this; it also depends on the type of constituents used and the type of method used for the preparation. Nanocomposites are also dependent on the processing factors and method for synthesis.

There are various different types of methods which are employed for the preparation of the polymer nanocomposites including that of the carbon nanotubes and that of the layered materials. Important methods include:

I. In situ polymerization
II. Introduction of the prepolymer or polymers from the solution
III. Injection of the melts
IV. Solution blending
V. Template synthesis
VI. Sol-gel process
VII. Particulates and the polymers having direct mixture

Commonly used methods are in situ polymerization, melt processing and solution blending.

9.2.3.1 In Situ Polymerization

In this method, usually the solvents are avoided, and it produces the uniformly distributed polymer matrix with fillers in it. The preparation includes the typical techniques of polymerization such as stirring, UV curing, ultrasonic, etc. which are used to polymerize the mixture-containing fillers and monomer particles. This method is ideal for the production of nanocomposite constituents having covalent bonds in them. Those polymers can also be prepared by this method, which cannot be prepared by the methods such as melt blending or solution mixing. This method is also useful for the preparation of thermally unstable and insoluble polymers.

9.2.3.2 Melt Processing

Every method needs the use of solvents which are generally toxic in nature. But this method is one of the simplest methods which do not requires the use of such toxic solvents. It is a cost-effective technique. Melt processing is also known as the solution-blending technique. In this method, the fillers are directly scattered in the melted polymers. Techniques such as moulding, extrusion or injection are used to blend the melts. Thermoplastic polymers have high viscosity due to which high temperature is required in this method to melt the polymers. High shear force along with high temperature is required to scatter the fillers uniformly.

9.2.3.3 Solution Blending

It is one of the most used and simple methods for the synthesis of nanocomposites. Solution blending technique includes the solvent in which the scattered solution of fillers and polymers are mixed. The most common type of solvents includes toluene, water, chloroform, etc. Production of the nanocomposite includes the mixing of the solution and further removal of the solvent. Simple filtration and precipitation can result in the preparation of the nanocomposites. This technique includes the limitation of firstly seeking and then removing the solvent in end stage. The solvent used in this method is toxic in nature. The preparation of the graphene oxide or graphene is done by combining it with the polymers such as polyimides, polyacrylamides, polystyrene, etc. with the help of blending method of their solution (Potts et al. 2011). To encourage covalent bonding and effective scattering with the matrix, the graphene is fabricated acidically. In these nanocomposites, one of the important steps is to exfoliate the fabricated graphene.

9.2.3.4 Other Techniques

There are various other different types of techniques which can be employed for the preparation of polymer nanocomposites. The synthesis of polymer nanocomposites which is reinforced by carbon nanotubes can be synthesized by various methods such as melt mixing, solution mixing, in situ polymerization, direct mixing, etc. In the same way for the synthesis of the conducting polymer nanocomposites, various methods such as electrochemical and chemical methods are employed (Sanchez et al. 1999).

Table 9.8 shows the various techniques for the processing of polymer-based nanocomposites; Table 9.9 shows the benefits and drawbacks of polymer-based nanocomposite synthesis approaches, whereas Table 9.10 shows the methods for the synthesis of polymer-CNT nanocomposites.

Table 9.8 Techniques for the processing of polymer-based nanocomposites (Henrique et al. 2009)

Method	Structure	Procedure	Reference
In situ intercalative polymerization	Montmorillonite through PCL/Epoxy/N_6/PMMA	(I) Firstly, wrapping of the silicate of layered stage in the liquid monomer solution is done (II) After that the polymers are prepared in the introduced sheets (III) With the help of radiations or heat, the polymerization is carried away (IV) In the interlayer, the suitable catalyst is diffused prior than step of swelling	Kojima et al. (1993), Messersmith and Giannelis (1994), Yao et al. (2001)
Template synthesis	Hectorite with HPMC, PDDA, PVPR, PANI, PAN	Within the aqueous solution having polymer in it, the in situ preparation of inorganic material is done. This inorganic structure consists of the layered structure. In the preparation of layers, the polymers, which are water soluble in nature, act as the models. This is less used for the preparation of the layered silicates but is widely used in the preparation of the LDH nanocomposites	Tomasko et al. (2003), Carrado and Xu (1998), Watkins and McCarthy (1995)
Sol-gel process	2-hydroxyethyl acrylate (HEA)/SiO_2 Poly(amide-imide)/TiO_2 PMMA/SiO_2 Polycarbonate/SiO_2 Polyimide/SiO_2 Polyimide/silica Polyethylacrylate/SiO_2	(I) Inserting the monomers and the organic molecules over the sol-gel matrices (II) The creation of the chemical bonds leads to the initiation of the organic group (III) In the polymer, the sol-gel matrices are formed by in situ action (IV) Other process of formation of organic/inorganic networks can also be done by simultaneous action	Liu et al. (2002), Ver Avadhani et al. (1997), Kickelbick (2003)
Prepolymer/intercalation by solution	Clay with PCL, PLA, HDPE, PEO, PVA, PVP, PVA, etc.	(I) It is used in the material having layered reinforcement and into which the introduction of polymer is possible (II) It is generally applied for the layered silicates, which allow the introduction of prepolymer/polymer from the solution (III) In it only those solvents are used, in which the silicate layers get swells and the solubility of prepolymer/polymer is good	Alexandre and Dubois (2000), Jimenez et al. (1997)

(continued)

Table 9.8 (continued)

Method	Structure	Procedure	Reference
Melt intercalation	Montmorillonite by using PEO/PVP/PS/Clay-PVPH/PP	(I) The process of annealing is carried out of the mixture of the layered reinforcement and the polymer. The softening point for this process under the shear or at static mode must be high than that of the polymer (II) During the annealing process, the polymer chains from the melt of the polymer get dispersed into the passages of the reinforcement layers	Hasegawa et al. (1998), Gilman (1999)
Mixing and in situ polymerization	TiO_2/PMMA/Pd polyester γ-Fe_2O_3/Epoxy vinylester $CuSO_4$, $AgNO_3$ and $NiSO_4$ Ag/PVA Fe_3O_4/Epoxy vinylester, $CaCO_3$/PET PAA/Cu, Ag/PAA (acrylic acid), PAA/Ni	(I) Firstly the monomers or either the polymers are mixed with the reinforcements (II) The inorganic particles are then scattered into the precursor made of the monomers (III) The relevant catalyst is then used in the mixture for the polymerization process (IV) The traditional moulding process is then carried out in order to process the material further (V) The ultrasonic are used for the scattering of the epoxies (VI) In order to encourage the formation of the metal nanoparticles and instantaneous polymerization, the exposition of the AG systems to 60 Co γ-ray is carried out	Aymonier et al. (2003), Evora and Shukla (2003), Di Lorenzo et al. (2002)

9.2.3.5 Synthesis of Polymer-CNT Nanocomposites

Some of the techniques for the synthesis of polymer-CNT nanocomposites are shown in Table 9.10. The structure and method for their synthesis are also shown along with remarks.

9.3 Conclusion

It is concluded that technology is emerging day by day, and hence it requires the material that should have novel properties and have the improved performance as compared with the existing materials. Therefore, nanocomposites are the materials that are the need of the today's technological world. There are various techniques and methods available to us for the synthesis of these nanocomposite materials, but still there is need for the improvement in these techniques. Nanocomposites show improved performance as compared to microcomposites or monolithics and therefore

Table 9.9 Benefits and drawbacks of polymer-based nanocomposite synthesis approaches (Henrique et al. 2009)

Method	Benefits	Drawback	Reference
In situ intercalative polymerization	(I) This method depends upon the scattering of fillers inside the precursors of polymers (II) It is an easy method	(I) The control for the intragallery polymerization is difficult (II) It has very limited applications	Kojima et al. (1993), Messersmith and Giannelis (1994), Okamoto et al. (2001)
Synthesis of template	(I) Procedure is easy (II) Production can be done at large scale	(I) Its applications are very limited (II) It can be polluted by the side products (III) It is generally based upon the polymers that are soluble in water	Tomasko et al. (2003), Watkins and McCarthy (1995), Carrado and Xu (1998)
Sol-gel process	(I) It is a simple process (II) It has the ability to produce products with maximum purity (III) It is a versatile process (IV) It requires low processing temperature (V) The chemical homogeneity in this process is high	(I) The amount of cavities is low (II) It has more shrinkage	Reuter (1991), Vorotilov et al. (1999), Ennas et al. (1998)
Prepolymer/intercalation from solution	(I) In it the preparation of intercalated nanocomposites having less or no polarity can be obtained which is established on the basis of polymers (II) Fillers with homogeneous scattering can be synthesized	This process utilizes the huge quantity of the solvents in term of industrial level	Kamigaito (1991), Alexandre and Dubois (2000), Jimenez et al. (1997)
Melt intercalation	(I) This process is well suited for the industrial polymer processes (II) It is environment-friendly (III) The practice of using the polymers for other processes is not suitable	It has a very restricted application for the use in polyolefin, which characterizes most of the used polymers	Vaia and Giannelis (1997a, b), Gilman (1999), Vaia et al. (1995)

Table 9.10 Methods for the synthesis of polymer-CNT nanocomposites (Henrique et al. 2009)

Technique	Structure	Method	Remarks	Reference
Solution mixing	Thermoplastic resins (Epoxy/PS)	(I) Scattering of the 0.2–1% of the CNTs having 10 mm length and 100 nm of diameter (II) The solvent of the polymer is then removed (III) Curing	(I) The behaviour of the polymer is firstly modified (III) The effect of synergistic is done (II) The shape memory nanocomposites can be formed	Qian et al. (2000), Koerner et al. (2004)
In situ polymerization	Polyether-polyester; Epoxy/CNT; Polyaniline/CNT; MMA/CNT	(I) The scattering in the matrix or monomers is done by using the ultrasonic technique (II) Curing	(I) The CNTs are used for the synthesis of the polymers (II) It has better chemical bonding	Park et al. (2003), Andrews et al. (2002)
Direct mixing	Thermoset resins are used	(I) Scattering of CNTs is done (II) Curing		Sandler et al. (1999)
Melt mixing	N6 Polymers	(I) The CNTs are mixed with the melts of prepolymer by the mechanical action (II) After that the extrusion process is carried out (III) The compression or injection moulding process can be applied	(I) The 0.2–2.0% of the multiwalled carbon nanotubes are used (II) The mixer having twin screw is used	Tang et al. (2003), Andrews et al. (2002), Liu (2004)
Others	PANI/SWCNT; PP-CNT; IPP-SWCNT; PVK-SWCNT	(I) The mechano-chemical pulverization process of the solid state takes place (II) The sonication and blending together take place (III) The melt blending is done (IV) VDP	0–10% of CNTs by weight is used	Xia et al. (2004a, b), Valentini et al. (2003)

are the best materials for today's industry to be incorporated in modern machinery and devices. There are various applications of these materials in today's world, but we should not limit the use of nanocomposite materials to only these applications; nanocomposite materials can do great in future devices and machinery. Due to the unique properties and vast field of applications, nanocomposite materials are becoming the interest of the modern industry and proving itself a great field of research for the researchers.

References

Alexandre M, Dubois P (2000) Polymer-layered silicate nanocomposites: preparation, properties and uses of a new class of materials. Mater Sci Eng R Rep 28(1):1–63. https://doi.org/10.1016/S0927-796X(00)00012-7

Andrews R et al (2002) Fabrication of carbon multiwall nanotube/polymer composites by shear mixing NA. Macromol Mater Eng 287(6):395–403. https://doi.org/10.1002/1439-2054(20020601)287:6<395::AID-MAME395>3.0.CO;2-S

Aymonier C et al (2003) Poly (methyl methacrylate)/palladium nanocomposites : synthesis and characterization of the morphological, thermomechanical, and thermal properties. Chem Mater 15(17):4874–4878

Baker C, Shah SI, Hasanain SK (2004) Magnetic behavior of iron and iron-oxide nanoparticle/polymer composites. J Magn Magn Mater 280(2–3):412–418. https://doi.org/10.1016/j.jmmm.2004.03.037.

Balázsi C et al (2005) Processing of carbon nanotube reinforced silicon nitride composites by spark plasma sintering. Composit Sci Technol 65(5 SPEC. ISS):727–733. https://doi.org/10.1016/j.compscitech.2004.10.006

Bhattacharya V, Chattopadhyay K (2004) Microstructure and wear behaviour of aluminium alloys containing embedded nanoscaled lead dispersoids. Acta Mater 52(8):2293–2304. https://doi.org/10.1016/j.actamat.2004.01.020

Branagan DJ, Tang Y (2002) Developing extreme hardness (>15 GPa) in iron based nanocomposites. Compos A: Appl Sci Manuf 33(6):855–859. https://doi.org/10.1016/S1359-835X(02)00028-3

Carpenter EE et al (2000) Synthesis and magnetic properties of gold-iron-gold nanocomposites. Mater Sci Eng A 286(1):81–86. https://doi.org/10.1016/S0921-5093(00)00681-X

Carrado KA, Xu L (1998) In situ synthesis of polymer−clay nanocomposites from silicate gels. Chem Mater 10(5):1440–1445. https://doi.org/10.1021/cm970814n

Chen X et al (2000) Carbon-nanotube metal-matrix composites prepared by electroless plating. Compos Sci Technol 60(2):301–306. https://doi.org/10.1016/S0266-3538(99)00127-X

Chen WX, Lee JY, Liu Z (2002) Electrochemical lithiation and de-lithiation of carbon nanotube-Sn2Sb nanocomposites. Electrochem Commun 4(3):260–265. https://doi.org/10.1016/S1388-2481(02)00268-0.

Chen WX, Lee JY, Liu Z (2003) The nanocomposites of carbon nanotube with Sb and SnSb0.5 as Li-ion battery anodes. Carbon 41(5):959–966. https://doi.org/10.1016/S0008-6223(02)00425-6

Choa YH et al (2003) Preparation and characterization of metal/ceramic nanoporous nanocomposite powders. J Magn Magn Mater 266(1–2):12–19. https://doi.org/10.1016/S0304-8853(03)00450-5.

Chow GM et al (1990) Alternative approach to nanocomposite synthesis by sputtering. Appl Phys Lett 56(19):1853–1855. https://doi.org/10.1063/1.103067

Choy KL (2003) Chemical vapour deposition of coatings. Prog Mater Sci 48(2):57–170. https://doi.org/10.1016/S0079-6425(01)00009-3

Cushing BL, Kolesnichenko VL, O'Connor CJ (2004) Recent advances in the liquid-phase syntheses of inorganic nanoparticles. Chem Rev 104(9):3893–3946. https://doi.org/10.1021/cr030027b

Dalton AB et al (2003) Super-tough carbon-nanotube fibres. Nature 423:703

Di Lorenzo M, Errico M, Avella M (2002) Thermal and morphological characterization of poly (ethylene terephthalate)/calcium carbonate nanocomposites. J Mater Sci 7:2351–2358. https://doi.org/10.1023/A:1015358425449

Ennas G et al (1998) Sol–gel preparation and characterization of Ni–SiO 2 nanocomposites. J Non-Cryst Solids 232–234(3):587–593

Evora VMF, Shukla A (2003) Fabrication, characterization, and dynamic behavior of polyester/TiO2nanocomposites. Mater Sci Eng A 361(1–2):358–366. https://doi.org/10.1016/S0921-5093(03)00536-7.

Gilman JW (1999) Flammability and thermal stability studies of polymer layered-silicate (clay) nanocomposites. Appl Clay Sci 15(1–2):31–49. https://doi.org/10.1016/S0169-1317(99)00019-8

Gleiter H (1992) Materials with ultrafine microstructures: retrospectives and perspectives. Nanostruct Mater 1(1):1–19. https://doi.org/10.1016/0965-9773(92)90045-Y

Gupta G et al (2018) Effect of SiC reinforcement on mechanical behavior of aluminum alloys – a review, in, p. 030051. https://doi.org/10.1063/1.5051307

Hasegawa N et al (1998) Preparation and mechanical properties of polypropylene – clay hybrids using a maleic anhydride-modified polypropylene oligomer. J Appl Polym Sci 67(96):87–92. https://doi.org/10.1002/(SICI)1097-4628(19980103)67:1<87::AID-APP10>3.3.CO;2-9

Henrique P et al (2009) Nanocomposites : synthesis, structure, properties and new application opportunities. Mater Res 12(1):1–39. https://doi.org/10.1590/S1516-14392009000100002

Holtz RL, Provenzano V (1994) Enhanced microhardness of copper-niobium nanocomposites. Nanostruct Mater 4(2):241–256. https://doi.org/10.1016/0965-9773(94)90082-5

Hussain F et al (2006) Review article: polymer-matrix nanocomposites, processing, manufacturing, and application: an overview. J Compos Mater 40(17):1511–1575. https://doi.org/10.1177/0021998306067321

Jiang L, Gao L (2003) Carbon nanotubes – magnetite nanocomposites from solvothermal processes : formation, characterization, and enhanced electrical properties. Chem Mater 15:2848–2853

Jimenez G et al (1997) Structure and thermal/mechanical properties of poly (e -caprolactone) – clay blend. J Appl Polym Sci 64(July):2211–2220. https://doi.org/10.1002/(SICI)1097-4628(19970613)64:11<2211::AID-APP17>3.0.CO;2-6

Kamalakaran R et al (2003) I n-situ formation of carbon nanotubes in an alumina–nanotube composite by spray pyrolysis. Carbon 41:2737–2741. https://doi.org/10.1016/S0008-6223(03)00380-4

Kamigaito O (1991) What can be improved by nanometer composites? J Jpn Soc Powder Powder Metall 38(3):315–321. https://doi.org/10.2497/jjspm.38.315

Khalid FA et al (2003) Study of microstructure and interfaces in an aluminium-C60composite material. Acta Mater 51(15):4575–4582. https://doi.org/10.1016/S1359-6454(03)00294-5

Kickelbick G (2003) Concepts for the incorporation of inorganic building blocks into organic polymers on a nanoscale. Prog Polym Sci (Oxford) 28:83. https://doi.org/10.1016/S0079-6700(02)00019-9

Koerner H et al (2004) Remotely actuated polymer nanocomposites – stress-recovery of carbon-nanotube-filled thermoplastic elastomers. Nat Mater 3(2):115–120. https://doi.org/10.1038/nmat1059

Kojima Y et al (1993) One-pot synthesis of nylon 6-clay hybrid. J Polym Sci Part A Polym Chem 31(5):1755–1758

Kori SA, Biradar SL, Auradi V (2014) Synthesis of Al-Al3Ti in-situ metal matrix composites by salt route and evaluation of their mechanical properties. Adv Mater Res 984–985:280–284. https://doi.org/10.4028/www.scientific.net/AMR.984-985.280

Laurent C et al (1998) Carbon nanotubes-Fe-alumina nanocomposites. Part II: microstructure and mechanical properties of the hot-pressed composites. J Eur Ceram Soc 18(14):2005–2013. https://doi.org/10.1016/S0955-2219(98)00142-3

Li X, Yang Y, Cheng X (2004) Ultrasonic-assisted fabrication of metal matrix nanocomposites. J Mater Sci 39(9):3211–3212. https://doi.org/10.1023/B:JMSC.0000025862.23609.6f

Liu et al (2004) Morphology and mechanical properties of multiwalled carbon nanotubes reinforced Nylon-6 composites. Macromolecules 37(19):7214–7222. https://doi.org/10.1021/ma049132t

Liu J et al (2002) Preparation and characteristic of a new class of silica/polyimide nanocomposites. J Mater Sci 37:3085–3088

Mathur S et al (2002) Structural and optical properties of NdAlO 3 nanocrystals embedded in an Al 2 O 3 matrix. Chem Mater 14(2):568–582. https://doi.org/10.1021/cm0111433

Messersmith PB, Giannelis EP (1994) Synthesis and characterization of layered silicate-epoxy nanocomposites. Chem Mater 6(10):1719–1725. https://doi.org/10.1021/cm00046a026

Ning J et al (2003) Fabrication and mechanical properties of SiO2matrix composites reinforced by carbon nanotube. Mater Sci Eng A 357(1–2):392–396. https://doi.org/10.1016/S0921-5093(03)00256-9.

Noguchi T et al (2004) Carbon nanotube/aluminium composites with uniform dispersion. Mater Trans 45(2):602–604. https://doi.org/10.2320/matertrans.45.602

Okamoto M, Morita S, Kotaka T (2001) Dispersed structure and ionic conductivity of smectic clay/polymer nanocomposites. Polymer 42:2685–2688

Palmero P (2015) Structural ceramic nanocomposites: a review of properties and powders' synthesis methods. Nano 5(2):656–696. https://doi.org/10.3390/nano5020656

Pandey JK et al (2005) Recent Advances in Biodegradable Nanocomposites. J Nanosci Nanotechnol 5(4):497–526. https://doi.org/10.1166/jnn.2005.111.

Park SJ et al (2003) Synthesis and dispersion characteristics of multi-walled carbon nanotube composites with poly(methyl methacrylate) prepared by in-situ bulk polymerization. Macromol Rapid Commun 24(18):1070–1073. https://doi.org/10.1002/marc.200300089

Potts JR et al (2011) Graphene-based polymer nanocomposites. Polymer. Elsevier Ltd 52(1):5–25. https://doi.org/10.1016/j.polymer.2010.11.042

Qian D et al (2000) Load transfer and deformation mechanisms in carbon nanotube-polystyrene composites. Appl Phys Lett 76(20):2868–2870. https://doi.org/10.1063/1.126500

Ravichandran K et al (2018) Synthesis of nanocomposites, synthesis of inorganic nanomaterials. Elsevier Ltd. https://doi.org/10.1016/B978-0-08-101975-7.00006-3

Ray SS, Bousmina M (2005) Biodegradable polymers and their layered silicate nanocomposites: in greening the 21st century materials world. Prog Mater Sci 50(8):962–1079. https://doi.org/10.1016/j.pmatsci.2005.05.002

Reuter H (1991) Sol-gel processes. Adv Mater 3(5):258–259. https://doi.org/10.1002/adma.19910030510

Riedel R et al (1995) Polymer-derived Si-based bulk ceramics, Part I: preparation, processing and properties. J Eur Ceram Soc 15(8):703–715. https://doi.org/10.1016/0955-2219(95)00041-R

Roy R, Roy RA, Roy DM (1986) Alternative perspectives on "quasi-crystallinity": non-uniformity and nanocomposites. Mater Lett 4(8–9):323–328. https://doi.org/10.1016/0167-577X(86)90063-7

Sanchez C, Ribot F, Lebeau B (1999) Molecular design of hybrid organic-inorganic nanocomposites synthesized via sol-gel chemistry. J Mater Chem 9(1):35–44. https://doi.org/10.1039/a805538f

Sandler J et al (1999) Development of a dispersion process for carbon nanotubes in an epoxy matrix and the resulting electrical properties. Polymer 40(21):5967–5971. https://doi.org/10.1016/S0032-3861(99)00166-4

Schadler LS, Giannaris SC, Ajayan PM (1998) Load transfer in carbon nanotube epoxy composites. Appl Phys Lett 73(26):3842–3844. https://doi.org/10.1063/1.122911

Schmidt D, Shah D, Giannelis EP (2002) New advances in polymer/layered silicate nanocomposites. Curr Opinion Solid State Mater Sci 6(3):205–212. https://doi.org/10.1016/S1359-0286(02)00049-9

Sealy C (2004) Stronger by a hair COMPOSITES. Mater Today 7(10):15

She J et al (2000) Mechanical properties and fracture behavior of ® brous Al 2 O 3/SiC ceramics. J Eur Ceram Soc 20:1–5

Shiva Shankar GS, Basavarajappa S (2015) Synthesis and characterization of aluminiun 2219 reinforced with boron carbide and molybdenum disulfide metal matrix hybrid composites. Adv Mater Res 1101:28–31. https://doi.org/10.4028/www.scientific.net/AMR.1101.28

Stearns LC, Zhao J, Harmer MP (1992) Processing and microstructure development in Al2O3-SiC "nanocomposites". J Eur Ceram Soc 10(6):473–477. https://doi.org/10.1016/0955-2219(92)90022-6

Sternitzke M (1997) Review : stru. ctural ceramic nanocomposites. J Eur Ceram Soc 17:1061–1082

Tang W et al (2003) Melt processing and mechanical property characterization of multi-walled carbon nanotube/high density polyethylene (MWNT/HDPE) composite films. Carbon 41(14):2779–2785. https://doi.org/10.1016/S0008-6223(03)00387-7

Thostenson ET, Ren Z, Chou T-W (2001) Advances in the science and technology of carbon nanotubes and their composites: a review. Compos Sci Technol 61(13):1899–1912. https://doi.org/10.1016/S0266-3538(01)00094-X

Tjong SC, Wang GS (2004) High-cycle fatigue properties of Al-based composites reinforced with in situ TiB2 and Al2O3 particulates. Mater Sci Eng A 386(1–2):48–53. https://doi.org/10.1016/j.msea.2004.07.021.

Tomasko DL et al (2003) Supercritical fluid applications in polymer nanocomposites. Curr Opinion Solid State Mater Sci 7(4–5):407–412. https://doi.org/10.1016/j.cossms.2003.10.005.

Vaia RA, Giannelis EP (1997a) Lattice model of polymer melt intercalation in organically -modified layered silicates. Macromolecules 30(25):7990–7999. https://doi.org/10.1021/ma9603488.

Vaia RA, Giannelis EP (1997b) Polymer melt intercalation in organically-modified layered silicates: model predictions and experiment. Macromolecules 30(25):8000–8009. https://doi.org/10.1021/ma9603488

Vaia RA et al (1995) New polymer electrolyte nanocomposites: melt intercalation of poly(ethylene oxide) in mica-type silicates. Adv Mater 7(2):154–156. https://doi.org/10.1002/adma.19950070210

Valentini L et al (2003) Morphological characterization of single-walled carbon nanotubes-PP composites. Compos Sci Technol 63(8):1149–1153. https://doi.org/10.1016/S0266-3538(03)00036-8

Ver Avadhani C et al (1997) Polyamide-silica gel hybrids containing metal salts: preparation via the sol-gel reaction. Polym Bull 38(5):501–508. https://doi.org/10.1007/s002890050079

Vorotilov KA et al (1999) Sol-gel derived ferroelectric thin films: avenues for control of microstructural and electric properties. J Sol-Gel Sci Technol 16(1):109–118. https://doi.org/10.1023/A:1008776120997.

Watkins JJ, McCarthy TJ (1995) Polymerization of styrene in supercritical CO2-Swollen poly(chlorotrifluoroethylene). Macromolecules 28(12):4067–4074. https://doi.org/10.1021/ma00116a004

West R, Wang Y, Goodson T (2003) Nonlinear absorption properties in novel gold nanostructured topologies. J Phys Chem B 107(15):3419–3426. https://doi.org/10.1021/jp027762w

Xia H et al (2004a) Preparation of polypropylene/carbon nanotube composite powder with a solid-state mechanochemical pulverization process. J Appl Polym Sci 93(1):378–386. https://doi.org/10.1002/app.20435

Xia Z et al (2004b) Direct observation of toughening mechanisms in carbon nanotube ceramic matrix composites. Acta Mater 52(4):931–944. https://doi.org/10.1016/j.actamat.2003.10.050

Xu CL et al (1999) Fabrication of aluminum/carbon nanotube composites and their electrical properties. Carbon 37(5):855–858. https://doi.org/10.1016/S0008-6223(98)00285-1

Yang J, Schaller R (2004) Mechanical spectroscopy of Mg reinforced with Al2O3 short fibers and C nanotubes. Mater Sci Eng A 370(1–2):512–515. https://doi.org/10.1016/j.msea.2003.08.124

Yao KJ et al (2001) Polymer/layered clay nanocomposites: 2 polyurethane nanocomposites. Polymer 43(3):1017–1020. https://doi.org/10.1016/S0032-3861(01)00650-4.

Ying D, Zhang D (2000) Processing of Cu–Al2O3 metal matrix nanocomposite materials by using high energy ball milling. Mater Sci Eng A 286(1):152–156. https://doi.org/10.1016/S0921-5093(00)00627-4

Yoon ES et al (2002) Microstructure and sintering behavior of W-Cu nanocomposite powder produced by thermo-chemical process. Int J Refract Met Hard Mater 20(3):201–206. https://doi.org/10.1016/S0263-4368(02)00003-3

Chapter 10
Appraisal of Chitosan-Based Nanomaterials in Enzyme Immobilization and Probiotics Encapsulation

Subham Rakshit, Suman Kumar Halder, and Keshab Chandra Mondal

10.1 Chitosan

Chitin is most plenteous amino-polysaccharide in biosphere, found in the shells of crustaceans (such as lobsters, crabs, shrimp, shellfish), fungal cell wall, insect cuticles and exoskeleton, radula of mollusks, and the cephalopods beaks (Gooday et al. 1986; Choi et al. 2004). Henri Braconnot, a French pharmacist and chemist, first discovered chitin in 1811. It is a linear polymer of N-acetyl-D-glucosamine (GlcNAc) linked by β-(1 → 4) bond, and chitosan is the deacetylated form of chitin. The main dissimilarity between chitin and chitosan is that in the latter acyl groups are absent which are attached by the amide bond in the chain of chitin. The free amines in the chitosan polymer act as functional group. The reactivity and solubility of the chitosan polymer are better than chitin and nontoxic in nature. Practically chitosan is β-(1 → 4)-linked D-glucosamine (GlcN) along with few GlcNAc. The physicochemical and functional properties of chitosan can be modified by changing the degree of deacetylation (DD), derivatization, and chemical hydrolysis (depolymerization) of the polymer (Halder and Mondal 2018). Chitosan is water insoluble but soluble in acidic pH. This is due to the fact that at low pH most amino groups transformed into protonated form and the positive charge reacts with water to solubilize (Sorlier et al. 2001). In case of commercially produced chitosan, molecular weight varies between 3800 and 20,000 dalton. Viscosity of chitosan varies between 10 and 1000 centipoise. High superiority polymers of known molecular weight are essential in biomedical uses of chitosan. It has wide varieties in degradation mechanism like oxidation-reduction of free radical polymerization and enzymatic-catalysis. Chitosan has numerous employments in pharmaceutical, medical, agricultural, cosmetics, biotechnology, food technology, textile sectors because of its (i) nontoxic,

S. Rakshit · S. K. Halder (✉) · K. C. Mondal
Department of Microbiology, Vidyasagar University, West Bengal, India
e-mail: sumanmic@mail.vidyasagar.ac.in

I. Bhushan et al. (eds.), *Nanomaterials and Environmental Biotechnology*, Nanotechnology in the Life Sciences,
https://doi.org/10.1007/978-3-030-34544-0_10

(ii) biodegradable, (iii) biocompatibile, (iv) cationic, (v) bioactive, and (vi) bioadhesive natures, and it is also easily commercially available at low cost (Chen et al. 2006; Klotzbach et al. 2006; Halder and Mondal 2018).

10.2 Why Chitosan Is Useful in Enzyme Immobilization

According to many literature surveys, chitosan is one of the most common biopolymer used for enzyme immobilization. Chitosan has a special property to control over the discharge of the active mediator and eliminate the hazardous solvents while constructing the particles in aqueous condition (Sheldon 2007). It is a very useful property to make immobilized enzymes. It is a polyamine comprising of free amino groups which are freely available for cross-linking with multivalent ions. The reactive hydroxyl groups exert a better enzyme coupling effectiveness in adsorption and covalent bonding of enzyme immobilization methods (Cao et al. 2012). At the absorption site of enzyme, the residual time increases because of chitosan's mucoadhesive character. Owing to its membrane-forming ability, low immunogenicity, good adhesion property, and hydrophobicity, chitosan is a potential immobilization matrix which provides excellent stability of the immobilized biocatalyst (de Moura et al. 2008). Chitosan is also used in immobilization techniques because of its unique characteristics like enhanced resistance in degradation of chemical and elimination of metal ion disturbances to enzyme (Yang et al. 2010). It also can act in numerous shapes and forms. For example, in nuclease immobilization cross-linked with bifunctional glutaraldehyde with chitosan provides the support through interaction of amino group with the aldehyde groups. Chitosan acts through other functional groups like phenols, thiols, and imidazoles.

10.3 Nanoparticles

Nanoparticles are organic or inorganic particles which range varying between 1 and 100 nm. "Nano" word originated from Greek literature known as "Nanos" which means dwarf or extremely short (Pal et al. 2011). Nanoparticles have various special properties than other particles, which made a bridge between bulk materials and molecular structures. When a particle size reduces in nanometers, their properties and behavior (e.g., melting point, chemical reactivity, fluorescence, electrical conductivity, magnetic permeability, etc.) change from those at high scales, known as quantum effects, and they can imprison their electrons. The unique properties of nanoparticles are higher surface/volume ratio and melting temperature which provides tremendous driving force for diffusion. Due to size-dependent properties of nanoparticles, surface plasmon resonance is observed in some metal particles. It is used as enzyme immobilization because of its higher permeability properties through barriers (e.g., biological membrane, blood-brain barrier) and improved biocompatibility and some drug or active materials can be encapsulated, entrapped, adsorbed, or attached in micromolecular particles on the matrix surface (Janssen et al. 2002; Sheldon 2007). Nanoparticles acquire the various sites for enzyme molecules to make them a rigid particle. Due to its high surface to volume ratio, it is encouraging for enzyme immobilization. Nanoparticles on biocatalytic system provide elevated retention activity, as enzymes attached to the particle surface make them available for substrates. Numerous inorganic materials such as nanogold and graphene are used as matrices for immobilization of enzymes. Not only that, various biopolymeric supports are widely used for this purpose such as starch, agarose, cellulose, gelatin, albumin, chitosan, etc. (Katchalski-Katzir and Kraemer 2000; Lozinsky et al. 2003; Salemi 2010). The main ambition of the utilization of nanoparticles for releases of enzymes is to preserve the structures and performance of the attached substance for a long time (Jia et al. 2003). With the application of nanoparticles, it produces metabolites which effortlessly degrade in biological environment.

10.3.1 Methods of Preparation of Nanoparticles

Nanoparticles are synthesized by both (i) top-down method and (ii) bottom-up method.

Top-down method In this method the desirable material is formed from the bulk material after cutting down into nano sizes.

Bottom-up method In this method desirable material is formed by gathering the building block particles.

10.3.2 Methods of Preparation of Chitosan Nanoparticles for Enzyme Immobilization

There are several methods to prepare the chitosan nanoparticles. These are (i) reverse micelle method, (ii) ionic cross-linking method, (iii) coprecipitation method, (iv) emulsion cross-linking method, and (v) ionotropic gelation method.

10.3.2.1 Reverse Micelle Method

A narrow range of polymeric nanoparticles are prepared following this method. In the solvent, the surfactant is liquefied to form reverse micelles. In this process low molecular weight of chitosan has greater control in particle size distribution. Following this method, bovine serum albumin (BSA)-loaded chitosan nanoparticles are synthesized, and 143–428 nm particle size was obtained. It also established that high chitosan concentration decreases the BSA release. Magnetic nanoparticles of chitosan can also be prepared by reverse micelle method by employing glutaraldehyde and Span-80, respectively, as cross-linking agent and emulsifier (Brunel et al. 2008).

10.3.2.2 Ionic Cross-Linking Method

This is based on the ionic interaction of positively charged chitosan with oppositely charged (negative) molecules. Tripolyphosphate (TPP) has frequently used for this owing to its multivalent, nontoxic, and gel-forming properties. The preparation of chitosan/TPP depends on pH, component ratio, concentrations, and process of mixing. TPP concentration does not affect the surface amino groups of chitosan nanoparticles. Ironic chitosan nanoparticle can be easily prepared by this process having a diameter that varies between 50 and 100 nm (Nasti et al. 2009; Wu et al. 2009a).

10.3.2.3 Coprecipitation Method

Single step coprecipitation can be adopted for preparation of chitosan-coated magnetic nanoparticles. Electron microscopy, X-ray diffraction, and magnetic measurements confirmed that the nanoparticles diameter stands between 10 and 11 nm. It is very essential for removal of heavy metal ions from water (Gregorio-Jauregui et al. 2012).

10.3.2.4 Emulsion Cross-Linking Method

Acidic solution of chitosan is treated with n-hexadecane and sunflower oil mixture for emulsification. This method revealed that wide range of particles size from 20 to 500 μm may be produced (Nagpal et al. 2010).

10.3.2.5 Ionotropic Gelation Method

In this method various sizes of chitosan nanoparticles can be obtained by using different gelation agent. The mean nanoparticle size of 162 nm and 517 nm were obtained when sodium tripolyphosphate and sodium sulfate were used respectively as gelation agent (Avadi et al. 2010).

10.4 Enzyme Immobilization

Enzymes are biological catalysts which are accelerating the chemical reaction within the body to maintain life. Enzymes are large in size and very delicate molecules unlike inorganic catalysts. Conventional soluble enzymes have many disadvantages due to the instability and sensitivity in reaction conditions. So, the uses of solid phase catalysts have become very important. Enzyme immobilization is a method of imprisonment of the enzyme molecules to a solid support that are insoluble in the reaction mixture. There are many benefits for using immobilized enzyme.

These are (i) it prevents the contamination of the substrate to reaction with enzyme which indirectly decreases the purification cost; (ii) it provides longer half-life and less degradation of enzymes; (iii) it also make possible the efficient retrieval and reuse of enzymes for the same reaction; (iv) it helps to control reaction rate and time; (v) product stability is very high; (vi) after binding of enzyme to substrate, enzyme changes into heterogeneous form to homogeneous form which results in simple separation, and higher purified product is formed; (vii) the matrix or solid support protects the enzyme in harsh reaction condition like pH, temperature, etc.; (viii) and enzyme immobilization gives high catalytic activity.

10.4.1 Methods of Preparation of Immobilized Enzymes

10.4.1.1 Support Binding

It can be physical such as salt linkages, hydrophobic interaction, van der Waals interaction, ionic, or covalent. Adsorption of enzyme is made by physical bonding. For this bonding enzyme is dried on the electrode surface. Proteolysis of enzyme is protected by contact with hydrophobic interfaces through adsorption (Spahn and Minteer 2008). Hydrogen bonding also gives successful immobilization on the pore walls silanols by use of silanized molecules sieves (Mitchell and Pérez-Ramírez 2011). The physical bonds are too weak and incapable to keep the enzyme fixed to the supports in high temperature in industries. Ionic and covalent binding have higher strength than physical bonding to support the enzyme and better performance (Sheldon 2007). Covalent binding also helps magnetic nanoclusters to interact with various active sites of enzymes (Patel et al. 2009). There are some limitations in this type of immobilization like if somehow enzyme deactivates the carrier then it will lost its function and become unusable. For this, support should be a synthetic polymer such as resin, silica, etc.

10.4.1.2 Cross-Linking

Cross-linking acts a vital role to conserve both structural and functional properties by using bifunctional chemicals. The most stable bifunctional cross-linking reagent is glutaraldehyde because of its covalent bond subunits and solubility in aqueous solution (Cirillo et al. 2014). Cross-linking is very valuable technique because in carrier-dependent techniques, the enzyme lost its native property up to 50% (Roy and Emilia Abraham 2004). Moreover, carrier bound technique is very laborious and time based. Some carrier-free approaches have been developed for highly concentrated enzymes activity, high stability, and low production costs. These approaches are cross-linked enzyme crystals (CLECs) (Roy et al. 2005) and cross-linked enzyme aggregates (CLEAs) (Sheldon et al. 2005, 2011). In nanotechnology, cross-linking technique is very important because of enhanced porosity and more surface area to support nano-diameters. Cross-linked enzyme has been composed of ionic polymer or organic solvents to make precipitation of aggregates (Sheldon and van Pelt 2013).

10.4.1.3 Entrapment

Imprisonment of enzyme within the polymeric network (like organic polymer, sol-gel, and microcapsule) with the help of covalent and non-covalent bonds is known as entrapment. In this process mechanical stability is enhanced; leakages of enzyme from carrier are less and easily can operate in the target (Shen et al. 2011). In this method enzyme is trapped on the external/internal surface of the prefabricated carrier. In different field application such as biofuels, fine chemistry, biomedicine, and biosensors, the entrapment method has great impact through the nanostructure carrier (Wang et al. 2009; Wen et al. 2011). Important particle of entrapment is mesoporous silica for its large surface area, high adsorption capacity of enzyme, etc. characteristics (Chen et al. 2011). For instance, the lipase activity increases when entrapped in sol-gel matrix of calixarene. Entrapment of both lipase and magnetic nanoparticles in biomimetic silica improved biocatalytic activity.

10.4.2 Supports to the Enzymes

To enhance the efficacy of the immobilized enzyme, various materials can be used as supports. These materials could be inorganic, organic, or composite. A support can significantly increase the physiochemical properties as well as kinetic properties of an enzyme. The support protects the structure of enzymes in harsh industrial conditions. It also provides high mechanical strength as well as high catalytic activity. The support should be nontoxic for the immobilized enzyme and for the product and has to consist of various functional groups for binding with enzyme. The support should have high loading capacity of enzyme. The structural characteristics such as porosity, compression, particle behavior, biodegradability, attract to protein are very essential for a good support (Krajewska 2005). If a support contains small pore size, it leads to diffusional limitation that can cause subsequent inactivity of enzyme (Wong et al. 2009). The particles used as supports are classified as (i) classic materials and (ii) new materials.

10.4.2.1 Classic Materials

Inorganic Materials

Silica is the most frequent inorganic resources employed in enzyme immobilization due to its chemical and high thermal resistance. Silica is also used as sol-gel silica, colloidal silica nanoparticles, and fumed silica. Other inorganic materials like aluminum, titanium, and zirconium oxides are widely used (Narwal et al. 2014; Zou et al. 2014; Foresti et al. 2010; Vallés et al. 2011; Reshmi et al. 2007; Yang et al. 2008).

Mineral Materials

Enzyme stability enhanced by using mineral materials as a support. Minerals are used because of its easy availability and several reactive groups present on the surface which helps formation of the covalent bond (Mbouguen et al. 2006).

Carbon-Based Materials

Due to high porous structure and surface area, carbon-based materials such as charcoals is used as a robust support in enzyme immobilization (Daoud et al. 2010; Rani et al. 2000).

Organic Materials

Due to some limitations of inorganic materials like restricted biocompatibility, lower affinity, and inability to create geometrical shapes, organic support is applied. Organic support is divided into synthetic and renewable materials.

10.4.2.2 New Materials

Synthetic Materials

The main advantage of this is that support material that constructs the polymeric chain can be decided according to the requirement of enzymes. In a polymeric structure, many chemical functional groups may be seen like carboxyl, hydroxyl, amine, carbonyl, and diol groups as well as trialkyl ammine and alkyl groups (Kirk and Christensen 2002; Ferrario et al. 2011). Polymer supports can be very effective because of its improved properties like reusability and thermal stability. The layer of polymer plays an essential role for protecting the active sites of enzyme in the reaction mixture. Various polymers having different functional groups are used as matrix for enzyme immobilization. For example, tyrosinase immobilized via –NH and C=O groups on nylon 66 and via –NH group of polyaniline alpha-amylase is immobilized (Ashly et al. 2011).

Biopolymers

Besides synthetic polymer, biopolymers are also used as an alternative matrix material for enzyme immobilization. Carbohydrate and proteins are used as biopolymer support. For immobilization collagen, keratins, cellulose, carrageenan, alginate, as well as chitin and chitosan are used. For various advantages like availability, abundance, and simple process in synthesis, these materials play an essential role as support for enzyme immobilization. The biopolymers have various special properties like biocompatibility, nontoxicity, and high affinity to proteins. Biopolymer reduces the structural deformation and properties of enzyme and helps to maintain high catalytic activity. Moreover, biopolymers are less expensive and therefore decreases the cost of immobilization process (Horchani et al. 2012; Vijayaraghavan et al. 2009; Homaei et al. 2013; Krajewska 2004).

Alginates act as biopolymeric supports for its capability for gelation by using calcium and sodium ions and also for production of capsules which is capable of immobilization of single or multiple enzymes (Hwang and Gu 2013; Coradin et al. 2003). So, alginates are used for entrapment and encapsulation methods. But due to some limitations such as diffusional limitations and mechanical stability, its application is hampered. Immobilization of lipase through entrapment in calcium alginate beads and polyphenol oxidase encapsulated in copper alginate beads shows fine catalytic activity but poor reusability. Agarose is widely used as support because of its heteropolysaccharide chain composed of β-D-galactose and 3,6-anhydro-alpha-L-galactose linked by glycosidic bonds. It has high gelation capability like alginates. It is used as capsules, beads, and fibers, leading great attention in industries. Agarose beads were used for entrapment of alpha-amylase and increase five times reusability of biocatalyst (Ball and Morell 2003; Delattre et al. 2011). Chitosan is the most used biopolymer because it can be used in different forms and shapes.

10.4.3 Immobilization of Enzyme Through Chitosan Nanoparticles

There are many researchers that tried to establish the suitability of chitosan nanoparticles on enzyme immobilization by using various methods. Few researches had assessed the optimum conditions for the enzymatic reaction like temperature, pH, and kinetic parameters as well as enzyme stability, maximum yielding, reusability, etc. In the following sections successful immobilization of different industrially valuable enzymes on chitosan based matrix were documented.

10.4.3.1 β-Galactosidase

β-Galactosidase commonly used to prepare low lactose milk from galactose and lactose through the transglycosylation reaction, and this enzyme has various industrial applications. For this purpose the reusability, production of product, and stability should be higher for real-life application at low cost. For immobilization of β-galactosidase, various methods like covalent linking, cross-linking, and adsorption techniques have been industrialized. It was reported that magnetic (Fe_3O_4)-chitosan nanoparticles preparation through electrostatic adsorption with Fe^{2+} and Fe^{3+}coprecipitation and glutaraldehyde-mediated coupling is a simple and quite efficient technique (Pan et al. 2009). This study shows that there is improvement in stability and provides structure rigidity which reduces the conformation changes. The study also revealed that immobilized β-galactosidase shows higher enzymatic activity in wide range of pH and temperature than the free enzyme. After 15 times of utilization, immobilized enzyme retains its 92% initial activity and also increases the storage capacity. Similarly, immobilization of β-galactosidase of *Kluyveromyces lactis* was carried out through covalent binding with chitosan nanoparticles which enhances the thermal stability and reused up to 50 times without loss of activity (Klein et al. 2012). Križnik et al. (2018) successfully synthesized magnetic maghemite nanoparticles with chitosan and amino-organosilanes for the immobilization of β-galactosidase of *Aspergillus oryzae*.

10.4.3.2 Cellulase

Cellulase is essential to produce glucose by enzymatic hydrolysis used for conversion of reducing sugar from cellulose. Zang et al. (2014) and Lin et al. (2017) documented that Fe_3O_4 nanoparticles coated with chitosan can be used in the immobilization of cellulase by covalent binding using glutaraldehyde as cross-linker. After 10 successive reuses, half of the initial enzyme was upheld in the immobilized matrix. The production rate of glucose remains higher than 80% with enhanced thermal stability. Cellulase is immobilized in chitosan-coated magnetic nanoparticles (10 nm in diameter) through covalent binding where glutaraldehyde was used as coupling reagent (Sánchez-Ramírez et al. 2017). The immobilized biocatalyst retained 37% of initial activity and 80% after 15 cycles of carboxymethylcellulose hydrolysis reaction. It can be easily separated from the reaction mixture. This also shows enhanced storage and thermal stability. It has various industrial applications like bioethanol production.

10.4.3.3 Glucose Oxidase

Chitosan nanoparticle is also applied in glucose sensing. Kang et al. (2007) successfully immobilized glucose oxidase by cross-linking with chitosan gold/platinum nanoparticles on carbon nanotubes (CNTs). This component shows great advantages like broad range of glucose sensing, fast response to glucose (<5 seconds), and low detection limit (0.21 M) with reasonable stability, reproducibility, and selectivity. Nanocomposite of glucose oxidase/platinum/functional graphene sheets/chitosan for sensing of glucose was prepared by Wu et al. (2009b). It shows very good sensitivity of glucose up to 0.6 μM. The nanocomposite biosensor has long-term stability and reproducibility and protects the signal from noise molecules. Onay et al. (2018)

immobilized glucose oxidase on electrode modified with Fe_3O_4 chitosan silver nanoparticles. This helps to find out 5–30 mM glucose concentration. The stability of electrode increases and shows good shelf life, and repeatability. It also provides fast amperometric response.

10.4.3.4 Invertase

Invertase is a catalyst found in *Saccharomyces cerevisiae* which plays an important role in sucrose hydrolysis and convert the sucrose in invert syrup which contains equal amount of β-D-fructose and α-D-glucose. It is a very essential enzyme in food industry to make sweeter product as fructose is sweeter than sucrose and helps to maintain the softness of product. Valerio et al. (2013) prepared chitosan nanoparticles and immobilized invertase by covalent modification. The immobilized invertase retains its maximum activity after 59 cycles. However, the storage and thermal stability of both free and immobilized enzyme remain unaltered. Invertase was immobilized on chitosan coated γ-Fe_2O_3 magnetic nanoparticles which increases the conversion rate and magnetic separation (Waifalkar et al. 2016). The immobilized invertase was more stable in a broad range of temperature for 20 cycles with high activity.

10.4.3.5 Glucoamylase

Glucoamylase also known as amyloglucosidase is an essential enzyme for digestion, and it has wide industrial-scale applications, viz., textile, beverage, food, detersive, pharmaceutical, papermaking, and brewing. Glucoamylase from *Aspergillus niger* immobilized into 1,3,5-triazine functionalized chitosan-coated superparamagnetic nanoparticles through covalent attachment which retained 70% of its initial activity after ten cycle (Amirbandeh et al. 2017).

10.4.3.6 Glucosidase

Liu et al. (2017) successfully immobilized α-glucosidase in chitosan-enriched magnetic composites by embedding method. Immobilized enzyme shows greater range of tolerance of pH and temperature. Reusability of enzyme was increased, and 62.2% initial activity was retained after ten times of use. The storage stability is also increased up to ~83%.

10.4.3.7 Xylanase

Xylanase (xylan degrading enzyme) is a very important enzyme in food industry because it can produce xylooligosaccharide and also produce reducing sugar from agricultural by-product. The bifunctional enzyme was incorporated onto carbon-coated chitosan nanoparticle by covalent bonding (Liu et al. 2015). It also retains 82.6% of its activity after seven reactions and exhibited increased thermal and storage stability.

10.4.3.8 α-Amylase

It is glycosyl hydrolase enzyme which breaks down the α-1,4-glycosidic bond of starch and produces glucose, maltose, maltotriose, etc. It is widely applied in paper, baking, textile, and food industries. Alpha-amylase is successfully immobilized onto chitosan-TiO_2 nanocomposite via adsorption method which increases the stability of the enzyme and encourages its industrial applicability (Bindu and Mohanan 2017).

10.4.3.9 Pectinase

Pectinase is an essential enzyme in fruit juice technology employed to break pectin. Sojitra et al. (2017) immobilized pectinase onto magnetic chitosan nanoparticles using cross-linker dextran polyaldehyde. Thermal stability of immobilized enzyme was improved two times than free enzyme. The K_m and V_{max} values are approximately the same as free enzyme. Immobilized enzyme retains its conformational flexibility. About 85% and 89% of the residual activity was retained by using the immobilized biocatalyst for seven successive reuses and after 15 days of storage, respectively.

10.4.3.10 Laccase

Laccase (dioxygen oxidoreductase) is blue copper oxidase enzyme found in bacteria, insects, plant, and various fungus families. It degraded and detoxifies various non-aromatic and aromatic compounds by using molecular oxygen. Fang et al. (2009) synthesized reversed-phase suspension-based magnetic chitosan nanoparticles and immobilized laccase onto it by means of adsorption. The immobilized laccase showed good operational stability and storage capability. Laccase of *Trametes versicolor* was immobilized into chitosan-coated magnetite nanoparticles covalently by adsorption (Kalkan et al. 2012). Here, chitosan binds with cyanuric chloride (CC) or carbodiimide (EDAC) through hydroxyl group. The kinetic parameters revealed that all combination of nanoparticles with CC and EDAC has lower K_m value than free enzymes, while chitosan nanoparticles with laccase show highest catalytic efficiency. After 30 cycle uses, immobilized enzyme retains 71% of its initial activity. Likewise, immobilized laccase in chitosan nanoparticles exhibited reusability, enhanced half-life, and thermal stability for Congo Red decolorization, and 98% of decolorization ability was retained in the immobilized matrix after 25 uses (Sadighi and Faramarzi 2013). Alver and Metin (2017) synthesized chitosan nanoparticles chelated with Cu (II) and immobilized laccase into it by adsorption, and the immobilized biocatalyst was applied to remove phenol from aqueous solution. The immobilized enzyme displayed activity in a wide range of temperature and pH and retains 50% of its initial activity after eight times of uses. The immobilized enzyme removes 96% of phenol from aqueous solution. Tarasi et al. (2018) synthesized chitosan-coated Fe@PA-CD (Fe@PACD-Cs) and immobilized laccase by physical adsorption. The immobilized biocatalyst showed 100% of its activity in temperature between 15 and 55 °C at pH 4.0. It also increases storage ability up to 70% after 12 days at 25 °C. Fe@PACD-Cs removes phenol and bisphenol A by 96.4% and 85.5%, respectively. Therefore, this method can be useful in waste treatment.

10.4.3.11 Lipase

Lipase (triacylglycerol acylhydrolase) is a hydrolase enzyme which generates free fatty acids, glycerol, monoacylglycerols, diacylglycerols from fats, and oil. It is found in fungi, bacteria, some plants, and animals. It has many biological roles like esterification, enantioselective hydrolysis, and enantio-enriched monomer synthesis and polymerization of macromolecules. But the free lipase has poor stability and reusability. Moreover, after enzymatic reaction, separation of enzyme is very difficult. Therefore, to improve activity of lipase, immobilization is the best approach. The enzyme lipase is immobilized in magnetic Fe_3O_4 chitosan nanoparticles with the help of sodium tripolyphosphate (TPP) by covalent binding which shows higher loading capacity and better stability of catalysts and decreases the loss of enzymes. In this technique 129 mg/g enzyme adsorption and 20.02 μmol/min/mg enzyme activity were acheived with improved retention capacity (Wu et al. 2009a). Liu et al. (2011) synthesized magnetic Fe_3O_4 chitosan nanoparticles, and lipase is immobilized through adsorption. The immobilized biocatalyst showed greater benefits like enhanced storage stability and higher activity than free form. Kuo et al. (2012) immobilized lipase of *Candida rugosa* into magnetic Fe_3O_4 chitosan nanoparticles covalently using N-(3-dimethylaminopropyl)-N′-ethylcarbodiimide (EDC) and N-hydroxysuccinimide(NHS) as coupling agents. It was found that highest immobilization can be achieved by performing immobilized reaction time 2.14 hours at pH 6.3. It shows wide range of pH and thermal stability, and higher operational and storage stability was also increased. Immobilized enzyme retains 83% of its initial activity after 20 times of uses. Liu et al. (2012) successfully developed alginate/chitosan nanospheres immobilized with lipase where ethylene glycol (oxidic poly) was used for functionalization. The alginate-chitosan nanosphere diameter was 60 nm and showed superparamagnetic properties. Alginate-chitosan contains positive charges and directly adsorbed lipase. *Candida rugosa* lipase (CRL) successfully immobilized through covalent binding and adsorption onto chitosan-alginate nanospheres, respectively. Layer-by-layer (LbL) assembly process was utilized for coat made up of chitosan and alginate to immobilized CRL (ICRL). Kinetic study revealed that K_m of alginate-ICRL was higher than chitosan-ICRL but V_{max} of chitosan-ICRL was higher than alginate-ICRL. The reusability of chitosan-ICRL is far better than alginate-ICRL. Lipase of *Thermomyces lanuginosus* was also immobilized in magnetic Fe_3O_4 chitosan nanoparticles through covalent modification (Wang et al. 2015). Immobilized lipase can tolerate broad pH and high temperature. After ten cycles of reaction, immobilized lipase retains its 70% initial activity. On the contrary, chitosan nanoparticle was synthesized by ionic gelation method and lipase was immobilized by cross-linking method (Devnani and Bahadur 2017). The concentration of preparatory reagents (like chitosan, acetic acid, glutaraldehyde, sodium) were checked and observed that changes in concentration of reagents it can alter the activity and stability of immobilized enzymes. Ghadi et al. (2015) prepared chitosan magnetic core-shell nanoparticles (CMNPs) in which lipase of *Burkholderia cepacia* was immobilized. Total relative activity and stability was increased in CMNP immobilized lipase. So, it is considered as suitable carrier for immobilization of lipase. An innovative support for the cross-linking of chitosan shell coated on magnetic nanoparticles with squaric acid (SqA) was developed in which lipase of *C. rugosa* was immobilized by EDC and NHS. In immobilized state the activity of lipase increased up to two times (Ziegler-Borowska et al. 2017). Immobilized lipase in magnetic chitosan/chitin nanoparticles was employed to produce biofuels, and 97.8% yield was achieved only in 5 hours. Moreover, the immobilized biocatalyst can be used successively up to eight esterification cycle (Rahman et al. 2018). In recent years for better activity of immobilized enzyme, the immobilization technique is improved and upgraded to nanocomposites. Magnetic chitosan nanocomposites were also designed for immobilization of porcine pancreas lipase (PPL) by Suo et al. (2018). The immobilized biocatalyst has better reusability and enhanced operational stability than free enzyme. Kinetic parameters also showed the improvement of affinity between substrate and immobilized enzyme. In another study, Xiang et al. (2018) synthesized SBA-CIL-chitosan nanocomposite as a carrier for immobilization of PPL. It also provides higher immobilization efficiency with tolerance at varied pH and temperature, stability, and reusability. The enzymatic activity increased up to 8.3 times and 82% of initial activity was retained.

10.4.3.12 Protease

Neutral proteinase is an essential catalyst in the industry like food, medicine, tanning, etc. because it produces peptides and free amines through hydrolyses of protein. So, its stability should increase by immobilization. To improve the activity of immobilized neutral proteinase, it was immobilized into chitosan nanoparticles by ionization gelation method (Tang et al. 2006). After immobilization storage stability, operational activity and thermal stability were increased, and it increased the enzyme activity up to 13.17%. In acidic condition its stability was also increased. Shojaei et al. (2017) immobilized protease from *Penaeus vannamei* in chitosan nanoparticles by ionic and electrostatic interaction which showed board range of pH and temperature tolerance. Immobilization also improved reusability as well as storage stability. Trypsin was immobilized in carboxymethyl chitosan-functionalized magnetic nanoparticles, cross-linked with 1-ethyl-3-(3-dimethylaminopropyl) carbodiimide (EDC) and glutaraldehyde (GA) (Sun et al. 2017). Immobilized enzyme retained 71% and 88.5% residual activity when cross-linked with EDC and GA, respectively, after six times of uses. After 14 days of storage, 80% and 88% of initial activity was retained in the immobilized biocatalyst cross-linked with EDC and GA, respectively. Demirkan et al. (2018) synthesized chitosan-blended cellulose monoacetate nanofibers by electrospinning method and immobilized protease by physical adsorption technique. The activity of enzyme was measured by utilized casein as substrate. After seven cycles of reaction, 20% of activity of enzyme still remain after glutaraldehyde activation. After glutaraldehyde and NaOH activation, the operational stability of immobilized enzyme was increased and hence can be utilized repeatability in textile and detergent industries.

10.4.3.13 Alcohol Dehydrogenase

Alcohol dehydrogenase oxidizes alcohols and converts into ketones and aldehydes. Li et al. (2008) immobilized alcohol dehydrogenase from *Saccharomyces cerevisiae* onto magnetic Fe_3O_4 chitosan nanoparticles. Thermal stability was increased and durability was also enhanced.

10.4.3.14 Penicillin G Acylase

This enzyme converts penicillin G to 6-aminopenicillanate and aminodeacetoxy-cephalosporanate. Both are essential for the construction of β-lactam antibiotics. This enzyme was immobilized onto magnetic (Fe_3O_4) chitosan nanoparticles via Schiff base reaction. Through this, higher thermal stability and wide range of pH tolerance was achieved. After 10 times of uses, immobilized enzyme retains its 75% of native activity (Ling et al. 2016).

10.4.3.15 Serratiopeptidase

It is a very important enzyme for the treatment of inflammation and pain. It is widely used in treatment of atherosclerosis, arthritis, carpal tunnel syndrome, chronic bronchitis, sinusitis, fibrocystic breast disease, postoperative swelling, etc. (Tiwari 2017). Serratiopeptidase was also immobilized onto chitosan-assisted magnetic nanoparticles (Kumar et al. 2014). It shows better storage stability than free form of the enzyme.

Not only that, there are numerous number of enzymes which were immobilized onto chitosan nanoparticles for improvement of their quality, stability, reusability, etc. That is summarized in Table 10.1.

Table 10.1 Enzyme immobilized onto chitosan nanoparticles, optimization, and application

Enzyme	Support	Mode of immobilization	Optimum pH and temperature	Reusability	Storage stability	Applications	Reference
Arsenite oxidase (AOase) from (*Exiguobacterium* sp. *As-9*)	Chitosan nanoparticles	Physical adsorption	Wide range of pH (5–10) and temperature (4–37 °C) can sustain		It retains its normal activity about 57.7% and 39.7% at 40 °C–60 °C, respectively	Conversion of As(III) to As(V)	Pandey and Bhatt (2018)
β-Fructofuranosidase	Chitosan-coated magnetic nanoparticles	Covalent bonding	Optimum pH and temperature were 5.5 °C and 60 °C, respectively	After 10 times of FOS reaction, it sustains its 55% activity		Conversion of sucrose to fructooligosaccharides	Chen et al. (2014)
α-Chymotrypsin	Fe_3O_4-chitosan nanoparticles	Covalent bonding	Optimum pH and temperature were 7.6 °C and 21.7 °C, respectively	After 12 times of dipeptide synthesis, it sustains its 60% activity		The developed immobilization model provides a simple, effective, and inexpensive process for industry application	Ju et al. (2012)
Pullulanase	Magnetic chitosan/Fe3O4 nanoparticles	Covalent bonding	Optimum temperature above 60 °C	After 6 times of reuses, co-immobilized via cross-linking retains its 85% activity	Free enzyme retains its activity by 11% whereas sol-gel encapsulation and co-immobilized retains 52% and 69%, respectively, at 62 °C after 5 hours	Due to high storage stability and easy recovery, it is applied in food industry	Long et al. (2017)

(continued)

Table 10.1 (continued)

Tannase from *Aspergillus niger*	Amino-functionalized magnetic Fe_3O_4–chitosan nanoparticles		Optimum pH and temperature were 4.5 °C and 30 °C, respectively	After 8 times of utilization, it sustains its 50% initial activity		Application in tea infusion	Li et al. (2018)
Acetylcholinesterase enzyme (AChE)	Chitosan magnetic iron nanoparticles					Malathion detection	Rodrigues et al. (2018)
Horseradish peroxidase (HRP) enzyme	Gold nanoparticle customized chitosan membrane (Au–chitosan)			After 34 times of use, the biosensor shows its normal activity		Applied for the detection of H_2O_2 in biosensor incorporates in indium tin oxide (ITO) electrode	Luo et al. (2005)

10.5 Probiotics

From the very ancient times the living microorganisms in fermented products were used for maintaining of our health. In the last century the Russian scientist Elie Metchnikoff delivered the concept of consumption of live microorganisms for health benefits (Gordon 2008). Generally probiotics are nonpathogenic live microorganisms which exert health beneficial effects when consumed in sufficient amounts. There are various health-promoting properties of probiotics, like (a) improves immune system in which it will supply, (b) maintains the natural gut microflora, (c) increases the lactose tolerance, (d) decreases cholesterol level, (e) exerts anti-carcinogenic properties, (f) inhibits pathogen by obstructing them to adhere the activation sites, (g) provides available supply of nutrient, and (f) removes toxin and its receptors (Prakash et al. 2011; Sullivan and Nord 2005). The various functions of probiotics assist in the decrease of the gene expression related with colonic cancer, prevent acute diarrhea, reduce the abdominal pain, defend the assault of pouchitis, and decrease the chance of ulcerative colitis and necrotizing enterocolitis. Probiotics can prevent or reduce the intestinal diseases such as (a) infectious diarrhea, (b) antibiotic-associated diarrhea (AAD), (c) irritable bowel syndrome (IBS), and (d) inflammatory bowel disease (IBD). *Lactobacillus* and *Bifidobacterium* genera were employed as probiotic product (Hoffmann et al. 2013). Probiotic microorganisms belong to the genus *Lactobacillus*, *Bifidobacterium*, *Enterococcus*, *Propionibacterium*, and certain yeasts like *S. boulardii*. Probiotics can exert their effect through various mechanisms. With the gut bacteria, probiotics can interact through three different portions such as epithelial layer, mucus layer, and GALT or gut-associated lymphoid tissue (Johansson et al. 2011). Probiotics act on mucus layer and intestinal layer and make team with gut microbiota for the interference of pathogenic microorganisms through increases of the production of SCFA, blocking the adhesion of pathogens to lumen. Moreover, it has also produced some toxin like bacteriocin. Probiotics also enhance the barrier functions of Paneth cells, colonocytes, goblet cells, and enterocytes to increase the production of mucin, defensin, and IgA (Quevrain et al. 2016; Hong et al. 2011). It also acts upon M cells, macrophages, dendritic cells, and T cells to regulate cytokine production and T_{reg} responses as well as some receptors situated on epithelium such as toll-like, C-type lectin, Nod-like and RIG-I-like receptors, etc. Some probiotics bacteria like *Lactobacillus* can exert their effects of immunomodulation through enhancing the production of some peptides (Bouskra et al. 2008; Wang et al. 2014).

10.5.1 Probiotic Encapsulation

To ensure the favorable role, probiotics should survive in the digestion pattern of host. After colonization and growth of the probiotics in the GI tract, they will perform the beneficial activities. The major hindrances are acidic pH of stomach, the immune system of host, competition with commensal, and pathogenic microorganisms (Mawad et al. 2018). There are many obstacles until or unless the probiotics bacteria reach the target portion. Therefore, the protection in the passage of digestive system should be given to probiotics through encapsulation (Ramos et al. 2018). Encapsulation of probiotics provides improved survivality of the organism in the digestive system and provides protection from acidic condition, high oxygen levels, throughout the gastrointestinal passage. The most essential advantages of encapsulation are the following: (1) increase the adherence to mucosal tissue of the intestine, (2) enhance the resistance of bacteria, and (3) keep the cell viable for a long time (Chandramouli et al. 2004; Ramos et al. 2018).

10.5.2 Methods of Encapsulation

Encapsulation is the incorporation of essential ingredients in food to protect the function and manage the discharge of ingredients. In microencapsulation solids, liquids, and gaseous substances are packed in small capsules which release the contents at controlled rates through long time (Champagne and Fustier 2007). The main characteristics of encapsulation are: (a) should be water soluble, (b) keep their structure after incorporation, and (c) release probiotics in the target portion. There are mainly three types of technique for encapsulation, these are (1) spray drying (Burgain et al. 2011), (2) extrusion (de Vos et al. 2010; Rokka and Rantamäki 2010), and (3) emulsion (Heidebach et al. 2012).

10.5.3 Techniques of Coated Capsules

For better performance of encapsulated capsules, they need to be coated by edible polymers. Coating provides various advantages like (a) higher stability in acidic pH and wide range of temperature, (b) reduces capsules permeability, (c) proper release of micronutrient, (d) proper delivery to the target. There are mainly two coating techniques: (i) layer-by-layer (LbL) (Decher 1997) and (ii) coacervation (Eratte et al. 2015). Among the various materials for coating of microcapsules, alginate, chitosan, poly-l-lysine (PLL), whey protein, etc. are extensively utilized.

10.5.4 Probiotics Encapsulation in Chitosan-Based Nanomaterials

In the digestive tract and the whole time between processing of food and consumption, the viability of probiotics becomes low and failed to exert their effects. Encapsulation provides storage (such as increase shell life of foods) and packaging (from oxygen, temperature, humidity). Li et al. (2011) reported that high viable cell count of 7.38 and 7.91 log cfu/g of *Lactobacillus casei* was maintained at immobilized state in alginate-chitosan and alginate-chitosan-carboxymethyl chitosan microcapsules after 2 hours in simulated gastric fluid. Similarly survival rate of *Bifidobacterium breve* is reasonably good (6.6 ± 0.5 log cfu/mL) in immobilized state in alginate-chitosan after 1 hour in simulated gastric fluid (Cook et al. 2011). *Saccharomyces boulardii* yeast cells were encapsulated in chitosan and dextran sulfate which protect it from acidic condition in simulated gastric fluid (Thomas et al. 2014; Shori 2017). Gandomi et al. (2016) successfully encapsulated probiotic bacteria *Lactobacillus rhamnosus* GG incorporated in chitosan-alginate with inulin and examine their survival effect under stimulated gastric fluid and through storage of apple juice. The storage study was done during storage at 25 and 4 °C for 90 days. After the 90 days, the free bacterial survival diminished to 13.6%, but encapsulated *L. rhamnosus* provides 4.5-fold improved survival rate than free cells. Sensory scores revealed that encapsulation bringing the improvement of all characteristics of probiotics. This study advocated that probiotic encapsulation enhanced the bacteria survival rate both during storage of apple juice and in stimulated GI fluid without any adverse effect. Two probiotics bacteria *Lactobacillus acidophilus* and *Lactobacillus rhamnosus* were microencapsulated in calcium alginate-chitosan and Eudragit S 100 nanoparticles (Ansari et al. 2017). In the study free cells (unencapsulated), single coated (only with chitosan), and double-coated beads (chitosan and Eudragit nanoparticles) were applied in gastric juice without pepsin at pH 1.55 and in intestinal juice with 1% bile sat at pH 7.5. The outcome of the study showed that free bacterial counts diminished significantly ($P < 0.001$). The rate of decrease of free *L. acidophilus* and *L. rhamnosus* was significantly faster than single- and double-coated formation of encapsulated probiotics. Due to the lack of strength of single chitosan-coated beads under pH 1.55, the counts reduced 5-log and 4-log of *L. rhamnosus* and *L. acidophilus*, respectively. But double-coated beads showed more viability of bacteria than that of single coated. Therefore, this study proved that encapsulation with chitosan and Eudragit S 100 nanoparticles showed improved survivability of *L. acidophilus* and *L. rhamnosus* probiotics in the acidic environment of stomach. Kim et al. (2017) successfully encapsulated probiotics bacteria *L. acidophilus* through ionic gelation method and examine the encapsulated probiotics in stimulated gastric condition and storage. At first *L. acidophilus* encapsulated in biocompatible and biodegradable capsules constructed by ionic gelation between chitosan (CS) and phytic acid (PA). To increase the stability of capsules, calcium carbonate ($CaCO_3$) and starch were utilized. After 2 hours in acidic condition (pH 1.5, stimulated gastric fluid), the cells of *L. acidophilus* decreased only 0.64 log CFU/g. Among them, PA-CS-$CaCO_3$ capsules are more stable than other two capsules in pH 1.5 and 2.0. This study proved that encapsulation provides more viability and storage capacity of probiotics bacteria. Likewise, Ebrahimnejad et al. (2017) demonstrated that nanoencapsulation of probiotic *Lactobacillus acidophilus* in chitosan endorse good probiotic viability and stability in intestinal juice.

10.6 Conclusion

As renewable and biodegradable nanoparticles, chitosan continuously draws attention of both academicians and industrialists. In recent ages, chitosan-based nanomaterials have acquired a remarkable advantage over their conventional counterparts. The amazing and unique physicochemical properties of chitosan advocate its widespread application in immobilization of many important enzymes as well as encapsulation of probiotic microorganisms. In the arena of enzyme immobilization, chitosan-based nanomaterials endorsed enhance stability and activity of the immobilized biocatalyst which attested their long-term use and reuse. Encapsulated probiotic microbes in chitosan-based nanomaterial enhance the stability and viability of the microbes especially in simulated gastrointestinal environment. In conclusion, chitosan-based nanomaterials can be considered to be a promising outlook for long-term repetitive use of industrially viable enzymes as well as introducing of viable probiotics for enhanced activity and stability.

References

Alver E, Metin AÜ (2017) Chitosan based metal-chelated copolymer nanoparticles: Laccase immobilization and phenol degradation studies. Int Biodeterior Biodegrad 125:235–242

Amirbandeh M, Taheri-Kafrani A, Soozanipour A et al (2017) Triazine-functionalized chitosan-encapsulated superparamagnetic nanoparticles as reusable and robust nanocarrier for glucoamylase immobilization. Biochem Eng J 127:119–127

Ansari F, Pourjafar H, Jodat V et al (2017) Effect of Eudragit S100 nanoparticles and alginate chitosan encapsulation on the viability of *Lactobacillus acidophilus* and *Lactobacillus rhamnosus*. AMB Express 7(1):144

Ashly PC, Joseph MJ, Mohanan PV (2011) Activity of diastase α-amylase immobilized on polyanilines (PANIs). Food Chem 127(4):1808–1813

Avadi MR, Sadeghi AM, Mohammadpour N et al (2010) Preparation and characterization of insulin nanoparticles using chitosan and Arabic gum with ionic gelation method. Nanomedicine 6(1):58–63

Ball SG, Morell MK (2003) From bacterial glycogen to starch: understanding the biogenesis of the plant starch granule. Annu Rev Plant Biol 54(1):207–233

Bindu VU, Mohanan PV (2017) Enhanced stability of α-amylase via immobilization onto chitosan-TiO_2 nanocomposite. Nanosci Technol 4(2):1–9

Bouskra D, Brézillon C, Bérard M et al (2008) Lymphoid tissue genesis induced by commensals through NOD1 regulates intestinal homeostasis. Nature 456(7221):507

Brunel F, Véron L, David L et al (2008) A novel synthesis of chitosan nanoparticles in reverse emulsion. Langmuir 24(20):11370–11377

Burgain J, Gaiani C, Linder M et al (2011) Encapsulation of probiotic living cells: from laboratory scale to industrial applications. J Food Eng 104(4):467–483

Cao M, Li Z, Wang J (2012) Food related applications of magnetic iron oxide nanoparticles: enzyme immobilization, protein purification, and food analysis. Trends Food Sci Technol 27(1):47–56

Champagne CP, Fustier P (2007) Microencapsulation for the improved delivery of bioactive compounds into foods. Curr Opinion Biotechnol 18(2):184–190

Chandramouli V, Kailasapathy K, Peiris P et al (2004) An improved method of microencapsulation and its evaluation to protect *Lactobacillus spp.* in simulated gastric conditions. J Microbiol Methods 56(1):27–35

Chen XG, Liu CS, Liu CG et al (2006) Preparation and biocompatibility of chitosan microcarriers as biomaterial. Biochem Eng J 27(3):269–274

Chen GC, Kuan IC, Hong JR et al (2011) Activity enhancement and stabilization of lipase from *Pseudomonas cepacia* in polyallylamine-mediated biomimetic silica. Biotechnol Lett 33(3):525–529

Chen SC, Sheu DC, Duan KJ (2014) Production of fructooligosaccharides using β-fructofuranosidase immobilized onto chitosan-coated magnetic nanoparticles. J Taiwan Inst Chem Eng 45(4):1105–1110

Choi YJ, Kim EJ, Piao Z et al (2004) Purification and characterization of chitosanase from *Bacillus* sp. strain KCTC 0377BP and its application for the production of chitosan oligosaccharides. Appl Environ Microbiol 70(8):4522–4531

Cirillo G, Nicoletta FP, Curcio M et al (2014) Enzyme immobilization on smart polymers: catalysis on demand. React Funct Polym 83:62–69

Cook MT, Tzortzis G, Charalampopoulos D et al (2011) Production and evaluation of dry alginate-chitosan microcapsules as an enteric delivery vehicle for probiotic bacteria. Biomacromolecules 12:2834–2840

Coradin T, Nassif N, Livage J (2003) Silica–alginate composites for microencapsulation. Appl Microbiol Biotechnol 61(5–6):429–434

Daoud FB, Kaddour S, Sadoun T (2010) Adsorption of cellulase *Aspergillus niger* on a commercial activated carbon: kinetics and equilibrium studies. Colloid Surf B Biointerfaces 75(1):93–99

de Moura MR, Aouada FA, Mattoso LH (2008) Preparation of chitosan nanoparticles using methacrylic acid. J Colloid Interface Sci 321(2):477–483

de Vos P, Faas MM, Spasojevic M et al (2010) Encapsulation for preservation of functionality and targeted delivery of bioactive food components. Int Dairy J 20(4):292–302

Decher G (1997) Fuzzy nanoassemblies: toward layered polymeric multicomposites. Science 277(5330):1232–1237

Delattre C, Fenoradosoa TA, Michaud P (2011) Galactans: an overview of their most important sourcing and applications as natural polysaccharides. Braz Arch Biol Technol 54(6):1075–1092

Demirkan E, Avci T, Aykut Y (2018) Protease immobilization on cellulose monoacetate/chitosan-blended nanofibers. J Ind Tex 47(8):2092–2111

Devnani H, Bahadur A (2017) Variation in activity of lipase immobilized on chitosan and alginate nanoparticles by changing concentration of the preparatory reagents. Res Rev: J Life Sci 7(3):14–22

Ebrahimnejad P, Khavarpour M, Khalili S (2017) Survival of *Lactobacillus acidophilus* as probiotic Bacteria using chitosan nanoparticles. Int J Eng Trans A: Basics 304:456–463

Eratte D, McKnight S, Gengenbach TR et al (2015) Co-encapsulation and characterisation of omega-3 fatty acids and probiotic bacteria in whey protein isolate–gum Arabic complex coacervates. J Funct Foods 19:882–892

Fang H, Huang J, Ding L et al (2009) Preparation of magnetic chitosan nanoparticles and immobilization of laccase. J Wuhan Univ Technol-Mater Sci Ed 24(1):42–47

Ferrario V, Ebert C, Knapic L et al (2011) Conformational changes of lipases in aqueous media: a comparative computational study and experimental implications. Adv Syn Cat 353(13):2466–2480

Foresti ML, Valle G, Bonetto R et al (2010) FTIR, SEM and fractal dimension characterization of lipase B from *Candida antarctica* immobilized onto titania at selected conditions. Appl Surface Sci 256(6):1624–1635

Gandomi H, Abbaszadeh S, Misaghi A et al (2016) Effect of chitosan-alginate encapsulation with inulin on survival of *Lactobacillus rhamnosus* GG during apple juice storage and under simulated gastrointestinal conditions. LWT-Food Sci Technol 69:365–371

Ghadi A, Tabandeh F, Mahjoub S et al (2015) Fabrication and characterization of core-shell magnetic chitosan nanoparticles as a novel carrier for immobilization of *Burkholderia cepacia* lipase. J Oleo Sci 64(4):423–430

Gooday GW, Aruchami M, Gowri N et al (1986) Chitin deacetylases in invertebrates. In chitin in nature and technology. Springer, Boston

Gordon S (2008) Elie Metchnikoff: father of natural immunity. Eur J Immunol 38(12):3257–3264

Gregorio-Jauregui KM, Pineda MG, Rivera-Salinas JE et al (2012) One-step method for preparation of magnetic nanoparticles coated with chitosan. J Nanomater 2012:4

Halder SK, Mondal KC (2018) Microbial valorization of chitinous bioresources for chitin extraction and production of chito-oligomers and N-acetylglucosamine: trends, perspectives and prospects. In: Microbial biotechnology. Springer, Singapore, pp 69–107

Heidebach T, Först P, Kulozik U (2012) Microencapsulation of probiotic cells for food applications. Crit Rev Food Sci Nutr 52(4):291–311

Hoffmann DE, Fraser CM, Palumbo FB et al (2013) Probiotics: finding the right regulatory balance. Science 342(6156):314–315

Homaei AA, Sariri R, Vianello F et al (2013) Enzyme immobilization: an update. J Chem Biol 6(4):185–205

Hong YS, Hong KS, Park MH et al (2011) Metabonomic understanding of probiotic effects in humans with irritable bowel syndrome. J Clin Gastroenterol 45(5):415–425

Horchani H, Aissa I, Ouertani S et al (2012) *Staphylococcal lipases*: biotechnological applications. J Mol Catal B: Enzym 76:125–132

Hwang ET, Gu MB (2013) Enzyme stabilization by nano/microsized hybrid materials. Eng Life Sci 1:49–61

Janssen MH, van Langen LM, Pereira SR et al (2002) Evaluation of the performance of immobilized penicillin G acylase using active-site titration. Biotechnol Bioeng 78(4):425–432

Jia H, Zhu G, Wang P (2003) Catalytic behaviors of enzymes attached to nanoparticles: the effect of particle mobility. Biotechnol Bioeng 84(4):406–414

Johansson ME, Larsson JM, Hansson GC (2011) The two mucus layers of colon are organized by the MUC2 mucin, whereas the outer layer is a legislator of host–microbial interactions. Proc Natl Acad Sci U S A 108(Supplement 1):4659–4665

Ju HY, Kuo CH, Too JR et al (2012) Optimal covalent immobilization of α-chymotrypsin on Fe3O4-chitosan nanoparticles. J Mol Catal B: Enzym 78:9–15

Kalkan NA, Aksoy S, Aksoy EA et al (2012) Preparation of chitosan-coated magnetite nanoparticles and application for immobilization of laccase. J Appl Pol Sci 123(2):707–716

Kang X, Mai Z, Zou X et al (2007) A novel glucose biosensor based on immobilization of glucose oxidase in chitosan on a glassy carbon electrode modified with gold–platinum alloy nanoparticles/multiwall carbon nanotubes. Anal Biochem 369(1):71–79

Katchalski-Katzir E, Kraemer DM (2000) Eupergit® C, a carrier for immobilization of enzymes of industrial potential. J Mol Catal B Enzym 10(1–3):157–76

Kim JU, Kim B, Shahbaz HM et al (2017) Encapsulation of *probiotic Lactobacillus acidophilus* by ionic gelation with electrostatic extrusion for enhancement of survival under simulated gastric conditions and during refrigerated storage. Int J Food Sci Technol 52(2):519–530

Kirk O, Christensen MW (2002) Lipases from *Candida antarctica*: unique biocatalysts from a unique origin. Org Process Res Dev 6(4):446–451

Klein MP, Nunes MR, Rodrigues RC et al (2012) Effect of the support size on the properties of β-galactosidase immobilized on chitosan: advantages and disadvantages of macro and nanoparticles. Biomacromolecules 13(8):2456–2464

Klotzbach T, Watt M, Ansari Y et al (2006) Effects of hydrophobic modification of chitosan and Nafion on transport properties, ion-exchange capacities, and enzyme immobilization. J Membr Sci 282(1–2):276–283

Krajewska B (2004) Application of chitin-and chitosan-based materials for enzyme immobilizations: a review. Enzym Microb Technol 35(2–3):126–139

Krajewska B (2005) Membrane-based processes performed with use of chitin/chitosan materials. Sep Purif Technol 41(3):305–312

Križnik L, Vasić K, Knez Ž et al (2018) Hyper-activation of ß-galactosidase from *Aspergillus oryzae* via immobilization onto amino-silane and chitosan magnetic maghemite nanoparticles. J Clean Prod 179:225–234

Kumar S, Jana AK, Dhamija I et al (2014) Chitosan-assisted immobilization of serratiopeptidase on magnetic nanoparticles, characterization and its target delivery. J Drug Target 22(2):123–137

Kuo CH, Liu YC, Chang CM et al (2012) Optimum conditions for lipase immobilization on chitosan-coated Fe_3O_4 nanoparticles. Carbohydr Polym 87(4):2538–2545

Li GY, Huang KL, Jiang YR et al (2008) Preparation and characterization of *Saccharomyces cerevisiae* alcohol dehydrogenase immobilized on magnetic nanoparticles. Int J Biol Macromol 42(5):405–412

Li XY, Chen XG, Sun ZW et al (2011) Preparation of alginate/chitosan/ carboxymethyl chitosan complex microcapsules and application in *Lactobacillus casei* ATCC 393. Carbohydr Polym 83:1479–1485

Li R, Fu G, Liu C (2018) Tannase immobilisation by amino-functionalised magnetic Fe_3O_4-chitosan nanoparticles and its application in tea infusion. Int J Biol Macromol 114:1134–1143

Lin Y, Liu X, Xing Z et al (2017) Preparation and characterization of magnetic Fe_3O_4–chitosan nanoparticles for cellulase immobilization. Cellulose 24(12):5541–5550

Ling XM, Wang XY, Ma P et al (2016) Covalent immobilization of penicillin G acylase onto Fe_3O_4@ chitosan magnetic nanoparticles. J Microbiol Biotechnol 26(5):829–836

Liu Y, Jia S, Wu Q et al (2011) Studies of Fe_3O_4-chitosan nanoparticles prepared by co-precipitation under the magnetic field for lipase immobilization. Cat Com 12(8):717–720

Liu X, Chen X, Li Y et al (2012) Preparation of superparamagnetic Fe_3O_4@ alginate/chitosan nanospheres for *Candida rugosa* lipase immobilization and utilization of layer-by-layer assembly to enhance the stability of immobilized lipase. ACS Appl Mater Interfaces 4(10):5169–5178

Liu MQ, Huo WK, Xu X et al (2015) An immobilized bifunctional xylanase on carbon-coated chitosan nanoparticles with a potential application in xylan-rich biomass bioconversion. J Mol Cat B: Enzym 120:119–126

Liu DM, Chen J, Shi YP (2017) α-Glucosidase immobilization on chitosan-enriched magnetic composites for enzyme inhibitors screening. Int J Biol Macromol 105:308–316

Long J, Li X, Zhan X et al (2017) Sol–gel encapsulation of pullulanase in the presence of hybrid magnetic (Fe_3O_4–chitosan) nanoparticles improves thermal and operational stability. Bioprocess Biosyst Eng 40(6):821–831

Lozinsky VI, Galaev IY, Plieva FM et al (2003) Polymeric cryogels as promising materials of biotechnological interest. Trends Biotechnol 21(10):445–451

Luo XL, Xu JJ, Zhang Q et al (2005) Electrochemically deposited chitosan hydrogel for horseradish peroxidase immobilization through gold nanoparticles self-assembly. Biosens Bioelectron 21(1):190–196

Mawad A, Helmy YA, Shalkami AG et al (2018) *E. coli* Nissle microencapsulation in alginate-chitosan nanoparticles and its effect on *Campylobacter jejuniin vitro*. Appl Microbiol Biotechnol 9:1–6

Mbouguen JK, Ngameni E, Walcarius A (2006) Organoclay-enzyme film electrodes. Anal Chim Acta 578(2):145–155

Mitchell S, Pérez-Ramírez J (2011) Mesoporous zeolites as enzyme carriers: synthesis, characterization, and application in biocatalysis. Catal Today. 2011 168(1):28–37

Nagpal K, Singh SK, Mishra DN (2010) Chitosan nanoparticles: a promising system in novel drug delivery. Chem Pharm Bull (Tokyo) 58(11):1423–1430

Narwal SK, Saun NK, Gupta R (2014) Characterization and catalytic properties of free and silica-bound lipase: a comparative study. J Oleo Sci 63(6):599–605

Nasti A, Zaki NM, de Leonardis P et al (2009) Chitosan/TPP and chitosan/TPP-hyaluronic acid nanoparticles: systematic optimisation of the preparative process and preliminary biological evaluation. Pharm Res 26(8):1918–1930

Onay A, Dogan Ü, Ciftci H et al (2018) Amperometric glucose sensor based on the glucose oxidase enzyme immobilized on graphite rod electrode modified with Fe_3O_4-CS-Au magnetic nanoparticles. Ionics 2018:1–8

Pal SL, Jana U, Manna PK et al (2011) Nanoparticle: an overview of preparation and characterization. J Appl Pharm Sci 1(6):228–234

Pan C, Hu B, Li W et al (2009) Novel and efficient method for immobilization and stabilization of β-d-galactosidase by covalent attachment onto magnetic Fe_3O_4–chitosan nanoparticles. J Mol Catal B: Enzym 61(3–4):208–215

Pandey N, Bhatt R (2018) Improved biotransformation of arsenic by arsenite oxidase–chitosan nanoparticle conjugates. Int J Biol Macromol 106:258–265

Patel SR, Yap MG, Wang DI (2009) Immobilization of l-lactate dehydrogenase on magnetic nanoclusters for chiral synthesis of pharmaceutical compounds. Biochem Eng J 48(1):13–21

Prakash S, Tomaro-Duchesneau C, Saha S et al (2011) The gut microbiota and human health with an emphasis on the use of microencapsulated bacterial cells. Biomed Res Int 2:2011

Quevrain E, Maubert MA, Michon C et al (2016) Identification of an anti-inflammatory protein from *Faecalibacterium prausnitzii*, a commensal bacterium deficient in Crohn's disease. Gut 65(3):415–425

Rahman IN, Attan N, Mahat NA et al (2018) Statistical optimization and operational stability of *Rhizomucor miehei* lipase supported on magnetic chitosan/chitin nanoparticles for synthesis of pentyl valerate. Int J Biol Macromol 115:680–695

Ramos PE, Cerqueira MA, Teixeira JA et al (2018) Physiological protection of probiotic microcapsules by coatings. Crit Rev Food Sci Nutr 58(11):1864–1877

Rani AS, Das ML, Satyanarayana S (2000) Preparation and characterization of amyloglucosidase adsorbed on activated charcoal. J Mol Cat B: Enzym 10(5):471–476

Reshmi R, Sanjay G, Sugunan S (2007) Immobilization of α-amylase on zirconia: a heterogeneous biocatalyst for starch hydrolysis. Catal Commun 8(3):393–399

Rodrigues NF, Neto SY, Luz RD et al (2018) Ultrasensitive determination of malathion using acetylcholinesterase immobilized on chitosan-functionalized magnetic iron nanoparticles. Biosensors 8(1):16

Rokka S, Rantamäki P (2010) Protecting probiotic bacteria by microencapsulation: challenges for industrial applications. Euro Food Res Technol 231(1):1–2

Roy JJ, Emilia Abraham T (2004) Strategies in making cross-linked enzyme crystals. Chem Rev 104(9):3705–3722

Roy JJ, Abraham TE, Abhijith KS et al (2005) Biosensor for the determination of phenols based on cross-linked enzyme crystals (CLEC) of laccase. Biosens Bioelectron 21(1):206–211

Sadighi A, Faramarzi MA (2013) Congo red decolorization by immobilized laccase through chitosan nanoparticles on the glass beads. J Taiwan Ins Chem Eng 44(2):156–162

Salemi Z (2010) Tailor-made enzyme carriers: preparation and use of adsorbents specifically designed to immobilize allosteric enzymes in activated conformation. Am J Biochem Biotechnol 6:111–115

Sánchez-Ramírez J, Martínez-Hernández JL, Segura-Ceniceros P et al (2017) Cellulases immobilization on chitosan-coated magnetic nanoparticles: application for Agave Atrovirens lignocellulosic biomass hydrolysis. Bioprocess Biosyst Eng 40(1):9–22

Sheldon RA (2007) Enzyme immobilization: the quest for optimum performance. Adv Synth Catal 349(8–9):1289–1307

Sheldon RA (2011) Characteristic features and biotechnological applications of cross-linked enzyme aggregates (CLEAs). Appl Microbiol Biotechnol 92(3):467–477

Sheldon RA, van Pelt S (2013) Enzyme immobilisation in biocatalysis: why, what and how. Chem Soc Rev 42(15):6223–6235

Sheldon RA, Schoevaart R, Van Langen LM (2005) Cross-linked enzyme aggregates (CLEAs): a novel and versatile method for enzyme immobilization (a review). Biocatal Biotransform 23(3–4):141–147

Shen Q, Yang R, Hua X et al (2011) Gelatin-templated biomimetic calcification for β-galactosidase immobilization. Process Biochem 46(8):1565–1571

Shojaei F, Homaei A, Taherizadeh MR et al (2017) Characterization of biosynthesized chitosan nanoparticles from *Penaeus vannamei* for the immobilization of *P. vannamei* protease: an eco-friendly nanobiocatalyst. Int J Food Prop 20(sup2):1413–1423

Shori AB (2017) Microencapsulation improved probiotics survival during gastric transit. HAYATI J Biosc 24(1):1–5

Sojitra UV, Nadar SS, Rathod VK (2017) Immobilization of pectinase onto chitosan magnetic nanoparticles by macromolecular cross-linker. Carbohydr Polym 157:677–685

Sorlier P, Denuzière A, Viton C et al (2001) Relation between the degree of acetylation and the electrostatic properties of chitin and chitosan. Biomacromolecules 2(3):765–772
Spahn C, Minteer SD (2008) Enzyme immobilization in biotechnology. Recent Pat Eng 2(3):195–200
Sullivan Å, Nord CE (2005) Probiotics and gastrointestinal diseases. J Int Med 257(1):78–92
Sun J, Yang L, Jiang M et al (2017) Stability and activity of immobilized trypsin on carboxymethyl chitosan-functionalized magnetic nanoparticles cross-linked with carbodiimide and glutaraldehyde. J Chromatogr B 1054:57–63
Suo H, Xu L, Xu C et al (2018) Enhancement of catalytic performance of porcine pancreatic lipase immobilized on functional ionic liquid modified Fe_3O_4-chitosan nanocomposites. Int J Biol Macromol 119:624–632
Tang ZX, Qian JQ, Shi LE (2006) Characterizations of immobilized neutral proteinase on chitosan nano-particles. Process Biochem 41(5):1193–1197
Tarasi R, Alipour M, Gorgannezhad L et al (2018) Laccase immobilization onto magnetic β-cyclodextrin-modified chitosan: improved enzyme stability and efficient performance for phenolic compounds elimination. Macromol Res 2018:1–8
Thomas MB, Vaidyanathan M, Radhakrishnan K et al (2014) Enhanced viability of probiotic Saccharomyces boulardii encapsulated by layer-by-layer approach in pH responsive chitosan-dextran sulfate polyelectrolytes. J Food Eng 136:1–8
Tiwari M (2017) The role of serratiopeptidase in the resolution of inflammation. Asian J Pharma Sci 12(3):209–212
Valerio SG, Alves JS, Klein MP et al (2013) High operational stability of invertase from *Saccharomyces cerevisiae* immobilized on chitosan nanoparticles. Carbohydr Polym 92(1):462–468
Vallés D, Furtado S, Villadóniga C et al (2011) Adsorption onto alumina and stabilization of cysteine proteinases from crude extract of solanum granuloso-leprosum fruits. Process Biochem 46(2):592–598
Vijayaraghavan K, Yamini D, Ambika V et al (2009) Trends in inulinase production–a review. Crit Rev Biotechnol 29(1):67–77
Waifalkar PP, Parit SB, Chougale AD et al (2016) Immobilization of invertase on chitosan coated γ-Fe_2O_3 magnetic nanoparticles to facilitate magnetic separation. J Colloid Interface Sci 482:159–164
Wang ZG, Wan LS, Liu ZM et al (2009) Enzyme immobilization on electrospun polymer nanofibers: an overview. J Mol Cat B: Enzym 56(4):189–195
Wang Y, Begum-Haque S, Telesford KM et al (2014) A commensal bacterial product elicits and modulates migratory capacity of CD39+ CD4 T regulatory subsets in the suppression of neuroinflammation. Gut Microbes 5(4):552–561
Wang XY, Jiang XP, Li Y et al (2015) Preparation Fe_3O_4@ chitosan magnetic particles for covalent immobilization of lipase from *Thermomyces lanuginosus*. Int J Biol Macromol 75:44–50
Wen H, Nallathambi V, Chakraborty D et al (2011) Carbon fiber microelectrodes modified with carbon nanotubes as a new support for immobilization of glucose oxidase. Microchim Acta 175(3–4):283–289
Wong LS, Khan F, Micklefield J (2009) Selective covalent protein immobilization: strategies and applications. Chem Rev 109(9):4025–4053
Wu Y, Wang Y, Luo G et al (2009a) In situ preparation of magnetic Fe_3O_4-chitosan nanoparticles for lipase immobilization by cross-linking and oxidation in aqueous solution. Bioresour Technol 100(14):3459–3464
Wu H, Wang J, Kang X et al (2009b) Glucose biosensor based on immobilization of glucose oxidase in platinum nanoparticles/graphene/chitosan nanocomposite film. Talanta 80(1):403–406
Xiang X, Ding S, Suo H et al (2018) Fabrication of chitosan-mesoporous silica SBA-15 nanocomposites via functional ionic liquid as the bridging agent for PPL immobilization. Carbohydr Polym 182:245–253
Yang Z, Si S, Zhang C (2008) Study on the activity and stability of urease immobilized onto nanoporous alumina membranes. Microporous Mesoporous Mater 111(1–3):359–366

Yang K, Xu NS, Su WW (2010) Co-immobilized enzymes in magnetic chitosan beads for improved hydrolysis of macromolecular substrates under a time-varying magnetic field. J Biotechnol 148(2–3):119–127
Zang L, Qiu J, Wu X et al (2014) Preparation of magnetic chitosan nanoparticles as support for cellulase immobilization. Ind Eng Chem Res 53(9):3448–3454
Ziegler-Borowska M, Chelminiak-Dudkiewicz D, Siódmiak T et al (2017) Chitosan–collagen coated magnetic nanoparticles for lipase immobilization—new type of "enzyme friendly" polymer shell crosslinking with squaric acid. Catalysts 7(1):26
Zou B, Hu Y, Cui F et al (2014) Effect of surface modification of low cost mesoporous SiO_2 carriers on the properties of immobilized lipase. J Colloid Interface Sci 417:210–216

Chapter 11
Nano-Based Drug Delivery Tools for Personalized Nanomedicine

Shama Parveen, Neera Yadav, and Monisha Banerjee

11.1 Introduction

Recent chemotherapeutics are not so effective; therefore, the development of new alternative strategies is necessary to fight lethal diseases such as cancer, HIV, hepatitis, etc. Nanotechnology provides the manipulation of compounds at nanorange and has the possibility to increase response, compatibility, and reduce the expense of diagnosis. The generative biomolecules such as DNA or proteins can be analyzed by red-shifted absorbance of gold nanoparticles or alteration in the conductance of nanowires and declination of a nano-cantilever. Different types of nanocarriers, for example, quantum dots and metal oxides have shown adequate benefits over conventional detections, tagging of intracellular, and image of the target site. Nanotechnology has also provided many routes which enable increased detection of target cells. Proteins and DNA are used over a point for care of tool. Chips (protein or gene) can be developed by using nanomaterials which could be attached to convenient nanofluidic tools for better diagnosis of diseases. Nanomaterials have design flexibility and small size and can be modified easily for compatibility to the target site. Interactions of nanosized materials with the biological system will benefit from knowing their effects and development of differently sophisticated nanodevices. Different nanocarriers are being used for delivery of compounds like inorganic nanocarriers (metal nanoparticles, graphene, mesoporous silica, etc.), lipid-based nanovesicles (liposome, nanoemulsion, solid lipid nanocarrier, etc.), polymer-based nanocapsules (dendrimer, ligand decorated nanoparticles, core-shell nanocapsule, etc.), and molecular complexes (protein nanocomplexes, cyclodextrin nanocomplexes, etc.) (Fig. 11.1). The main goal of nanobiotechnology is to use this technique

S. Parveen · N. Yadav · M. Banerjee (✉)
Molecular & Human Genetics Laboratory, Department of Zoology, University of Lucknow, Lucknow, India

I. Bhushan et al. (eds.), *Nanomaterials and Environmental Biotechnology*, Nanotechnology in the Life Sciences,
https://doi.org/10.1007/978-3-030-34544-0_11

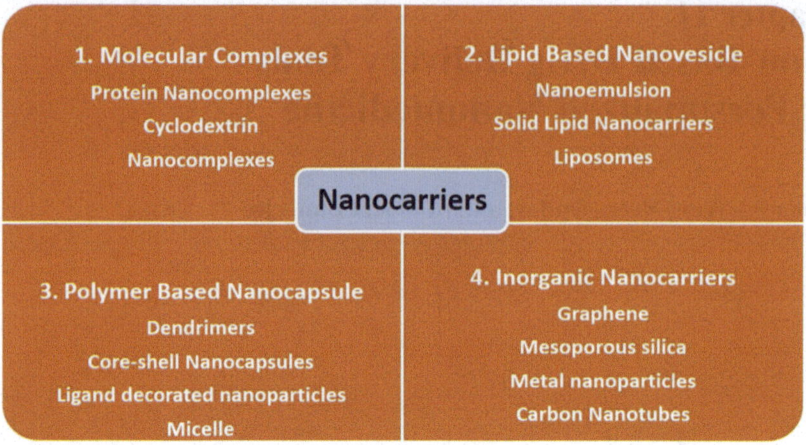

Fig. 11.1 Types of nanocarriers

to study how biological systems work and, in turn, can perk up existing nanotechnologies or create more beneficial nanomaterials.

11.2 Applications of Nanotechnology in Biological Sciences

11.2.1 Drug Delivery in Cancer

11.2.1.1 Gelatin Nanoparticle

A targeted drug-reaching system followed by reformed "decrease in size and hydrophobic/hydrophilic modification" was processed by the addition of gelatin nanoparticles (GNPs) with a drug Doxorubicin-Lactose (Dox-Lac). The effect of GNPs-Dox-Lac nanoformulation lasting in the circulatory system has a more effective accumulation toward the tumor region. Nanocarrier breakdown is stimulated by MMP2 (matrix metalloproteinase) present in tumor cells, and drugs are released to make an easy penetration in tumor tissues and cellular uptake. The pH change will result in dissociation of drug Dox-Lac in tumor cells, and free Dox (Mw 543 Da) will stimulate toxicity. GNPs-Dox-Lac promotes inhibition of tumor growth rate up to 90.4% by less in vivo toxicity, indicating that the activity was improved in the nanoform for hepatocellular carcinoma (HCC) (Liu et al. 2018).

11.2.1.2 PEGylated Liposomes

PEGylated liposome (PEG-LS) formulation mostly is used as a targeted drug delivery tool. The inflexibility and hydrophilicity of liposomal PEG corona membranes limited cellular uptake of the drug (Sadzuka et al. 2003). A new assembly for drug release was developed on lipid-based self-assembled nanoparticles that are competent for increased cellular uptake, penetration in tissues, and release of drug. Researchers characterized a better bicellar nanoformulation with the addition of a PEGylated phospholipid which presented a remarkable change of physicochemical property in morphology, i.e., the bicelles (disc-shaped) equally distributed into very small (~12 nm) circular micelles. The converted lipid nanoparticles were defined as hyper-cell-permeable micelles (HCPMi). The HCPMi confirmed not only extended stability in serum but also higher cellular and tumoral uptake than the usual PEGylated liposomal system (PEG-LS) (Saw et al. 2017). HCPMi exhibited fast cellular intake and release of cargo in the cytoplasm of cancerous cells. Cancerous cells treated from the drug (docetaxel) loaded with HCPMi compared to the DTX (docetaxel) loaded with PEG- LS showed about fivefold decrease in viability of cells, showing best effectiveness in HCPMi intake and discharge of drug. HCPMi entered deeply into the tumor cells and found more uptake than PEG-LS as shown by in vivo tumor imaging analysis. Studies showed that in vivo treatment with HCPMi (DTX) increased tumoral uptake and remarkable tumor growth inhibition, about 70% tumor weight reduction in an experimental model (Saw et al. 2017).

11.2.1.3 Nanovaccines

Nanoformulation is useful in the development of cancer vaccines. Earlier PLGA (poly lactic-co-glycolic acid) nanoformulations were modified with toll-like receptor 7 agonist, imiquimod. These, immune adjuvant nanoparticles (NP-R) was coated with membranes of cancer cell (NP-R@M). These formulations were further modified by mannose moiety (NP-R@M-M). The resulting nanovaccine exhibited increased cellular uptake through dendritic cells that are antigen-presenting cells, which could be further stimulated for enhanced antitumor immune response by delaying tumor development and thereby act as a preventive vaccine (Yang et al. 2018). Combination of vaccine (NP-R@M-M) by cellular checkpoint-regulator treatment provides us with more effective therapeutic impact to treat established tumors. It would be helpful to design cancer nanovaccines, and the principle may apply to a broad range of tumor types (Yang et al. 2018).

11.2.2 Phytochemical-Based Nanodrugs

11.2.2.1 Nanocurcumin

The anticancerous activity of curcumin has been reported in different studies including pancreatic cancer (Vallianou et al. 2015). Curcumin is of immense importance it can not be utilized completly as medicine due to certain drawbacks such as low bioavailabilty, solubility and stability. For enhancing its therapeutic effect, a polymeric form of nanoparticles shows an ideal delivery method. Moreover, for improving the potential of a drug, availability at target site, and circulation time in blood, modifications on the surface of nanocarriers by different macromolecules, such as chitosan and polyethylene glycol (PEG), hold tremendous potential. Curcumin loaded PLGA make it more effective by modifing the surface with chitosan and PEG can diminish the restrictions of native curcumin delivery for obtaining a maximum therapeutic response. In vitro studies of CNPs in pancreatic cancer showed higher cytotoxicity, increased anti-migratory, and ability to induce apoptosis in comparison to the native form of curcumin. Therefore, results of such studies can provide options for combating pancreatic cancer (Arya et al. 2018).

11.2.2.2 Nano-ginseng

Ginseng plant product having anticancerous activity was used in nanoformulation with ginsenoside Rb1/protopanaxadiol nanoparticles (Rb1/PPD NPs). The range of nanoparticles was about 110 nm with high drug loading capacity and stability in blood circulation, effective both in vitro and in vivo with better accumulation in tumor tissues and less toxic to normal tissue. The process of producing "nano-ginseng" formulation is easy, measurable, and economical (Dai et al. 2018).

11.2.2.3 Nano-quercetin

Quercetin is a potential chemotherapeutic drug. The drug was encapsulated in soluplus polymeric micelles by using the film dispersion method in order to improve its poor aqueous solubility and stability. In polymeric micelles (Qu-PMs) loaded with quercetin, more encapsulation efficiency and narrow size distribution led to complete drug dispersibility in water. Quercetin was in molecular form within the plasma membrane and was analyzed by X-ray diffraction (XRD) (Dian et al. 2014). Quercetin also formed intermolecular hydrogen bonding with carriers, analyzed by Fourier-transform infrared spectroscopy (FTIR). The stable formulation Qu-PMs exhibited significant sustained-release property. The pharmacokinetic study showed that absorption was also improved when compared with the free form of quercetin. Development of more appropriate formulations such as soluplus polymeric micelles were used for encapsulation of many drugs, i.e., poorly water soluble (Dian et al. 2014).

11.2.2.4 pH-Dependent Nanotools

siRNA-based nanotools in RISC (RNA-induced silencing complex) will undergo many changes in pH before showing their gene silencing mechanism (Nelson et al. 2014). They flood through the circulatory system, protrude into tumor tissue (pH 7.2–6.0), are introduced in the cell organelles such as endosomes with pH 5.0–6.5 and lysosomes (pH 4.0–5.0), and at last evade in the cytoplasmic environment of the cell. Many pH-based formulations were developed by decreasing the pH. Due to a decrease in pH, ionizable polymers like polyacids, cationic lipid, and poly amino acids can protonate to change solubility and electrostatic interaction or break the acid-cleavable bonds like hydrazine, ketal, ester, and nanocarriers for the release of cargo (Zhang et al. 2018).

11.3 Disease Diagnostics

11.3.1 Magnetic and Electrochemical-Based Nanoparticles

HIV-infected cells have been detected by magnetic nanoparticles (NPs) in vitro (Dodd et al. 2001). Magnetic NPs can be formulated according to target cells. Separation of treated cells from other cells is through a magnetic separation device. The formulation of anti-HIV drug-loaded magnetic nanoparticles is applied with changing magnetic field. Magnetic NPs such as magnetite (Fe_3O_4) have been studied due to their specific characteristics like stability, biocompatibility, and sensitivity toward the magnetic field. Magnetic-based carriers for targeted drug delivery are mostly used due to their less toxicity, biocompatibility, magnetic resonance imaging, and higher magnetic saturation (Mahajan et al. 2012). Magnetic properties of these NPs are dependent on their synthesis. Co-precipitation method was used to synthesize Fe_3O_4 NPs. Size of the particles can be controlled by coating of polyethylene glycol (PEG) at the surface. Evaluation of biocompatibility of these nanoparticles was done by cytotoxic assay and analysis of cell cycle in MCF-7 (human breast cancer cell line) (Kansara et al. 2018).

11.3.2 Gold Nanoparticles

Hepatitis is a viral disease which can cause chronic syndrome and many other diseases. Nanoformulation also performs a good role in diagnosing hepatitis caused by hepatitis B and C viruses. Previous recognition tools did not yield much efficiency and susceptibility. New detection systems were developed by researchers that are based on the electrochemical method, with a test of gold-enhanced nanoformulation conjugation of magnetic beads. This technique is very sensitive and specific for recognition of the DNA sequences of hepatitis B virus (HBV) (Hanaee et al. 2007). Electric potential of immunosensors increases with an increase in the amount of HBV (Tang et al. 2004). The study provides support that immunosensors can be used in the future as a detection system for hepatitis B virus. Gold nanoparticles are a preferred delivery system for immunosensors because of their compatibility with antibodies. Nanosized gold protein chips were created to detect and analyze antibodies of hepatitis B and C at the same time. Formulation of gold nanoparticles provides evenness and stability. Easier detection of antibodies is a result of stronger signals generated by gold nanoparticles (Duan et al. 2005). Significant amount of data shows that gold nanoparticles make the analysis and detection of hepatitis virus more successful.

Mycobacterium tuberculosis (MTB) bacteria cause the infectious disease tuberculosis that infects respiratory system (Shingadia and Novelli 2003) and considered as the second deadliest disease (WHO 2010). According to WHO report 2010, one third of the global population was infected by MTB, the strain which provides difficulty in detection and treatment. Gold nanoparticles detect MTB strands by DNA and RNA screening. MTB can be detected within few hours by nanoparticle probe (Veigas et al. 2010). Detection of MTB is based on pattern of color change due to aggregation of nanoparticles. The culture of MTB was important to analyze and detect bacterial strains. DNA fragments of MTB are broken down by electrochemical biosensors and are labeled by gold nanoparticles (Wang et al. 2013).

Nanotechnology is emerging as an advanced technique for the treatment of bone inflammation. Acton initiates as metallic nanoparticles offer a good surface for osteoblasts to connect to the bone. Metallic nanoparticles enable the growing of osteoblasts in a specific time interval and are helpful in successful osteo-regeneration. Nanocomposite materials induced specific protein for bone regrowth. Nanoformulated hydroxyapatite, the ceramic component of the bone, shows an increase in production of osteoblast (Tran and Webster 2009). Nanocomposites with carbon nanotubes also exhibit increasing osteoblast formation. Osteointegration process is accomplished in nanotubes by preventing the competition with other types of cells.

11.3.3 Nitric Oxide-Embedded Nanoparticles

Recently, the barrier of skin infection treatment has been overcome by nanotechnology. Nanoformulated therapeutic drugs are able to release the drug over a time period without any damage to the skin (Cevc and Vierl 2010). Nitric oxide-loaded nanoparticles are used in skin treatment, and the drug is released in a controlled time period. Different types of skin diseases were also treated by other formulations based on the same technique (Wang et al. 2008). Nitric oxide-loaded nanoparticles showed an improved response to treatment compared to the use of injectable needles for a skin infection (Englander and Friedman 2010).

11.3.4 Sunscreen

Ultraviolet (UV) radiation is a well-known cause of skin inflammation. The outer most layer, stratum corneum, provides protection to the body against harmful UV radiation. Sunscreens containing nanoparticles are used by humans to protect the skin from UV damage. Zinc oxide (ZnO)- and titanium dioxide (TiO2)-based nanoformulations are mostly used in the manufacturing of sunscreens. This nanoformulation reflects the harmful UV-A and UV-B radiations to the atmosphere in the form of heat. ZnO and TiO2 inhibit UV photons from reaching the skin cells through absorption and reflection mechanism (Popov et al. 2010).

11.3.5 Personalized Nanomedicine

Personalized medicine is a healthcare strategy that involves development of specific treatment at the individual level on the basis of their genetic makeup, phenotype, and environmental conditions (Ge et al. 2014). These days, use of nanomedicines is increasing exponentially and providing a platform for the treatment of every patient or group of patients (cohort) on the basis of a specific requirement in their genome (Zhang et al. 2012). Nanomedicines are used as therapeutic agents for personalized treatment, and their application is considered for a better treatment by pharma companies. Different types of nanomedicines exist, i.e., liposome, polymeric nanoparticles, magnetic-based nanoparticles, and micelles. All show advantages and improve the effect of conventional therapy (Fig. 11.2). These advantages are due to their small size, different nature, protected encapsulated material, controlled release

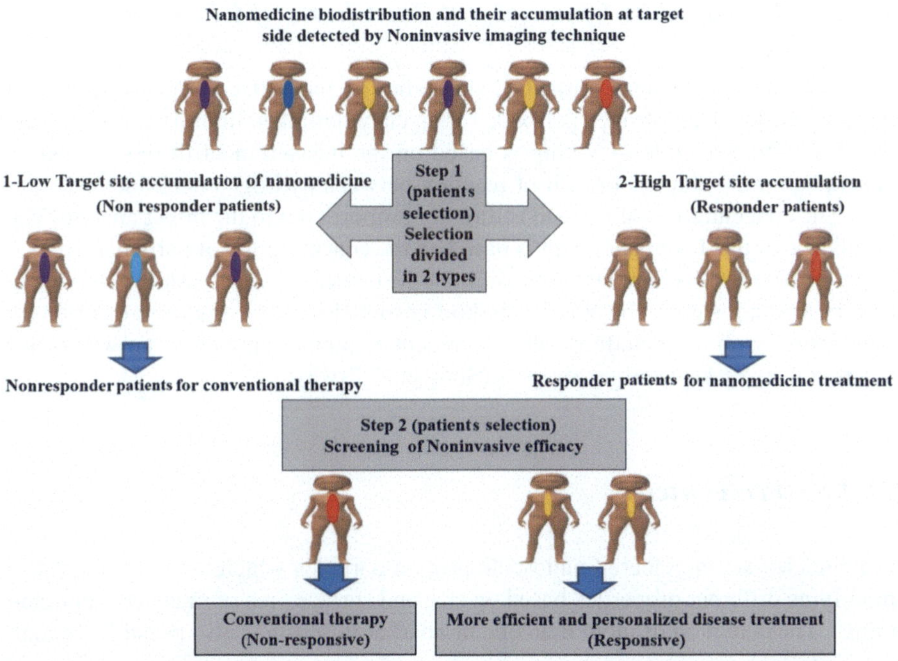

Fig. 11.2 Selection of patients selection after treatment with nanomedicines

for targeted delivery, increased therapeutic effect, and reduced side effects (Fornaguera and García-Celma 2017).

11.3.6 Personalized Nanodevices

Point of care (P-o-C) provides a beneficial platform for personalized medicine. It is useful in early diagnostic and more sensitive toward the detection of biomarkers in treatment monitoring. The infectious diseases can be controlled by a personalized diagnostic method by using P-o-C technologies in developing countries and environment where resource is limited. The liability of disease diagnostic protocols can be improved by using multiplexed assays by analyzing multiple biomarkers. Multiplexing of assays provides the multivariate analysis of huge number of the patient samples at one time (Romeo et al. 2016). In this way improved determination diagnosis of a particular biomarker of the disease can be achieved. The much more diagnostic power of multiplexed point-of-care biosensors is generated which are portable, cheap, and user-friendly devices. The first device of such kind, i.e., multiplexed volumetric bar-chart chip, was reported for analyzing protein biomarker of cancer (Song et al. 2012).

11.3.7 Microfluidic Channels on Bar Charts of Glass Chip

Channels contain covalently attached antibodies on the surface of glass walls with dye and pre-loaded hydrogen peroxide for an enzyme-linked immune sorbent assay (ELISA). The principle of V-chip is based on the measurement of oxygen that is generated on the chip as a result of reaction between hydrogen peroxide and catalase. The concentration of oxygen is directly proportional to the target analyte. The excellence of this inked bar result is based on the concentration of particular ELISA target in every well. Complete unification of the device provides results without any data processing steps. The V-chip method is rooted in the expression of different biomarkers such as including estrogen receptor, human epidermal growth factor receptor 2, and progesterone receptor (Song et al. 2012).

11.3.8 Proteinticles

Proteinticles are engineered nanoscale protein particles which are very helpful in modifying different properties based on size and surface area of many conventional things. The detection of more than one biomarker in serum is also possible through multiplexed viral detection (Chen et al. 2015).The method of disease detection is imperative in diseases like AIDS and hepatitis. This protocol is based on lateral flow assay (LFA) for protein nanoformulation. Proteinticles have better biodegradability and biocompatibility and are easily accessible to surface modifications (Lee et al. 2015). These nanoformulations are synthesized by using different proteins like gelatin, gliadin, elastin, zein, legumin, soy protein, albumin, and milk protein. Different methods used for the formulation include desolvation, emulsification, electrospray, and coacervation. Characterization parameters of these nanoforms comprise size of particle, morphology of particle, their surface charge, loading of drug, entrapment of drug, and structure of particle and drug release in vitro. Application of administration route of protein nanoparticles through different methods has been examined by renowned researchers (Verma et al. 2018).

11.3.9 Aptamers

The aptamer is an oligonucleotide-based nanoformulation. Advance characteristics of aptamer show high-binding affinity and specificity with target molecules in both intra- and extracellular environment. It works as an agonist or antagonist in a biological system. In recent years, several aptamers are used in the detection of disease, and curative purposes are under development to identify different molecules of HCC (hepatocellular carcinoma). The aptamer has been to increase the effect of conventional chemotherapies and reduces the growth of HCC cells in vitro. Aptamer induces antitumor activity and cell death in vivo. Overall data show that aptamer has reduced toxicity levels, and it may provide safer platform in the field of personalized medicine (Ladju et al. 2018).

11.4 Conclusion

Nanoscale materials allow nanodevices to enter new scientific and technological frontiers in disease diagnosis and treatment. Advance nanoscale characteristics of these devices especially their increased surface to volume ratio afford high-sensitivity and low-response times.

Acknowledgments SP and NY are thankful to Council of Scientific and Industrial Research (CSIR) and University Grants Commission (UGC), New Delhi, respectively, for providing financial assistance. The research grant provided to Molecular and Human Genetics Lab by Higher Education, Govt. of UP, Lucknow, under Centre of Excellence program is duly acknowledged.

References

Arya G, Das M, Sahoo SK (2018) Evaluation of curcumin loaded chitosan/PEG blended PLGA nanoparticles for effective treatment of pancreatic cancer. Biomed Pharmacother 102:555–566
Cevc G, Vierl U (2010) Nanotechnology and the transdermal route: a state of the art review and critical appraisal. J Control Release 141(3):277–299
Chen P, Chung MT, McHugh W, Nidetz R, Li Y, Fu J et al (2015) Multiplex serum cytokine immunoassay using nanoplasmonic biosensor microarrays. ACS Nano 9(4):4173–4181
Dai L, Zhu W, Si C, Lei J (2018) "Nano-ginseng" for enhanced cytotoxicity against cancer cells. Int J Mol Sci 19(2):627
Dian L, Yu E, Chen X, Wen X, Zhang Z, Qin L, Wu C (2014) Enhancing oral bioavailability of quercetin using novel soluplus polymeric micelles. Nanoscale Res Lett 9(1):684
Dodd CH, Hsu HC, Chu WJ, Yang P, Zhang HG, Mountz JD Jr et al (2001) Normal T-cell response and in vivo magnetic resonance imaging of T cells loaded with HIV transactivator-peptide-derived super paramagnetic nanoparticles. J Immunol Methods 256(1–2):89–105
Duan L, Wang Y, Li SSC, Wan Z, Zhai J (2005) Rapid and simultaneous detection of human hepatitis B virus and hepatitis C virus antibodies based on a protein chip assay using nano-gold immunological amplification and silver staining method. BMC Infect Dis 5(1):53
Englander L, Friedman A (2010) Nitric oxide nanoparticle technology: a novel antimicrobial agent in the context of current treatment of skin and soft tissue infection. J Clin Aesthet Dermatol 3(6):45
Fornaguera C, García-Celma MJ (2017) Personalized nanomedicine: a revolution at the nanoscale. J Pers Med 7(4):12
Ge Y, Li S, Wang S, Moore R (eds) (2014) Nanomedicine: Principles and Perspectives. Springer, New York
Hanaee H, Ghourchian H, Ziaee AA (2007) Nanoparticle-based electrochemical detection of hepatitis B virus using stripping chronopotentiometry. Anal Biochem 370(2):195–200
Kansara K, Patel P, Shukla RK, Pandya A, Shanker R, Kumar A, Dhawan A (2018) Synthesis of biocompatible iron oxide nanoparticles as a drug delivery vehicle. Int J Nanomedicine 13:79
Ladju RB, Pascut D, Massi MN, Tiribelli C, Sukowati CH (2018) Aptamer: a potential oligonucleotide nanomedicine in the diagnosis and treatment of hepatocellular carcinoma. Oncotarget 9(2):2951
Lee JH, Seo HS, Kwon JH, Kim HT, Kwon KC, Sim SJ et al (2015) Multiplex diagnosis of viral infectious diseases (AIDS, hepatitis C, and hepatitis A) based on point of care lateral flow assay using engineered proteinticles. Biosens Bioelectron 69:213–225
Liu Y, Li L, Li L, Zhou Z, Wang F, Xiong X, Huang Y (2018) Programmed drug delivery system based on optimized "size decrease and hydrophilicity/hydrophobicity transformation" for enhanced hepatocellular carcinoma therapy of doxorubicin. Nanomedicine 14(4):1111–1122
Mahajan SD, Aalinkeel R, Law WC, Reynolds JL, Nair BB, Sykes DE et al (2012) Anti-HIV-1 nanotherapeutics: promises and challenges for the future. Int J Nanomedicine 7:5301
Nelson CE, Kim AJ, Adolph EJ, Gupta MK, Yu F, Hocking KM et al (2014) Tunable delivery of siRNA from a biodegradable scaffold to promote angiogenesis in vivo. Adv Mater 26(4):607–614
Popov AP, Zvyagin AV, Lademann J, Roberts MS, Sanchez W, Priezzhev AV, Myllylä R (2010) Designing inorganic light-protective skin nanotechnology products. J Biomed Nanotechnol 6(5):432–451
Romeo A, Leung TS, Sanchez S (2016) Smart biosensors for multiplexed and fully integrated point-of-care diagnostics. Lab Chip 16(11):1957–1961
Sadzuka Y, Kishi K, Hirota S, Sonobe T (2003) Effect of polyethyleneglycol (PEG) chain on cell uptake of PEG-modified liposomes. J Liposome Res 13(2):157–172
Saw PE, Yu M, Choi M, Lee E, Jon S, Farokhzad OC (2017) Hyper-cell-permeable micelles as a drug delivery carrier for effective cancer therapy. Biomaterials 123:118–126

Shingadia D, Novelli V (2003) Diagnosis and treatment of tuberculosis in children. Lancet Infect Dis 3(10):624–632

Song Y, Zhang Y, Bernard PE, Reuben JM, Ueno NT, Arlinghaus RB et al (2012) Multiplexed volumetric bar-chart chip for point-of-care diagnostics. Nat Commun 3:1283

Tang DP, Yuan R, Chai YQ, Zhong X, Liu Y, Dai JY, Zhang LY (2004) Novel potentiometric immunosensor for hepatitis B surface antigen using a gold nanoparticle-based biomolecular immobilization method. Anal Biochem 333(2):345–350

Tran N, Webster TJ (2009) Nanotechnology for bone materials. Wiley Interdiscip Rev Nanomed Nanobiotechnol 1(3):336–351

Vallianou NG, Evangelopoulos A, Schizas N, Kazazis C (2015) Potential anticancer properties and mechanisms of action of curcumin. Anticancer Res 35(2):645–651

Veigas B, Machado D, Perdigao J, Portugal I, Couto I, Viveiros M, Baptista PV (2010) Au-nanoprobes for detection of SNPs associated with antibiotic resistance in *Mycobacterium tuberculosis*. Nanotechnology 21(41):415101

Verma D, Gulati N, Kaul S, Mukherjee S, Nagaich U (2018) Protein based nanostructures for drug delivery. J Pharm 2018:9285854. https://doi.org/10.1155/2018/9285854

Wang Q, Zhang N, Hu X, Yang J, Du Y (2008) Chitosan/polyethylene glycol blend fibers and their properties for drug controlled release. J Biomed Mater Res A 85(4):881–887

Wang S, Inci F, De Libero G, Singhal A, Demirci U (2013) Point-of-care assays for tuberculosis: role of nanotechnology/microfluidics. Biotechnol Adv 31(4):438–449

World Health Organization (2010) Global tuberculosis control: WHO report 2010. World Health Organization, Geneva

Yang R, Xu J, Xu L, Sun X, Chen Q, Zhao Y et al (2018) Cancer cell membrane-coated adjuvant nanoparticles with mannose modification for effective anticancer vaccination. ACS Nano 12(6):5121–5129

Zhang XQ, Xu X, Bertrand N, Pridgen E, Swami A, Farokhzad OC (2012) Interactions of nanomaterials and biological systems: implications to personalized nanomedicine. Adv Drug Deliv Rev 64(13):1363–1384

Zhang P, An K, Duan X, Xu H, Li F, Xu F (2018) Recent advances in siRNA delivery for cancer therapy using smart nanocarriers. Drug Discov Today 23(4):900–911

Chapter 12
Nanotechnology as Potential and Innovative Platform Toward Wastewater Treatment: An Overview

Akhilesh Kumar Singh and Paras Porwal

12.1 Introduction

Clean and safe water is a prerequisite for the survival of living creatures (flora and fauna) on earth. It is not only crucial for survival of living organisms but also drinking, domestic purposes (like cooking, washing and cleaning, etc.), and industrial as well as agricultural usages, e.g., irrigation for food production. In present scenario, several anthropogenic activities like revolutionary industrialization and urbanization, climate change, as well as the advanced agricultural practices have put massive strain on water resources including significant deterioration in water quality with increased water pollution and water crisis (Alcamo et al. 2017). Approximately, 844 million people lack basic drinking water facility, and 2 billion people drink fecal polluted water (WHO 2017). Also, it is anticipated that world's population will enhance up to approximately 2.9 billion by 2050 (Rockström 2003). Scarcity of freshwater is also because of the use of water sources toward domestic, industry, as well as irrigation purposes globally (Rockström 2003). Pollution of surface as well as groundwater resources is an additional reason toward decreased freshwater availability. Mostly, in developing countries, surface water is polluted by industrial and domestic wastewater containing microbial agents (pathogenic and nonpathogenic) with rich source of metals (toxic at above certain concentration) like plumbing (Zn, Pb, Cu, Al, etc.), cosmetics (Fe, Cu, Zn, etc.), pesticides (Hg, Zn, Cu, etc.), detergents (Fe, etc.), and pharmaceuticals (Ag, Fe, Al, etc.). Furthermore, As, Hg, Pb,

A. K. Singh
Department of Biotech and Genome, School of Life Sciences, Mahatma Gandhi Central University, Motihari, Bihar, India

P. Porwal (✉)
Amity Institute of Biotechnology, Amity University Uttar Pradesh, Lucknow Campus, Lucknow, India
e-mail: pporwal@lko.amity.edu

I. Bhushan et al. (eds.), *Nanomaterials and Environmental Biotechnology*, Nanotechnology in the Life Sciences,
https://doi.org/10.1007/978-3-030-34544-0_12

and Cd are toxic for all living beings, and they have no biological functions (Zinicovscaia et al. 2015). Pollutants in water not only cause waterborne diseases but also result in potable water crisis. It has been estimated that relatively 50% of the world's population will live in water-stressed areas by 2025 (WHO 2017). Thus, there is urgent need to develop economical as well as stable materials together with innovative approaches by water industries so as to yield the freshwater in abundant quantities. There are various physical and chemical strategies like chemical precipitation, oxidation or reduction, filtration, ion exchange, evaporation and electrochemical treatment, reverse osmosis, etc., with the help of which one can get rid of the contaminants like heavy metals present in the water. However, the majority of these approaches turn out to be unsuccessful with heavy metals concentrations <100 mg/l (Ahluwalia and Goyal 2007). Biological treatment systems like activated sludge as well as biological trickling filters are not capable to eliminate vast range of increasing contaminants. Moreover, the majority of these contaminants are found to be soluble in the effluent (Amin et al. 2017). Considering these, it is obvious that conventional water or wastewater treatment strategies are found to be inefficient toward yielding sufficient harmless water owing to enhancing need of water together with strict health guidelines as well as newly emerged pollutants/contaminants. Therefore, nanotechnology-based multifunctional as well as extremely competent approaches are found to come up with solutions of water or wastewater treatments, which require no vast infrastructures/centralized systems (Amin et al. 2017). Nanotechnology is found to be efficient toward solving aforementioned water issues concerning quality and quantity (Bottero et al. 2006). Among advanced water treatment systems, nanoparticles like carbon nanotubes (CNTs) as well as dendrimers are playing important role for the development of improved treatment strategies (Porwal and Sharma 2016; Obare and Meyer 2004).

Nanotechnology is a technology-based science in diverse directions and deals with the processing, deformation, amalgamation, and separation of materials by molecules or atoms (Ardelean 2015). Nanotechnology has induced and heralded a new industrial revolution mainly due to their biocompatibility (Sharma et al. 2016). In today's scenario, nanoscience due to the extraordinary properties of nanosized materials has become the thirst area for researchers. These nanosized materials are intervening macro and atomic or molecular level materials (Singh et al. 2018, 2019; Pandey et al. 2019; Rani et al. 2020). The unrestricted application of these nanoparticles in diverse field of science has boost their demand, and to meet the enhanced demand in a short time, researchers have utilized various physical and chemical methods to synthesize nanoparticles like arc-discharging liquid, lithography, laser ablation, gas phase deposition, ball milling (Blosi et al. 2011; Kim et al. 2006a, b), microemulsion, electrochemical deposition, coprecipitation sonochemical, and heat decomposition process (Chatterjee et al. 2012; Liu et al. 2012; Ramyadevi et al. 2012; Nakamura et al. 2007; Park et al. 2007). These methods produce metallic as well as other nanoparticles of diverse size and shape with numerous applications like agriculture, food processing, packaging, clinical diagnostics, wastewater treatment, and environmental remediation (Shan et al. 2009; Macaskie et al. 2010; Bradley et al. 2011; Zambrano-Zaragoza et al. 2011). But these environmentally unfriendly methods of nanoparticle synthesis give rise to environmental hazards and biohazard (Singh et al. 2016, 2017; Rawat et al. 2018; Du et al. 2018). Thus, there is an urgent need to develop alternate approaches/methods for synthesis of nanoparticles that are eco-friendly and cost-effective and pose negligible negative effect on human health and can be exploited at commercial scale. Therefore, green synthesis of nanoparticles is an attractive strategy in current scenario, which is gaining huge importance owing to its nontoxicity, exploitation of renewable sources of energy, consumption of lesser energy, and operability at low temperatures with environment-friendly procedures over traditional approaches (Sharma et al. 2016). Green synthesis of nanoparticles involves biological systems like plant or microorganisms (like bacteria and fungi), which are responsible for the transformation of metal salts into nanoparticles by playing the role of reducing, stabilizing, and capping agents. The biosynthesis of nanomaterial exemplifies a link between nanotechnology and biotechnology. Among various applications of nanoparticles, the treatment of wastewater/contaminated water using nanobiotechnology is gaining tremendous attention globally in current scenario. This is owing to their simplicity, economic feasibility, and disinfection efficacy.

12.2 Fabrication of Nanoparticles: Physical, Chemical, and Biogenic Approaches

Generally, the fabrication of nanoparticles can be categorized into two, i.e., "top-down" and "bottom-up" approaches. In top-down approach, any chemical or physical method is used to convert bulk material into nanosized particles, while bottom-up approach utilizes any biological or chemical method to transform into nanoparticles from atom/molecules through a series of step (Fig. 12.1). Nanotechnologist mostly prefers bottom-up approach as nanoparticle synthesized by this is more homogenous in structural and chemical composition and the chances of imperfections are also minimized. Although the physical, chemical, and mechanical methods are the predominant conventional methods for nanoparticle synthesis, currently the biological methods are becoming more popular because of their several advantages over conventional methods (Fig. 12.2). Normally, nanoparticles are of two types, i.e., organic (include carbon nanoparticles) and inorganic nanoparticles; they comprise of metal nanoparticles like Ag, Pt, Au, and Pd magnetic and semiconductor nanoparticles like TiO_2, SiO_2, and ZnO_2. Green synthesis of nanoparticles using different biomass like bacteria, algae, fungi, yeast, viruses, actinomycetes, plants and plant parts, pure or commercially available biomolecules, and biological waste materials is widely employed as a potential platform for fabrication of metallic nanoparticles (Mittal et al. 2013; Sharma et al. 2015). Mostly, a metal precursor is involved in the biosynthesis of nanoparticles, which is reduced or transformed either intracellularly or extracellularly using biological extract/biochemicals as a source of reductant present in the biomass (Asmathunisha and Kathiresan 2013). The extra-

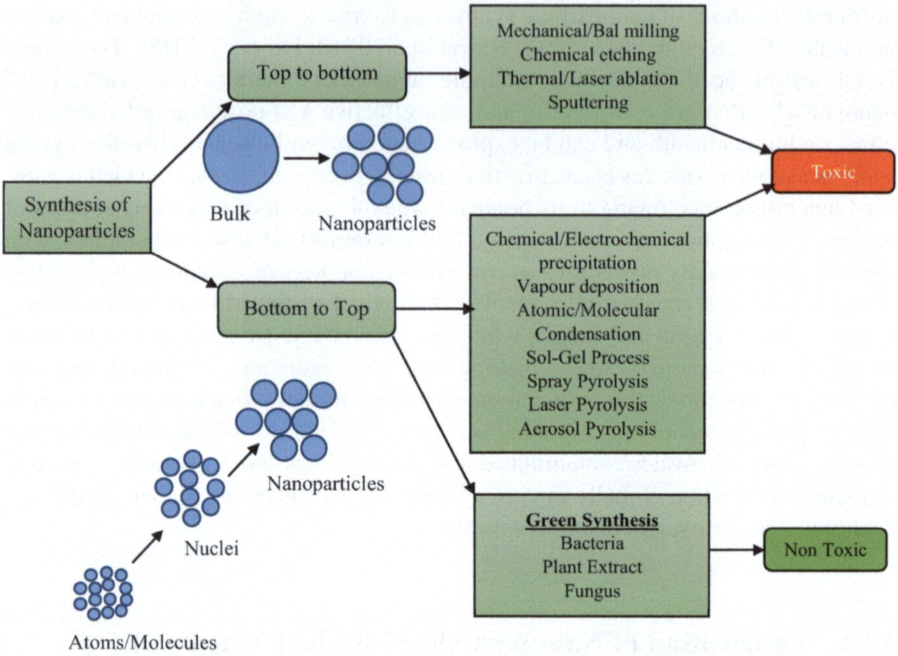

Fig. 12.1 Approaches for the synthesis of nanoparticles (Ahmed et al. 2016, 2017)

cellular approach exploits plant/food extracts (downstream processing) for the synthesis of metallic nanoparticles that produce highly energetic and more reactive spherical shape structure, while intracellular approach involves dissimilarities in reducing as well as stabilizing ability of phytochemicals present in different parts of the plants that are responsible to produce extremely disperse polymetallic nanoparticles of heterogeneous morphology. So, extracellular approach is desired over intracellular. Actual mechanism is yet to be established for the nanoparticle synthesis. However, scientists have proposed a hypothetical mechanism of nanoparticle synthesis (Fig. 12.3) based on the results obtained from specific biomass utilized (Velusamy et al. 2016). Furthermore, plant-mediated approach is favored over microbial-based nanoparticle synthesis because of single-step method for biosynthesis (takes less time), nontoxic, economical, and eco-friendly process (Kavitha et al. 2013; Kumar and Yadav 2009). Due to rich biodiversity, plants are considered as warehouse for the nanoparticles synthesis, and their potential is yet to be completely exploited (Table 12.1). At present, the scientists are working on individual plant extracts like polyphenols, flavonoids, terpenoids, glycosides, proteins, organic acids, etc., for the production of metallic nanoparticles to achieve more control over size along with their promising usages in various fields as well as morphology of nanoparticles (Seralathan et al. 2014; Tamuly et al. 2014). The waste products of food processing and agriculture-based industries are also used for fabrication of biogenic metallic nanoparticle synthesis like silver nanoparticles from aqueous corn

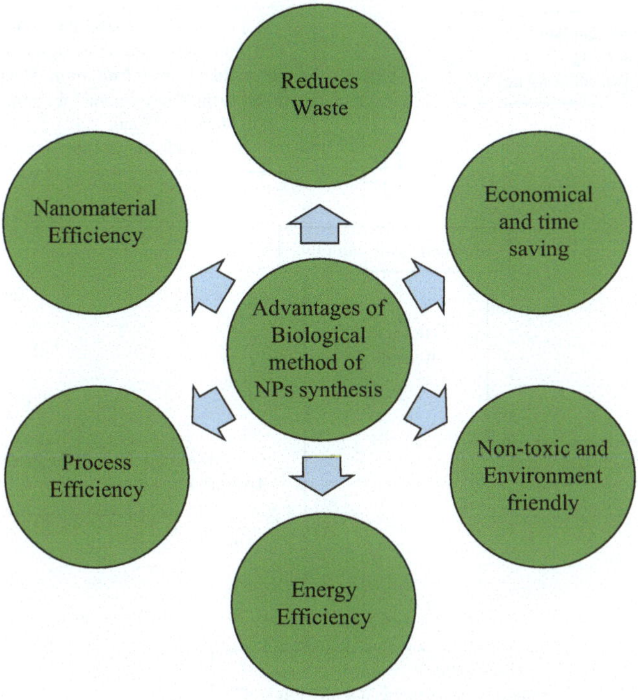

Fig. 12.2 Comparative account on the biological approaches of NP synthesis over chemical and physical approaches (Hasan et al. 2018)

husk extract and carboxymethyl sago pulp waste (Villanueva-Ibáñez et al. 2015; Deekonda et al. 2016); gold nanoparticles from palm oil mill effluent and sugar beet pulp (Gan et al. 2012; Castro et al. 2010); and platinum from waste peel extract of pomegranate (Dauthal and Mukhopadhyay 2015).

12.3 Characterization Techniques of Fabricated Nanoparticles

The characterization of synthesized nanoparticles is carried out so as to find out its various physicochemical properties like surface area, shape, size, electrical conductivity, quantum and charge parameters, hydrophobicity indices, crystalline structure, aqueous solubility, type of coating, UV/visible absorption, fluorescence, dermal penetration, and topological parameters (Xiao et al. 2017; Qu et al. 2017; Tuo et al. 2013). There are ample numbers of techniques available for the detection, measurement, and characterization of nanoparticles, but there is no single method which can be selected as ideal method for all; rather a method is to be chosen on the basis of sample type, information required, time taken, and economic feasibility. In

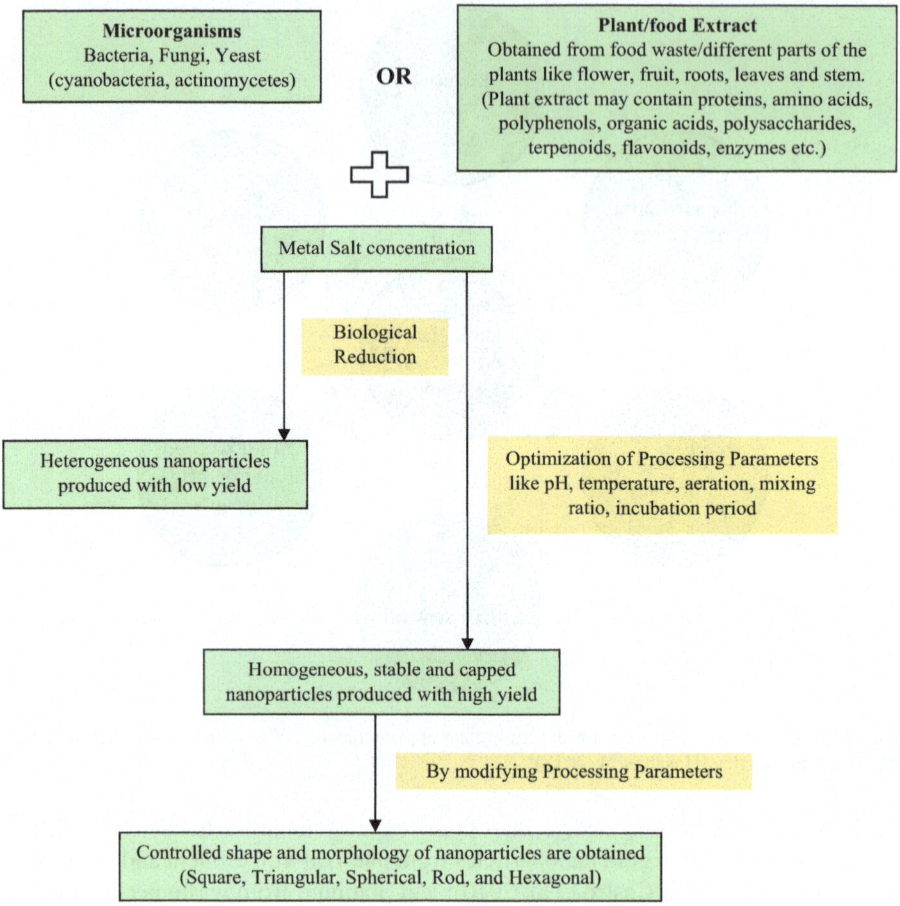

Fig. 12.3 Biogenic approach for the synthesis of nanoparticles. (Modified from Singh et al. 2016; Ghosh et al. 2017)

this context, Table 12.2 shows overview of frequently employed analytical and biochemical techniques used for the evaluation and characterization of physicochemical parameters for the fabricated nanoparticles with their strengths and limitations like transmission electron microscopy (TEM), scanning electron microscopy (SEM), high-resolution transmission electron microscopy (HRTEM), TEM with energy-dispersive X-ray spectroscopy (EDS), scanning tunneling microscopy (STM), atomic force microscopy (AFM), ultraviolet-visible (UV-Vis) spectroscopy, Fourier-transform infrared spectroscopy (FTIR), X-ray diffraction (XRD), dynamic or static light scattering (DLS and SLS), Brunauer-Emmett-Teller (BET) theory, quasi-elastic light scattering (QELS), nuclear magnetic resonance (NMR), Rutherford backscattering spectrometry, scanning mobility particle sizer (SMPS), photon correlation spectroscopy (PCS), nanoparticle tracking analysis (NTA), multiangle light scattering (MALS), and small-angle neutron scattering (SANS). Efforts are continuously

Table 12.1 An overview on the fabrication of plant-mediated nanoparticles

Plant system	Plant part used for NP synthesis	Type of NP synthesized	Size (nm)	Shape	References
Cucurbita pepo	Leaf extract	Silver	5–40	Roughly spherical	Krishnaraj et al. (2016)
Ceropegia thwaitesii	Leaf extract	Silver	100	Spherical	Muthukrishnan et al. (2015)
Butea monosperma	Bark extract	Silver	~35	Spherical	Pattanayak et al. (2017)
Aloe vera	Plant extract	Silver	35–55	Spherical and cubic	Dinesh et al. (2015)
Andrographis echioides	Leaf extract	Silver	68.06–91.28	–	Elangovan et al. (2015)
Azadirachta indica	Gum extract	Silver	12.09–29.65	Spherical	Velusamy et al. (2015)
Mukia maderaspatana	Leaf extract	Silver	13–34	Spherical	Chitra et al. (2015)
Phyllanthus niruri	Leaf extract	Silver	30–60	Spherical	Suresh et al. (2015)
Caulerpa scalpelliformis	Plant aqueous extract	Silver	20–35	Spherical and cubic	Murugan et al. (2015a)
Gmelina asiatica	Leaf extract	Silver	20–64	Spherical and triangle	Muthukumaran et al. (2015)
Adhatoda vasica	Leaf extract	Silver	10–50	Spherical	Latha et al. (2016)
Dillenia indica	Fruit	Silver	11–24	Spherical	Singh et al. (2013)
Andrographis paniculata	Leaves	Silver	67–88	Spherical	Suriyakalaa et al. (2013)
Withania somnifera	Leaves	Silver	5–40	Irregular, spherical	Nagati et al. (2012)
Citrullus colocynthis	Calli	Silver	5–70	Triangle	Satyavani et al. (2011)
Galaxaura elongata	Powder or extract	Gold	4–77	Spherical	Abdel-Raouf et al. (2017)
Carica papaya	Leaf extract	Gold	2–20	Spherical	Muthukumar et al. (2016)
Citrus maxima	Fruit extract	Gold	25 7 ± 10	–	Yu et al. (2016)
Couroupita guianensis	Flower extract	Gold	29.2–43.8	Cubic	Subramaniam et al. (2016)
Mimusops elengi	Bark extract	Gold	9–14	Spherical	Majumdar et al. (2015)
Ziziphus mauritiana	Leaf extract	Gold	20–40	Spherical	Sadeghi (2015)
Chloroxylon swietenia	Leaf extract	Gold	18–31	Spherical	Balasubramani et al. (2015)

(continued)

Table 12.1 (continued)

Plant system	Plant part used for NP synthesis	Type of NP synthesized	Size (nm)	Shape	References
Sterculia acuminata	Fruit extract	Gold	9.37–38.12	Spherical	Bogireddy et al. (2015)
Stevia rebaudiana	Leaf extract	Gold	17	Cubic	Sadeghi et al. (2015)
Cymbopogon citratus	Leaf extract	Gold	20–50	Triangular, spherical, and rod	Murugan et al. (2015b)
Cassia fistula	Stem	Gold	55–98	Spherical	Daisy and Saipriya (2012)
Trigonella foenum-graecum	Seed	Gold	15–25	Spherical	Aromal and Philip (2012)
Mirabilis jalapa	Flowers	Gold	100	Spherical	Vankar and Bajpai (2010)
Mentha piperita	Leaves	Gold	90–150	Spherical	MubarakAli et al. (2011)
Prunus × *yedoensis*	Gum extract	Platinum	10–50	Spherical and oval	Velmurugan et al. (2016)
Punica granatum	Peel extract	Platinum	16–23	Spherical	Dauthal and Mukhopadhyay (2015)
Barberry	Fruit extract	Palladium	18	Spherical	Nasrollahzadeh et al. (2016)
Euphorbia thymifolia L.	Leaf extract	Palladium	30–35	–	Nasrollahzadeh and Sajadi (2016)
Eclipta prostrata	Leaf extract	Palladium	27 ± 1 3	–	Rajakumar et al. (2015)
Catharanthus roseus	Leaf extract	Palladium	38	Spherical	Kalaiselvi et al. (2015)

carried out to standardize the way nanoparticles are measured for assessing occupational exposure, health, and environmental risks. As accuracy of the method varies, comparing the results of the same sample by different techniques can provide accuracy indication.

12.4 Nanoparticles: Potential Platform for the Removal of Water Contaminants

Polluted water/wastewater can be used for synthesis of biogenic nanoparticles (Ahluwalia et al. 2016), producing energy as well as recovering nutrients for agricultural uses (Mal et al. 2016). Carbon nanotubes (CNTs) are extensively used for treating water and wastewater as they have antimicrobial property with no side

Table 12.2 Various approaches employed toward the evaluation of physicochemical properties of nanomaterial (Rasmussen et al. 2017; Lin et al. 2014)

Characterization technique	Physicochemical parameter	Type of material	Strengths	Limitations
Transmission electron microscopy (TEM)	Size and size distribution, shape heterogeneity, aggregation, dispersion	Inorganic, organic, carbon-based, biological, complex materials (coated, core-shell). Spherical, equiaxial particles, tubes, flakes, rods, fiber	Direct measurement of the parameters with higher spatial resolution than SEM. Other analytical methods can be united with this to investigate chemical composition and electronic structure of nanomaterial	Costly equipment, samples required as ultrathin samples in nonphysiological condition
Scanning electron microscopy (SEM)	Size and size distribution, shape, aggregation	Organic, inorganic, carbon-based, biological, complex materials (coated, core-shell). Spherical, equiaxial particles, tubes, flakes, rods, fibers	The size/size distribution and shape of nanomaterial is directly measured	Dry samples, conducting sample, or coating conductive are materials required
Atomic force microscopy (AFM)	Size and size distribution, shape, structure, surface properties (modified AFM), surface charge density (mapping)	Organic, inorganic, carbon-based, biological, complex materials (coated, core-shell). Spherical, equiaxial particles, tubes, flakes, rods, fibers	3D sample surface topography. Sub-nanoscale topographic resolution. Direct measurement of samples in dry, aqueous, or ambient environment	Time taking and tough sample preparation. Tip effects often disturb measurements. Analysis in general limited to the external of nanomaterials
X-ray diffraction (XRD)	Size, shape, and crystal structure for crystalline materials	Inorganic, carbon-based, complex materials (coated, core-shell not organic)	Well-acknowledged technique. High spatial resolution at atomic scale	Confined to crystalline materials, low intensity compared to electron diffraction
Electrophoretic mobility	Zeta potential	Organic, inorganic, carbon-based, biological, complex materials (coated, core-shell). Spherical, equiaxial particles, tubes, flakes, rods, fiber	Simultaneous measurement of many particles (using ELS)	Very diluted sample required, not very specific method

(continued)

Table 12.2 (continued)

Characterization technique	Physicochemical parameter	Type of material	Strengths	Limitations
Infrared/ Fourier-transform infrared spectroscopy (IR/ FTIR)	Structure molecular composition surface properties (FTIR)	Organic, inorganic, carbon-based, biological, complex materials (coated, core-shell). Spherical, equiaxial particles, tubes, flakes, rods, fiber	Minimal or no sample preparation requirement. Fast and inexpensive measurement	Commonly used for dry powder samples, and molecule must be active in IR region. Interference and strong absorbance of H_2O (IR). Relatively low sensitivity in nanoscale analysis. Detection limits routine: 2% Special techniques: 0.01%
Ultraviolet-visible spectroscopy (UV-vis)	Surface chemistry composition, optical properties, particle size (for certain materials)	Organic, inorganic, carbon-based, biological, complex materials (coated, core-shell). Only for certain elements	Easy and low-cost technique. Solid or liquid, both types of samples can be evaluated	Material-dependent technique. Plasmonic samples can only be determined for size. Samples in powders require reflectance accessory. Only for diluted samples (not in powders), e.g., thin films, dispersions, etc.
Nuclear magnetic resonance (NMR)	Structure composition, size (indirect analysis), purity, conformational change	Organic, inorganic, carbon-based, biological, complex materials (coated, core-shell). Only for certain elements	Sample preparation is easy and nondestructive/ noninvasive method	Huge amount of sample is required, more time-consuming and less sensitivity
X-ray photoelectron spectroscopy (XPS)	Surface chemistry composition	Organic, inorganic, carbon-based, complex materials (coated, core-shell). Spherical, equiaxial particles, tubes, flakes, rods, fiber	Very sensitive for surface chemistry	Only detects the elements in a very thin layer close to surface (about 10 nm), needs careful sample preparation, very sensitive, ultrahigh vacuum

(continued)

Table 12.2 (continued)

Characterization technique	Physicochemical parameter	Type of material	Strengths	Limitations
Energy dispersive X-ray analysis (EDX)	Composition, surface chemistry	Organic, inorganic, carbon-based, complex materials (coated, core-shell). Spherical, equiaxial particles, tubes, flakes, rods, fiber	Gives information on elements from beryllium to einsteinium. Detection limit 0.1 at %	Not a real surface science technique so used in combination with TEM or SEM. Not suitable for nonconductive samples, low atomic number elements are difficult to detect (they overlap with X-ray peaks)
Raman spectroscopy	Composition, surface chemistry, particle size (for certain materials)	Organic, inorganic, carbon-based, biological, complex materials (coated, core-shell). Spherical, equiaxial particles, tubes, flakes, rods, fibers	Frequently available, easy and low-cost technique	Metals and alloys cannot be analyzed. Size for certain materials can only be determined like CNT, thickness of graphene flakes. Very sensitive to impurities. Samples heating through intense laser radiation can lead to sample decomposition
Single particle inductively coupled plasma mass spectrometry (SP-ICP-MS)	Particle size distribution, surface chemistry	Inorganic, complex materials (coated, core-shell if contains detectable element). Spherical, equiaxial particles	Measures individual particles. Cost-efficient, quick, and chemically precise method	Accurate size determination restricted to spherical particles only. Can detect only one element at a time
Scanning mobility particle sizer (SMPS)	Particle size, particle size distribution	Organic, inorganic, carbon-based, complex materials (coated, core-shell). Spherical, equiaxial particles	On-site technique frequently used for aerosols	Only aerosol samples in size range 2.5–1000 nm with resolution 2 nm particles need to be charged
Scanning tunneling microscopy (STM)	Size and size distribution, shape, aggregation, structure surface charge density (mapping)	Organic, inorganic, carbon-based, biological, complex materials (coated, core-shell). Spherical, equiaxial particles, tubes, flakes, rods, fibers	Direct measurement. High spatial resolution down to atomic scale	For conductive samples only

effects (Gehrke et al. 2015). The antimicrobial property of nanoparticles for wastewater treatment is mainly used for disinfection, control over membrane biofouling, and biofilm control on other relevant surfaces. Nanomaterials as an adsorbent are used in several forms such as absorptive, catalytic, catalytic membrane, bioactive nanoparticles, polymeric as well as nanocomposite membrane, etc., for treatment of wastewater. Nanometal oxides such as iron oxides like ferrous oxide, TiO_2, and Al_2O_3 are used as adsorbents for heavy metals and radio nuclides (Mayo et al. 2007). Owing to the low toxicity and chemical stability, TiO_2 nanoparticles are extensively employed as decontaminating and disinfecting agents, which remain stable throughout the process of decontamination (Savage and Diallo 2005). Sharma et al. (2018) reported that nanocomposite (TiO_2–Fe_2O_3) can be used as an active microbial disinfectant that can eradicate *E. coli* through photocatalytic activity. Silver nanocatalyst N-doped TiO_2 was used for the degradation of microbial contaminants (Prachi et al. 2013). Nano iron hydroxide and zinc oxide adsorbent permit adsorption of arsenic from drinking water and wastewater (Aredes et al. 2012). Nanotechnology also contributes in the field of agriculture by protecting the plants and enhancing their growth, e.g., colloidal solution of ZnO nanoparticles were used as a fertilizer (Sabir et al. 2014), silica nanoparticles (plant nanobiotechnology) as a bio-insecticides, vector for gene delivery, pesticide delivery (mesoporous silica), protects quenching of active reagent (Baeza et al. 2015). Nowadays, nanocatalysts are extensively being used for treating wastewater due to their high catalytic and high surface to volume ratio. Catalytic nanoparticles which find their application in removal of environmental contaminants are semiconductor materials, metallic/bimetallic catalyst, zero-valance metal (like zero-valent iron), etc. (Chokshi and Bora 2014). Silver nanoparticles embedded with zeolites are used for sanitary purposes by inhibiting the growth of microorganisms in water (Dutta 2011). Recently, a number of waste-derived as well as tailored waste-derived nanomaterials are being utilized in different fields such as silica nanopowder from rice husk, sugarcane waste, bamboo culm for energy storage (Praneetha and Murugan 2015), waste plastic-derived nanoparticles for bioimaging (Hu et al. 2014), elimination of heavy metals, organic pollutants, and disintegration of toxic organic materials from water/wastewater, e.g., flue dust waste-derived catalyst has the potential to degrade 69% of the dye content from wastewater mainly due to their catalytic properties (Amorim et al. 2013; Buruga et al. 2018; Sarkar et al. 2018).

Silver (Ag) nanoparticles were used in water filtration membranes (e.g., polysulfone) owing to their low toxicity, biofouling reduction, and strong antimicrobial activity against gram-positive and gram-negative bacteria and viruses like *Pseudomonas aeruginosa*, *E. coli*, *Staphylococcus aureus*, *Klebsiella*, etc. The antimicrobial activity of Ag nanoparticles is due to the formation of free radicals that were responsible for damaging the bacterial membranes and alter the membrane properties by adhering to cell surfaces, DNA interactions, and enzyme destruction (Jain and Pradeep 2005; Panahi et al. 2018). The noble metals, magnetic, and dye-doped nanomaterials are constantly used to check the wastewater quality by detecting pathogens at a very low concentration (Chokshi and Bora 2014). The extracellular water-soluble polysaccharides of cyanobacterial strain *Nostoc commune* were exploited toward the biogenic fabrication of silver nanomaterials after interaction with $AgNO_3$ (spherical shape, size 15–45 nm). The synthesized nanoparticles depicted antimicrobial activity toward *E. coli* as well as outstanding antiseptic properties against pathogenic seed and grain-borne fungi (Morsy et al. 2014). Qing et al. (2017) reported that silver nanoparticles synthesized by using waste tea extract (as a reducing and stabilizing agent) had shown the catalytic degradation of methylene blue and ethyl violet solution in presence as well as absence of H_2O_2, restrain impact on gram-positive and gram-negative bacteria, and have the potential to be used for industrial wastewater treatment.

Copper (Cu) and copper oxide (CuO) nanoparticles can be synthesized biologically through plant extracts like magnolia leaf extract, clove extract, stem latex of *Euphorbia nivulia*, *Sterculia urens* extract (Karaya gum), and *P. stutzeri* bacteria from wastewater via electroplating route (Varshney et al. 2010) and *Morganella* bacteria (Din and Rehan 2017). Several research studies have revealed that these copper and copper oxide nanoparticles have antioxidant and antibacterial property toward *E. coli* as well as *S. aureus* (Subhankari and Nayak 2013; Lee et al. 2011). Biogenic Ag/Cu (multimetallic and combine) nanocatalysts synthesized from the leaf extract of *Carica papaya* with average size of 150 nm have catalytic activity for chlorpyrifos disintegration in water with 3,5,6-trichloro-2-pyridinol and diethylthiophosphate (Rosbero and Camacho 2017).

Palladium-coated gold nanoparticles are highly efficient catalyst for the removal of trichloroethane from groundwater than palladium alone. Palladium nanoparticles were used for the reduction of Cr(VI) to Cr(III) (Özlem et al. 2013). Nanoparticles were significantly used as key ingredients for makeup products like sunscreens, fungicides, rubber processing, and wastewater treatment (Gunalan et al. 2012). Palladium-incorporated ZnO nanocatalyst showed high photocatalytic activity for the removal of *E. coli* from wastewater (Prachi et al. 2013).

12.5 Limitations of Nanoparticle-Based Wastewater Treatment

In the era of commercialization, the success or failure of any nanoengineered material exploited toward the treatment of water/wastewater technology purely depends on their impact on the aquatic flora and fauna. Presently limited evidences are accessible toward the behavior, transport, as well as chemical alteration of nanoparticles in the surrounding atmosphere as well as the mechanism of their effect on biological systems. So, nanoparticles used for water treatment might interact with human health and environment at any stage of product life cycle and may have involuntary effects. The results yielded are contradictory from several studies together with the life cycle analysis, technology assessment, toxicity test, pathways, and dispersal of nanoparticles in aquatic bodies to evaluate health risks of particular nanomaterial. For instance, silver nanoparticles that are used to decrease bad smell from socks are being released in water during washing and may destroy beneficial effect of bacteria responsible for the breakdown of organic matter present in wastewater (Chokshi and Bora 2014). Since no standard laws and regulations are set for the general experimental tests, it slows down the necessary decision processes. Different procedures of nanotechnology have shown a great potential on the lab scale, but their promptness for commercialization varies extensively owing to technical obstacles, cost-efficacy, potential environmental, and human risks.

12.6 Conclusion

Harmless/safe water has turn out to be a competitive source worldwide because of growing population, severe droughts, climate change, etc. Nanoparticles depict exceptional features like large surface areas, size, shape, as well as dimensions, which enable them as eye-catching platform toward water/wastewater treatment including disinfection, adsorption, and membrane separations. Overall, water/wastewater treatment exploiting nanomaterials is a potential area toward present as well as upcoming investigations. Furthermore, some safety measures need to be taken so as to circumvent any risk to human health and environment owing to exploitation of these various nanoparticles for different applications including water/wastewater treatment. However, such issue of nanoparticles can be overcome as a result of association among research institutions, industry, as well as government.

References

Abdel-Raouf N, Al-Enazi NM, Ibraheem IBM (2017) Green biosynthesis of gold nanoparticles using *Galaxaura elongata* and characterization of their antibacterial activity. Arab J Chem 10:S3029–S3039

Ahluwalia SS, Goyal D (2007) Microbial and plant derived biomass for removal of heavy metals from wastewater. Bioresour Technol 98:2243–2257

Ahluwalia S, Prakash NT, Prakash R, Pal B (2016) Improved degradation of methyl orange dye using bio-co-catalyst Se nanoparticles impregnated ZnS photocatalyst under UV irradiation. Chem Eng J 306:1041–1048

Ahmed S, Ahmad M, Swami BL (2016) A review on plants extract mediated synthesis of silver nanoparticles for antimicrobial applications: a green expertise. J Adv Res 7:17–28

Ahmed S, Chaudhry SA, Ikram S (2017) A review on biogenic synthesis of ZnO nanoparticles using plant extracts and microbes: a prospect towards green chemistry. J Photochem Photobiol B Biol 166:272–284

Alcamo J, Henrichs T, Rösch T (2017) World water in 2025: global modeling and scenario analysis for the world commission on water for the 21st century. Report A0002. Centre for Environmental Systems Research, University of Kassel, Kurt Wolters Strasse 3, 34109 Kassel, Germany

Amin MT, Alazba AA, Manzoor U (2017) A review of removal of pollutants from water/wastewater using different types of nanomaterials. Adv Mater Sci Eng 2014:825910

Amorim CC, Leao MMD, Moreira RFPM, Fabris JD, Henriques AB (2013) Performance of blast furnace waste for azo dye degradation through photo-Fenton-like processes. Chem Eng J 224:59–66

Ardelean II (2015) Metallic nanoparticle synthesis by cyanobacteria: fundamentals and applications. In: The algae world. Springer, Dordrecht, pp 429–448

Aredes S, Klein B, Pawlik M (2012) The removal of arsenic from water using natural iron oxide minerals. J Clean Prod 29-30:208–213

Aromal SA, Philip D (2012) Green synthesis of gold nanoparticles using *Trigonella foenum-graecum* and its size dependent catalytic activity. Spectrochim Acta A 97:1–5

Asmathunisha N, Kathiresan K (2013) A review on biosynthesis of nanoparticles by marine organisms. Colloids Surf B Biointerfaces 103:283–287

Baeza A, Colilla M, Vallet-Regí M (2015) Advances in mesoporous silica nanoparticles for targeted stimuli-responsive drug delivery. Expert Opin Drug Deliv 12(2):319–337

Balasubramani G, Ramkumar R, Krishnaveni N, Sowmiya R, Deepak P, Arul D, Perumal P (2015) GC-MS analysis of bioactive components and synthesis of gold nanoparticle using Chloroxylon swietenia DC leaf extract and its larvicidal activity. J Photochem Photobiol B 148:1–8

Blosi M, Albonetti S, Dondi M, Martelli C, Baldi G (2011) Microwave-assisted polyol synthesis of Cu nanoparticles. J Nanopart Res 13:127–138

Bogireddy NKR, Anand KKH, Mandal BK (2015) Gold nanoparticles-synthesis by *Sterculia acuminata* extract and its catalytic efficiency in alleviating different organic dyes. J Mol Liq 211:868–875

Bottero JY, Rose J, Wiesner MR (2006) Nanotechnologies: tools for sustainability in a new wave of water treatment processes. Integr Environ Assess Manag 2(4):391–395

Bradley EL, Castle L, Chaudhry Q (2011) Applications of nanomaterials in food packaging with a consideration of opportunities for developing countries. Trends Food Sci Technol 22:604–610

Buruga K, Kalathi JT, Kim KH, Ok YS, Danil B (2018) Polystyrene-halloysite nano tube membranes for water purification. J Ind Eng Chem 61:169–180

Castro L, Blázquez ML, González F, Muñoz JA, Ballester A (2010) Extracellular biosynthesis of gold nanoparticles using sugar beet pulp. Chem Eng J 164:92–97. https://doi.org/10.1016/j.cej.2010.08.034

Chatterjee AK, Sarkar RK, Chattopadhyay AP, Aich P, Chakraborty R, Basu T (2012) A simple robust method for synthesis of metallic copper nanoparticles of high antibacterial potency against *E. coli*. Nanotechnology 23:100–103

Chitra G, Balasubramani G, Ramkumar R, Sowmiya R, Perumal P (2015) Mukia maderaspatana (Cucurbitaceae) extract-mediated synthesis of silver nanoparticles to control Culex quinquefasciatus and Aedes aegypti (Diptera: Culicidae). Parasitol Res 114(4):1407–1415

Chokshi N, Bora L (2014) An overview of nanotechnology in waste water treatment. http://www.researchgate.net/publication/264004597_An_Overview_of_Nanotechnology_in_waste_water_treatment

Daisy P, Saipriya K (2012) Biochemical analysis of *Cassia fistula* aqueous extract and phytochemically synthesized gold nanoparticles as hypoglycemic treatment for diabetes mellitus. Int J Nanomedicine 7:1189–1202

Dauthal P, Mukhopadhyay M (2015) Biofabrication, characterization and possible bio-reduction mechanism of platinum nanoparticles mediated by agro-industrial waste and their catalytic activity. J Ind Eng Chem 22:185–191

Deekonda K, Muniyandy S, Lim YY, Janarthanan P (2016) Electron beam radiation mediated green synthesis of silver nanoparticles using carboxymethyl sago pulp obtained from sago waste. Polymer (United Kingdom) 86:147–156. https://doi.org/10.1016/j.polymer.2016.01.048

Din MI, Rehan R (2017) Synthesis, characterization, and applications of copper nanoparticles. Anal Lett 50(1):50–62

Dinesh D, Murugan K, Madhiyazhagan P, Panneerselvam C, Kumar PM, Nicoletti M, Jiang W, Benelli G, Chandramohan B, Suresh U (2015) Mosquitocidal and antibacterial activity of green-synthesized silver nanoparticles from Aloe vera extracts: towards an effective tool against the malaria vector Anopheles stephensi? Parasitol Res 114(4):1519–1529

Du W, Tan W, Yin Y, Ji R, Peralta-Videa JR, Guo H, Gardea-Torresdey JL (2018) Differential effects of copper nanoparticles/microparticles in agronomic and physiological parameters of oregano (*Origanum vulgare*). Sci Total Environ 618:306–312

Dutta P (2011) Silver nanoparticles embedded in zeolite membranes: release of silver ions and mechanism of antibacterial action. Int J Nanomedicine 15:1833

Elangovan K, Elumalai D, Anupriya S, Shenbhagaraman R, Kaleena PK, Murugesan K (2015) Phyto mediated biogenic synthesis of silver nanoparticles using leaf extract of *Andrographis echioides* and its bio-efficacy on anticancer and antibacterial activities. J Photochem Photobiol B 151:118–124

Gan PP, Ng SH, Huang Y, Li SFY (2012) Green synthesis of gold nanoparticles using palm oil mill effluent (POME): a low-cost and eco-friendly viable approach. Bioresour Technol 113:132–135. https://doi.org/10.1016/j.biortech.2012.01.015

Gehrke I, Geiser A, Somborn-Schulz A (2015) Innovations in nanotechnology for water treatment. Nanotechnol Sci Appl 8:1–17

Ghosh PR, Fawcett D, Sharma SB, Poinern GEJ (2017) Production of high-value nanoparticles via biogenic processes using aquacultural and horticultural food waste. Materials 10(8):852

Gunalan S, Sivaraj R, Rajendran V (2012) Green synthesized ZnO nanoparticles against bacterial and fungal pathogens. Prog Nat Sci Mater Int 22:693–700

Hasan M, Ullah I, Zulfiqar H, Naeem K, Iqbal A, Gul H, Ashfaq M, Mahmood N (2018) Biological entities as chemical reactors for synthesis of nanomaterials: Progress, challenges and future perspective. Mater Today Chem 8:13–28

Hu Y, Yang J, Tian J, Jia L, Yu JS (2014) Green and size-controllable synthesis of photoluminescent carbon nanoparticles from waste plastic bags. RSC Adv 4(88):47169–47176

Jain P, Pradeep T (2005) Potential of silver nanoparticle-coated polyurethane foam as an antibacterial water filter. Biotechnol Bioeng 90:59–63

Kalaiselvi A, Roopan SM, Madhumitha G, Ramalingam C, Elango G (2015) Synthesis and characterization of palladium nanoparticles using Catharanthus roseus leaf extract and its application in the photo-catalytic degradation. Spectrochim Acta Part A 135:116–119. https://doi.org/10.1016/j.saa.2014.07.010

Kavitha KS, Syed B, Rakshith D, Kavitha HU, Rao HCY, Harini BP, Satish S (2013) Plants as green source towards synthesis of nanoparticles. Int Res J Bio Sci 2(6):66–76

Kim YH, Lee DK, Cha HG, Kim CW, Kang YC, Kang YS (2006a) Preparation and characterization of the antibacterial Cu nanoparticle formed on the surface of SiO_2 nanoparticles. J Phys Chem B 110:24923–24928

Kim YH, Lee DK, Jo BG, Jeong JH, Kang YS (2006b) Synthesis of oleate capped Cu nanoparticles by thermal decomposition. Colloid Surf A Physicochem Eng Asp 445:231–238

Krishnaraj C, Ji BJ, Harper SL, Yun SI (2016) Plant extract-mediated biogenic synthesis of silver, manganese dioxide, silver-doped manganese dioxide nanoparticles and their antibacterial activity against food- and water-borne pathogens. Bioprocess Biosyst Eng 39:759–772

Kumar V, Yadav SK (2009) Plant-mediated synthesis of silver and gold nanoparticles and their applications. J Chem Technol Biotechnol 84(2009):151–157

Latha M, Priyanka M, Rajasekar P, Manikandan R, Prabhu NM (2016) Biocompatibility and antibacterial activity of the Adathoda vasica Linn extract mediated silver nanoparticles. Microb Pathog 93:88–94

Lee HJ, Lee G, Jang NR, Yun JH, Song JY, Kim BS (2011) Biological synthesis of copper nanoparticles using plants extract. Nanotechnology 1:371–374

Lin PC, Lin S, Wang PC, Sridhar R (2014) Techniques for physicochemical characterization of nanomaterials. Biotechnol Adv 32(4):711–726

Liu Q, Zhou D, Yamamoto Y, Ichino I, Okido M (2012) Preparation of Cu nanoparticles with $NaBH_4$ by aqueous reduction method. Trans Nonferrous Metals Soc China 22:117–123

Macaskie LE, Mikheenko IP, Yong P, Deplanche K, Murray AJ, Paterson-Beedle M, Coker VS, Pearce CI, Cutting R, Pattrick RAD, Vaughan D, van der Laan G, Lloyd JR (2010) Today's wastes, tomorrow's materials for environmental protection. Hydrometallurgy 104(3):483–487

Majumdar R, Bag BG, Ghosh P (2015) *Mimusops elengi* bark extract mediated green synthesis of gold nanoparticles and study of its catalytic activity. Appl Nanosci 6:521–528

Mal J, Nancharaiah YV, Van HED, Lens PNL (2016) Metal chalcogenide quantum dots:biotechnological synthesis and applications. RSC Adv 6(47):41477–41495

Mayo JT, Yavuz C, Yean S (2007) The effect of nano crystalline magnetite size on arsenic removal. Sci Technol Adv Mater 8:1–2

Mittal AK, Chisti Y, Banerjee UC (2013) Synthesis of metallic nanoparticles using plant extracts. Biotechnol Adv 31:346–356

Morsy FM, Nafady NA, Abd-Alla MH, Elhady DA (2014) Green synthesis of silver nanoparticles by water soluble fraction of the extracellular polysaccharides/matrix of the cyanobacterium Nostoc commune and its application as a potent fungal surface sterilizing agent of seed crops. Univers J Microbiol Res 2:36–43

MubarakAli D, Thajuddin N, Jeganathan K, Gunasekaran M (2011) Plant extract mediated synthesis of silver and gold nanoparticles and its antibacterial activity against clinically isolated pathogens. Colloid Surf B 85:360–336

Murugan K, Benelli G, Ayyappan S, Dinesh D, Panneerselvam C, Nicoletti M, Hwang JS, Kumar PM, Subramaniam J, Suresh U (2015a) Toxicity of seaweed-synthesized silver nanoparticles against the filariasis vector Culex quinquefasciatus and its impact on predation efficiency of the cyclopoid crustacean Mesocyclops longisetus. Parasitol Res 114:2243–2253

Murugan K, Benelli G, Panneerselvam C, Subramaniam J, Jeyalalitha T, Dinesh D, Nicoletti M, Hwang JS, Suresh U, Madhiyazhagan P (2015b) Cymbopogon citratus-synthesized gold nanoparticles boost the predation efficiency of copepod Mesocyclops aspericornis against malaria and dengue mosquitoes. Exp Parasitol 153:129–138

Muthukrishnan S, Bhakya S, Kumar TS, Rao MV (2015) Biosynthesis, characterization and antibacterial effect of plant-mediated silver nanoparticles using *Ceropegia thwaitesii*-an endemic species. Ind Crop Prod 63:119–124

Muthukumar T, Sambandam B, Aravinthan A, Sastry TP, Kim JH (2016) Green synthesis of gold nanoparticles and their enhanced synergistic antitumor activity using HepG2 and MCF7 cells and its antibacterial effects. Process Biochem 51(3):384–391

Muthukumaran U, Govindarajan M, Rajeswary M, Hoti SL (2015) Synthesis and characterization of silver nanoparticles using Gmelina asiatica leaf extract against filariasis, dengue, and malaria vector mosquitoes. Parasitol Res 114(5):1817–1827

Nagati VB, Alwala J, Koyyati R, Donda MR, Banala R, Padigya PRM (2012) Green synthesis of plant-mediated silver nanoparticles using *Withania somnifera* leaf extract and evaluation of their anti-microbial activity. Asian Pac J Trop Biomed 2:1–5

Nakamura T, Tsukahara Y, Sakata T, Mori H, Kanbe Y, Bessho H, Wada Y (2007) Preparation of monodispersed Cu nanoparticles by microwave-assisted alcohol reduction. Bull Chem Soc Jpn 80:224–232

Nasrollahzadeh M, Sajadi SM (2016) Green synthesis of Pd nanoparticles mediated by *Euphorbia thymifolia L.* leaf extract: catalytic activity for cyanation of aryl iodides under ligand-free conditions. J Colloid Interface Sci 469:191–195

Nasrollahzadeh M, Mohammad-Sajadi S, Rostami-Vartooni A, Alizadeh M, Bagherzadeh M (2016) Green synthesis of the Pd nanoparticles supported on reduced graphene oxide using barberry fruit extract and its application as a recyclable and heterogeneous catalyst for the reduction of nitroarenes. J Colloid Interface Sci 466:360–368

Obare SO, Meyer GJ (2004) Nanostructured materials for environmental remediation of organic contaminants in water. J Environ Sci Health A 39(10):2549–2582

Özlem Z, Ataklı K, Yürüm Y (2013) Synthesis and characterization of anatase nanoadsorbent and application in removal of lead, copper and arsenic from water. Chem Eng J 225:625–635

Panahi Y, Mellatyar H, Farshbaf M, Sabet Z, Fattahi T, Akbarzadehe A (2018) Biotechnological applications of nanomaterials for air pollution and water/wastewater treatment. Mater Today Proc 5:15550–15558

Pandey B, Singh AK, Singh SP (2019) Nanoparticles mediated gene knock out through miRNA replacement: recent progress and challenges. In: Mishra RK, Thomas S, Mohapatra S, Dasgupta N, Ranjan S (eds) Applications of targeted nano-drugs and delivery systems. Elsevier, Amsterdam, pp 469–497

Park BK, Jeong S, Kim D, Moon J, Lim S, Kim JS (2007) Synthesis and size control of monodisperse copper nanoparticles by polyol method. J Colloid Interface Sci 311:417–424

Pattanayak S, Mollick MMR, Maity D, Chakraborty S, Dash SK, Chattopadhyay S, Roy S, Chattopadhyay D, Chakraborty M (2017) *Butea monosperma* bark extract mediated green synthesis of silver nanoparticles: characterization and biomedical applications. J Saudi Chem Soc 21(6):673–684

Porwal P, Sharma A (2016) Improving water quality through nanotechnology. Int J Appl Res Technol 1(2):119–133

Prachi, Gautam P, Madathil D, Nair A (2013) Nanotechnology in waste water treatment a review. Int J Chem Tech Res 5:2303–2308

Praneetha S, Murugan AV (2015) Development of sustainable rapid microwave assisted process for extracting nanoporous Si from earth abundant agricultural residues and their carbon-based nanohybrids for lithium energy storage. ACS Sustain Chem Eng 3(2):224–236

Qing W, Chen K, Wang Y, Liu X, Lu M (2017) Green synthesis of silver nanoparticles by waste tea extract and degradation of organic dye in the absence and presence of H_2O_2. Appl Surf Sci. https://doi.org/10.1016/j.apsusc.2017.07.007

Qu Y, Pei X, Shen W, Zhang X, Wang J, Zhang Z, Li S, You S, Ma F, Zhou J (2017) Biosynthesis of gold nanoparticles by Aspergillum sp. WL-Au for degradation of aromatic pollutants. Physica E: Low Dimens Syst Nanostruct. https://doi.org/10.1016/j.physe.2017.01.010

Rajakumar G, Rahuman AA, Chung IM, Kirthi AV, Marimuthu S, Anbarasan K (2015) Antiplasmodial activity of eco-friendly synthesized palladium nanoparticles using *Eclipta prostrata* extract against *Plasmodium berghei* in Swiss albino mice. Parasitol Res 114:1397–1406

Rani H, Singh SP, Yadav TP, Khan MS, Ansari MI, Singh AK (2020) In-vitro catalytic, antimicrobial and antioxidant activities of bioengineered copper quantum dots using Mangifera indica (L.) leaf extract. Mater Chem Phys 239:122052

Ramyadevi J, Jeyasubramanian K, Marikani A, Rajakumar G, Rahuman AA (2012) Synthesis and antimicrobial activity of copper nanoparticles. Mater Lett 71:114–116

Rasmussen K, Rauscher H, Mech A (2017) Chapter 2: Physicochemical characterization. In: Fadeel B, Pietroiusti A, Shvedova AA (eds) Adverse effects of engineered nanomaterials, 2nd edn. Academic Press, London, pp 15–49

Rawat S, Pullagurala VL, Hernandez-Molina M, Sun Y, Niu G, Hernandez-Viezcas JA, Peralta-Videa JR, Gardea-Torresdey JL (2018) Impacts of copper oxide nanoparticles on bell pepper (*Capsicum annum L.*) plants: a full life cycle study. Environ Sci Nano. https://doi.org/10.1039/C7en00697g

Rockström J (2003) Water for food and nature in drought-prone tropics: vapour shift in rain-fed agriculture. Philos Trans R Soc B Biol Sci 358:1997–2009

Rosbero TMS, Camacho DH (2017) Green preparation and characterization of tentacle-like silver/copper nanoparticles for catalytic degradation of toxic chlorpyrifos in water. J Environ Chem Eng 5:2524–2532

Sabir S, Arshad M, Chaudhari S (2014) Zinc oxide nanoparticles for revolutionizing agriculture: synthesis and applications. Sci World J 2014:1–8

Sadeghi B (2015) *Zizyphus mauritiana* extract-mediated green and rapid synthesis of gold nanoparticles and its antibacterial activity. J Nanostruct Chem 5:265–273

Sadeghi B, Mohammadzadeh M, Babakhani B (2015) Green synthesis of gold nanoparticles using Stevia rebaudiana leaf extracts: characterization and their stability. J Photochem Photobiol B 148:101–106

Sarkar B, Mandal S, Tsang YF, Kumar P, Kim KH, Ok YS (2018) Designer carbon nanotubes for contaminant removal in water and wastewater: a critical review. Sci Total Environ 612:561–581

Satyavani K, Gurudeeban S, Ramanathan T, Balasubramanian T (2011) Biomedical potential of silver nanoparticles synthesized from calli cells of *Citrullus colocynthis (L.)* Schrad. J Nanobiotechnol 9:43–51

Savage N, Diallo M (2005) Nanomaterials and water purification: opportunities and challenges. J Nanopart Res 7:331–342

Seralathan J, Stevenson P, Subramaniam S, Raghavan R, Pemaiah B, Sivasubramanian A, Veerappan A (2014) Spectroscopy investigation on chemo-catalytic, free radical scavenging and bactericidal properties of biogenic silver nanoparticles synthesized using Salicornia brachiata aqueous extract. Spectrochim Acta Part A 118:349–355

Shan G, Surampalli RY, Tyagi RD, Zhang TC (2009) Nanomaterials for environmental burden reduction, waste treatment, and nonpoint source pollution control: a review. Front Environ Sci Eng China 3(3):249–264

Sharma D, Kanchi S, Bisetty K (2015) Biogenic synthesis of nanoparticles: a review. Arab J Chem. https://doi.org/10.1016/j.arabjc.2015.11.002

Sharma A, Sharma S, Sharma K, Chetri SPK, Vashishtha A, Singh P, Kumar R, Rathi B, Agrawal V (2016) Algae as crucial organisms in advancing nanotechnology: a systematic review. J Appl Phycol 28(3):1759–1774

Sharma B, Boruah PK, Yadav A, Das MR (2018) TiO_2-Fe_2O_3 nanocomposite heterojunction for superior charge separation and the photocatalytic inactivation of pathogenic bacteria in water under direct sunlight irradiation. J Environ Chem Eng 6(1):134–145

Singh S, Saikia JP, Buragohain AK (2013) A novel 'green' synthesis of colloidal silver nanoparticles (SNP) using *Dillenia indica* fruit extract. Colloid Surf B 102:83–85

Singh P, Kim YJ, Zhang D, Yang DC (2016) Biological synthesis of nanoparticles from plants and microorganisms. Trends Biotechnol 34(7):588–599

Singh S, Vishwakarma K, Singh S, Sharma S, Dubey NK, Singh VK, Liu S, Tripathi DK, Chauhan DK (2017) Understanding the plant and nanoparticle interface at transcriptomic and proteomic level: a concentric overview. Plant Gene. https://doi.org/10.1016/j.plgene.2017.03.006

Singh AK, Pal P, Gupta V, Yadav TP, Gupta V, Singh SP (2018) Green synthesis, characterization and antimicrobial activity of zinc oxide quantum dots using *Eclipta alba*. Mater Chem Phys 203:40–48

Singh AK, Yadav TP, Pandey B, Gupta V, Singh SP (2019) Engineering nanomaterials for smart drug release: recent advances and challenges. In: Mishra RK, Thomas S, Mohapatra S, Dasgupta N, Ranjan S (eds) Applications of targeted nano-drugs and delivery systems. Elsevier, Amsterdam, pp 411–449

Subhankari I, Nayak PL (2013) Synthesis of copper nanoparticles using *Syzygium aromaticum* (*Cloves*) aqueous extract by using green chemistry. World J Nano Sci Technol 2:14–17

Subramaniam J, Murugan K, Panneerselvam C, Kovendan K, Madhiyazhagan P, Dinesh D, Kumar PM, Chandramohan B, Suresh U, Rajaganesh R, Alsalhi MS, Devanesan S, Nicoletti M, Canale A, Benelli G (2016) Multipurpose effectiveness of Couroupita guianensis-synthesized gold nanoparticles: high antiplasmodial potential, field efficacy against malaria vectors and synergy with Aplocheilus lineatus predators. Environ Sci Pollut Res 23(8):7543–7558

Suresh U, Murugan K, Benelli G, Nicoletti M, Barnard DR, Panneerselvam C, Kumar PM, Subramaniam J, Dinesh D, Chandramohan B (2015) Tackling the growing threat of dengue: Phyllanthus niruri-mediated synthesis of silver nanoparticles and their mosquitocidal properties against the dengue vector Aedes aegypti (Diptera: Culicidae). Parasitol Res 114(4):1551–1562

Suriyakalaa U, Antony JJ, Suganya S, Siva D, Sukirtha R, Kamalakkannan S, Pichiah PBT, Achiraman S (2013) Hepatocurative activity of biosynthesized silver nanoparticles fabricated using *Andrographis paniculata*. Colloid Surf B 102:189–194

Tamuly C, Hazarika M, Bordoloi M, Bhattacharyya PK, Kar R (2014) Biosynthesis of Ag nanoparticles using pedicellamide and its photocatalytic activity: an eco-friendly approach. Spectrochim Acta Part A 132:687–691

Tuo Y, Liu G, Zhou J, Wang A, Wang J, Jin R, Lv H (2013) Microbial formation of palladium nanoparticles by Geobacter sulfureducens for chromate reduction. Bioresour Technol 133:606–611

Vankar PS, Bajpai D (2010) Preparation of gold nanoparticles from *Mirabilis jalapa* flowers. Indian J Biochem Biophys 47:157–160

Varshney R, Bhadauria S, Gaur MS, Pasricha R (2010) Characterization of copper nanoparticles synthesized by a novel microbiological method. JOM 62(12):102–104

Velmurugan P, Shim J, Kim K, Oh BT (2016) Prunus × yedoensis tree gum mediated synthesis of platinum nanoparticles with antifungal activity against phytopathogens. Mater Lett 174:61–65

Velusamy P, Das J, Pachaiappan R, Vaseeharan B, Pandian K (2015) Greener approach for synthesis of antibacterial silver nanoparticles using aqueous solution of neem gum (*Azadirachta indica L.*). Ind Crop Prod 66:103–109

Velusamy P, Kumar GV, Jeyanthi V, Das J, Pachaiappan R (2016) Bio-inspired green nanoparticles:synthesis, mechanism, and antibacterial application. Toxicol Res. https://doi.org/10.5487/TR.2016.32.2.095

Villanueva-Ibáñez M, Yañez-Cruz MG, Álvarez-García R, Hernández-Pérez MA, Flores-González MA (2015) Aqueous corn extract – mediated green synthesis of AgCl and Ag nanoparticles. Mater Lett 152:166–169. https://doi.org/10.1016/j.matlet.2015.03.097

WHO (World Health Organization) (2017) Drinking-water. Fact sheet. Available at http://www.who.int/media-centre/factsheets/fs391/en/. Accessed 23.01.17

Xiao X, Liu QY, Lu XR, Li TT, Feng XL, Li Q, Liu ZY, Feng YJ (2017) Self-assembly of complex hollow CuS nano/micro shell by an electrochemically active bacterium Shewanella oneidensis MR-1. Int Biodeterior Biodegrad 116:10–16

Yu J, Xu D, Guan HN, Wang C, Juang LK (2016) Facile one-step green synthesis of gold nanoparticles using *Citrus maxima* aqueous extracts and its catalytic activity. Mater Lett 166:110–112

Zambrano-Zaragoza ML, Mercado-Silva E, Gutiérrez-Cortez E, Castaño-Tostado E, Quintanar-Guerrero D (2011) Optimization of nanocapsules preparation by the emulsion diffusion method for food applications. LWT Food Sci Technol 44(6):1362–1368

Zinicovscaia I, Duca G, Cepoi L, Chiriac T, Rudi L, Mitina T, Frontasyeva MV, Pavlov S, Gundorina SF (2015) Biotechnology of metal removal from industrial wastewater: zinc case study. Clean Soil Air Water 43:112–117

Chapter 13
Solid Lipid Nanoparticles

Akhlesh Kumar Jain and Suresh Thareja

13.1 Introduction

The area of Novel Drug Delivery System is getting wider day by day in expanded area of biomedical science, bioengineering and nanotechnology (Ekambaram et al. 2012). Most of the latest delivery techniques explore nanosize-based particles, i.e. nanocarriers having the API (Shah et al. 2011). Few important drug carriers developed using nanotechnology-based approaches are nanoemulsion, nanosuspension, nanocrystals, nanoparticles and solid lipid nanoparticles (Jain 1997). Recent advances in the development of nanocarriers have started a new era in Formulation Science. Solid lipid nanoparticles (SLNs) were reported in 1991 as an unconventional carrier system to typical colloidal carriers such as emulsions, microemulsions, self micro-emulsifying drug delivery system, micellar systems, liposomes, polymeric microparticles and nanoparticles (Ramteke et al. 2012).

SLNs mingle advantages of the conventional carriers along with circumventing some of their major disadvantages. SLNs showed potential applications in drug, gene and vaccine delivery along with controlled and site-specific drug targeting. SLNs are effortlessly made nanoparticles composed of biodegradable polymers of high stability devoid of significant toxicity as well as commercially economic and could incorporate wide variety of drugs for effective targeting. SLNs are novel lipid-based formulations constituted exclusively of biodegradable lipids such as highly purified triglycerides, monoglycerides, complex glyceride mixtures, hard fats or even waxes, which turn solid at room temperature. Solid lipid nanoparticles are nanometre-sized particles that range from 50 to 200 nm and made of solid hydrophobic core which are suspended in aqueous phase containing surfactant.

A. K. Jain (✉) · S. Thareja
School of Pharmaceutical Sciences, GGV Central University, Bilaspur, Chhattisgarh, India

I. Bhushan et al. (eds.), *Nanomaterials and Environmental Biotechnology*,
Nanotechnology in the Life Sciences,
https://doi.org/10.1007/978-3-030-34544-0_13

The drug is dissolved or dispersed in solid core contains the solid high melting fat matrix. Both kinds of lipophilic or hydrophilic therapeutics and diagnostics could be incorporated into the SLN (Shah et al. 2011). SLNs not only unite the advantages of emulsion, liposomes and solid polymeric nanocarriers together but also eliminate few of their disadvantages. Major advantages included are biocompatibility and biodegradability, avoidance of drug leakage, stability against coalescence, nontoxicity, hydrolysis, physical stability and being an excellent carrier for lipophilic drugs (Cavalli et al. 2002). Lipid emulsion and liposomes are entirely different. Oil core made of a neutral lipid, covered by monolayer of amphiphilic lipid, makes lipid emulsion, whereas liposomes are bilayer lipid vesicles made of amphiphilic phospholipid having an interior aqueous cavity (Jain 1997). On the other hand, SLNs are designed from solid lipids and stabilized with an aqueous suspension of emulsifying agents. They look a lot like nanoemulsion; the only difference is that liquid lipid is replaced with a solid lipid, hence providing an outstanding opportunity for controlled drug release as solid lipid lowers the movement of encapsulated drug drastically compared to liquid oil phase (Martins et al. 2007). Also, encapsulation in solid lipids improves the stability of incorporated chemically sensitive lipophilic ingredients in contrast to liquid lipids of nanoemulsion. These prospective benefits of physicochemical properties associated with the physical state of the lipid phase are as follows:

(i) Movement of reactive radicals in solid material is slower compared to liquid medium and hence limits the degradation pathways.
(ii) Phase partition of the API and lipid phase into the solid lipid thus prevents leaching of drugs at the surface of SLN.
(iii) Enhanced absorption of inadequately absorbed drugs is reported after administration using SLN.

Large-scale production of SLNs could be achieved out in a cost-effective and relatively simple manner using high-pressure homogenization technique. Another approach for the production of SLNs is microemulsions or simply suspending liquid lipid in a solution of surfactant with stirring and sonication. SLNs made using various methods are present in suspension form; hence storing for prolonged period of time showed instability due to hydrolysis reactions. However conversion of SLNs into dry powder which can be reconstituted in order to improve stability of SLNs, with the help of lyophilization or spray drying, is an excellent way (Sinha et al. 2010). SLNs provide an excellent opportunity as an advanced drug carrier for oral delivery, topical administration, pulmonary administration, parenteral administration, gene delivery and potential adjuvant for vaccines. In a nutshell, they propose an extremely versatile platform for second- and third-order targeting of drugs.

The major advantages associated by SLNs are as follows:

(a) Suitable for controlled drug release and drug targeting.
(b) Suitable for delivery of both hydrophilic and lipophilic drugs.
(c) Reduced toxicity compared to polymeric nanoparticles as SLNs are made of biocompatible lipids.

(d) Provide protection to labile drugs from chemical, photochemical and oxidative degradation.
(e) Water-based technology (organic solvents can be avoided).
(f) SLN could be administered through various routes such as oral, pulmonary, intravenous, ophthalmic and dermal (Rabinarayan and Padilama 2010).

13.2 Composition of Solid Lipid Nanoparticles

Common ingredients used for preparation of solid lipid nanoparticles are solid lipid(s), stabilizers mainly surfactants and water. The word lipid is having broader sense here and also inclusive of fatty acids (e.g. stearic acid), steroids (e.g. cholesterol), triglycerides (e.g. tristearin), partial glycerides (e.g. Imwitor) and waxes (e.g. cetyl palmitate). All the varieties of emulsifying agents, irrespective of their molecular weight and net charge, could provide the desired stability to the SLN suspension. Sometimes, single surfactant unable to stop particle agglomeration effectively hence the combination of emulsifiers might be required for proper stabilization. The selection of emulsifiers is generally made as per the application and mode of administration of SLN. An obvious benefit of SLN is that the lipid mixture contains physiological lipids and hence limits the toxicity. Most of the additives used for preparation of SLNs are not harmful and falls in the generally recognized as safe (GRAS) category list of the Food and Drug Administration (FDA). SLN is composed of lipids and of stabilizers.

13.2.1 Lipids

In the last two and half decades, SLN proved to be a runaway success among nanocarriers because SLN not only "combine the few benefits of fatty emulsifying agents, liposomes and polymeric nanoparticles, but also have the ability to overcome major drawbacks of these carriers" (Muller and Runge 1998). Few problems associated with SLNs such as crystallinity and low entrapment have been overcome by nanostructured lipid carriers (NLC) (Patlolla et al. 2010). Lipids composed of mainly oily or waxy materials (fatty acids, their alcohols, their glycerol esters, i.e. phospholipids, glycolipids and sphingophospholipids) are having low solubility in polar solvents and freely miscible in organic solvents. The core of the SLN comprises of the lipid; this core itself determines pharmaceutical properties of the particles such as drug stability and drug release.

Generally, two or more lipid combinations are useful because it allows manipulation of encapsulation efficiency and release kinetics and also the physical characteristics, i.e. melting point, crystallinity and polymorphism. A list of various types of lipids can be chosen for development of SLN, and NLC is reported in Table 13.1.

Table 13.1 Commonly used lipids in the preparation of SLN and NLC

S. no.	Triglycerides	Free fatty acids	Free fatty alcohols	Waxes	Miscellaneous
1	Glyceryl tristearate	Behenic acid	Stearyl alcohol	Cetyl palmitate	Castor oil
2	Glyceryl tripalmitate	Palmitic acid and palmitoleic acid	Cetyl alcohol	Beeswax	Hydrogenated castor oil
3	Glyceryl trimyristate	Oleic acid	Myristyl alcohol	Carnauba wax	Hydrogenated palm oil
4	Medium chain triglycerides	Myristic acid	Lauryl alcohol	–	Anhydrous milk fat
5	Glyceryl trioleate	Stearic acid	–	–	Cacao butter
6	Tricaprylin	Arachidic acid and arachidonic acid	–	–	Goat fat
7	Tricaprin	Lauric acid	–	–	–
8	Trilaurin	Linoleic acid or linolenic acid	–	–	–

13.2.2 Surface-Active Compounds (SACs)

Once two phases are immiscible into each other (e.g. oil and aqueous) and mixed together, one phase is dispersed into the other phase, and an interfacial boundary is formed. Surface-active agents reduce this interfacial tension and stabilize the dispersion. Nevertheless, surfactant properties are associated by their chemical structure, and they could reduce the surface tension only up to a specific level which is sometimes not enough to stabilize the dispersion. Additional decrease can be carried out by the addition of cosurfactants (Solans and Sole 2012). The climbing of temperature diminishes the surface tension and hence could be utilized to manufacture hot emulsions with small energy involvement (Webster and Cates 2001). In addition to main constituents, a variety of other additives can also be required to be added such as cryoprotectants (trehalose, sorbitol) (Soares et al. 2013) and stabilizers which act through other mechanisms (polyvinyl alcohol) (Xie et al. 2011). Table 13.2 shows a list of excipients used for preparation and stabilization of SLN.

13.3 Techniques Used for Preparation

13.3.1 High-Pressure Homogenization

The basic method used for production of SLN is high-pressure homogenization which is a consistent and dominant practice, first of all explored for synthesis of SLNs. High-pressure homogenization can be carried out at higher temperature (hot HPH technique) or at under room temperature (cold HPH technique) (Muller and Lucks 1996). In both techniques, the lipid and drug are melted approximately

Table 13.2 List of various excipients used for preparation and stabilization of SLN

S. no.	Nonionic				Anionic	Cationic	Amphoteric	Cosurfactant
	Polyoxyethylene sorbitan fatty esters	Polyoxyethylene alkyl/aryl ethers	Poloxamers	Others				
1	Polysorbate 20	Tyloxapol	Poloxamer 407	Polyglyceryl-6 distearate	Sodium lauryl sulphate	Dimethyldioctadecy-lammonium bromide	L-α-phosphatidylcholine	1-Butanol
2	Polysorbate 60	Polyoxyethylene (20) stearyl ether	Poloxamer 188	Polyglyceryl-3 methylglucose distearate	Sodium dehydrocholate	Cetrimonium bromide	Egg lecithin	Diethylene glycol monoethyl ether
3	Polysorbate 80	Polyoxyethylene (20) cetyl ether	–	PEG caprylic/capric triglycerides	Sodium taurocholate	DOTAP	Soya lecithin	Low molecular weight PEG
4	Polysorbate 85	Polyoxyethylene (20) isohecadecyl ether	–	Macrogol (15)-hydroxystearate	Sodium cholate	Chlorhexidine salts	–	Propylene glycol
5	–	–	–	Polyoxyethylene glyceryl monostearate	Sodium glycocholate	DOTMA	–	Sorbitan monostearate

5–10 °C above the melting point of the lipid. High-pressure homogenizers push liquid through a narrow gap at high pressure (100–2000 bar) which accelerates the liquid particles to extremely higher speed (over 1000 km/h). Such higher forces and cavitation stress break the lipid droplets to the nanometric level. Usually, 5–10% lipid ratio is good enough to prepare the SLN, but up to 40% lipid content can be processed using this technique.

13.3.1.1 Hot Homogenization

In hot homogenization technique, the lipid melt containing the drug is dispersed under stirring in aqueous solution of surfactant at similar higher temperature to the melt. The resulted pre-emulsion is homogenized through a temperature-directed high-pressure homogenizer; usually three cycles (500 bar) are sufficient, followed by cooling of hot O/W nanoemulsion to room temperature, which leads to recrystallization of lipids in the form of solid lipid nanoparticles. Figure 13.1 provides a graphical illustration of SLN production using hot homogenization method. Generally, hot homogenization method can be used for hydrophobic and poorly soluble drugs. As the exposure period to elevated temperature is comparatively low, hence, majority of thermolabile drugs could be encapsulated into the SLNs before being degraded. However, this technique is not suitable for hydrophilic drugs because higher partitioning of drugs in to the aqueous phase during homogenization

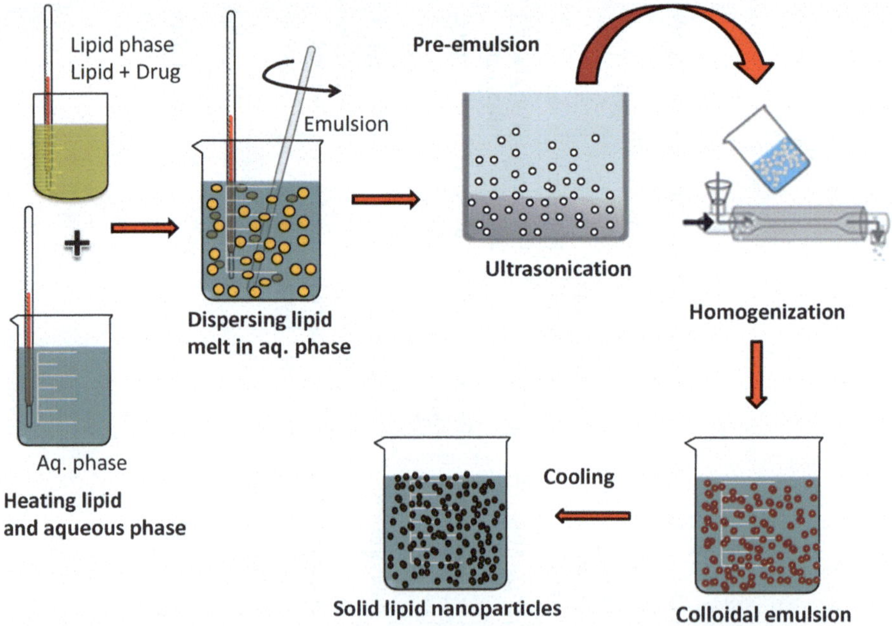

Fig. 13.1 Schematic depiction of various steps involved in the hot homogenization technique

process results in low entrapment efficiency (Gohla and Dingler 2001; Schwarz et al. 1994).

The properties of the final product are affected by the conditions at which pre-emulsion is processed, and it is desirable to obtain droplets in submicron range. Usually, higher temperatures produce smaller particle size because of the reduced viscosity of the internal phase. Nevertheless, elevated temperatures fasten up the degradation process of the drug as well as of matrix. Also, homogenization process can be repeated several times. In majority, 3–5 homogenization rounds at 500–1500 bar are enough to obtain the desired size. It should be remembered that the high homogenization pressure or greater number of cycles often results in an increase of the particle size because of the occurrence of particle coalescence due to high kinetic energy of the particles. Moreover, high-pressure homogenization produces a rise in the temperature of the sample (approximately 10 °C for 500 bar).

13.3.1.2 Cold Homogenization

The cold HPH is a suitable technique for processing temperature-sensitive drugs and/or hydrophilic drugs. During cold homogenization method, the lipid mixture containing the drug is cooled down to the melting temperature of the lipids; therefore, the solid lipid ground to lipid microparticles which are dispersed in a cold surfactant solution yield a pre-suspension. Pre-suspension of microparticles is homogenized at or below room temperature; the cavitation forces produced by the homogenization process are strong enough to rupture the particles into solid lipid nanoparticles. Strict control of temperature is required in order to make sure that the lipid remains in the unmolten state even after increase in temperature during homogenization. Initial step is similar to hot homogenization that comprises the dispersing or solubilization of the medicament into the lipid mixture. Second step is rapidly cooling of the melt which leads to uniform pattern of API in the lipid particles. Also, reduced temperatures enhance the delicateness of the lipid and ultimately particle comminution. Figure 13.2 provides a stepwise production of SLNs using cold homogenization method. Another way to minimize the loss of hydrophilic drug to aqueous phase is to replace water with other suitable media in which the drug is having low solubility (e.g. oil or PEG 600) (Rabinarayan and Padilama 2010).

Cold homogenization process has an edge over hot homogenization as it eliminates or limits the three major problems associated with the hot homogenization technique.

(a) Heat-induced drug degradation.
(b) Partitioning of the drug into the aqueous phase during homogenization.
(c) Crystallization step of the nanoemulsion leads to complexity and several modifications.

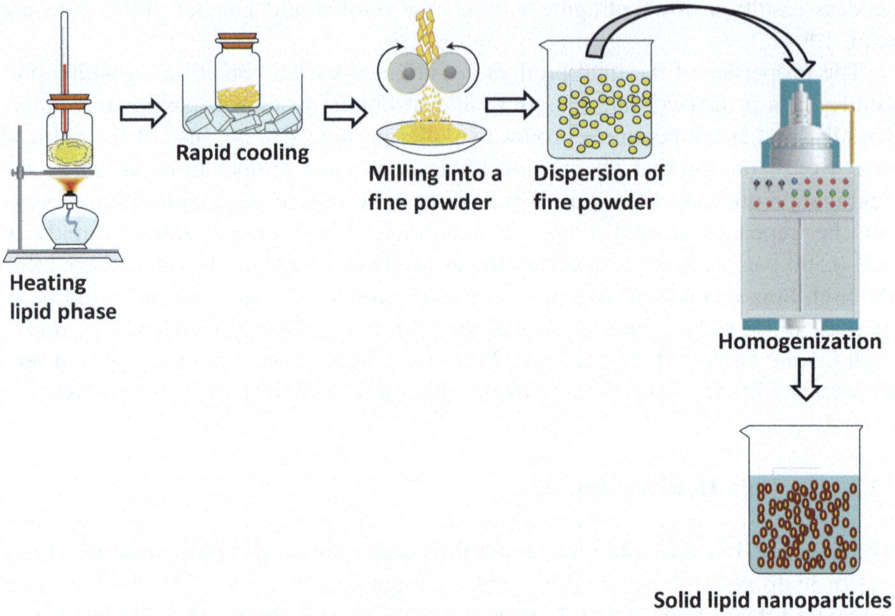

Fig. 13.2 Schematic depiction of various steps involved in the cold homogenization technique

13.3.2 Ultrasound Dispersion/Ultrasonication

Ultrasonication depends on the cavitations in hydrophilic phase that break down lipid matrix into finer particles. The acquired hot microemulsion is allowed to cool to solidify the droplets. The drug is added to the melted lipid; the heated aqueous phase is added to the melted lipid and subjected to high-speed stirring to prepare a nanoemulsion. Pre-emulsion of liquified lipid and heated surfactant mixture is formed, and ultrasound is supplied through probe which is in direct touch of the liquid. SLNs of lower size are formed using ultrasonication and high-speed homogenizer techniques. This technique has the possibility for scale up with flow cells, without moving elements (ease in cleaning), and allows to manage the procedure by changing the energy supplied by sonication. Major shortcoming is the chances of metal adulteration that goes higher with prolonged sonication times and physical instability due to growing solid particles while stirring the SLNs.

13.3.3 Solvent Emulsification/Evaporation

In this technique, water-immiscible organic solvents like cyclohexane, dichloromethane, toluene and chloroform are used for dissolving the hydrophobic drug and the lipid components. Further, the mixture is emulsified with a hydrophilic solution using a high-speed homogenizer. In order to get smaller-sized SLN, the coarse

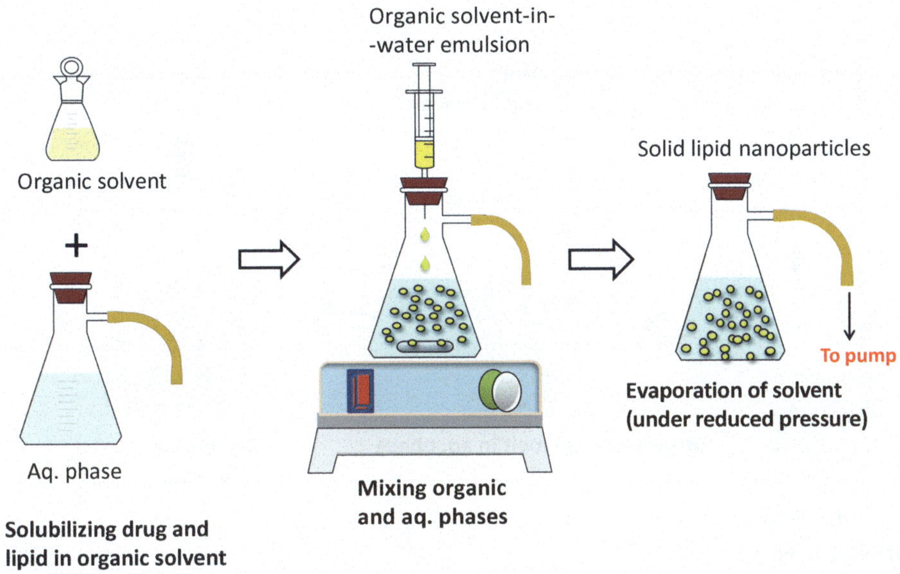

Fig. 13.3 Schematic depiction of various steps involved in the solvent evaporation technique

emulsion is immediately processed through the microfluidizer which produces nanoemulsion. Later on, this nanoemulsion was mechanically stirred at room temperature using rotary evaporator which evaporates the organic liquid evaporation under reduced pressure (40–60 mbar) and leaves lipid precipitates of SLNs (Siekmann and Westesen 1996). Figure 13.3 provides a stepwise demonstration of various aspects involved in the solvent evaporation method. The particle size of the SLNs can be modified by varying the concentration of lipid in organic phase. Generally, finer particle size could be obtained by keeping the lipid load below (Pinto and Muller 1999). Avoidance of high heat during the method is advantageous, and the produced SLN suspension can be centrifuged to collect the SLN.

13.3.4 Microemulsion-Based Technique

This method is based on the dilution of microemulsions. Microemulsions are clear or transparent biphasic liquid composed of a lipid phase and aqueous phase which are made stable using surfactant and also a cosurfactant in most of the cases. They are made by mixing optically transparent mixture composed of fatty acid (i.e. stearic acid), an emulsifier (i.e. polysorbate 20), co-emulsifiers (i.e. butanol) and water at 65–70 °C. The lipid chosen should have a low melting point. Additives like butanol are less preferred with respect to regulatory aspects. The microemulsions show properties of emulsions as well as a real solution. Addition of a microemulsion to water leads to the formation of fine particles and hence could be exploited as method of preparation of SLN (Gasco 1993). Because of the involvement of a dilution step,

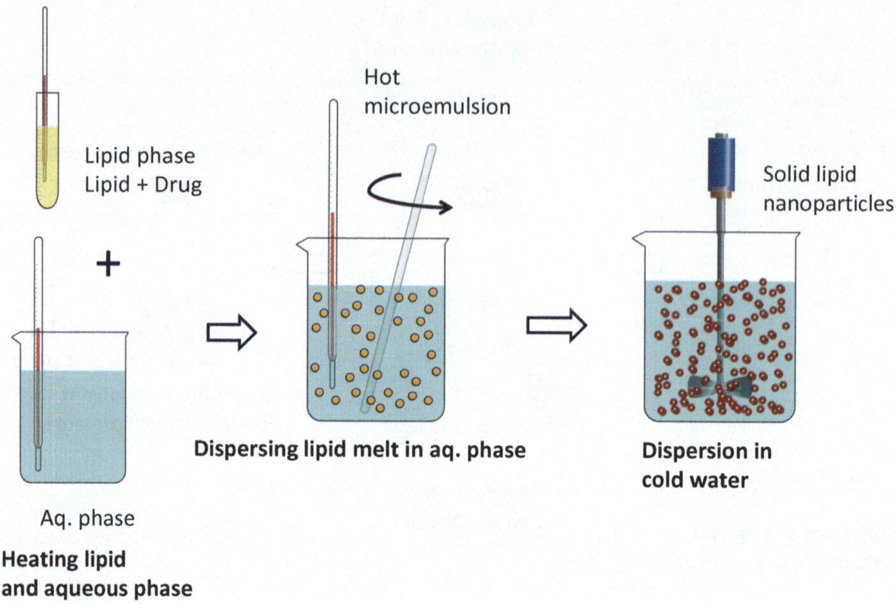

Fig. 13.4 Schematic depiction of various steps involved in the microemulsion-based technique

noticeably inferior concentration of SLN has been produced compared to HPH method. First of all, a hot microemulsion is prepared by stirring 10% molten solid lipid, 15% surfactant and up to 10% of cosurfactant with water. This hot microemulsion is diluted with cold water (1:50) through an especially designed thermostated syringe. The mixture is kept on stirring while dilution which leads to nanoprecipitation of lipids. Figure 13.4 provides a stepwise illustration of production of SLNs using microemulsion-based method. Large-scale production of SLN is possible using microemulsion technique. The microemulsion is produced in a large tank which is heated through a temperature-controlled device. Later on, microemulsion is diluted by transferring it to a cold water tank where lipid precipitation takes place (Carli 1999). Excess water could be taken by ultrafiltration or by lyophilization in order to make concentrated SLN. Also, microwave-assisted technique for formation of microemulsion is another way of preparing SLN based on microemulsion technique (Fig. 13.5).

13.3.5 Double Emulsion Method

So far, numerous methods have been developed to prepare SLN, for example, high-speed homogenization, ultrasound divergence and high-pressure homogenization. In the last few years, microemulsion method has emerged quickly due to its simple operations and so on, compared to previous method. Nevertheless, SLNs prepared

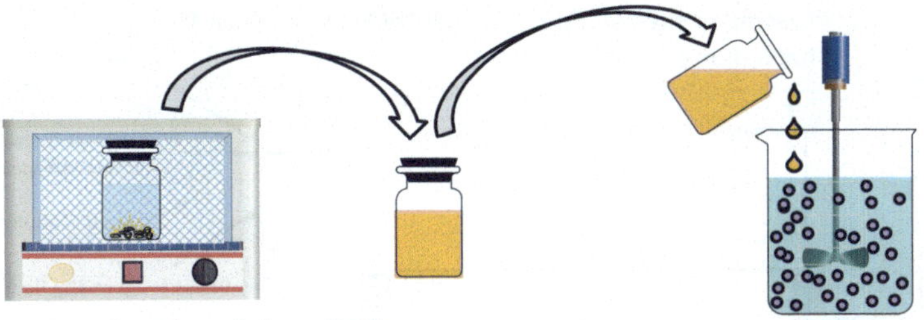

Fig. 13.5 Schematic depiction of various steps involved in microwave-assisted technique of microemulsion formation for SLN production

by microemulsion method have few disadvantages, particularly limited loading capacity of some water-soluble drugs because water-soluble drugs diffuse out from oil phase to aqueous phase while in emulsification process (Shi et al. 2011). The double emulsion technique was established to solubilize agents having high aqueous solubility in the internal aqueous phase of a W/O/W emulsion, with the aid of a stabilizer to prevent partitioning of the aqueous soluble drug to the external aqueous phase during solvent evaporation.

In this method the drug is dissolved into an aqueous solution which is emulsified with melted lipid mixture in order to form primary W/O emulsion stabilized with suitable excipients (e.g. gelatin, poloxamer 407). Further, the primary W/O emulsion is re-emulsified with hydrophilic solution of emulsifying agents to produce a double W/O/W emulsion. Stirring of W/O/W emulsion allows formation of SLNs which are isolated by filtration. Using this method, relatively large particles are produced, however, allow incorporation of higher quantity of hydrophilic molecules and also provide the opportunity of surface alterations (e.g. with PEGs). Double emulsion method can also keep away the need of melting the lipid for the development of peptide/protein-loaded lipid particles.

13.3.6 Membrane Contactor Technique

This method based on simple step of passing lipid through pores of a membrane with pressure at above the melting point of the lipid leads to formation of small droplets. The holes in the membrane work as parallel capillaries for entrance of the lipid phase. Simultaneously, hydrophilic phase is stirred constantly and flows tangentially inside the membrane module and brush off the droplets being produced at the hole openings. SLNs were produced by the cooling of the formulation at the RT. Initially, materials were kept in the thermostated bath to uphold the desired temperature, and N_2 was purged to generate the pressure for the liquid phase. Figure 13.6 provides a stepwise graphical view of SLN production using membrane

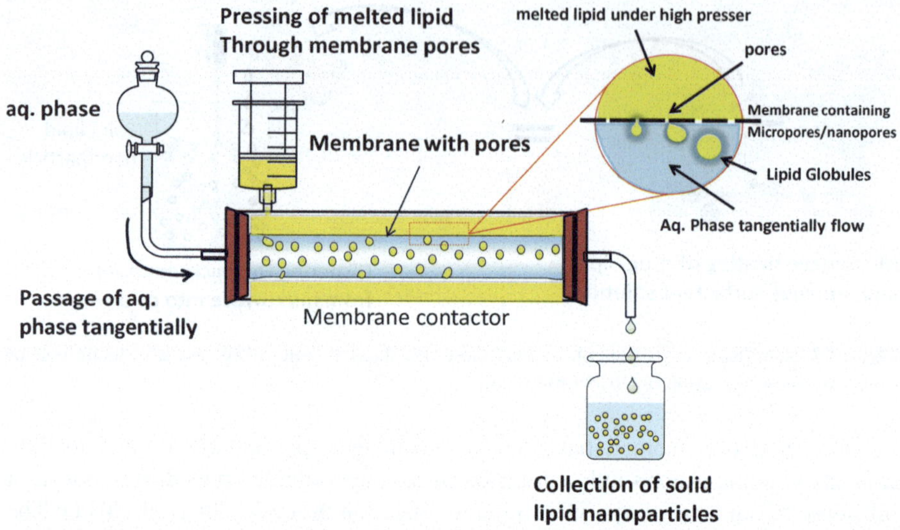

Fig. 13.6 Schematic depiction of various steps involved in the in the membrane contactor-based technique

contactor. The various process parameters (aqueous and lipid phase temperature, the lipid phase pressure, aqueous phase cross flow velocity, lipid phase amount and membrane pore size) affect the size of SLN formed.

The membrane contactor method can also be used for the preparation of polymeric nanoparticles, by methods involving a polymerization of dispersed monomers (interfacial polymerization method) or a dispersion of preformed polymers (nanoprecipitation method). The advantage of membrane contactor method is to manage the size of SLN by a suitable selection of membrane pore size and process parameters as well as it's scaling-up ability (Rabinarayan and Padilama 2010). Higher lipid content not only increases the mean particle diameters but also deteriorates membrane performance. The temperature and flow speed of the hydrophilic phase affect the diameter of SLN (Charcosset et al. 2005). The selection of emulsifying agents and their concentration influences the lipid flux and the diameter of SLNs (El-Harati et al. 2006).

13.3.7 Supercritical Fluid (SCF) Technology

Supercritical fluid (SCF) technique is a comparatively novel method used for preparation of SLN and however has been used for preparation of microparticles and drug powder (Yasuji et al. 2008). It has been proved to be useful alternatives for drying protein formulations as it has major advantage of solvent-less processing in addition to its low toxicity, relatively low critical temperature as well as cost-effectiveness. A fluid is considered as supercritical at the stage once its temperature and pressure go above their correspondence critical value. A gas does not liquefy by increasing the pressure above the critical temperature. Upon increasing the pressure, the density of the gas raises lacking any noteworthy enlargement in viscosity, whereas the capability of the fluid to solubilize the agents too enhances.

The gas which has minute solubility property to dissolve a compound under ambient condition can behave as super solvent for no. of compound under supercritical state. Various gases, i.e. CO2, ethane CH2FCF3 and ammonia, were investigated and tried, but CO2 has proved itself as the best for SCF technique. The principle behind SCF technology is precipitating out of drug or particles by compressed anti-solvent, i.e. supercritical carbon dioxide (Yang and Ciftci 2016). The lipid and drug are dissolved in a solvent; the supercritical fluid, selected to act as anti-solvent to the solutes, is completely immiscible or incompletely soluble with the solvent. The dissolved solid in solvent is sprayed into the flowing SCF and leads to precipitation of solid solute particles (Byrappa et al. 2008).

Another method of SLN production using SCF method is considered as "supercritical extraction of emulsions" (SFEE) (Chattopadhyay et al. 2007). The lipoidic substances and the drug are solubilized in an lipophilic liquid like chloroform. The organic solution containing surfactant is suspended with a hydrophilic solvent containing cosurfactant. The resultant product is subjected to high-pressure homogenizer in order to produce O/W emulsion which enters from the top of the extraction column at a steady speed, and countercurrently the SCF is entered at a steady flow rate. Solid lipid nanoparticles are produced as dispersion into the SCF.

13.4 Characterization of Solid Lipid Nanoparticles

As discussed previously, SLNs are generally produced by homogenizing a suspension of lipids and aqueous phase. Characterization of SLN is a challenge due to the smaller size of the particles and the complexity of the system. Many important aspects have to be considered which might affect the stability and drug release kinetics of SLN. Particle size, size distribution and components of the nanocarriers, morphology and minute structure are vital characteristics of SLNs which affect encapsulation efficiency, stability and release pattern. All these factors point out that the performance of lipid nanoparticles—drug incorporation, release and stability—is influenced by particle size, shape and structure.

13.4.1 Physical Properties

13.4.1.1 Size and Its Distribution

Particle size is a main factor which ensue the in vivo fate of SLNs. Particles larger than 5 μm are restricted for parenteral route as these might block capillary and ultimately can cause embolisms. Clearance of particles from the circulation by the reticuloendothelial organs is also a size-based phenomenon (Wu et al. 2012). Formulation-related variables (lipid nature, dispersing medium, surfactant, cosurfactant and other additives) as well as process variables (preparation method, extent of homogenization and sonication, temperature, equipment) are generally referred to as quality constraints. Size determination study of SLNs is carried out by scattering-based sizing techniques like photon correlation spectroscopy and/or laser diffraction. Particle size and morphology, therefore, may affect the entrapment and release kinetics of the therapeutics from the lipid nanoparticles.

Photon Correlation Spectroscopy

Photon correlation spectroscopy (PCS) or dynamic light scattering (DLS) is the mostly preferred technique to study the particle diameter of solid lipid nanoparticles in a suspension (Obeidat et al. 2010). It measures the fluctuation of the intensity of the scattered light which is caused by particle movement. PCS generally requires a tiny amount of sample, and that does not require any laborious sample preparation. It is nondestructive, faster technique which can determine the size range from 3 to 3000 nm approximately. PCS analyses the Brownian motion or random thermal of the particles in the suspended state in a dispersion medium. Particles in dispersion are irradiated by a laser light, and the changes in the intensity of scattered light are measured after multiple collisions of particles as a function of time. The PCS device has a laser source, which falls on the dispersion in a sample cell (temperature controlled), and scattered light is detected using a photomultiplier. Smaller particles cause high-intensity fluctuations compared to larger particles, because of their high diffusion coefficient. A statistical relationship is applied to correlate the scattering intensity vs time curve to obtain particle diameter and its distribution.

The technique gives the particle size, in terms of z-average diameter and polydispersity index (PDI) that is an indicative of the total length of the particle size distribution. The lower PDI is considered to be narrow particle size distribution, i.e. monodisperse suspension. Higher PDI values indicated poor uniformity of size distribution pattern. Most of the scientists recognize PDI values less than 0.3 as optimum values; however, 0.5 value of PDI is also sometimes accepted. Even though PCS is the extensively established technique, it is based on the translational diffusion coefficient D of the particles; hence it is an indirect method for measurement of particle size. Although PCS is a consistent method for analysing particles having tight distribution pattern, it is not recommended for multimodal size distributions. As presence of larger aggregates or microparticles may hinder the movement of nano range particles hence have a noteworthy effect on measurement process. SLN suspension is generally diluted in order to decrease manifold scattering effects. In addition PCS assumes that all the particles are of spherical shape; however, this is of least importance for SLNs but is a matter of concern for other colloidal carriers.

Laser Diffraction (LD) Spectroscopy

LD is an effective tool which has a broader detection range (20 nm–2000 μm). Laser diffraction (LD), which is also known as laser light scattering, can be used alone or in combination with PCS to obtain an entire population of particle size from tiny ones to large particles. LD is depending on the composite phenomenon of Fraunhofer, Mie and Rayleigh scattering from an illuminated particle in dispersion state. Principle behind LD is that once a laser light falls on a sample having suspended particles, the bigger particles scatter light at narrow angles whereas the minor ones at broader angles (Stoye et al. 1998). Nevertheless the technique deficient in precision for particle dispersions has drastically smaller diameter compared to laser wavelength.

In addition, laser diffractometer provides a good estimation of polydispersity of particles; hence LD covers a broader size range from nanometre to 100 micron. The main limitation of this theory is that its application to SLN demands knowledge refractive index at the measurement wavelength as results are strongly dependent on these optical parameters. The development of PIDS technology (Polarization Intensity Differential Scattering) greatly enhanced the sensitivity of LD to smaller particles (Jores et al. 2004). This technique merges wavelength dependence and polarization effects together and hence significantly increases the accurateness of LD towards tiny nanoparticles.

13.4.2 Microscopic Methods

13.4.2.1 Shape and Surface Morphology

Microscopic techniques such as electron microscopy (EM) and atomic force microscopy can also be used to get information on particle size in the nanometric range. In contrast to light scattering-dependent method, individual particles are measured by the Coulter counter and microscope techniques; hence these provide the benefit of a closer straight analysis. In addition, the results generated by PCS and LD could be affected by the anisometric morphology of the particles. Therefore, the data produced by PCS and LD could be corroborated with EM to determine the real size of the SLNs.

Morphology usually narrates to the external structure and surface pattern of the particle, while ultrastructure narrates to the internal structure of the particle. Ultrastructure indicates the existence and orientation of the different portions of the formulation and nanoparticle as a whole as well as associated structures like micelles, a core-shell structure which may or may not coexist. Generally, spheres have the minimum likely surface area out of other possible shape and, hence, could be stabilized with minimum quantity of surfactant. In addition, sphere particles have the greatest diffusion length and allow the opportunity of sustained and predetermined release of encapsulated therapeutics. In contrast, anisometric shape desires a larger quantity of surfactant for stabilization. It is highly required that the drug is to be entrapped into the surfactant layer or adsorbed onto outer layer.

Electron Microscopy

Electron microscopy provides the direct information on the particle shape. However, special concern is possible artefacts which may arise by the sample preparation procedures. For example, solvent removal may influence the particle shape. These techniques utilize the electron beam as an alternative of light to visualize nanoparticles.

Scanning electron microscopy (SEM) measures electrons sprinkled from the outer layer of the sample to largely create the surface structure, whereas transmission electron microscopy (TEM) measures electrons transmitted through the sample and allows to look beyond the surface. Both techniques have some drawbacks associated with processing conditions (e.g. vacuum, heating) as well drawbacks associated with sample preparation, i.e. dehydration, conductive coating staining, etc.; sometimes changes can be induced within the nanoparticles, while sample preparation and analysis might produce invalid results.

TEM is also being used to determine the morphology and ultrastructure of SLNs (Friedrich et al. 2010). Sample preparation steps involve negative staining, freeze-fracture and vitrification to study the nanocarriers by TEM. Negative staining is carried out by utilizing heavy metal salt solutions on to the colloidal lipid dispersions (Blasi et al. 2013; Silva et al. 2011). A tiny droplet of SLN dispersion is kept on a TEM, grid stained with osmium tetroxide, phosphotungstic acid or uranyl acetate. Also, the resulting TEM photomicrographs generally may not be fair representations of the whole SLN population. Mainly TEM techniques provide information of two-dimensional projection of SLN; hence in order to get three-dimensional anisometric particles and to get additional information on the ultrastructure, freeze-fracture technique is utilized. The TEM micrographs of SLN after coating with uranyl acetate is shown in Fig. 13.7, which shows that SLNs have spherical shape and size from 80 to 200 nm (Li et al. 2006).

Atomic Force Microscopy

Atomic force microscopy is a superior method which can be used to study the realistic shape and surface morphology of the SLN. AFM offers few additional advantages compared to TEM and SEM. As vacuum is not required, the sample needs not to be coated by a conductive substance; however, the sample should be immobilized prior to scanning which can be easily done by removal of the aqueous media from SLN dispersion. Precaution needs to be taken as process of water removal can bring

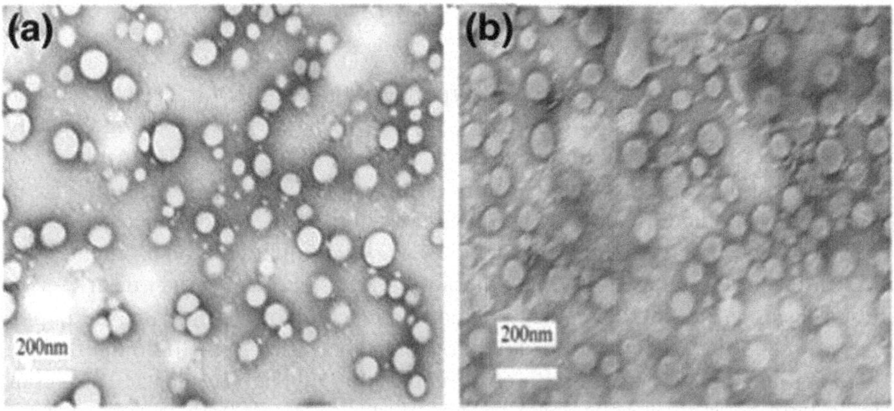

Fig. 13.7 Transmission electron microscopy photographs of solid lipid nanoparticles after staining with 1% uranyl acetate solution. (**a**) 5-FU loaded Cetyl palmitostearate SLNs (CP) (**b**) 5-FU loaded glyceryl monostearate (GMS) SLNs (Reprinted from Shenoy et al. (2013), with permission from Springer)

alterations linked with shrinkage, clustering and crystallization of the lipid (Shakesheff et al. 1994). Another strong benefit of AFM is that it can also provide the important details about soft surfactant layering surrounding the particles that is not possible with other similar techniques (Gref et al. 1994).

13.4.3 Surface Charge

The measurement of electric charge permits forecasting the long-term storage stability of SLNs. Generally, particle clustering occurs slightly for charged particles (high zeta potential) because electrical repulsion takes place between them. Though it is not tightly followed by the systems stabilized with steric stabilizers as adsorption of steric stabilizer around the particles diminishes the zeta potential but prevents aggregation (Müller 1996), zeta potential provides an idea about the amplitude of the electric attraction, or repulsion takes place among SLNs. Generally, greater zeta potential (e.g. more than +30 mV or less than −30 mV) stabilizes the SLN dispersion (Saupe et al. 2006) which prevents the contact between the particles. Particle cell interactions can also be affected by surface. However, colloidal particles having zeta potential closer to zero can express good long-term stability if they are sterically stabilized with the help of stabilizers (Okonogi and Riangjanapatee 2013). Zeta potential is affected by the pH, type and concentration of ions present; hence, the effect of pH and electrolyte on the stability of SLNs has been studied (Choi et al. 2014).

13.4.4 *Drug Encapsulation and Loading Capacity*

Huge varieties of drugs have been incorporated into the SLN. The most critical parameter to analyse the appropriateness of a drug carrier is its loading capacity which is generally expressed in percent related to the lipid phase. The various factors which influence the loading capacity of drug in the lipid are drug solubility in melted lipid, mixing efficiency of lipid and polymorphs.

The precondition to attain a considerable drug encapsulation is an adequate solubility in the lipid matrix, and drug solubility should be greater than desired as it diminishes upon cooling down the lipid matrix and could be lesser in the solid lipid. To encourage drug solubilization, solubilizers can be added to the lipid melt. More complex lipids rather simple mixtures of mono-, di- and triglycerides and also containing fatty acids of various chain lengths produce crystals with many imperfections allowing space to accommodate the drugs (Westesen and Siekmann 1997). Crystalline structure although property of the chemical nature of the lipid is of vital importance in determining whether a drug will tightly incorporate for the long-term or expel out easily.

13.4.4.1 Determination of Incorporated Drug

Quantity of encapsulated agents in SLNs manipulates the release kinetics; therefore it is of utmost crucial importance to determine the encapsulated agents. The quantity of drug entrapped per unit wt. of SLN is estimated after removal of the free drug. Separation of free drug can be carried out by ultracentrifugation and extensive dialysis. The concentration of drug after separation of free drug is determined by using standard analytical technique such as spectrophotometer, spectroflourophotometry or HPLC. Alternatively, encapsulation efficiency can be assessed by dissolving the SLN in a suitable solvent followed by measurement using respective analytical method. Another way to determine encapsulation efficiency is by extraction of the drug into appropriate solvent (Huang et al. 2008). EE is calculated in %:

$$\text{EE\%} = \frac{\text{Amount of drug into the particles}}{\text{Amount of drug added to the formulation}} \times 100$$

13.4.5 *Drug Localization and Drug Release*

A major concern with lipid nanoparticles is the burst release observed with these systems. A prolonged drug release was first obtained when studying the incorporation of prednisolone. Even it is possible to alter the release profiles as a function of lipid matrix, stabilizer concentration and production condition (e.g. temperature) (Muhlen and Mehnert 1998). It is possible to achieve in vitro drug release up to 5–7 weeks with SLN. Moreover, release profiles could be altered to avoid any burst release; however this burst effect could be useful as loading dose.

Even though the drug is place inside the SLN but not always distributed homogenously throughout the matrix (Ren et al. 2013) and can be discreted in different regions of SLNs as well as in associated structures i.e. micelles, liposomes, drug nanocrystals. Methods used to find out the release can inherently manipulate the in vitro drug release pattern of SLN/NLC; hence critical consideration of methodologies should be performed. Regrettably, release experiments are carried out through several ways such as separation by centrifugation and dialysis; hence it is difficult to make comparison. The SLN dispersion is filled in pretreated dialysis bag and immersed in an appropriate dissolution fluid at RT under sink or nonsink conditions with gentle stirring. In vitro release profiles of carvedilol from SLN composed of various lipids (Compritol, lecithin) are shown in Fig. 13.8 and found that SLN made of higher lecithin concentration has shown a faster drug release compared to SLN made of Compritol (Aboud et al. 2016).

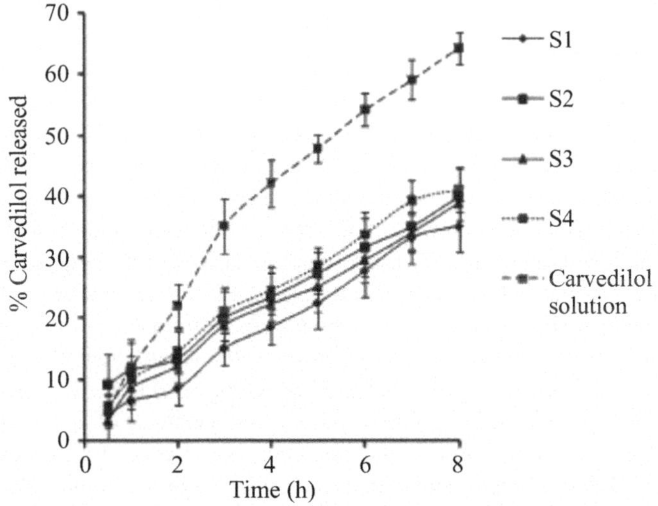

Fig. 13.8 In vitro release profiles of carvedilol from various SLNs made from different lipids. (Reproduced with kind permission from Springer, Aboud et al. (2016))

13.5 Applications of Solid Lipid Nanoparticles

13.5.1 Parenteral Delivery

SLNs are originally discovered to sort out few inherent difficulties of the parenteral nanoemulsions. During last two decades, SLN was explored for a diversity of novel applications and has been shown very useful to solve the problems associated to the faster drug leakage from the parenteral nanoemulsion. Modification of SLN surface of poloxamer- or poloxamine-stabilized SLN showed in vivo sustained and controlled effect (Scholer et al. 2001). Also, in vivo chemical stability of camptothecin increased after its encapsulation into the SLN (Joshi and Muller 2009). Many of the side effects of docetaxel have been reduced by parenteral administration of SLN-encapsulated docetaxel compared to plain drug (Liu et al. 2011). SLNs have proved to increase cellular uptake and targeting potential and ultimately improved the cancer therapy after encapsulation of cytotoxic drugs into the SLN (Liu et al. 2011). Also, SLNs have potential to overcome the challenges linked with parenteral administration of major diseases like malaria by increasing poor solubility of antimalarial drugs, for example, artemether and primaquine, after encapsulation in NLC by reducing haemolytic potential and increasing antimalarial response (Nayak et al. 2010; Mosallaei et al. 2013; Omwoyo et al. 2014). SLN increased drug levels in the lung, spleen and brain compared to plain drug solution which leads to higher distribution into the liver and kidneys.

13.5.2 Oral Delivery

Although the oral route is the most natural choice of drug administration, low solubility and bioavailability of diversity of therapeutics restrict its use. After orally administered, triglycerides (one of the components of SLN) are broken into mono- and diglycerides. These simple glycerides interact with bile salts present in the GIT and form micelles containing the drug which absorbed via chylomicron into the lymphatic system (Patil et al. 2012). Additionally, SLN may be taken up by M-cells of Peyer's patches (He et al. 2005). SLN may be administered orally as aqueous dispersions as well as traditional dosage forms such as tablets, pellets or capsules. Also, presence of food greatly affects SLN behaviour in GIT. The atmosphere of the stomach favours particle aggregation because of the presence of acid and high ionic strength. Improved bioavailability and sustained blood concentration of cyclosporine were found through lipid nanodispersions after per oral administration (Martins et al. 2007). In a latest study, SLNs have protected insulin against in vitro proteolytic enzymes upon encapsulation into the SLN (Almeida and Souto 2007). Also, SLNs have demonstrated the improvement in the bioavailability and protect the proteins against degradation after oral administration (Aungst 2000; Basu et al. 2010).

13.5.3 Transdermal and Topical Use

The use of SLN for transdermal and topical purpose expanded great attention. SLNs present higher chemical stability of the drugs, easily make film on skin surface, greatly hydrate skin and modulate drug release. Also, as they are made of nonirritant and nontoxic lipids, they can be applied to damaged and inflamed skin (Attama et al. 2009). Enhanced safety and therapeutic effectiveness after entrapment into the lipid particles have been reported for coenzyme *Q10* (Lin et al. 2010), celecoxib (Del Pozo-Rodriguez et al. 2010), ketoprofen (Obeidat 2012), vitamin A (Cirri et al. 2012) and many more. Also when systemic absorption is undesired after dermal application, SLN achieved epidermal targeting of podophyllotoxin which was confirmed with fluorescence microscopy (Dingler et al. 1999). Moreover SLN decreased the irritation effect of retinoic acid and also enhanced its stability (Chen et al. 2006). Additionally, SLNs have its application in the cosmetics as the incorporation of novel technologies has led to successful marketing. Various anti-ageing creams, sunscreen products and many more cosmetics have been using SLN and NLC and found that lipid particles in the form of cream formulation did not have irritant prospective with a greater hydration compared to traditional creams (Junyaprasert et al. 2009). Generally, smaller particle SLNs are more useful for dermal application. In the majority of cases, the SLN dispersion is added to an ointment or gel in order to design a formulation which can be easily applied to the skin. Enhancement of the SLN quantity in to the suspension comes out in a semisolid, gel-like consistency that is suitable for direct application on the skin surface (Farboud et al. 2011).

The greater advantage of using SLNs for topical products is that they protect the labile drugs against degradation, for example, retinol or vitamin C, which cannot be added directly into the creams because of the chemical instability which can be overcome by using SLN (Siddiqui et al. 2014).

13.5.4 Pulmonary, Nasal and Ocular Administration

SLNs are appropriate delivery system with excellent acceptability and higher safety for pulmonary treatment (Pardeike et al. 2010). However, SLNs as dry powders are not suitable for delivery to the lungs as the particle size is too small to be retained inside. Simple approach of aerosolization of aqueous SLN dispersions provides as an attractive approach. Also, tiny size of SLNs facilitates their incorporation into the microparticles and drops which can deliver the content to the alveoli. In this regard, SLN and NLC have improved the pharmacokinetic parameters of itraconazole (Pardeike et al. 2011), phenethyl isothiocyanate (Pilcer and Amighi 2010), celecoxib (Patlolla et al. 2010) and thymopentin (Jaafar-Maalej et al. 2011), after pulmonary administration (Pardeike et al. 2010). Also, incorporation of antituberculosis drugs into the SLN showed convincing findings after pulmonary administration (Liu et al. 2008).

Nasal administration is also proved to be promising using SLNs for variety of drugs, i.e. ropinirole (Chavan et al. 2013), alprazolam (Pardeshi et al. 2013) and budesonide (Joshi et al. 2012). Brain targeting is also reported using SLN after intranasal administration for risperidone and demonstrated enhanced brain bioavailability after nasal administration to the mice (Li et al. 2009). The major limitation of ophthalmologic formulation is low retention time into the eye cavity. Diclofenac encapsulated into the SLN demonstrated prolonged release and greater translocation to artificial corneal surface (Attama et al. 2008). Also, SLN-based tobramycin formulations proved to be more effective in the treatment of ocular infections (Zhang et al. 2006).

13.6 SLNs as a Carrier for Site-Specific Delivery

13.6.1 Application in Gene Delivery

In recent time, lipid-based nanocarriers have received greater interest in the area of gene therapy as convincing substitute of positively charged lipids as of protection ability of the incorporated agents against chemical degradation and their faster uptake by cells. In addition production method of SLNs requires low mechanical force; hence processing steps do not degrade DNA and RNA strands. However, cationic charge is a prerequisite for the efficient contact with the anionic cell surface and could be easily induced by incorporating suitable cationic surfactants into lipid matrix. Cationic lipid carriers have been investigated for transfection efficiency in African green monkey kidney fibroblast-like cells (Cos-1) and human bronchoepithelial cell (Olbrich et al. 2001) and found that SLNs efficiently bind and transfected plasmid DNA for the first time using solid lipid nanoparticles. In the recent past, several reports about successful incorporation of genetic materials such as DNA, plasmid DNA and other nucleic acid into the SLN have been published (Zhuang et al. 2010). Gene delivery of a diametric HIV-1 HAT peptide (TAT 2) by incorporation into SLN is demonstrated. Nucleic acid-encapsulated nanoparticles were prepared from a liquid phase composed of aqueous phase containing DNA and a water-miscible organic solvent containing lipid. The obtained particles are called genospheres (70–100 nm) which targeted specifically by insertion of an antibody-lipo polymer conjugated in the particle. Cationic SLNs have shown comparable in vivo transfection efficacy to the liposome made of similar cationic lipids. Recently Carrillo and coworkers used cationic SLN gene delivery to the brain (Carrillo et al. 2013).

13.6.2 SLN as Carriers for Peptides and Protein Drugs

Exception to most polymeric micro- and nanocarrier production techniques used for SLN does not require highly toxic organic liquids, which might be harmful for protein bioactives. Moreover, SLNs can be made to accumulate hydrophobic or water-soluble drugs for parenteral and non-parenteral routes.

SLN production is based on solidified emulsion (dispersed phase) methodologies; hence due to hydrophilic nature of proteins, they are expected to be poorly microencapsulated into the hydrophobic matrix of SLN. A large number of proteins and peptides, e.g. calcitonin, insulin, LHRH, somatostatin and protein antigens, have been studied for drug release kinetics, protein stability and in vivo performance after incorporation into the SLNs. Peptide ligand-modified SLNs are evaluated to increase oral bioavailability of proteins and found that absolute bioavailability of peptides was increased by 2.45 to 1.98 times compared to unmodified SLNs, indicating the scope and effectiveness for the improvement of the oral bioavailability of proteins (Fan et al. 2014). Another research group has prepared peptide-loaded SLNs by solvent diffusion method. The gonadorelin (a model peptide) was incorporated and evaluated for various parameters. The average diameter and zeta potential of SLN were found to be 421.7 nm and −21.1 mV, respectively. However, in vitro release followed biphasic pattern with slow release of gonadorelin (Hu et al. 2004). SLNs were prepared by solubilizing lysozyme (a model peptide) into the melted lipid phase. Results revealed solubility-dependent entrapment efficiency of the peptide in the lipid phase of the finished preparation.

13.6.3 Lipid Nanoparticle as a Carrier for Vaccine

Adjuvants are of great importance in vaccination in order to elicit enhanced immune response. Safer new subunit vaccines have poor immunization potential when used alone, and hence effective adjuvants are required (Copland et al. 2005). Improving the quantity of antigen transported is not a resolution as it enhances the vaccination cost drastically particularly for the developing countries. Conventional adjuvant such as alum is to be employed; however they suffer many disadvantages and have limited immunization to boost up potential. Alternatively various biodegradable polymeric microparticles, niosomes, liposomes and nanoparticles have been explored as an adjuvant for parenteral or mucosal administration routes (Eyles et al. 2003). However, most of the additives used for production of these colloidal carriers as well as manufacturing conditions have deterioration effect on vaccine stability. PLA/PLGA microspheres have shown potential as useful antigen delivery systems (Storni et al. 2005). Advantages of SLN as adjuvant compared to traditional adjuvants are the excellent tolerability by the body and biodegradation. The lipid portion of SLN degrades gradually allowing an extended contact of antigens to the immune cells. Also, degradation of the antigen by the enzyme complexes is restricted due to steric stabilization by the surfactants which hamper the direct exposure (Olbrich and Muller 1999). First report of SLN as adjuvant displayed 43 and 73% effective protection (antibody levels) (Muller et al. 2000). Malaria protein antigen R32NS1 was incorporated into lipospheres and demonstrated increased specific IgG response in serum after intramuscular injection which extended for a minimum of 12-week period after primary immunization (Amsteel et al. 1992). Proteolytic degradation at various mucosae makes peptide or protein vaccines less effective for mucosal delivery; formulation into SLNs is a documented approach to enhance mucosal immunization. Intranasal administration of particles showed deposition of the particles into the lungs, and its lung mucosal layer takes a crucial part in generating immune response (Eyles et al. 2003).

13.7 Stability

SLN stability can be improved by removal of the water as the powder form is usually more stable than the suspensions. Lyophilization is a preferred and appropriate technique used to get rid of the water in the formulation. However, while in lyophilization the protection efficiency of the surfactant might be transformed as a result, the lyophilized SLNs may lead to bigger particles after redispersion (Del Pozo-Rodriguez et al. 2009). Drug leakage and modification in the surface charge are likely during lyophilization. Therefore, addition of appropriate cryoprotectant such as mannose, maltose and trehalose (10–15%) is essential before lyophilization (Cavalli et al. 1997). The properties of the redispersed powder are largely affected by the selection of the cryoprotectant.

Alternatively, spray drying is also another approach used to form powders from SLN dispersions. As heat is required in the process, hence lipids are preferred which have melting points higher than 70 °C. Sometimes it results in particle aggregation. Spray drying cannot be used for more diluted SLN dispersions. An additional method, of stabilizing SLN dispersions, is to make it viscous, as viscous dispersions have reduced rate of sedimentation and collision of the particles (Freitas and Muller 1998). In spite of all these problems observed during stabilization and storage of SLNs, these systems might be stable up to 3 years in dispersion state (Freitas and Muller 1999).

13.8 Conclusions

SLNs unite the advantage of polymeric nanoparticles, fatty emulsions as well as lipid vesicles. Moreover, possibility of accumulation of hydrophobic and aqueous soluble agents, enhanced stability, economy and simplicity of large-scale production process makes them attractive carriers for drug delivery. Production techniques based on commercial production of emulsions and nutrition parenteral was primarily used for large-scale production of SLNs because necessary set of instrument simply existed and can be engaged at both lab and industrial scale. The scalability of methods of preparation of lipid nanoparticles leads to no difficulty in commercialization. In addition site-specific delivery and prolonged effect of drug could be healthier attained with SLNs. The other aspects of SLN related to commercial production such as sterilization, freeze drying and shelf life have already been developed to good standard.

References

Aboud HM, El Komy MH, Ali AA, El Menshawe SF, Elbary AA (2016) Development optimization, and evaluation of carvedilol-loaded solid lipid nanoparticles for intranasal drug. AAPS PharmSciTech 2016(17):1353–1365

Almeida AJ, Souto E (2007) Solid lipid nanoparticles as a drug delivery system for peptides and proteins. Adv Drug Deliv Rev 59:478–490

Amsteel S, Domb AJ, Laving CR (1992) Lipospheres as a vaccine carrier system: effects of size, charge, and phospholipids composition. Vaccin Res 1:383–395

Attama AA, Reichl S, Muller-Goymann CC (2008) Diclofenac sodium delivery to the eye: in vitro evaluation of novel solid lipid nanoparticle formulation using human cornea construct. Int J Pharm 355:307–313

Attama AA, Reichl S, Muller-Goymann CC (2009) Sustained release and permeation of timolol from surface-modified solid lipid nanoparticles through bioengineered human cornea. Curr Eye Res 34:698–705

Aungst BJ (2000) Intestinal permeation enhancers. J Pharm Sci 89:429–442

Basu B, Garala K, Bhalodia R, Joshi B, Mehta K (2010) Solid lipid nanoparticles :a promising tool for drug delivery system. J Pharm Res 3(1):84–92

Blasi P, Schoubben A, Romano G, Giovagnoli S, Di Michele A, Ricci M (2013) Lipid nanoparticles for brain targeting II. Technological characterization. Colloid Surf B 110:130–137

Byrappa K, Ohara S, Adschiri T (2008) Nanoparticles synthesis using supercritical fluid technology - towards biomedical applications. Adv Drug Deliv Rev 60:299–327

Carli F (1999) Physical chemistry and oral absorption of the nanoparticulate systems. Rencentre Pharmapeptides 5:158–160

Carrillo C, Hernandez N, Garcia-Montoya E, Perez-Lozano P, Sune-Negre JM, Tico JR, Sune C, Minarro M (2013) DNA delivery via cationic solid lipid nanoparticles (SLNs). Eur J Pharm Sci 49:157–165

Cavalli R, Caputo O, Carlotti ME, Trotta M, Scarnecchia C, Gasco MR (1997) Sterilization and freeze-drying of drug-free and drug-loaded solid lipid nanoparticles. Int J Pharm 148:47–54

Cavalli R, Gasco MR, Chetoni P, Burgalassi S, Saettone MF (2002) Solid lipid nanoparticles (SLN) as ocular delivery system for tobramycin. Int J Pharm 238:241–245

Charcosset C, El-Harati AA, Fessi H (2005) Preparation of solid lipid nanoparticles using a membrane contactor. J Control Release 108(1):112–120

Chattopadhyay P, Shekunov BY, Yim D, Cipolla D, Boyd B, Farr S (2007) Production of solid lipid nanoparticle suspensions using supercritical fluid extraction of emulsions (SFEE) for pulmonary delivery using the AERx system. Adv Drug Deliv Rev 59:444–453

Chavan SS, Ingle SG, Vavia PR (2013) Preparation and characterization of solid lipid nanoparticle-based nasal spray of budesonide. Drug Deliv Transl Res 3:402–408

Chen H, Chang X, Du D, Liu W, Liu J, Weng T, Yang Y, Xu H, Yang X (2006) Podophyllotoxin-loaded solid lipid nanoparticles for epidermal targeting. J Control Release 110:296–306
Choi KO, Aditya NP, Ko S (2014) Effect of aqueous pH and electrolyte concentration on structure, stability and flow behavior of non-ionic surfactant based solid lipid nanoparticles. Food Chem 147:239–244
Cirri M, Bragagni M, Mennini N, Mura P (2012) Development of a new delivery system consisting in drug–in cyclodextrin–in nanostructured lipid carriers for ketoprofen topical delivery. Eur J Pharm Biopharm 80:46–53
Copland MJ, Rades T, Davies NM, Baird MA (2005) Lipid based particulate formulations for the delivery of antigen. Immunol Cell Biol 83:97–105
Del Pozo-Rodriguez A, Solinis MA, Gascon AR, Pedraz JL (2009) Short- and long-term stability study of lyophilized solid lipid nanoparticles for gene therapy. Eur J Pharm Biopharm 71:181–189
Del Pozo-Rodriguez A, Delgado D, Solinis MA, Pedraz JL, Echevarria E, Rodriguez JM, Gascon AR (2010) Solid lipid nanoparticles as potential tools for gene therapy: in vivo protein expression after intravenous administration. Int J Pharm 385:157–162
Dingler A, Blum RP, Niehus H, Muller RH, Gohla S (1999) Solid lipid nanoparticles (SLN/Lipopearls)--a pharmaceutical and cosmetic carrier for the application of vitamin E in dermal products. J Microencapsul 16:751–767
Ekambaram P, Hasansathali AA, Priyanka K (2012) Solid lipid nanoparticles: a review. Sci Rev Chem Commun 2(1):80–102
El-Harati AA, Charcosset C, Fessi H (2006) Influence of the formulation for *solid lipid nanoparticles* prepared with a membrane contactor. Pharm Dev Technol 11(2):153–157
Eyles JE, Carpenter ZC, Alpar HO, Williamson ED (2003) Immunological aspects of polymer microsphere vaccine delivery systems. J Drug Target 11:509–514
Fan T, Chen C, Guo H, Xu J, Zhang J, Zhu X, Yang Y, Zhou Z, Li L, Huang Y (2014) Design and evaluation of solid lipid nanoparticles modified with peptide ligand for oral delivery of protein drugs. Eur J Pharm Biopharm 88:518–528
Farboud ES, Nasrollahi SA, Tabbakhi Z (2011) Novel formulation and evaluation of a Q10-loaded solid lipid nanoparticle cream: in vitro and in vivo studies. Int J Nanomedicine 6:611–617
Freitas C, Muller RH (1998) Spray-drying of solid lipid nanoparticles (SLN TM). Eur J Pharm Biopharm 46:145–151
Freitas C, Muller RH (1999) Correlation between long-term stability of solid lipid nanoparticles (SLN) and crystallinity of the lipid phase. Eur J Pharm Biopharm 47:125–132
Friedrich H, Frederik P, de With G, Sommerdijk N (2010) Imaging of self-assembled structures: interpretation of TEM and Cryo-TEM images. Angew Chem Int Ed 49:7850–7858
Gasco MR (1993) Method for producing solid lipid microspheres having a narrow size distribution. US Patent 5 250 236
Gohla SH, Dingler A (2001) Scaling up feasibility of the production of solid lipid nanoparticles (SLN). Pharmazie 56:61–63
Gref R, Minamitake Y, Peracchia MT, Trubetskoy V, Torchilin V, Langer R (1994) Biodegradable long-circulating polymeric nanospheres. Science 263:1600–1603
He J, Hou SX, Feng JF, Cai BQ (2005) Effect of particle size on oral absorption of silymarin-loaded solid lipid nanoparticles. China J Chin Mater Med 30:1651–1653
Hu FQ, Hong Y, Yuan H (2004) Preparation and characterization of solid lipid nanoparticles containing peptide. Int J Pharm 273:29–35
Huang G, Zhang N, Bi X, Dou M (2008) Solid lipid nanoparticles of temozolomide: potential reduction of cardial and nephric toxicity. Int J Pharm 355:314–320
Jaafar-Maalej C, Andrieu V, Elaissari A, Fessi H (2011) Beclomethasone-loaded lipidic nanocarriers for pulmonary drug delivery: preparation, characterization and in vitro drug release. J Nanosci Nanotechnol 11:1841–1851
Jain NK (1997) Controlled and novel drug delivery, 1st edn. CBS, New Delhi

Jores K, Mehnert W, Drechsler M, Bunjes H, Johann C, Mader K (2004) Investigations on the structure of solid lipid nanoparticles (SLN) and oil-loaded solid lipid nanoparticles by photon correlation spectroscopy, field-flow fractionation and transmission electron microscopy. J Control Release 95:217–227

Joshi MD, Muller RH (2009) Lipid nanoparticles for parenteral delivery of actives. Eur J Pharm Biopharm 71:161–172

Joshi AS, Patel HS, Belgamwar VS, Agrawal A, Tekade AR (2012) Solid lipid nanoparticles of ondansetron HCl for intranasal delivery: development, optimization and evaluation. J Mater Sci Mater Med 23:2163–2175

Junyaprasert VB, Teeranachaideekul V, Souto EB, Boonme P, Muller RH (2009) Q10-loaded NLC versus nanoemulsions: stability, rheology and in vitro skin permeation. Int J Pharm 377:207–214

Li Y, Dong L, Jia A, Chang S, Xue H (2006) Preparation and characterization of solid lipid nanoparticles loaded traditional Chinese medicine. Int J Biol Macromol 38:296–299

Li H, Zhao X, Ma Y, Zhai G, Li L, Lou H (2009) Enhancement of gastrointestinal absorption of quercetin by solid lipid nanoparticles. J Control Release 133:238–244

Lin YK, Huang ZR, Zhuo RZ, Fang JY (2010) Combination of calcipotriol and methotrexate in nanostructured lipid carriers for topical delivery. Int J Nanomed 5:117–128

Liu J, Gong T, Fu H, Wang C, Wang X, Chen Q, Zhang Q, He Q, Zhang Z (2008) Solid lipid nanoparticles for pulmonary delivery of insulin. Int J Pharm 356:333–344

Liu D, Liu Z, Wang L, Zhang C, Zhang N (2011) Nanostructured lipid carriers as novel carrier for parenteral delivery of docetaxel. Colloids Surf B 85:262–269

Martins S, Sarmento B, Ferreira DC, Souto EB (2007) Lipid-based colloidal carriers for peptide and protein delivery-liposomes versus lipid nanoparticles. Int J Nanomedicine 2:595–607

Mosallaei N, Jaafari MR, Hanafi-Bojd MY, Golmohammadzadeh S, Malaekeh-Nikouei B (2013) Docetaxel-loaded solid lipid nanoparticles: preparation, characterization, in vitro, and in vivo evaluations. J Pharm Sci 102:1994–2004

Muhlen Z, Mehnert W (1998) Drug release and release mechanism of prednisolone loaded solid lipid nanoparticles. Pharmazie 53:552

Müller RH (1996) Zetapotential und Partikelladung in der Laborpraxis – Einführung in die Theorie, praktische Meßdurchführung, Dateninterpretation. Wissenschaftliche Verlagsgesellschaft, Stuttgart, 254 S

Muller RH, Lucks JS (1996) Arzneistofftrager aus festen Lipidteilchen, Feste Lipidnanospharen (SLN). European Patent No. 0605497

Muller RH, Runge SA (1998) Solid lipid nanoparticles (SLN) for controlled drug delivery. In: Benita S (ed) Submicron emulsions in drug targeting and delivery. CRC, Amsterdam, pp 219–234

Muller RH, Mader K, Gohla S (2000) Enzymatic degradation of Dynasan 114 SLN – effect of surfactants and particle size. Eur J Pharm Biohgarm 50:161–170

Nayak AP, Tiyaboonchai W, Patankar S, Madhusudhan B, Souto EB (2010) Curcuminoids-loaded lipid nanoparticles: novel approach towards malaria treatment. Colloids Surf B Biointerfaces 81:263–273

Obeidat WM (2012) Investigation of temperature-induced physical instability of preserved coenzyme Q10-loaded (NLC): a comparative study at different temperatures. Afr J Pharm Pharmacol 6:2413–2423

Obeidat WM, Schwabe K, Muller RH, Keck CM (2010) Preservation of nanostructured lipid carriers (NLC). Eur J Pharm Biopharm 76:56–67

Okonogi S, Riangjanapatee P (2013) Potential technique for tiny crystalline detection in lycopene loaded SLN and NLC development. Drug Dev Ind Pharm 40(10):1–8

Olbrich C, Muller RH (1999) Enzymatic degradation of SLN-effect of surfactant and surfactant mixtures. Int J Pharm 180:31–39

Olbrich C, Bakowsky U, Lehr CM, Muller RH, Kneuer C (2001) Cationic solid-lipid nanoparticles can efficiently bind and transfect plasmid DNA. J Control Release 77:345–355

Omwoyo WN, Ogutu B, Oloo F (2014) Preparation, characterization, and optimization of primaquine-loaded solid lipid nanoparticles. Int J Nanomedicine 9:3865–3874

Pardeike J, Schwabe K, Muller RH (2010) Influence of nanostructured lipid carriers (NLC) on the physical properties of the Cutanova Nanorepair Q10 cream and the in vivo skin hydration effect. Int J Pharm 396:166–173

Pardeike J, Weber S, Haber T, Wagner J, Zarfl HP, Plank H, Zimmer A (2011) Development of an itraconazole-loaded nanostructured lipid carrier (NLC) formulation for pulmonary application. Int J Pharm 419:329–338

Pardeshi CV, Rajput PV, Belgamwar VS, Tekade AR, Surana SJ (2013) Novel surface modified solid lipid nanoparticles as intranasal carriers for ropinirole hydrochloride: application of factorial design approach. Drug Deliv 20:47–56

Patil S, Joshi M, Pathak S, Sharma S, Patravale V (2012) Intravenous β-artemether formulation (ARM NLC) as a superior alternative to commercial artesunate formulation. J Antimicrob Chemother 67:2713–2716

Patlolla RR, Chougule M, Patel AR, Jackson T, Tata PN, Singh M (2010) Formulation, characterization and pulmonary deposition of nebulized celecoxib encapsulated nanostructured lipid carriers. J Control Release 144:233–241

Pilcer G, Amighi K (2010) Formulation strategy and use of excipients in pulmonary drug delivery. Int J Pharm 392:1–19

Pinto JF, Muller RH (1999) Pellets as carriers of solid lipid nanoparticles (SLNk) for oral administration of drugs. Pharmazie 54:506–509

Rabinarayan P, Padilama S (2010) Production of solid lipid nanoparticles-drug loading and release mechanism. J Chem Pharm Res 2:211–227

Ramteke KH, Joshi SA, Dhole SN (2012) Solid lipid nanoparticle: a review. IOSR J Pharm 2(6):34–44

Ren J, Zou M, Gao P, Wang Y, Cheng G (2013) Tissue distribution of borneol-modified ganciclovir-loaded solid lipid nanoparticles in mice after intravenous administration. Eur J Pharm Biopharm 83:141–148

Saupe A, Gordon KC, Rades T (2006) Structural investigations on nanoemulsions, solid lipid nanoparticles and nanostructured lipid carriers by cryo-field emission scanning electron microscopy and Raman spectroscopy. Int J Pharm 314:56–62

Scholer N, Olbrich C, Tabatt K, Muller RH, Hahn H, Liesenfeld O (2001) Surfactant, but not the size of solid lipid nanoparticles (SLN) influences viability and cytokine production of macrophages. Int J Pharm 221:57–67

Schwarz C, Mehnert W, Lucks JS, Muller RH (1994) Solid lipid nanoparticles (SLN) for controlled drug delivery: I. Production, characterization and sterilization. J Control Release 30:83–96

Shah C, Shah V, Uphadhyay U (2011) Solid lipid nanoparticles: a review. Curr Pharm Res 1(4):351–368

Shakesheff KM, Davies MC, Domb A, Glasbey TO, Jackson DE, Heller J, Roberts CJ, Shard AG, Tendler SJB, Williams PM (1994) Visualizing the degradation of polymer surfaces with an Atomic Force Microscope. Proc Int Symp Control Release Bioact Mater 21:1343–1344

Shenoy VS, Gude RP, Murthy RSR (2013) In vitro anticancer evaluation of 5-fluorouracil lipid nanoparticles using B16F10 melanoma cell lines. Int Nano Lett 3:36

Shi L, Li Z, Yu L, Jia H, Zheng L (2011) Effects of surfactants and lipids on the preparation of solid lipid nanoparticles using double emulsion method. J Dispers Sci Tech 32(2):254–259

Siddiqui A, Alayoubi A, El-Malah Y, Nazzal S (2014) Modeling the effect of sonication parameters on size and dispersion temperature of solid lipid nanoparticles (SLNs) by response surface methodology (RSM). Pharm Dev Technol 19:342–346

Siekmann B, Westesen K (1996) Investigations on solid lipid nanoparticles prepared by precipitation in o/w emulsions. Eur J Pharm Biopharm 43:104–109

Silva AC, Gonzalez-Mira E, Garcia ML, Egea MA, Fonseca J, Silva R, Santos D, Souto EB, Ferreira D (2011) Preparation, characterization and biocompatibility studies on risperidone-loaded solid lipid nanoparticles (SLN): high pressure homogenization versus ultrasound. Colloid Surf B 86:158–165

Sinha VR, Srivastava S, Goel H, Jindal V (2010) Solid lipid nanoparticles (SLN'S) – trends and implications in drug targeting. Int J Adv Pharm Sci 1:212–238

Soares S, Fonte P, Costa A, Andrade J, Seabra V, Ferreira D, Reis S, Sarmento B (2013) Effect of freeze-drying, cryoprotectants and storage conditions on the stability of secondary structure of insulin-loaded solid lipid nanoparticles. Int J Pharm 456:370–381

Solans C, Sole I (2012) Nano-emulsions: formation by low-energy methods. Curr Opin Colloid Interface Sci 17:246–254

Storni T, Kundig TM, Senti G, Johansen P (2005) Immunity in response to particulate antigen-delivery systems. Adv Drug Deliv Rev 57:333–355

Stoye I, Schroder K, Muller-Goymann CC (1998) Transformation of a liposomal dispersion containing ibuprofen lysinate and phospholipids into mixed micelles – physico-chemical characterization and influence on drug permeation through excised human stratum corneum. Eur J Pharm Biopharm 46:191–200

Webster AJ, Cates ME (2001) Osmotic stabilization of concentrated emulsions and foams. Langmuir 17:595–608

Westesen K, Siekmann B (1997) Investigation of the gel formation of phospholipid-stabilized solid lipid nanoparticles. Int J Pharm 151:35–45

Wu M, Tia Z, Yuan S, Huang Z (2012) Magnetic and optical properties of the Aurivillius phase $Bi_5Ti_3FeO_{15}$. Mater Lett 68:190–192

Xie S, Zhu L, Dong Z, Wang Y, Wang X, Zhou W (2011) Preparation and evaluation of ofloxacin loaded palmitic acid solid lipid nanoparticles. Int J Nanomedicine 6:547–555

Yang J, Ciftci ON (2016) Formation of hollow solid lipid micro- and nanoparticles using supercritical carbon dioxide. Food Bioprod Process 98:151–160

Yasuji T, Takeuchi H, Kawashima Y (2008) Particle design of poorly water-soluble drug substances using supercritical fluid technologies. Adv Drug Deliv Rev 60:388–398

Zhang N, Ping Q, Huang G, Xu W, Cheng Y, Han X (2006) Lectin-modified solid lipid nanoparticles as carriers for oral administration of insulin. Int J Pharm 327:153–159

Zhuang CY, Li N, Wang M, Zhang XN, Pan WS, Peng JJ, Pan YS, Tang X (2010) Preparation and characterization of vinpocetine loaded nanostructured lipid carriers (NLC) for improved oral bioavailability. Int J Pharm 394:179–185

Chapter 14
Nanotechnology Applications and Synthesis of Graphene as Nanomaterial for Nanoelectronics

Anil Bhardwaj, Gaurav Sharma, and Sumeet Gupta

14.1 Introduction

At present the tremendous development in the field of integrated circuit technology (IC technology) is based on the continuous scaling of silicon-based CMOS transistor devices (Roberts 1984). The scaling helps to maintain the law proposed by Moore in 1965, which states that "the number of devices or transistor per square inch on the integrated circuit doubles every year." The scaling in the nanometer (nm) region is posing serious problems in the integration of devices and maintaining performance of the device's parameters (Kilby 2000; Mollick 2006). At the nanometer (nm) scale level, the effect of quantum tunneling is very much prevalent as it makes device fabrication difficult (Horowitz et al. 2005; Wu et al. 2013).

At nanometer (nm) scale region, the secondary effects come into existence. There is more power dissipation in the system along with subthreshold leakage current flow, which mainly affects the performance of the whole system. Therefore, the problem that arises due to scaling has made researchers search for alternative materials. Hence, carbon nanotubes (CNTs) and graphene are the prime contenders that can potentially replace silicon (Si) in various conventional devices. This will also be helpful in the electronics industry in the near future. The various properties of these devices include thermal, mechanical, optical, and electrical properties which are very much effected by using the nanometer designs.

In this chapter the detailed study of various properties of graphene as well as carbon nanotubes (CNTs) which are useful in the present day-to-day scenario in the nanotechnology regime has been reported. This chapter will also illustrate the implementation and the conceptualization of the term nanotechnology (nm) in numerous number

A. Bhardwaj (✉) · G. Sharma · S. Gupta
Shri Mata Vaishno Devi University, Katra, Jammu and Kashmir, India
e-mail: anil.bhardwaj@smvdu.ac.in

I. Bhushan et al. (eds.), *Nanomaterials and Environmental Biotechnology*, Nanotechnology in the Life Sciences,
https://doi.org/10.1007/978-3-030-34544-0_14

of fields like electronics, medical, chemical, robotics, mechanical, aeronautics, and information technology along within the computers.

The term nanotechnology has its origin from the two Greek words "nano" and "technology." The Greek prefix "nano" means "to a billionth of any entity," and the word "technology" implicates the concept of innovation and also the art and science which thus investigate the manipulation of the matter at nanoscale (Roberts 1984; Kilby 2000; Mollick 2006). Because of which, nanotechnology or nanotech or nanoscaled technology is customarily weighed below 0.1 um or 100 nm (so nm scale is equivalent to 10^{-9} m). Nanoscale science (or nanoscience) investigates the phantasm and properties and also measures the varied responses of materials at various scales including atomic, molecular, and macromolecular levels along with the sizes falling universally in the range of 1–100 nm.

In the current scenario which is nowadays preeminently under 5 nm, the materials properties diverge incomparably (i.e., vital role being played by quantum-scale effects) vis-a-vis a bigger particulate scale. Nanotech involves various aspects which include design, manipulation, building, production, and application, which is achieved by regulating the shape and size, formulating the properties, responses, and performance of devices and systems up to < than 100 nm.

Nanotechnology being a branch of science and technology deals with the materials having the dimensional size of the order of nanometer scale. The term nanometer means that the scale unit is having 10^{-9} m value. These materials have the size less than 100 nm. Nanotech also denotes the manipulation of the matter on supramolecular scale, atomic scale, and molecular scale (Ganji and Kachapi 2015).

Another more authenticated and generalized definition of nanotech is imparted by the National Nanotechnology Initiative (NNI) where it includes the manipulations of the element dimensions by measurably varying one of the dimensions from 1 to 100 nm. The basis of the above fact is that the quantum mechanical effects become significant in the quantum realm scale.

To conceive this idea of nanotechnology, one should observe the nail of the little finger. The dimensional width of the nail on the little finger is round about 10 million nm; another live example of the nanotechnology is the strand of human hair which is of the range 75,000–100,000 nm in diameter. This term was first coined in the late 1950s of the twentieth century especially around 1959. Being an umbrella phrase that beset all the fields of Science & Engineering at nanometer scale. Nanotech is immensely distinct and multidisciplinary field extending from the novel addendum of conventional physics devices to absolutely a new approach planted on the molecular self-assembly.

Nanotechnology being new was mainly talked by the then physicist and Nobel laureate Richard Feynman in which he stated that "There is plenty of room at the bottom" (Feynman 1992).

Nanotech being an enabling technology is expected to revolutionize the major portions of the industrial sectors. The earliest widespread description of nanotechnology was primarily focused on the particular goal of precise manipulation of atoms and molecules which is required for the fabrication of macroscopic products.

The NNI states any material to be in the nanotechnology state only if the concerned material fulfills all the conditions as [EPA, US 2007]:

- Research and technological evolution centered at the atomic, molecular, or macromolecular levels ranging approx. from 1 to 100 nm
- Creating and using systems, structures, and devices containing innovative properties and functions due their small or intermediate size
- Manipulation and control ability at atomic scale

Nanotech being an appearing technology because of feasibility of put forwarding through going materials and to forge state-of-the-art materials having utterly unique features and functionality instilled with colossal abilities for wider spectrum of applications. Supplemental to its utility in industrial arena, pioneering innovations can be forecasted in various fields like ICT, medical and allied technology, metrology, etc. (Seeman 2003).

Symbolic utilizations of nanosciences and nanoengineering field can be seen in various arenas like pharmaceuticals, cosmetic products, canned food, chemical engineering, materials with high performance, electronic items, precision mechanics, light energy models, energy generation, and environmental sciences.

Nanoscience (nanotechnology) is the most energized discipline of science and technology as it promises superior health, wealth, defense, and closing ranks between science and fiction. It has grown rapidly over the years and has already gained considerable international and national interest. It has abeyance for making enormous impact on the society, humanity, and ecology apart from their immediate uses in hardware and electronics system (Kumar 2006; Phillips 2006).

Their direct and indirect impact would be widespread, and it would give a new boost to various emerging technologies such as information technology, data storage, semiconductor technology, sensors and actuators, pharmaceutical, biomedicine, biotech, chemical industry, aerospace, and automotive industry.

As the particle size reduces and reaches the nanoscale regime, the surface area to volume ratio of the particle rapidly increases. The number of atoms on the surface of the particle in this size range will dramatically increase compared to the atoms within the particle. A particle of 5 nm in diameter has nearly 50% of its atoms on the surface compared to the 100 nm particle where less than 5% are surface atoms. This affects the internal structure of the particle including the arrangement of atoms and the electronic configuration, and quantum effects dominate (Ferrari 2007).

In the 1980s era, two major discoveries stimulated the growth of nanotechnology in modern era:

- The first comprised of scanning via tunneling microscope discovery in the year 1981 which put forward the visualization of individual atoms along with their bonds. In year 1989 the above said technique was successfully put into action to maneuver distinctive atoms. Gerd Binnig and Heinrich Rohrer who were the developers of the microscope at IBM's research lab at Zurich got Nobel Prize in the field of physics in 1986 (Meyyappan 2004; Novoselov et al. 2004). In the same year, the discovery of analogous atomic force microscope made by them led to the revolution in the field of nanotech.

- The second comprised of the discovery of fullerenes in 1985 by Kroto, Smalley, and Curl for which they won the Nobel Prize in the field of chemistry in 1996. C_{60} was initially not considered in the nanotech field but was then considered with the discovery of graphene tubes (called carbon nanotubes and every now and then known as the buckytubes) with numerous utilizations in the field of electronics (Meng et al. 2012).

The nanotech field during the 2000s called for upsurge in scientific, political, and commercial attention that resulted in development and controversy at the same time. The major skepticisms were related to the definitions and utilities of nanotech. In 2001 and 2003, apprehensions surfaced about the feasibility of applications forecasted by supporters of molecular nanotech, which ultimately ended up with a public discussion between Drexler and Smalley. Further the profit-oriented products started to appear with the progression of nanotech in the world. These materials are narrowed down to a wide-scale usage of these nanomaterials without the inclusion of any atomic control of matter (Wang et al. 2012).

Various instances of these encompass the usage of nanoparticles of silver as antibiotics, sunscreens composed of nanoparticle components, silica nanoparticles to strengthen carbon fiber, and clothing made of CNTs with stain resilience properties.

Governments nowadays have started promoting nanotechnology and invested funds in the research field of nanotech; for example, the USA started with the NNI that gave a size-dependent description of nanotech and facilitated monetary support for explorations and analysis on the nanoscale. Similarly the European Framework Program was established to give a boost in the development of nanotech. The mid-2000s saw the prospering of innovative and serious scientific concerns in these fields.

Various projects evolved for making nanotech road maps based on meticulous maneuvering of matter with discussions on both present and future capacities and goals.

14.1.1 Types of Nanomaterials

The nanomaterials are the foundations on which nanoscience and nanotechnology are based.

In the past few years, nanostructure science-based technology is recognized as an interdisciplinary area for researchers to collaborate. This leads to the development activity in the worldwide researcher's fraternity. Nanotechnology is having potential to revolutionize the methods in which the materials and other products are created. There are some nanomaterials that occur naturally to be utilized in various applications.

The various types of nanomaterials which have been manufactured and developed are:

- *Nanotubes*: Being a tubelike structure in nm scale, nanotubes are the kind of nanoparticles within which other smaller nanoparticles can be channeled with a property of an electric conductor as well as insulator. The various kinds of nanotubes are (Lakhtakia 2012):
  - *BCN nanotubes*
  - *Boron nitride nanotubes*
  - *CNTs*
  - *DNA nanotubes*
  - *Gallium nitride nanotubes*
  - *Silicon nanotubes*
  - *Inorganic nanotubes*
  - *Tunneling and titania nanotubes*
- *Nanorods*: Physiology here refers to branching (part of particular field) one physiology of nanoscale particle with each comprising the structural range from 1 to 100 nm after being synthesized from the metals or the semiconductors. They have standard aspect ratio to be in the range of 3–5. They are manufactured through chemical synthesis after the combination of ligands which are acting as shape-controlling agents bonding to obtain different facets of nanorods of varying strengths. They are majorly implemented in MEMS-based devices and accelerometers (Lakhtakia 2012; Respaud et al. 2016).
- *Nanocomposite*: These are the multiphase solid materials in which one phase is provided with one, two, or three dimensions of less than 100 nm. In some cases more than one phase dimensional size is in nm range, and they are thus embedded in the metal, ceramic, or polymer matrix.
- *Nanocoatings*: These are the nanosized thin films which are applied to the surface of various materials so as to enhance the various characteristic features of the materials such as the corrosion protection, protection against the degradation caused by the water and ice element, frictional minimization, antifouling and antibacterial properties, self-cleaning property, and resistance to heat and radiations along with thermal management. Nanocoatings are used for aerospace, defense, medical, oil, marine, and metallic industries (Kolyadina et al. 2016).
- *Graphenes*: They are nowadays a very attractive material for post-CMOS or silicon electronics because of their high carrier mobility and planar structure, thereby making the integration with the standard technology easier than that of CNTs. Moreover, nowadays the 2D material systems have gained a great amount of attention among various global research communities. They are majorly 2D material with only one-atom thickness (Imperiale et al. 2010).
- *Thin films*: These comprise of the layers of materials having thickness in the range of few nm to several mm.

14.1.2 Applications of Nanotechnology

Nanotechnology may be defined as the technique to manipulate matter at atomic level or molecular level. The design obtained is utilized for a particular application. The structures developed are worked in the range between 1 and 100 nm in size. This technique refers to development of structures by selection either of top-down or bottom-up design processes for individual components (Wu et al. 2008; Shahabadi et al. 2016).

From the definition, it acts on the emerging technologies associated with the novel classes of therapeutics. Using the improved technology of nanotechnology, it may be possible to improve upon targeted drug delivery (Hu et al. 2012), detection (Goenka et al. 2014), medical instrumentation (Goenka et al. 2014), and other specific areas. Many materials will be acted upon where the nanomaterial being used has a large surface area to volume ratio and is size dependable for many applications as stated above. Before extensively discussing about the advantages of the Nanotechnology, there should be discussion about why the nanotech is becoming the most favored technology among the researchers. The nanotech is the most widely illustrated field because of the following:

- *Engineers*: Development of processes and techniques to handle materials at nanoscale and fabrication of devices
- *Electronics*: Development of new types of semiconductor materials having enhanced features

14.1.3 Advantages of Nanotechnology

Among the engineers it is most widely privileged because of the following factors:

- Materials manufactured are stronger, lighter, cheaper, durable, and precise.
- With the help of nanotech, the computers manufactured can become billion times faster and million times smaller. The manufacturing cost is also negligible.
- Nanotechnology is having great impact on various electronics products; thus they refashion/remodel their procedures and applications. The areas which are benefitted most include diodes, nanotransistors, OLED, quantum computers, plasma displays, etc. (Yeole and Padole 2015).
- Nanotech is also profitable for the energy sector. With the advancement of smaller and energy-efficient products, especially the energy-absorbing and the energy-storing devices, the energy consumption and wastage have been reduced dramatically. Production of solar cells, fuel cells, and hot solar cells has revolutionized the energy sector.
- Materials like nanotubes, aerogels, nanoparticles, etc. are very much stronger, durable, and lighter than some of the conventional tools and equipment like steel, aluminum, and other heavy earth metals (Prokhorov and Botsula 2010).

- With the advent of nanotech in the medical sector, smart drugs have been created for faster, side effect-free, and effective treatment of human ailments like diabetes and cancer and for tissue regeneration along with bone repair (Jelicic et al. 2015).
- In the defense sector, this technology is mainly useful for the development of weapons of mass destruction with ease and with less cost.

14.2 Graphene as Nanotechnology Material

The demand for fast response time devices is increasing tremendeously in the past few decades. With the reduction in the area of devices, the conventional semiconductor devices are not suitable because they tend to exhibit the second-order effects. This leads to the technology scaling of the devices. This leads to shift in the technology and made researchers to search for alternative materials to be used in electronics industry. Out of many materials, graphene and CNT have commanding presence in the electronics industry.

There are several areas for applications of graphene due to its properties such as high conductivity. Its large surface area makes it a promising material for batteries and for energy storage applications (Bae et al. 2010), while its transparency, conductivity, and flexibility make it an interesting material for flexible displays and touch screens (Novoselov et al. 2004). Other applications include the use of graphene as a conductive ink, chemical sensor, barrier layer, or heat spreader. Yet, the initial interest in graphene was sparked by its potential for post-silicon optical and electronic applications, due to its high intrinsic mobility, scalability, and observed electric field effect (Novoselov et al. 2004; Schwierz 2010).

There is limitation while looking forward for miniaturization of the devices there is limitation for scaling of device technology while following the Moores law (Schaller 1997). The time limit for reaching the limit has been postponed with more innovations in structures and architectures utilized by researchers throughout the world. The limit is about to reach its full restriction when the length of the transistors becomes less than 10 nm.

At this duration the potential of graphene as electronic material can be utilized. Thus the research on improving and modifying various techniques related to graphene need to be improved. Since the last decade, there is tremendous research related to problems associated with graphene-based electronics reported by researchers across the globe. There is definitely some more time requirement in the future to fully develop graphene-based transistors so as to replace silicon-based electronics.

The structures developed using graphene for logic transistors are probably different than the contemporary MOS structure. This is attributed to the difference in properties of graphene and silicon.

Because of its high mobility, graphene is considered to be an important material for devices that require ultrahigh speed such as analog and digital logic circuits. Use of devices in RF applications with graphene is nowadays very much prevalent. As graphene possesses remarkable optical property of linear dispersion. This property of Dirac electrons drives the use of graphene-based optoelectronic devices in broadband applications. Graphene is known to exhibit high optical damage threshold, and high third-order optical nonlinearities of graphene devices are also implemented in photonic devices. The application of graphene can be further extended to interconnect and to achieve that there is requirement of techniques so as to realize various applications as reviewed. Graphene being an extraordinary material used in the channel of devices is very much useful in high-frequency RF applications. On the basis of available facts, it can be clearly stated that graphene possess ultrahigh mobility that leads to ballistic transport at mm scale, along with large saturation velocity. In addition to high-frequency applications, they are also used in the design of LNA (low-noise amplifier). As it is well known fact that the RF transistors have maximum frequency of oscillation (f_{max}) and the cutoff frequency (f_T). Currently the highest reported cutoff frequency for graphene MOSFET is 427 GHz, and the maximum f_T is 45 GHz (Wu et al. 2016). Figure 14.1 describes the cutoff frequency for graphene transistors and other competing transistor types.

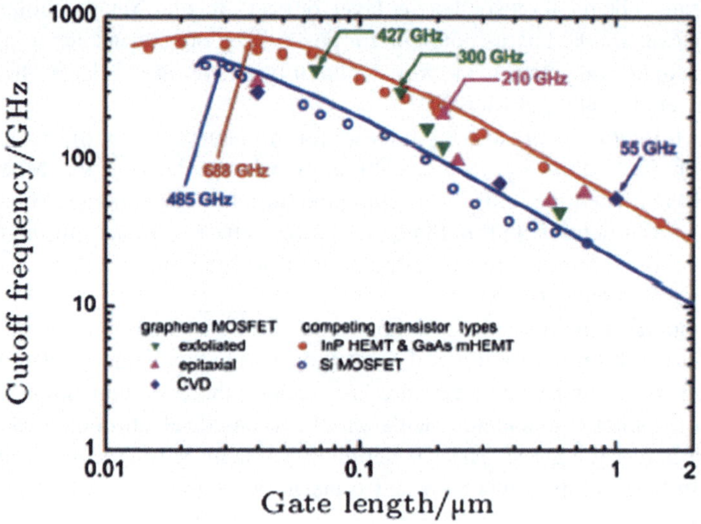

Fig. 14.1 Cutoff frequency of graphene MOSFETs, Si MOSFETs, and III-V HEMTs versus gate length in um (Schwierz 2011)

Table 14.1 Advent of high-frequency GFET devices (Hua-Qiang et al. 2013; Lin et al. 2010)

Achievement	F_T/GHz	Gate length/nm
Field effect in graphene	–	–
1st top-gated GFET	–	500
1st GHz GFET	14.7	500
1st GFET (submicrometer gate length)	26	150
1st epitaxial (GFET on SiC)	4.4	2000
1st 100GHz (epitaxial GFET on SiC)	100	240
1st nanowire gate GFET	300	144
1st CVD GFET with deep-submicrometer gate length	155	40
1st 427 GHz GFET with transferred gate stacks	427	67

Table 14.2 Comparison of different material characteristics

Characteristics	Si	SiC	InAlAs/ InGaAs	AlGaAs/ InGaAs	AlGaN/ GaN	Graphene
Peak electron velocity ($\times 10^7$ cm/s)	1.0	2.0	1.0	1.3	1.3	5–7
Electron mobility at 300 K (cm^2/V-s)	1500	700	5400	8500	1500–2200	>100,000
Thermal conductivity (W/cm-K)	1.5	4.5	0.7	0.5	>1.5	48.4–53

The various improvements in the cutoff frequency are depicted in Table 14.1. The comparison analysis of various characteristics is shown in Table 14.2.

Another material that has also found numerous applications in the field of electronics is carbon nanotubes (CNTs). Because of the rapid scaling of electronic devices, the energy consumption is a major cause of concern for researchers (Lin et al. 2011). Most of the researchers are now focusing their concern onto the ultralow-power sensors. Out of most of the sensors, gas sensors for air quality detection consume maximum power when used, so CNTs are used instead for energy-saving process. CNT-based gas sensor devices can detect NO_2 with power consumption of less than 50 nW. CNTs are innovative devices with good transistor scalability with growth in performance, and CNTFETs have outstanding current capability. CNTFETs have been extensively used at 32 nm technology and also below 28 nm. As it is well-known that the modern-day wireless receivers have to support a variety of wireless standards working at different frequency bands, thus there is an immediate requirement of power conservation devices. The perfect solution to these problems is the CNTFETs which have transit frequency (f_T) in the range of terahertz (THz). The CNT model parameter description is illustrated in Fig. 14.2.

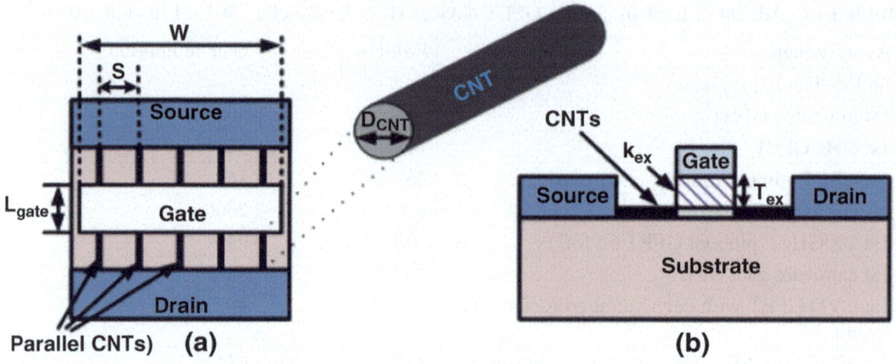

Fig. 14.2 (a) CNTFET as Channel (b) CNTFET structure

14.3 Graphene and Its Future Aspects

The structure of graphene can be described as a 2D honeycomb lattice of carbon atoms having thickness of one atom. Graphene being an allotropic form of carbon is 2D (two-dimensional). It has a hexagonal lattice packing at the atomic scale. Its structure is similar to other allotropes of carbon, namely, CNTs, graphite, charcoal, and fullerenes (C_{60} buckminsterfullerene). The aromatic molecule is considered to be of indefinitely large size. The combination of the words graphite and "ene" was first coined by scientist Hanns-Peter Boehm in 1962. It was discovered by Konstantin Novoselov and Andre Geim in 2004 at the University of Manchester (Hancock 2011). Graphene is a layered material with single-atom-thick layer of graphite (McClure 1956; Slonczewski et al. 1958; Wallace 1947). Although its existence is theoretically known for many years, single-atom layer is considered as thermodynamically unstable.

Novoselov et al. were the initial researchers who experimentally identified the single-atomic-layer structure of graphene. The findings are not in contradiction to the theory as the layer was well supported on the substrate (Novoselov et al. 2004). Also demonstration of high mobility is done experimentally during examination of electrical properties.

In 2010, Nobel Prize in physics was awarded for revolutionary experiments on 2D graphene materials. Graphene is a flat monolayer of carbon atoms which are tightly packed in order to achieve honeycomb structural lattice. One carbon atom is covalently bonded with three adjacent carbon atoms as shown in Fig. 14.3 through sp^2 hybridization, while the rest of the p_z carbon orbitals form an extended electron system (Thakur and Thakur 2015). Due to the unique electronic state of graphene, it possesses appreciable electrical properties.

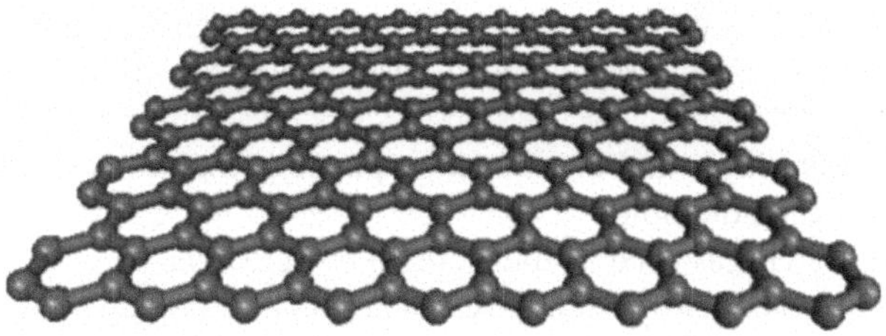

Fig. 14.3 Graphene atomic structure

Graphene has been known for excellent properties that include high mobility (10^5 cm^2 V^{-1} s^{-1} at room temperature), high current density (10^9 A/cm^2), high thermal conductivity, and good mechanical properties.

14.3.1 Properties of Graphene

Graphene falls under the category of zero-gap semiconductors having high electrical conductance. The charge carriers in these materials consisting of several layers are confined naturally in plane. This results in a weak van der Waals interaction between the nanomaterial and its surroundings (Mas-Balleste et al. 2011).

Due to this reason, 2D nanomaterials are sometimes classified as van der Waals materials. This weak interaction of graphene aids in the isolation of single-layer materials, with thickness of single atomic layer. The single-layer thickness of 2D nanomaterials like graphene makes them a suitable option to help meet the scaling demand requirement of future technology scenarios. These unique properties of graphene have attracted researcher's interest throughout the research community, thus elevating graphene's status to that of a true wonder material.

Graphene is being considered as a wonder material nowadays because of its exhibition of remarkable properties (Burtscher 2005):

- *Electronic properties*: Electron mobility possessed by graphene materials is surprisingly very high; the value of mobility reported at room temperature is noted to be around 15,000 cm^2 V^{-1} s^{-1}.
- *Optical properties*: Graphene possess high value of opacity in the case of atomic monolayer; in the initial stages, it is capable of absorption with $\pi\alpha \approx 2.3\%$ fraction of white light, the portion which is incident on its surface.
- *Thermal properties*: The thermal conductivity of graphene at room temperature is in the range of $4.84 \pm 0.44 \times 10^3$ to $5.30 \pm 0.48 \times 10^3$ Wm^{-1} K^{-1} (Burtscher 2005).
- *Mechanical properties*: The breaking strength of graphene as compared to that of steel is 200 times more, with a reported tensile modulus (stiffness) of 1 TPa (150,000,000 psi).

- Graphene is lighter, harder, stronger, and more flexible than steel. Graphene can also be recycled and is considered a sustainable, eco-friendly, cost-effective product.
- Containing single-layer carbon atoms, graphene is the thinnest material known and is stronger than diamond.
- Used for storing H_2 gas used as fuel cells in powered cars.
- Being highly chemically sensitive material, it is capable of detecting explosives.
- Containing nanometer-sized particles, graphene is helpful in removing ions in the desalination of water.

14.3.2 Different Types of Nanostructures and Methods of Graphene Preparation

Researchers have to deal with very fine structures during synthesis of nanomaterials. This characteristic of nanomaterials led researchers to come up with approaches for its synthesis. The two approaches used for the synthesis of nanomaterials can be bottom up or top down. In these approaches, atoms are either assembled together or disassembled. Bulk solids can be converted into finer pieces till there are few atoms remaining. The various techniques for the fabrication of different types of nanostructures are (Prokhorov and Botsula 2010):

- *Nanoprocessing*
- *Nanotooling*
- *Nanomachining*
- *Surface engineering*

In the field of electronics, the various materials fabricated are (Kolyadina et al. 2016; Yeole and Padole 2015; Prokhorov and Botsula 2010):

- *NEMS and MEMS devices*: These are basically the sensor and actuator circuits implanted at the nano- and microscale, thus offering opportunity for integration into application that requires high performance as well as high precision but at the same time containing some space constraints.
- *Nanosensors, actuators, and displays*: CNTs are used for the design of actuators, thereby leading to greater strength. SWNTs (single-walled nanotubes) are used for sensor applications as well as for the development of FED (field emission displays).
- *Double electron layer tunneling transistors (DELTTs)*: The most influencing quantum effect in transistor-like devices is tunneling which signifies that the particle plunges into the barrier which would not have been possible in the real world. They are built with a planar process using conventional semiconductor deposition. DELTT has an insulating barrier between two 2D walls. In this device

the operation is accomplished by the modulation of 2D-to-2D tunneling between the two coupled quantum wells and doesn't require lateral depletion or lateral gating.

- *Resonance tunneling devices (RTDs)*: These devices are specifically two-terminal devices comprising very high switching speed. RTDs for logic applications include resonant tunneling diode and hybrid devices in which resonant tunneling diodes are incorporated with single- or multi-FETs (RTD-FETs). Although they are not strictly transistor devices as they don't have a third terminal, these devices are used in addition with transistors to increase the performance of conventional circuits.
- *Single-electron transistors (SETs)*: These devices are three-terminal switching devices that transfer the electrons from source to drain. SETs include quantum dots, small conducting materials, as basic building material.
- *Spin logic devices (SLDs)*: The category includes a novel device, spin metal-oxide-semiconductor field-effect transistor (spin MOSFET). The device possesses a MOS gate structure and contacts having half metallic ferromagnetic (HMF) property for the source and drain. The other variations of the device include property called spin-torque transistor which provides new devices that include spin gain transistor (SGT) and magnetoresistive element (MRE) hybrid hall effect (HHE) devices (Hua-Qiang et al. 2013).
- *Hybrid hall effect (HHE) devices*: This particular class of devices includes single- or multiple-input wires passing over a region consisting of ferromagnetic material. The magnitude of the current in the direction of the input wire is large, and the magnetic field thus generated results in the magnetization of the ferromagnetic element, in right- or left-hand side direction. This direction absolutely depends on the direction of the current flow.
- *CNT-based field emission devices (FEDs)*: Field emission devices are basically electron emitters, which emit electrons on application of high electric field. Conventional cathode-ray tube and other electron emitters employ thermionic emission process. In the preceding process, the electron overcomes the work function barrier by thermally heating the cathode. But the field emission is different from the thermionic emission as in this process a very high electric field is applied at the tip of the emitter due to which the surface potential becomes very narrow and the electron tunnels through the barrier potential (Yeole and Padole 2015).

Graphene being a wonder material can be manufactured with the help of well-equipped and highly advanced environment instilled with the world's most advanced technology. The synthesis and application of graphene oxide nanosheets for the removal of toxicants from water is as shown below (Khurana et al. 2018):

From the above method shown in Fig. 14.4 for the formation of the graphene material, we define the following steps:

Step 1: Take some amount of graphite which is a composite of carbon, and add a few ml of sulfuric acid, i.e., H_2SO_4. A little amount of nitric acid, i.e., HNO_3, is also added to the mixture.

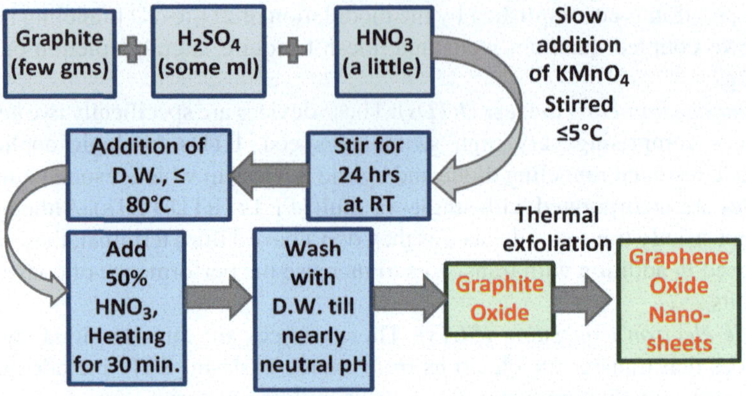

Fig. 14.4 Preparation method for graphene formation (Khurana et al. 2018)

Step 2: Afterward the slow addition of potassium permanganate, i.e., $KMnO_4$, in the mixture is accomplished which is followed by stirring done at a temperature less than 5 °C, i.e., temp ≤5 °C.

Step 3: Continuously stir the mixture for 24 hours at room temperature, and then add 140 ml of distilled water at a temperature of less than 80 °C.

Step 4: Add 50% of HNO_3 and subsequently heat for 30 min or half an hour.

Step 5: The mixture thus obtained is washed with D.W. till the pH returns to its neutral value of 7. This then leads to the formation of graphite oxide.

Step 6: After the formation of the graphite oxide by performing thermal exfoliation, graphene oxide nanosheets can be formed.

Outcome of the process yields 15 g/batch with optimization and ease.

14.3.3 Characterization of Graphene Material

The characterization of graphene material is a vital part of graphene research, and it involves measurements that are based on various spectroscopic and microscopic techniques. The process of characterization is required for the determination of layers and chemical purity of sample. The entire information is obtained by getting information about the absence and presence of defects. The simplest and most effective method that is mostly used for the identification of the number of graphene layers is optical contrast method. Characterization of the graphene material can be described through FTIR response (Fig. 14.5), Raman shift (Fig. 14.6), and X-ray diffraction (Fig. 14.7).

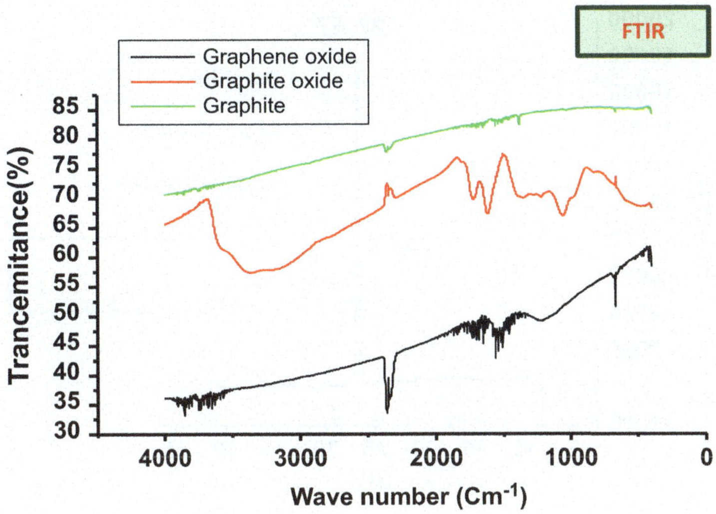

Fig. 14.5 FTIR response of graphene (Khurana et al. 2018)

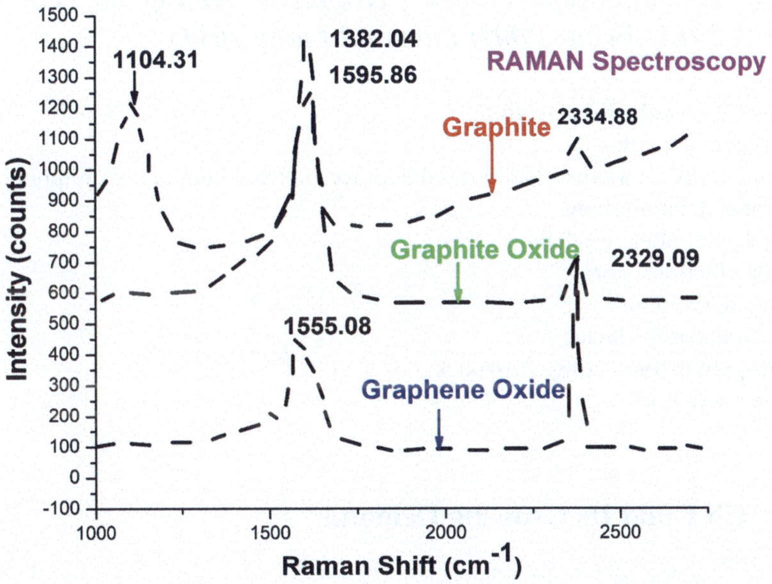

Fig. 14.6 Raman shift vs intensity in graphene using Raman spectroscopy (Khurana et al. 2018)

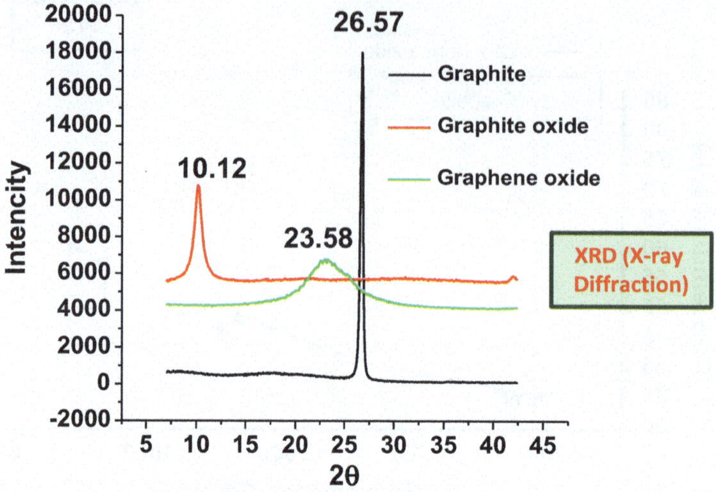

Fig. 14.7 X-ray diffraction method (Khurana et al. 2018)

14.3.4 Potential Applications of Graphene (Hua-Qiang et al. 2013; Awano 2009; Lam and Liang 2011)

- Single-molecule gas detection
- Sensors/solar cell
- Room temperature distillation of ethanol for fuel and human consumption
- Graphene nanoribbons
- Graphene optical modulator
- Graphene transistors
- Integrated circuits
- Electrochromic devices
- Transparent conducting electrodes
- Ultra-capacitors

14.4 CNT and Its Growing Demand

A number of nanomaterials are being developed for structural application; out of those nanomaterials, CNTs have been extensively researched around the world. They have numerous remarkable properties such as exceptionally high specific mechanical properties and physical strength with remarkable thermal properties (Dai 2001).

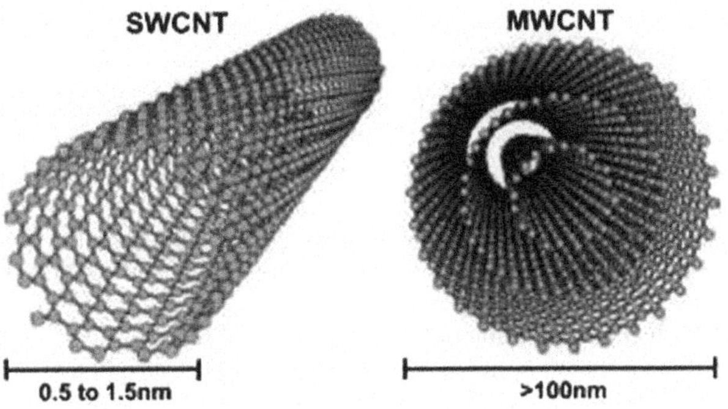

Fig. 14.8 Single-walled and multi-walled CNTs (Dai 2001; Saito et al. 1998)

The various types of CNTs (shown in Fig. 14.8) are:

- *Single-walled carbon nanotubes (SWCNTs)*: CNTs exist as single-walled (SWCNTs) or multi-walled structures (MWCNTs) which are composed of the concentric SWCNTs. Single-walled carbon nanotubes can be shown as a sheet of graphene rolled onto a cylindrical tube with the ends closed by the hemispheres of fullerene. In other words they can be considered as a structure obtained by elongating fullerenes along one axis. The strength and the stiffness of CNTs are derived from the C-C SP^3 bond, the strongest bond in nature. The result is the strongest and the stiffest material ever known. The diameters of SWNTs range from 0.5 to 3 nm (Dai 2001; Saito et al. 1998).
- *Multi-walled carbon nanotubes (MWCNTs)*: They are made up of concentric SWCNTs with different diameters. The equilibrium separation distance between them is considered to be equal to that of the equilibrium separation distance of graphene sheets which is 0.342 nm. The diameter of MWCNTs ranges from approximately 2–20 nm with the upper bound poorly defined. While MWCNTs are as stiff as SWCNTs, they are not as strong as SWCNTs due to the defects produced during the growth of the outer cylinder.

14.5 Conclusion

Nanotechnology because of its smaller size and unique properties has immense potential in the field of electronics, health care, and many other branches of science and technology. Nanotech being an emerging technology has a vast variety of futuristic applications in the day-to-day scenario of human life. It brings new aspects of various inert materials by modifying various optical, chemical, mechanical, and electronic properties at nanoscale. The degree of influence of nanotech on the daily lifestyle of humans depends on the humans entirely. Nanotech is the unavoidable future of humanity. Various nano-by-products such as graphene and CNTs are analyzed; these are wonder materials that have revolutionized entirely the electronics industry.

References

Awano Y (2009) Graphene for VLSI: FET and interconnect applications. In: Electron devices meeting (IEDM), 2009 IEEE international. IEEE, pp. 1–4
Bae S, Kim H, Lee Y et al (2010) Roll-to-roll production of 30-inchgraphene films for transparent electrodes. Nat Nanotechnol 5(8):574–578
Burtscher L (2005) Electrical and mechanical properties of carbon nanotubes
Dai H (2001) Carbon nanotubes: synthesis, structure, properties, and applications. In: Topics in applied physics, vol 80. Springer, Berlin
EPA, US (2007) Nanotechnology white paper. SP Council
Ferrari AC (2007) Raman spectroscopy of graphene and graphite: disorder, electron–phonon coupling, doping and nonadiabatic effects. Solid State Commun 143(1–2):47–57
Feynman RP (1992) There's plenty of room at the bottom. J Microelectromech Syst 1(1):60–66
Ganji DD, Kachapi SHH (2015) Introduction to nanotechnology, nanomechanics, micromechanics, and nanofluid
Goenka S, Sant V, Sant S (2014) Graphene-based nanomaterials for drug delivery and tissue engineering. J Control Release 173:75–88
Hancock Y (2011) The 2010 Nobel Prize in physics—ground-breaking experiments on graphene. J Phys D Appl Phys 44(47):473001
Horowitz M, Alon E, Patil D, Naffziger S, Kumar R, Bernstein K (2005) Scaling, power, and the future of CMOS. In: Electron devices meeting. IEDM technical digest. IEEE international. IEEE, p 7
Hu Y, Li F, Han D et al (2012) Simple and label-free electrochemical assay for signal-on DNA hybridization directly at undecorated graphene oxide. Anal Chim Acta 753:82–89
Hua-Qiang W et al (2013) Graphene applications in electronic and optoelectronic devices and circuits. Chin Phy B 22(9):098106
Imperiale I, Bonsignore S et al (2010) Computational study of graphene nanoribbon FETs for RF applications. In: Electron devices meeting (IEDM), 2010 IEEE international. IEEE, pp 32–33
Jelicic V, Magno M et al (2015). Towards Internet of Things for event-driven low-power gas sensing using carbon nanotubes. In: Advances in sensors and interfaces (IWASI), 2015 6th IEEE international workshop on. IEEE, pp 271–276
Khurana I, Shaw AK, Saxena A, Khurana JM, Rai PK (2018) Removal of trinitrotoluene with nano zerovalent iron impregnated graphene oxide. Water Air Soil Pollut 229(1):17
Kilby JS (2000) The integrated circuit's early history. Proc IEEE 88(1):109–111
Kolyadina EY, Matveeva LA et al (2016) Nanotechnology, features of surfaces and interfaces in the nanostructures with C 60 fullerenes and carbon composite films. In: Nanomaterials: application & properties (NAP), International conference on. IEEE, p 01PCSI03-1
Kumar V (2006) Alchemy at the nanoscale: magic heteroatom clusters and assemblies. Comput Mater Sci 36(1–2):1–11
Lakhtakia A (2012) Nanotechnology and metamaterials: conceptualization and intersection for new opportunities. In: Semiconductor electronics (ICSE), 2012 10th IEEE international conference on. IEEE, p A1
Lam KT, Liang G (2011) Electronic structure of bilayer graphene nanoribbon and its device application: a computational study. In: Graphene nanoelectronics. Springer, Berlin, Heidelberg, pp 509–527
Lin YM et al (2010) Science 327:662. See also Supporting Online Material, available online
Lin YM, Valdes-Garcia A et al (2011) Wafer-scale graphene integrated circuit. Science 332(6035):1294–1297
Mas-Balleste R, Gomez-Navarro C, Gomez-Herrero J, Zamora F (2011) 2D materials: to graphene and beyond. Nanoscale 3(1):20–30
McClure JW (1956) Diamagnetism of graphite. Phys Rev 104(3):666
Meng L, Zhang X, Lu Q, Fei Z, Dyson PJ (2012) Single walled carbon nanotubes as drug delivery vehicles: targeting doxorubicin to tumors. Biomaterials 33(6):1689–1698

Meyyappan M (2004) Carbon nanotubes: science and applications. CRC Press, Boca Raton
Mollick E (2006) Establishing Moore's law. IEEE Ann Hist Comput 28(3):62–75
Novoselov KS, Geim AK et al (2004) Electric field effect in atomically thin carbon films. Science 306(5696):666–669
Phillips JM (2006) Up close: nanoscale science research centers. MRS Bull 31(1):45–49
Prokhorov ED, Botsula OV (2010) Negative differential conductivity of semiconductor diode with resonance-tunnel border. In: Physics and engineering of microwaves, millimeter and submillimeter waves (MSMW), 2010 international Kharkov symposium on. IEEE, pp 1–3
Respaud M, Tap H, Grisolia C et al (2016) Nanotechnology practical teaching at school and university. In: 2016 IEEE nanotechnology materials and devices conference (NMDC). IEEE, pp 1–2
Roberts DH (1984) Silicon integrated circuits: a personal view of the first 25 years. Electr Pow 30(4):282–284
Saito R, Dresselhaus G, Dresselhaus MS (1998) Physical properties of carbon nanotubes. Imperial College Press, London
Schaller RR (1997) Moore's law: past, present and future. IEEE Spectr 34(6):52–59
Schwierz F (2010) Graphene transistors. Nat Nanotechnol 5(7):487–496
Schwierz F (2011) VLSI technology, systems and applications (VLSI-TSA)
Seeman NC (2003) DNA in a material world. Nature 421(6921):427
Shahabadi N, Jamshidbeigi M, Falsafi M (2016) Functionalization of Fe3O4@ SiO2 magnetic nanoparticles with nicotinamide and in vitro DNA interaction. J Mol Liq 224:227–233
Slonczewski JC et al (1958) Band structure of graphite. Phys Rev 109(2):272
Thakur VK, Thakur MK (2015) Carbon allotropes and fascinated nanostructures: the high-impact engineering materials of the millennium. In: Chemical functionalization of carbon nanomaterials. CRC Press, Boca Raton, pp 25–50
Wallace PR (1947) The band theory of graphite. Phys Rev 71(9):622
Wang X, Wang C et al (2012) Noble metal coated single-walled carbon nanotubes for applications in surface enhanced Raman scattering imaging and photothermal therapy. J Am Chem Soc 134(17):7414–7422
Wu W, He Q, Jiang C (2008) Magnetic iron oxide nanoparticles: synthesis and surface functionalization strategies. Nanoscale Res Lett 3(11):397
Wu J, Shen YL, Reinhardt K, Szu H, Dong B (2013) A nanotechnology enhancement to Moore's law. Appl Comput Intell Soft Comput 2013:2
Wu Y, Zou X, Sun M et al (2016) 200 GHz maximum oscillation frequency in CVD graphene radio frequency transistors. ACS Appl Mater Interfaces 8(39):25645–25649
Yeole P, Padole DV (2015) Design of basic logic gates using carbon nano tube field effect transistor and calculation of figure of merit. In: Emerging trends in engineering and technology (ICETET), 2015 7th international conference on. IEEE, pp 28–33

Chapter 15
Efficiency Enhancement of Renewable Energy Systems Using Nanotechnology

Neelu Raina, Preeti Sharma, Parvez Singh Slathia, Deepali Bhagat, and Atin Kumar Pathak

15.1 Introduction

In keeping with some expert evaluations, the world population will escalate to 9 billion by 2050 (Zekić et al. 2018). With the increase in population, universal energy use is expected to escalate in the succeeding 20 years by approximately 40% (Fig. 15.1) and almost two-fold by 2050 (Brinker and Ginger 2011). The worldwide energy demand is expected to remain roughly around 30 TW (terawatt) by the year 2050 and 46 TW (terawatt) by the year 2100 (Sahaym and Norton 2008; Hussein 2015). The most urgent and primary technological challenge our society is encountering these days is meeting the energy needs of the world's growing population in a geopolitically and environmentally sustainable manner (Lewis and Nocera 2006; Lewis 2007). Since there is a substantial worldwide competition for these limited energy resources, it is likely to have significant social and geopolitical consequences, making energy a security issue for every nation. Energy powers our life. Thus, life requires energy. An essential approach of all life forms is to retain energy. Therefore, subconsciously, every living being always opts for the ways and means of least resistance to energy (Ghernaout et al. 2018).

Today, the greatest attention in the world is devoted to energy sources because their use is irrevocable. In the present times, the fundamental energy supply for human survival comes primarily from fossil fuels (natural gas, crude oil, and coal). The supplies of traditional fossil fuels are running out at a very rapid pace. The overall potential of fossil fuels accessible on earth is measured at approximately 5.500 MTOE (million tons oil equivalents) with 60% credited to coal, nearly 30%

N. Raina (✉) · P. Sharma · P. S. Slathia · D. Bhagat
School of Biotechnology, Shri Mata Vaishno Devi University,
Katra, Jammu and Kashmir, India

A. K. Pathak
School of Energy Management, Faculty of Engineering, Shri Mata Vaishno Devi University,
Katra, Jammu and Kashmir, India

I. Bhushan et al. (eds.), *Nanomaterials and Environmental Biotechnology*,
Nanotechnology in the Life Sciences,
https://doi.org/10.1007/978-3-030-34544-0_15

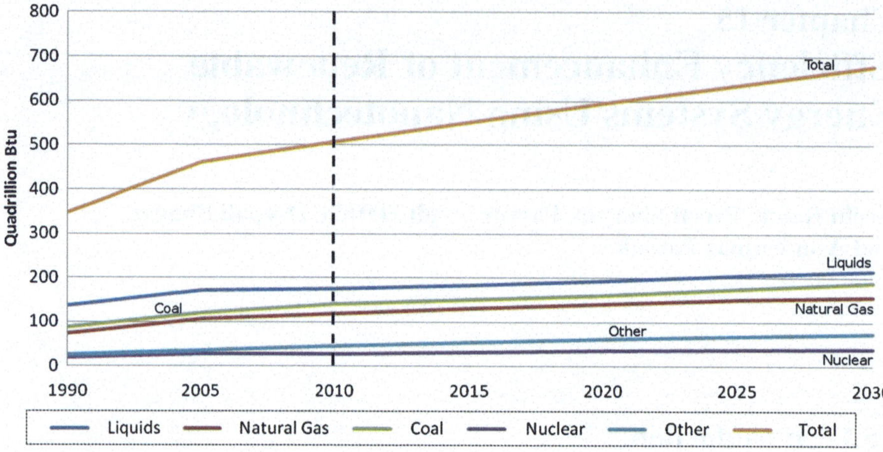

Fig. 15.1 Worldwide energy consumption from 1990 to 2030 in quadrillion BTUs (Brinker and Ginger 2011. Source: http://www.eia.doe.gov/oiaf/forecasting.html)

to natural gas, and approximately 10% to crude oil (Luther et al. 2015). It is stated that the current petroleum spending is 10^5 times quicker than that nature has the capacity to generate. Therefore, at this enormous speed of spending, the global fossil fuel reserves are likely to shrink before 2050 (Hussein 2015; Satyanarayana et al. 2011). One of the most imperative concerns for countries world over is dealing with the degrading environment combined with the rising economic baggage of feeding expanding populations. Excessive greenhouse gas (GHG) emissions in addition to the insufficient fossil fuel supplies are the current global energy problems resulting from increased fossil fuel consumption.

Oil price troble, environmental worries emerging out of the upsurge in global warming, exhausting economies, and dropping reservoir of non-renewable resources have put alternative sources into the limelight. With the aim of enabling an overall sustainable development, there is an urgent requirement to look for alternative, clean energy sources which are renewable and can replace fossil fuels (Deng et al. 2016). Because of the huge energy demand and low future availability of fossil fuels, there is a worldwide drift toward sustainability which is believed to rely heavily on renewable energy resources. Absence of fuel requirement decreases the release of carbon dioxide (CO_2) from renewable energy sources. Renewable energy can endow us with light, heat, and electricity devoid of environmental contamination. Solar energy, wind, tide, geothermal energy, etc. are potential renewable resources as they do not get exhausted with continuous usage and are environmentally benign (Deng et al. 2016; Bruce et al. 2008; Panwar et al. 2011).

"Energy efficiency" fuels the universal sustainable energy domain by incorporating technologies which fulfill the global energy outburst. Many technologies have made handsome strides in notable discipline/s of sustainability era. Nanotechnology is one of them. It is one of the topical technologies worldwide that is drawing immense research by scientists universally. These technologies comprise discovering sustainable energy sources, reducing pollution, furnishing unpolluted drinkable water (Anjum et al. 2016), withdrawing the use of pesticides and herbicides, and improving agricultural outputs to nourish an escalating population and variations in world climate ensuing from global warming (Ghernaout et al. 2018). However, few have made exponential gains in enhancing energy efficiency. Among them, prominent breakthroughs are still in pipeline in the renewable energy sector that will allow us to sustain our rising desire for energy in order to fulfill growing demands of the ever-increasing world population.

The twenty-first century presents mankind with an immense technical challenge which has been the expansion of technologies related to renewable energy resulting from some grave problems related to the utilization of energy and creation of energy from substrates. Renewable energy is poised to play a significant part in the global electricity mix by the year 2030. Going by the Alternative Policy Scenario, renewable energy sector will be supplying over two-thirds of the total electricity globally by that time (IEA 2007). Hence, nanotechnology can be viewed as a novel candidate for clean energy applications since it holds the key for several technological developments within the energy sector especially storage and transmission. Nanotechnology and nanoscience have the competence to offer resolution to quite a lot of these worldwide challenges (Ghernaout et al. 2018; Poinern 2014; Hussain et al. 2017; Hussein 2016; Quandt and Özdoğan 2010; Santos et al. 2015). The implementation of nanoscience to "practical" gadget/apparatus is called "nanotechnology" (Kandil 2016). It has the potential to be the basis of the next industrial development. It deals with some of the distinct mechanical, chemical, electrical, and physical parameters that arise due to restructuring matter at the nanometer scale. Materials can present diverse chemical, physical, and biological properties at the "nano" phase than they do at larger dimensions (Mathew et al. 2018). As a matter of fact, every individual present on earth has come across nano-sized particles. Every individual has been vulnerable to nanometer-ranged foreign elements. We draw them in with every breath and consume them with every gulp.

It is the science of investigating and exploring the distinctive properties of matter by fabricating new structures at an atomic scale of 1–100 nm. Mathematically, it is the billionth part of a meter (10^{-9} m) (Alkahtani 2018) in most classifications (Fig. 15.2). It deals with clusters of atoms of 1 nm at least in one dimension. Nanoscience gives materials distinctive properties such as size, nature, and performance, all due to change in their scale.

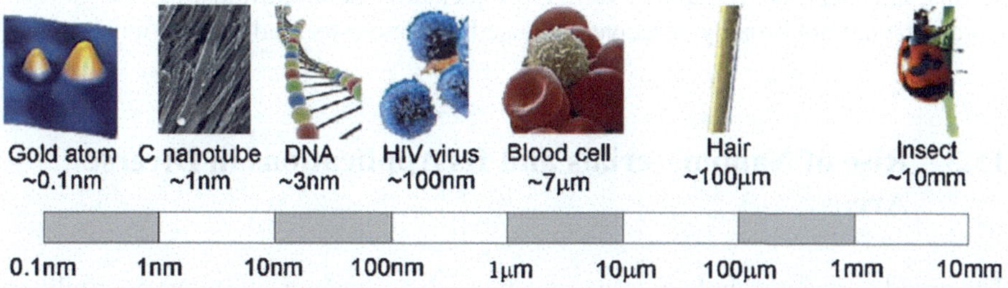

Fig. 15.2 Understanding the nanometer scale in comparison to the length of real materials (Abdin et al. 2013; Serrano et al. 2009)

15.2 Origin of Nanotechnology: The Science of Small Where Small Is Effective

Etymologically, the prefix "nano" is derived from the Greek word "nanos" implying "dwarf" (Booth and Baker 2017). "Nano" is currently a popular feature for the present-day science. The science of nanotechnology revolves around the basic idea of carrying out nanoscale manipulations and creating nanoscale objects. The origin of the concept is usually related to a discourse at the December 1959 congregation of the American Physical Society by world renowned Nobel laureate and physicist Richard Feynman. Feynman, during his well-known 1959 speech at California Institute of Technology (Caltech) entitled "There's Plenty of Room at the Bottom" (Feynman 2012; Khan et al. 2017), asked "What would happen if we could arrange the atoms one by one the way we want them?" (Goddard III et al. 2012; Buzea et al. 2007). His discourse described various stimulating experiments and the infinite possibilities afforded by "tininess." Since then there have been diverse groundbreaking expansions in this domain such as homeland security, medicine, information technology, energy, environmental science, food safety, transportation, and many others.

The term "nanotechnology" was first brought up in history in the late nineteenth century (1867) when James Clerk Maxwell published his primary remarks about this technology and offered potential of maneuvering distinct particles (Zekić et al. 2018; El Saliby et al. 2008). A Japanese scientist of the Tokyo University of Science, named Norio Taniguchi, was the first to use the term "nanotechnology" at a conference in 1974 (Kandil 2016; Aviram 1988). He used the term for describing semiconductor processes (for instance, ion beam milling and thin film deposition) which displayed characteristic control over the nanometer array. He defined "nanotechnology" as mainly consisting of the processing of separation, consolidation, and deformation of materials by one atom or "one molecule". However, in 1980, Eric Drexler in his first paper on nanotechnology again used the term "nanotechnology" being totally unaware of Taniguchi's prior use of the same term (Kandil 2016).

Many "nano" words have showed up in glossary, for instance, nanogram, nanometer, nanorobotics, nanoscience, nanosecond, nanostructure, nanotechnology, nanotube, nanowires, and many more. Some more uncommon words consist of nanoantenna, nanoarrays, nanocavity, nanocrystal, nanoelectronics, nanoencapsulation, nanofibers, nanolithography, nanomagnet, nanoprecursor, nanopatterning, nanoporous, nanoseparation, nanovalve, nanosequencing, nanoscaffolds, etc. Numerous words that are not broadly perceived are used in some esteemed publications.

15.3 Rise of Nanomaterials and Its Applications in Diverse Areas

The world of nanotechnology is ever-evolving. It is showing enormous possibilities in countless fields of science and technology by functioning at an elementary level. It is creating a lot of vigilance nowadays. It is not just a catchphrase but the future

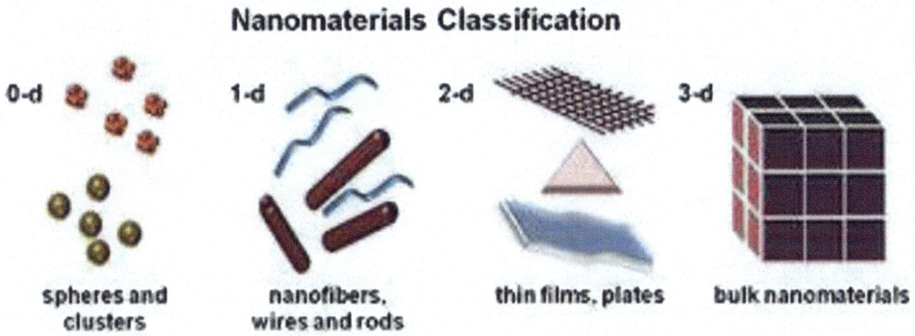

Fig. 15.3 Classification of nanomaterials (Raghav and Dinesh 2016)

of technology since it impacts every sector. It is the combined study of engineering and chemistry fields and is classified as the arrangement, assembly, and application of resources and appliances whose dimensions have been engineered at the nanoscale. Widely, the word "nano" is employed as prefix to illustrate matter, structures, or processes with characteristics coming from nano-sized structures. On the basis of overall shape, these materials can be classified as 0D (quantum dots, nanoclusters, and nanoparticles), 1D (nanotubes and nanorods), 2D (nano-sized thin films), or 3D (nanomaterials) (Khan et al. 2017; Tiwari et al. 2012) (Fig. 15.3).

Nanoscience creates an enormous prospect not only within the intellectual society but also among sponsors, administration, and industry due to its unique ability to engineer elements at an atomic scale. It has created novel equipment/devices with potential relevance in an extensive number of disciplines. Nanoscience and nanotechnology facilitate vast array of technologies that attain abundant and low-cost alternative energy production, efficient energy use, transmission, and storage. This section demystifies the topic for anybody engrossed on how nano-sized molecules and the nanotechnology involved affects our everyday lives. The relevance of nanomaterials is rising in diverse zones (Fig. 15.4), together with public health, agriculture, clothing, medicine, cosmetics, and food owing to their exceptional capability to boost bioavailability and solubility and to safeguard bioactive compounds while they are being prepared and then stored for use (Bajpai et al. 2018). Nanoparticles as coating materials have been the center of attraction over the previous few decades with their exceptional relevance in different spheres (Khezri et al. 2018). The study of these nanoparticle materials is an important element of nanoscience. It is regarded worldwide as the key technology for innovations and technological advancement in nearly all domains of economy (Table 15.1). Some of the improvements using scientific conception regarding the benefits of controlling materials at nanoscale have resulted into commercially viable products and technologies based on nanomaterials as given below:

1. *Influence of nanotechnology on transportation and sports equipment*: Increase in the toughness of lightweight sporting instruments/gadgets, extending from baseball bats, tennis and badminton racquets, hockey sticks, racing bicycles, golf balls/clubs, skis, fly fishing rods and archery arrows, etc. are some of the sporting

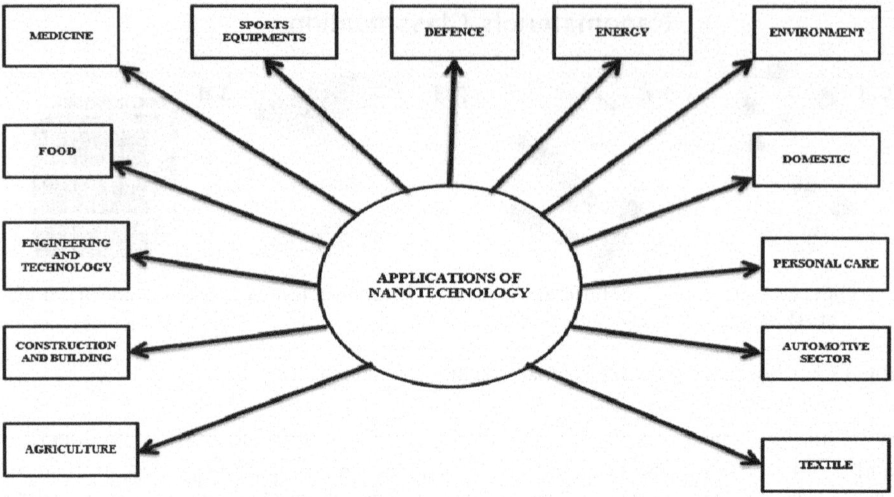

Fig. 15.4 Applications of nanotechnology

equipment, whose performance and durability are being enhanced with the assistance of nanotechnology (Nano werk 2013). In automobile sector, nanopolymers are used in paint coatings, lightweight body parts, self-cleaning nanofilms, scratch- and abrasion-resistant coatings, tires, mirrors and glasses, nanofilters for air cleaning, exhaust emission reduction, cooling systems, and many more. In aerospace, it is being applied to engines (improves heat transfer, lubrication) and tires (reduced friction, durability). In marine transport, nanostructured coatings are scratch resistant, and body parts made of nanoclay are light and tough (Mathew et al. 2018). Nanoscale sensors and instrument may possibly provide cost-effective constant monitoring of the structural integrity and functioning of bridges, tunnels, rails, parking structures, and pavements in due course (National Nanotechnology Initiative 2012).

2. *Food safety and as cleaning agent*: Nanotechnology has come to light as a scientific development tool to transform the agrifood sector, with the potential to improve global food production, in addition to the nutritional use, quality, and safety of food (Peters et al. 2016; King et al. 2018). Nanomaterials have been used to eradicate contaminants (Sun et al. 2013; Zhu et al. 2012), as antimicrobials (Kim et al. 2007; Morones et al. 2005) and as antioxidants (Deligiannakis et al. 2012). Besides being used in sensors for detection of pesticides, chemicals, heavy metals, foodborne pathogens and toxins, preservation, and food packaging (Gestal and Zurita 2015; De Francisco and GarcÍA-Estepa 2018), nanotechnology is also used in the removal of pesticides (Yu et al. 2017; Yu and Yang 2017; Krishna et al. 2018). Food industry includes food nano-preservation (encapsulation, humidity treatment), food nano-functionality (enhancement of physical qualities and nutritive value), and food nano-packaging (intelligent sensing/active antimicrobial/bio-based edible packaging) (Bajpai et al. 2018). Nanomaterials are used as cleaning agents in industrial solvents to remove groundwater contamination

Table 15.1 Nanomaterial particles commonly in use as coating materials

Feature	Nanomaterials commonly in use	Nanoparticle supplier/s
Antimicrobial activity/therapy	Fe_3O_4, NO, CuO, TiO_2, ZnO	Creative Diagnostics
Polymeric systems as gas barrier (decrease in gas permeability)	Polyimide (PI) nanotubes (silica, clay)	Southern Clay
Corrosion control	Li_2TiO_3, Al_2O_3,Cr_2O_3, Nanoclays (hydrotalcite, montmorillonite, etc.), boehmite nanofillers	Altairnano
Electrical conductivity, static charge dissipation	SnO_2, indium tin oxide (ITO), iron, antimony tin oxide (ATO), cobalt	Ishihara (ISK)
Fire retardant	Carbon nanotubes, nanoclays	Nanocor
IR absorption/reflection	C, ITO, Pt, ATO, SnO_2	Nanophase
Magnetic	Fe_2O_3,Fe_3O_4	TurboBeads
Mechanical, scratch resistance	TiO_2, Al_2O_3, ZnO, SiO_2, ZrO_2	Nissan Chemical
Photocatalytic/self-cleaning technology	Fe_3O_4,TiO_2/SiO_2, ZnO	Nanoscale
UV stability	Montmorillonite, SiO_2, TiO_2, ZnO, $BaSO_4$, CeO_2	Antaria
Nanotubes	Carbon, silica, cellulose	Adnano technologies, Cheap Tubes, EMFUTUR Technologies
Diesel additive to reduce fuel consumption	Cerium oxide	Energenics
Targeted delivery of drugs to tumors	Gold nanoparticles	CytImmune
Nanoparticles to treat groundwater pollutants	Iron nanoparticles	SiREM
Fabric treatments	Silver nanoparticles	NanoHorizons

(Yunus et al. 2012). Inexpensive, portable water purification systems using nanoscience can provide universal access to safe water (Pulizzi and Sun 2018; Jassby et al. 2018). Commonly used materials using nanotechnology in water purification systems have been documented (Suthar and Gao 2017). Nano-based films that can be applied to countertops or windows makes it much easier to keep those surfaces clean. Decontamination of industrial sites through environmental remediation by removal of numerous contaminants is another field where nanotechnology is working wonders (Guerra et al. 2018).

3. *Medicine and cosmetics*: Glucose-sensing tattoos, contact lenses, and skin patches examine diabetics' blood sugar levels and cautions when its level reaches upper or lower extreme (Aslan et al. 2005). In drug delivery systems, nanoparticles are effective due to site-specific targeting, high drug loading, side effect reduction, controlled and sustained drug release, manipulation of drug particle size, and surface characteristics (Simonazzi et al. 2018). In healthcare sector-inexpensive flexible and durable "electronic skin" sensors are used in robotic

systems and prosthetic limbs. Also nanotechnology helps in the field of visual realities, professional sports and entertainment sector to detect changes in pressure, humidity, and temperature thus enhancing experience (Lu et al. 2018). In wound dressings, nanotechnology-based drug delivery systems improve wound healing (Kalashnikova et al. 2015). In bandages, nanopolymers aid in increasing the surface area available to kill the germs, thus increasing the ability of the wound to absorb drug more effectively. New nanoscience-based prevention, detection, and treatment technologies could substantially reduce death and suffering from cancer and other deadly illnesses (National Cancer Institite 2017). Nanomedicine covers a range of uses, including nanomaterials, nanoelectronic biosensors, and molecular nanotechnology (Tan et al. 2016). Sunscreens provide protection from UVA and UVB rays without leaving a white residue on the skin (Lohani et al. 2014). An article "Cosmeceuticals market to 2018" projected that the worldwide cosmeceutical market will get to $42.4 billion by 2018 (Lohani et al. 2014; GBI Research 2013). Cosmetic products that have been successfully used are moisturizer, anti-wrinkle and antiaging creams, and face mask (Khezri et al. 2018). A large number of patents have been filed in this domain, and the income generated out of these market products has been bountiful.

4. *In fabric protection, safeguarding wood, paints, and varnishes*: Nanoparticles act as protective shield in fabrics against microorganisms and flammability and robustness against ultraviolet radiation (Rivero et al. 2015). Clothing that protects against toxins and pathogens (Bajpai et al. 2018; Rivero et al. 2015; Guardian 2014, February 14), seaweed capped zinc oxide nanoparticles having antibacterial and UV protection have been made having antibacterial and UV protection have been made (Pandimurugan and Thambidurai 2017). Nanotechnology-based protective coatings on wood have antimicrobial (Teng et al. 2018), fire-resistant, and chemical-resistant properties saving time, effort, and money. Paints and varnishes avert corrosion and microbial expansion as well as provide insulation (AZo Nano 2017).
5. *Areas where nanotechnology has left an indelible mark*: High-memory density devices are capable of storing the entire Library of Congress collection on a device of the size of a sugar cube. It improves banknote security with polymer-based ferroelectric memory (Nano Werk 2013). Nanomaterials enhance definite mechanical, physical, and chemical properties in addition to serving as carriers for some active substances, for instance, antioxidants and antimicrobial agents (Amirabad et al. 2018).
6. *Electronics and energy sector*: It aids in making integrated circuits with features that can be measured on nanometer (nm) scale, thus allowing companies to manufacture computer chips containing billions of transistors that are smaller, cheaper, and faster therefore consuming less power. Currently, flash memory manufacturers use nanolithographics to assemble memory chips as small as 20 nm (UnderstandingNano 2013). Nanotechnology develops clean, inexpensive, renewable power through energy creation, storage, and transmission technologies (Eureka 2018). Energy-efficient, low-emission systems help in reducing energy emissions (Abdin et al. 2018). Nanomaterials as catalyst in cost-effective chemical manufacturing save energy and minimize waste products (Boysen and

Muir 2011). Using nanoparticles in soap makes it work better while producing environmentally less harmful by-products. It is moreover being related to oil and gas withdrawal using nanotechnology-empowered gas lift valves in seaward activities or to the application of nanoparticles to sense infinitesimal down-well oil pipeline breaks (National Nanotechnology Initiative 2012). Applied nanotechnology enhances the functionality of traditional construction materials, reduces material carbon emissions, and increases infrastructure durability in addition to numerous other benefits bestowed by nanoscience technology (Oke et al. 2017).

15.4 Nanotechnology: The Future of Renewable Energy

Ranging from solar power to supercapacitors, nanotechnology has the capability to help resolve a large number of vital energy hurdles. This chapter focuses only on the efficiency enhancement of renewable energy systems using nanotechnology.

15.4.1 Benefits and Applications of Nanotechnology in the Renewable Energy Sector

Energy has become a major challenge in the present times since universal energy requirement is rising constantly. Energy is possibly influenced in a countless number of means by the spur of nanotechnology which thus has the chance to contribute in several ways to the heterogeneous array of hurdles in this sphere (Ramsden 2018). Replacing leading non-renewable energy sources by renewable forms of energy depends on the expansion of competent technologies. In consistent with prediction by the International Energy Agency (IEA), it will escalate from currently approximately 12,000 MTOE up to more than 18,000 MTOE till 2030 (Luther et al. 2015). Going by the current speed, the yearly consumption of energy will be twice of the present demand (Ampelli et al. 2014) by the year 2050 (Pandey 2018). It has thus established as the mainstream and cost-competitive technology in the renewable energy arena. Renewable energy sources, for instance, solar power, hydropower, bioenergy, and geothermal power/heat, have not remained untouched by the spur of nanotechnology boom. Especially, nanomaterials are progressively playing a dynamic role either by enhancing the efficiency of the energy storage system and conversion route or by upgrading the device design and performance (Chen et al. 2015).

15.4.2 Solar Energy

Energy from sun is a paramount renewable energy source. Since the sun is a changeable energy source which produces energy in a diffused form, therefore organizing incident solar radiations is not an easy task. Given that the solar energy is completely natural, it is believed to be a clean energy source (Chen 2015). Its creation is quickly becoming an essential source of renewable energy, being evolved as an alternative to conventional supply of energy. Energy sector especially solar energy storage and generation have therefore been significantly benefitted. Decreased manufacturing costs and improved material effectiveness are precisely two of the possible economic advantages that nanotechnology brings to these sectors (Raghav and Dinesh 2016). The most commonly used semiconductor material in generating energy using solar energy is silicon, and some other semiconductor materials are the basic substances of solar energy. At present, roughly 90% of the globally installed solar cells are made using silicon. There are four generations of PV cells which include (a) first-generation crystalline silicon PV (b) second-generation called thin-film PV, (c) third-generation – these are emerging PV, and (d) fourth generation – called high-efficiency PV (single and multi-junction concentrator). Third-generation PV devices are based upon nanocomposites, i.e., nanostructured semiconductors, organic-inorganic hybrid assemblage, as well as molecular assemblies. In addition to these, there is the field of solar thermal energy where the focus is on solar thermal technologies for electricity generation (Ralph Seitz et al. 2013). Solar thermal energy uses silicon as nanoparticle. Silicon is used in constitution with heavy metal alloys. This segment takes on the application of nanotechnology in the energy-associated processes that deals with the use of radiations from the sun as the source of energy. It can be proficiently used for many practical purposes such as solar cells, solar power plants, solar collectors, seawater desalination, etc. (Hussein 2015). Dye-sensitized solar cells, fuel cells, and solar photovoltaic cells are some other devices under the broad spectrum of systems using solar energy.

15.4.3 Solar Photovoltaic Cells

In this contemporary period of global awareness for sustainable energy development, countries are searching for viable solutions for energy crisis. Electricity formed from photovoltaic (PV) cells is one such example. These cells allow us to exploit the enormous quantity of energy in the form of radiations that the sun sends toward our planet earth (Wong et al. 2014). A solar photovoltaic cell is a device that converts photons from solar radiations by electrons into electricity (Lewis 2007). By solar photovoltaic (PV) electricity, we mean electricity created through the photovoltaic route, i.e., by the straight conversion of photons from solar radiation into electrical current using semiconductor tools (solar cells). It is a combined physical and chemical phenomenon. However, solar cells are termed photovoltaic whether the supply of light is coming from the sun or an artificial source. The surface of photovoltaics consists of an oxide layer which is conducting and a catalytic platinum layer that straightaway converts light from the sun into electricity. Although, solar energy is abundantly available and is free of cost. Still, just near about 0.04% of the fuel energy out of world's total primary energy supply is made available by the photovoltaic technology (IEA 2007). Photocatalysis studies are gaining considerable attention among researchers these days due to the prospects of higher energy savings and less environmental pollution. Near about 40 million tons of carbon dioxide emissions may be saved per annum, if a distributed solar grid is able to fulfill 1% electricity requirement of the world (Hussein 2015). Also, solar photovoltaic is anticipated to have a lesser LCOE (levelized cost of electricity) compared to coal or natural gas-fired generation world over by 2020 (Bleich and Guimaraes 2016). Among other elements such as silicon or additional material systems like selenium/copper/indium, further development is required in alternative cell types, for instance, polymer solar cells, thin-layer solar cells, or dye solar cells.

There are a number of applications making use of energy from sun; photovoltaic (PV) application is the most common one among them. PV market has stretched out considerably since the 1990s. The first-generation PV devices are based upon single or multiple, crystalline p-n junction silicon cells. Presently, PV market is founded on crystalline silicon wafer-based thick solar cells measuring ~150–300 nm. These cells are categorized as the first generation of photovoltaic cells which encompass greater than 86% of the world solar cell market. Current leading market technology, which employs monocrystalline or multicrystalline silicon wafers, scarcely leads to cost decrease through technological development and bulk production. Introducing a thin film layer of 1–2 nm size made of a semiconductor material leads to the introduction of second generation of photovoltaic substances. a-Si has a comparatively higher absorption rate in comparison to crystal silicon making it a viable option for thin film technology in P-V cells. Second generation of solar cells constitutes thin film-based solar cells. The high cost of the conventional photovoltaic cells is the main elementary obstacle to their widespread use. Inkjet printing technology incorporated with devices integrated with particles (e.g., dye-sensitized solar cells or Grätzel cells) present possibly much lower manufacturing price (Bella et al. 2015; Ramsden 2018).

Nanocoatings favor their relevance in PV applications. Multiple reflections thus attained lead to an amplification of the absorption path in turn leading to reduced recombination loss due to decreased absorber layer width. Thus, the energy bandgap of solar cells can be modified. Nanocoatings have valuable self-cleaning features too (Ralph Seitz et al. 2013; Kadırgan 2006). Gold nanoparticles boost the efficiency of PV cells. Fullerenes are widely used as electron acceptors in organic PVs (third generation). Novel fullerene acceptors (e.g., C60 including various organic acceptor materials) are projected to extend to elevated efficiencies. Nanoparticles (e.g., metal particles, metal oxide particles, or carbon-based nanostructures) if added to base fluids like propylene glycol, water, ethylene glycol, etc. augment the thermal conductivity. Crystalline silicon is a brilliant conductor, but it is expensive. The efficiency of energy conversion of roughly 25–30% in a photovoltaic cell is supplied by the crystalline silicon. Alternative conducting materials like titanium oxide are affordable to manufacture. However, in comparison to crystalline silicon, they have even lower conversion efficiency. Nanotechnology can be of immense use in manufacturing complex nanostructured lenses and mirrors to augment collection of solar energy. Aerogels containing nanopores are helpful as thermally insulating and transparent substances to be used as cover materials for solar collectors. The nanocomposite technology can be seen in numerous energy sector applications and is vital to both energy harvesting and storage market. In silicon-based solar cells, nanocomposites contribute in capturing energy generally in the form of nanowires. Predominantly, nanocomposites can be found in photovoltaic technology (inorganic polymer PV cells or as polymer gel electrolyte in dye-sensitized solar cells (DSSC); nanostructured solar cells or quantum dot-sensitized solar cells (QDSSC)) (Schueler 2011; Hashim 2011; Kamat 2007). Nanocomposites can also be sited in higher solar transmittance layers on collector glazing (e.g., Ti-Si-O films) and antireflection casing. In solar energy making and transformation, nanocomposites are applied to overcome interaction of electrons and chemicals (catalyst engineering), limits of single materials in solar spectrum (bandgap engineering), and passage of electrons in the material (defect engineering) as well as to trim down prices. These features result in better charge separation and as a result in improved photovoltaic efficiency (Hashim 2011; Kamat 2007).

15.5 Nanofluids for Solar Energy Applications

PV technology is a combination of conducting polymers or mesoporous metal oxides in addition to an enlarged surface area to augment internal reflections using nanoparticles and therefore makes a solo multi-spectrum layer. One of the economical techniques is to substitute the working fluid with a liquid which includes nanometer-sized molecules called nanofluid. A nanofluid is a colloidal mixture of nanometer-sized substances such as nanofibers, nanotubes, nanoparticles, nanorods, nanowires, or nanosheets in a fluidic system. The base material is a liquid phase, whereas the solid phase contains the nanomaterial. They are used to enhance the heat transfer coefficients, thermal conductivity, viscosity, and diffusivity resulted due to convection. The function of nanotechnology in focusing solar energy like parabolic dish, solar tower, or parabolic trough is because of advanced and novel heat transfer fluids for absorber tubes, e.g., nanocoatings for tubes especially antireflective coatings for collectors (Ralph Seitz et al. 2013). The total market of emerging PV technology involving organic dye cells, CIGS (copper indium gallium selenide) solar cells, and PV cells is expected to jump by 2021 to 7.7 billion US$ illustrating that nanotechnology-based PV market is powerfully associated with the upcoming PV technologies (Ralph Seitz et al. 2013). Better thermal conductivity was reported by Choi in 1995 (Choi et al. 1995) by using nanofluids on adding nanometer-sized particles to a fluid. According to literature study, the improved thermal conductivity of nanofluids is the chief basis of efficiency rise in solar systems (Raghav and Dinesh 2016). Nanometer-sized electrodes enhance electrical conductivity by lithium inclusion and augment the elevated rate potential. It assists in increasing lithium-ion battery's capacity, energy and power density, life cycle, and protection. It decreases the specific current density of active elements. It also reduces the volumetric variations and lattice stress and permits high rate of discharge. In case of anode, nanocomposite materials (e.g., Si- or Sn-based nanomaterials dispersed in a carbon-based matrix) as well as nanostructures (3D porous matter, nanowires, nanotubes, and nanorods) are currently looked into. As regards cathode, nanostructures are explored frequently as polyanion-based complexes or as transition metal oxides. Nanomaterial cathode electrodes present superior energy storage capacity along with discharge kinetics, superior cyclic stabilities due to an extensive surface area, small distances for mass and charge dispersal, and added liberty for volume alteration along with lithium intercalation and release (Ralph Seitz et al. 2013).

15.5.1 Solar Cells

In addition to the employment of electrodes for energy storage, electrodes are also exercised in photovoltaic sector in the solar energy quarter. In this area, indium tin oxide (ITO) commonly used as a transparent electrode can be replaced by several nanoscale materials like metal oxide nanoparticles, CNT, graphene, etc. In solar cells, flat screens, etc., ITO has an additional benefit to be employed as a transparent as well as conductive substance. However, indium is considered as a significant raw material in the solar cell market. Thus, the technologically sound industries are strongly researching for materials in order to substitute indium tin oxide (Ralph Seitz et al. 2013). Dye-sensitized solar cells (DSSCs) chiefly put to use nanocoatings, for instance, over transparent (carbon nanotubes/graphene) electrodes (McWilliams 2010). Ductility, wear and tear resistance, and defense against corrosion and erosion are some of the properties bestowed by thermal spray coatings such as chrome oxide, alumina/titania, tungsten carbide, yttria-stabilized zirconia, and nanoceramics (McWilliams 2010). In addition, titanium nanonets containing Li-ion batteries are coated with SiO_2 or silicon particles in order to prevent reaction of $TiVO_2$ with Ti^+ (Zhou et al. 2011). Ultra-black nanostructured, surface coatings utilized in the solar thermal sector are metal containing nano-sized oxide materials. Solar PV cells are nowadays coated with antireflection surface layer and are known to decrease the surface reflection drastically from 40% to 1% (Ralph Seitz et al. 2013; McWilliams 2010). Metal oxides in solar cells are nowadays substituted by ITO nanoparticles in the medium. Nanocarbon materials (third generation) like carbon nanotubes and graphene are used to enhance the counter electrodes in dye-sensitized solar cells by substituting platinum. The effectiveness of photovoltaic cells can be enhanced drastically by employing an intermediary band in between the conduction band and the valence band within the semiconductor. In this manner, energy levels which don't get placed in the energy bandgap can still be absorbed and converted into current. To achieve desired properties, the energy levels/bandgap of the semiconductor may be reframed by constructing quantum dots to be used in solar cells. Hybrid solar cells (organic-inorganic cell structures) use ZnO nanostructures as electron acceptors. Bad light absorption in thin-film solar cell absorption layers can be resolved by plasmonic metal nanoparticles that use light-scattering and absorption-enhancing layer. Solar cells having nanoparticles made of gold embedded in the plasmonic layer are projected to boost absorption and therefore the efficiency of the cell.

Nanostructures, such as quantum dots, optimally regulate bandgaps of semiconductors to the incident light radiation or emit numerous charge carriers per photon, thus enhancing conversion efficiency. Also this results in better light entrapment, for instance, by averting reflection loss at the front cover from nanostructured anti-reflection layers or by up- or down-conversion of light wave lengths via special nanostructures, which could facilitate better exploitation of the light spectrum.

The phenomenon of solar thermal (conversion of solar energy into heat such as steam and power generation) is commonly used to provide decentralized heat supply in buildings through roof-mounted solar collectors by using parabolic mirrors. The impact of nanotechnology is evident in spheres such as anti-reflection coatings for enhanced energy produce, optimized phase-change stores, and heat exchangers/carbide layers of collectors, which expand energy incorporation plus thermal and mechanical stability.

Contrary to the silicon wafer technology, thin-layer solar cells provide possibility for savings in the solar cell manufacturing because of material savings, low-temperature procedures, integrated cell insulation, and an elevated mechanization level in series production. Besides silicon, combination of elements such as copper/indium/gallium/sulfur/selenium (CIGS cells) as well as III–V semiconductors (e.g., gallium arsenide) is employed, which allow productivity of up to about 20% (Luther et al. 2015).

Different solar cell types can be summarized as below.

15.5.2 Dye-Sensitized Solar Cells (DSSC/DSC/DYSC/Grätzel Cell)

An electron transfer sensitizer (photo-sensitized electrode) is employed to take up the visible light and introduce charge carriers in the juncture between semiconductor and an electrolyte containing a wide and stable band width substrate. In the year 1991, this idea led to the emergence of a dye-sensitized solar cell (DSSC) (Hashim 2011). Dye-sensitized cells employ a sensitized semiconductor element connecting another strongly absorbing material (dye) and multi-junction solar cells to enhance the absorption of solar irradiation by using a layer of thin film semiconductors having bandgaps of variable energies. Thus, this combines a dye sensitizer (light absorbing matter) by using wide bandgap semiconductor (made up of a nanocrystalline structure) as the photo anode enabling the process of optical absorption and charge separation (Fig. 15.5). Semiconductor films such as TiO_2, ZnO, SnO_2, CdSe, etc. are formed of an interconnected range of small crystals (amounting to a couple of

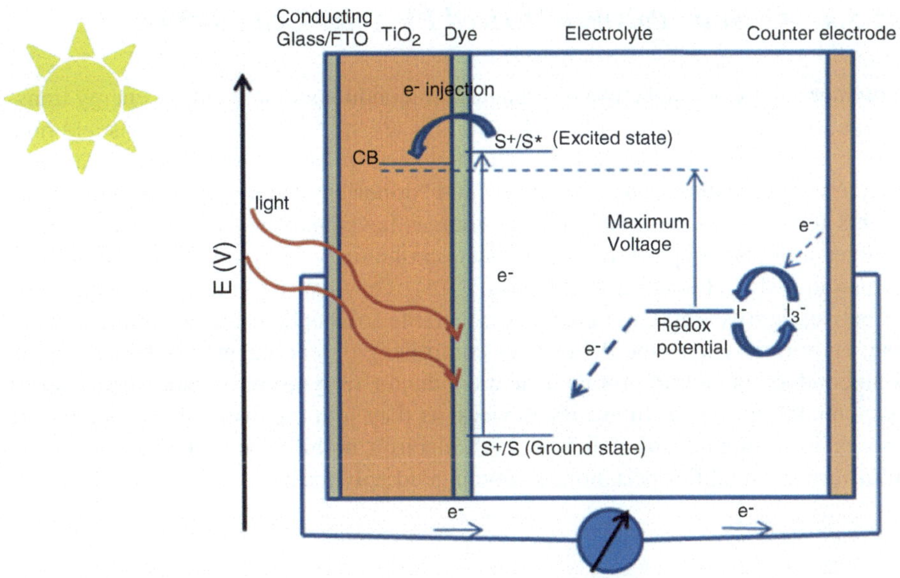

Fig. 15.5 Diagrammatic representation and working of a nanocrystalline dye-sensitized solar cell. (Sharma et al. 2018)

nanometers) in order to permit the process of electronic conduction. Compared to an unstructured electrode, this structure provides an enhanced surface area (nearly more than thousand times) accessible for dye chemisorption. An increase of 10^3–10^4 times in photocurrent and a total power conversion efficiency of 10.4% has been recorded when the sunlight passes using a single crystal into a nanocrystalline electrode for DSSC (Green et al. 2009).

15.5.3 Dye-Sensitized Nanocrystalline Solar Cells

The transport of charge carrier is completely separated from the absorption material in these types of cells. Fluorine-doped tin oxide (FTO) and tin-doped iron oxide (ITO) glass are some of the conducting transparent substrates. To these, a mesoporous nanocrystalline TiO_2 film with a monolayer of the charge transfer dye attached to its surface is pasted. A redox electrolyte is kept in contact to the film. Using photoexcitation, electrons in the TiO_2 conduction band can be introduced from the dye and carried onto the outer circuit in order to push the load and create electricity (Chen et al. 2012). This increases the percentage of energy from the sun (Hussein 2015).

15.5.4 Organic Polymer-Derived PV Solar Cell (OPV)

Polymer solar cells make use of organic semiconductors intended for energy transformation. In this type of solar cell, excitons are split into free electrons and hole pair by the effective electric field. This field is formed throughout the expanse of heterojunction called as the "acceptor" and "donor" molecules. Ascending requirements for low-cost renewable energy sources have enabled OPV solar cells to be utilized as a viable option for the production of energy from light at a minimal cost (Hussein 2015; Hussein and Reviews 2015). They can be used to engineer cost-effective, lightweight, large-area, and adaptable solar cells using economical painting. Hybrid polymer-type PV devices comprising of a conjugated polymer (p-type semiconductor) and an inorganic semiconductor (n-type) have captivated interest as potential option for future PV devices, as they join the distinctive properties of inorganic semiconductors such as high electron mobility and stability with fine film-forming qualities pertaining to conjugated polymers.

15.5.5 Hot Carrier Solar Cells

A highly energetic photon is used to knock a free d-electron higher in the conduction band, thereby raising the electronic temperature to a reasonably high scale, i.e., ~3000 K. However, within some hundred femtoseconds, the hot electron reverts back to the base of the conduction band, in the process passing on the heat energy to the lattice (Hussein 2015).

Nanoparticles present the following advantages to solar power plants (Hussein 2015; Taylor et al. 2011):

1. Small-sized nanoparticles make their way easily using pumps and plumbing of the particles without harmful effects.
2. Nanofluids have the capacity to absorb energy straightway which is greater than the energy lost in the intermediate steps.
3. Nanofluids are capable of being optically selective, that is, low emittance in the infrared band and high absorption within the solar scale.
4. An added, even temperature of the receiver may be attained within the solar collector.
5. Better transfer of heat using elevated thermal conductivity and convection can increase functioning of the receiver.
6. Nanoparticle dimensions may be enhanced to the required function for improving the absorption competence.

Nanomaterials significantly modify the characteristic features of the materials, i.e., biological, physical, and chemical properties. In comparison to macroscopic substances, nanocluster electrons can only assume specific "quantisized" states of energy affected by the quantity of atoms interacting in the process. Metal nanoclusters (e.g., Cu, Ni, Fe, etc.) comprised of a minute quantity of atoms are composed

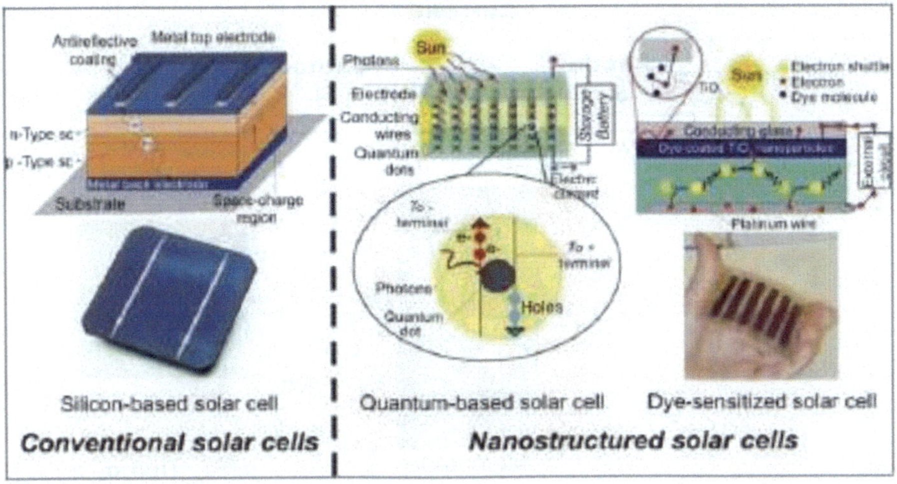

Fig. 15.6 Evolution of the solar cells (Raghav and Dinesh 2016)

either of multiple or a single element and measure characteristically fewer than 2 nm. They exhibit attractive optical, chemical, and electronic properties in comparison to their larger counterparts. The result is a distinctive fluorescence which differs strongly with the cluster size. A 2-nm-sized cadmium telluride particle fluoresces in green light and in red when an element is of 5 nm size. These quantum dots chiefly let a major improvement in the quantum yield and the conversion efficiency of solar cells (Luther et al. 2015).

Some of the benefits of 1D nanomaterials for enhancing solar cell efficiencies are summarized as follows (Yu and Chen 2009):

1. Charge transport follows a direct path.
2. The shape of the nanoparticles provides large surface area for maximum light yield.
3. In 1D nanomaterials, the movement of electrons is typically quite greater than the semiconductor films.

The evolutionary path of solar cells as revolutionized by nanotechnology is presented in Fig. 15.6.

15.6 Hydrogen Energy

15.6.1 Fuel Cells

Fuel cells transform chemical energy with a high effectiveness straightaway into electric current. In addition to pure hydrogen, natural gas, methanol, benzene, or biogas may be utilized to run fuel cells. Hydrogen is not as such an energy source but the carrier of atomic energy. Hydrogen fuel cell technologies transpired as the

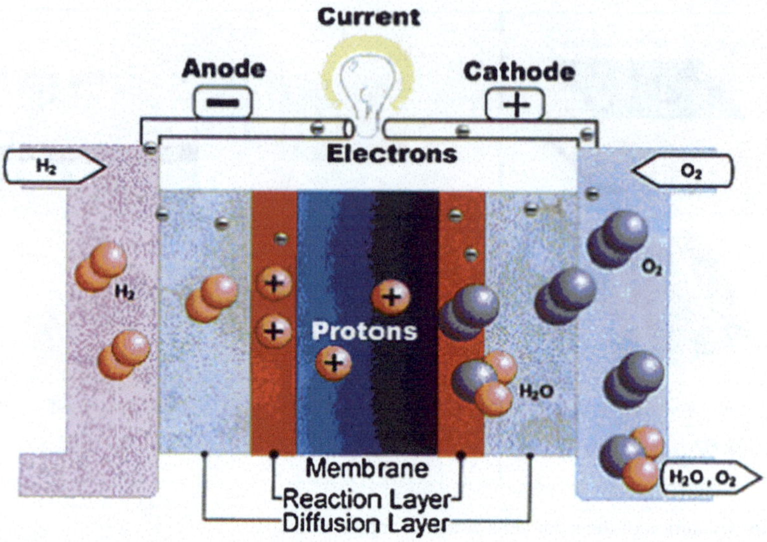

Fig. 15.7 Hydrogen fuel cell. (Hussein 2015; Danilov et al. 2008)

top promising solutions to sustainable energy and environment ("Bud" Peterson et al. 2010). Nanocoatings as titanium-coated nanotubes are used in hydrogen tanks (Ralph Seitz et al. 2013).

In a typical fuel cell, the union of hydrogen (Fig. 15.7) and oxygen molecules produces water. In this process, heat and electricity are released as by-products. This process takes place in an environment-friendly manner, without the release of any detrimental carbon dioxide (CO_2) emissions. However, fuel cells can also be produced using methane and methanol. Operability concerns, high cost, and resilience issues are some of the issues if using fuel cells in spite of their enormous utility. Nanotechnology comes to the rescue of fuel cells by resolving some of their limitations. Fuel cells, catalysts, and electrode membranes may be constructed by incorporating nanomaterials. One of the most proficient applications of nanotechnology in the hydrogen energy sector is concerned to the economical storage of hydrogen fuel since storing large volumes of hydrogen fuel is either highly priced or too immense. Nanoblades have been discovered by scientists to be having the capacity of possibly holding huge volumes of hydrogen fuel (G. 2008). In addition to this, it has the possibility of being put inside a nanomaterial, for instance, carbon nanofibers and carbon nanotubes (CNTs) (Matteo et al. 2012; Ye et al. 1999; Züttel et al. 2002; Nikitin et al. 2008; Gao et al. 2001; Danilov et al. 2008). These days, carbon nanotube fuel cells are being used to collect hydrogen gas which is considered to be environmentally safe. Since hydrogen gas has the best energy-to-weight ratio in comparison to any other fuel, therefore hydrogen energy is still being used widely in the space vehicles (Hussein 2015). Nanocomposite technology has a high potential to be used in devices for carbon dioxide conversion to hydrocarbon fuels like methanol (Kamat 2007), and nanocomposite sheets make available fresh type of merged hybrid device (Ducharme 2009; Pushparaj et al. 2007).

Proton exchange in membrane fuel cells has been enhanced using TiO_2/SnO_2 nanoparticles in the composite nafion membranes. This has augmented the ability of the modified fuel cell in comparison to conventional fuel cell with a nafion membrane predominantly at elevated temperatures (Abbaraju et al. 2008). Using photocatalysis, nanotechnology could be used as a probable tool for the economic production of hydrogen gas using solar energy in an environmentally friendly, clean, and economical manner (Jang et al. 2008). In fuel cells, carbon nanotubes (CNTs) due to their large surface area and thermal conductivity can be inducted into their components and in electro catalyst support system to upgrade its functioning and cut down its cost (Hussein 2015; Zhang et al. 2011).

15.6.2 Diesel Engine

In an internal combustion engine, clean water vapor is released utilizing hydrogen as the fuel. Nanofluids possess an important property of having a high capacity to absorb radiations. Therefore, nanofluids come handy in the way that they can be used ably in decreasing the release of radiations in the combustion process (Hussein 2015).

15.6.3 Biomass/Bioenergy

Biomass consists of all organic materials derived from plants. Biomass is currently projected to supply about 10–14% of the total energy reservoir present in the world (Hendriks and Zeeman 2009). With the stake of roughly 10% universally in the energy supply sector, biomass is presently a major vital renewable energy resource. On one hand, biomass assists in the production of heat and energy and on the other hand the supply of fuels. However, the segment of biofuels in the world fuel market is only approximately 1%, with bioethanol representing the most significant one. Nanotechnology might play an active role in the optimization of efficient biomass utilization by developing catalysts and new conversion protocols in addition to the nano-optimized development of bioresources, e.g., economical use of fertilizers and pesticides using nanoencapsulation and nanosensors (Luther et al. 2015).

15.6.4 Bio-oil

Nano-sized cerium oxide molecules as a supplement in biodiesel fuel significantly lower the release of NO_x constituents and several hydrocarbon compounds into the environment (Sajith et al. 2010). SWCNTs, i.e., single-walled carbon nanotubes having iron oxide nanoparticles imbedded in them, led to the rise of magnetic single-walled carbon nanotubes. This established the route for better efficacy in carbon nanotube-based enzyme bioreactors, thus decreasing economic expenses in industry-related enzyme systems and enhanced productivity of biofuel production (Goh et al. 2012).

15.6.5 Bio-diesel

The nanomaterial oxide-based magnetic catalysts such as strontium oxide, calcium oxide, titanium oxide, magnesium oxide, etc. are being applied for the preparation of biodiesel fuel. This gives an excellent possibility of the fuel being more economical (due to easy separation strength and reuse) by incorporating nanocatalysts having high efficiency, durability, and stability. Therefore, this paves the way for biofuel sector to achieve higher quality and optimal product yield for further development and application purposes (Hu et al. 2011). Mesoporous and carbon-based nanocatalysts like carbon nanofibers, biochar, carbon nanotubes, and graphene oxide have excellent potential for biodiesel production using an extensive range of feedstocks, specially nonfood substrates (Nizami and Rehan 2018).

15.6.6 Wind Energy

Wind energy uses wind as the source of energy. A wind turbine transforms kinetic energy of wind into electrical or mechanical energy which can be exploited for various application purposes. Turbines which run using wind energy don't require burning down the fuel for obtaining energy. As a result, they are considered environmentally friendly. Also, the wind is linked to the presence of sun; so long as the sun shines on this planet, the wind subsists on this planet. World wind power market is evaluated at approximately 27 bn $ (billion USD) (Luther et al. 2015). Nanotechnologies can play a determined part in the enhancement of wind power consumption owing to the presence of tribological coatings in addition to wear protection coat of gearboxes and bearings, conducting nanomaterials for superior lightning shield which permit added economic supply of wind power to the grid, and high-strength lightweight equipments for nanocomposite material-derived rotor blades.

Nanocomposite materials aid in the creation of longer and stronger blades since they have an outstanding capacity to handle stiffness and carry load to weight proportions very effectively. Tribological matters such as scuffing, micro-pitting, gearbox spalling, and wear hold a key aspect in terms of energy loss which can be alleviated using nanotechnology, thus enhancing the wind turbine efficiency. This can be executed via low-friction surface coatings and nano-lubricants (NanoConnectScandinavia 2018). A wear protective tribofilm is created using nano-colloid particle-based boron nitride lubricant additives, a corresponding technology in which it reacts with the boronized surface (Hussein 2015). An epoxy containing carbon nanotubes is utilized to make more grounded and lower-weight windmill sharp edges subsequently longer, and sharp edges enhance the measurement of power produced by every windmill (National Nanotechnology Initiative 2012, January 10). Wind turbine rotor blades are made lightweight, thereby increasing conductivity and strength of the blades by the use of carbon nanotubes (Pandey 2018).

15.6.7 Geothermal Energy

Geothermal energy is described as the heat energy dug out from inside the earth's surface which differs to an extent of 5–10 km in length. The temperature is extremely soaring deep within the earth's interior. Consequently, nanofluids may be employed in the form of a cooling agent to chill out electronic gadgets used in drilling machines such as sensors and pipes operating under extreme temperature and friction conditions. Energy sources derived from the earth's core can be defined in terms of a range of temperature. For example, temperature as low as $^{<}$100 °C to a medium range of 100–150 °C and as high as $^{>}$150 °C can be found when taken out by employing ground heat exchangers (Kalogirou 2005). Nanofluids act as working fluid to mine energy from the earth's interior and transfer it to a power plant system to generate huge amounts of workable energy (Nagar and Agrwal 2013). Nanofluids can absorb substantially higher amount energy in comparison to normal thermal fluids. This observation puts forward an array of future research projections by means of shallow borings to employ geothermal energy (Hussein 2015; Ganguly et al. 2012).

15.6.8 Tidal Energy

Waves are produced by the wind driving over the ocean surface. Moon causes the earth to pass through the course of ocean tidal projections when the earth rotates on its own axis. On earth, all throughout the sea coast, the impact of moon causes the rise and fall of tides almost at a regular interval of 12.5 h. This process goes on consecutively every day, thus making energy driven from tidal waves a prominent resource of renewable energy (Gillett 2002). Wave- powered devices obtain energy from the surface motion of ocean waves or from pressure oscillations below the surface.

Usually, energy from water may be regarded as a major resource of renewable energy due to some rational causes. One reason is that slightly more than two-thirds of the land area of the surface of the earth is surrounded with water. The other is that the world is capable of producing approximately 10,000–15,000TWh wave energy on an annual basis. The energy derived from sea waves can be utilized for producing hydrogen. It can also be used for mining ocean reserves, water purification, synthesizing liquid and solid state fuels, and in several other areas. A recent review (Qu et al. 2012) gives detailed discussion about how nanotechnology presents a future prospect for producing clean and affordable future water supply systems. This is due to many amazing features of nanotechnology like photosensitivity, antimicrobial and catalytic activity, high surface area; magnetic, electrochemical and optical properties thus offering valuable features like specialty adsorbents, high-performance membranes and sensors for monitoring water quality and solar decontamination. The authors also deduced that in order to build up a sustainable water management system, the expansion of nanotechnology must go hand in hand with the environmental safety and issues related to research in health sector (Hussein 2015).

15.7 Conclusions

History shows that science and technology has a profound impact on a society. Nanotechnology is the latest wonder of science whose arrival has created major buzz these days. It brings astonishing benefits to our everyday life. It is a rapidly evolving field and is making long strides in every discipline. Nanotechnology is touching lives through a wide spectrum of areas and is reshaping the future of mankind. It reduces materials to a very small (10^{-9}) scale with improved properties thereby enhancing its applicability. We are bound to see many nanotechnological wonders within the scientific world in the near future. But no one knows certainly what is in store for nanotechnology in the coming times.

Since the official commencement of the universal nanotechnology pursuit, it is one of the most quickly rising fields of technology in every sector. Renewable energy, in particular, is supposed to participate prominently in the contribution toward the energy sector world over in the coming hundred years as the society drifts against the use of fossil energy fuels. Renewable sources lead the way toward a sustainable economy by entering energy markets as a result of R&D. However, nanotechnology should focus more on using renewable energy as the primary energy source in an environmentally friendly manner for the advancement and development of humankind's bright future. These will unquestionably increase the value of life of our society. In an effort to enhance device efficiency and attain commercialization, additional research should be done in order to tailor the nanostructures, to manage charge separation and transport, to effectively utilize cost-effective co-catalysts, and to examine the reaction kinetics and their completion at a larger scale.

The utilization of nanotechnology-based materials in the renewable energy discipline can perform the pivotal task of escalating the effectiveness of renewable energy sources such as wind turbine engines, fuel cells, and solar cells for satisfying the growing appetite of the world for energy. This in turn reduces the hazardous effect of extracting fossil fuels in order to generate the huge amount of energy. It is also a key enabler of prominent decline in the expenditure of costly machinery of energy use and its storage. Additional investigations are therefore required to find out the prospective applications of nanotechnology in geothermal area as well as in the wind energy and tidal energy sector, with potential impacts on other types of the renewable energies.

References

"Bud" Peterson G, Li C, Wang M, Chen G (2010) Micro/nanotransport phenomena in renewable energy and energy efficiency. SAGE Publications Sage UK, London, UK

Abbaraju RR, Dasgupta N, Virkar AV (2008) Composite nafion membranes containing nanosize TiO_2/SnO_2 for proton exchange membrane fuel cells. J Electrochem Soc 155:B1307–B1313

Abdin Z, Alim M, Saidur R, Islam M, Rashmi W, Mekhilef S, Wadi A (2013) Solar energy harvesting with the application of nanotechnology. Renew Sust Energ Rev 26:837–852

Abdin AR, El Bakery AR, Mohamed MA (2018) The role of nanotechnology in improving the efficiency of energy use with a special reference to glass treated with nanotechnology in office buildings. Ain Shams Eng J 9(4):2671–2682

Alkahtani RN (2018) The implications and applications of nanotechnology in dentistry: a review. Saudi Dental J 30:107–116

Amirabad LM, Jonoobi M, Mousavi NS, Oksman K, Kaboorani A, Yousefi H (2018) Improved antifungal activity and stability of chitosan nanofibers using cellulose nanocrystal on banknote papers. Carbohydr Polym 189:229–237

Ampelli C, Perathoner S, Centi G (2014) Carbon-based catalysts: opening new scenario to develop next-generation nano-engineered catalytic materials. Chin J Catal 35:783–791

Anjum M, Miandad R, Waqas M, Gehany F, Barakat M (2016) Remediation of wastewater using various nano-materials. Arab J Chem

Aslan K, Lakowicz JR, Geddes CD (2005) Nanogold plasmon resonance-based glucose sensing. 2. Wavelength-ratiometric resonance light scattering. Anal Chem 77:2007–2014

Aviram A (1988) Molecules for memory, logic, and amplification. J Am Chem Soc 110:5687–5692

AZo Nano (2017) Nanotechnology in the paint industry [Online]. Available https://www.azonano.com/article.aspx?ArticleID=4710. Accessed 18 Dec 2018

Bajpai VK, Kamle M, Shukla S, Mahato DK, Chandra P, Hwang SK, Kumar P, Huh YS, Han Y-K (2018) Prospects of using nanotechnology for food preservation, safety, and security. J Food Drug Anal 26:1201

Bella F, Gerbaldi C, Barolo C, GrÄtzel M (2015) Aqueous dye-sensitized solar cells. Chem Soc Rev 44:3431–3473

Bleich K, Guimaraes R (2016) Renewable infrastructure investment handbook: a guide for institutional investors. World Economic Forum, Geneva

Booth T, Baker M (2017) Nanotechnology: building and observing at the nanometer scale. Pharmacognosy. Elsevier, Boston

Boysen E, Muir NC (2011) Electronics projects for dummies. Wiley Publishing Inc III. River Street Hoboken, NJ

Brinker CJ, Ginger D (2011) Nanotechnology for sustainability: energy conversion, storage, and conservation. In: Nanotechnology research directions for societal needs in 2020. Springer, Dordrecht

Bruce PG, Scrosati B, Tarascon JM (2008) Nanomaterials for rechargeable lithium batteries. Angew Chem Int Ed 47:2930–2946

Buzea C, Pacheco II, Robbie K (2007) Nanomaterials and nanoparticles: sources and toxicity. Biointerphases 2:MR17–MR71

Chen J (2015) Recent developments on silicon based solar cell technologies and their industrial applications. In: Energy efficiency improvements in smart grid components. IntechOpen, Croatia

Chen X, Li C, Grätzel M, Kostecki R, Mao SS (2012) Nanomaterials for renewable energy production and storage. Chem Soc Rev 41:7909–7937
Chen S, Li L, Sun H, Sun J, Lu B (2015) Nanomaterials for renewable energy. J Nanomater 2015:1
Choi S, Singer D, Wang H (1995) Developments and applications of non-Newtonian flows. Asme Fed 66:99–105
Danilov M, Melezhyk A, Kolbasov GY (2008) Carbon nanofibers as hydrogen adsorbing materials for power sources. J Power Sources 176:320–324
De Francisco EV, GarcÍA-Estepa RM (2018) Nanotechnology in the agrofood industry. J Food Eng 238:1–11
Deligiannakis Y, Sotiriou GA, Pratsinis SE (2012) Antioxidant and antiradical SiO2 nanoparticles covalently functionalized with gallic acid. ACS Appl Mater Interfaces 4:6609–6617
Deng J, Lu X, Liu L, Zhang L, Schmidt OG (2016) Introducing rolled-up nanotechnology for advanced energy storage devices. Adv Energy Mater 6:1600797
Ducharme S (2009) An inside-out approach to storing electrostatic energy. ACS Nano 3:2447–2450
El Saliby I, Shon H, Kandasamy J, Vigneswaran S (2008) Nanotechnology for wastewater treatment: in brief. EOLSS 7:1–15
Morgan TA. Eureka (2018) Nanotechnology could help hydrogen become a truly affordable, clean, green energy source. eurekamagazine.co.uk/design-engineering-news/nanotechnology-could-help-hydrogen-become-a-truly-affordable-clean-green-energy-source/180803/
Feynman RP (2012) There's plenty of room at the bottom: an invitation to enter a new field of physics. In: Handbook of nanoscience, engineering, and technology, 3rd edn. CRC Press, Boca Raton
G. G (2008) Public communication of science and technology; alternative energy and nanotechnology. [Online]. PCOSTproject. Available: https://communication.chass.ncsu.edu/pcost/prjects.php. Accessed
Ganguly S, Banerjee D, Kargupta K (2012) Nanotechnology and nanomaterials for new and sustainable energy engineering. Proceedings of the international conference nanomaterials: applications and properties. Sumy State University Publishing, 04NEA09-04NEA09
Gao X, Lan Y, Pan G, Wu F, Qu J, Song D, Shen P (2001) Electrochemical hydrogen storage by carbon nanotubes decorated with metallic nickel. Electrochem Solid-State Lett 4:A173–A175
Gbi Research (2013) Cosmeceuticals market to 2018 – Technological advances and consumer awareness boost commercial potential for innovative and premium-priced products [Online]. Available: http://gbiresearch.com/report-store/market-reports/therapy-analysis/cosmeceuticals-market-to-2018-technological-advances-and-consumer-awareness-boost-commercial-potential-for-innovative-and-pre. Accessed 15 June 2019
Gestal MC, Zurita J (2015) La nanotecnología en la producción y conservación de alimentos. Revista Cubana de Alimentación y Nutrición 25:24
Ghernaout D, Alghamdi A, Touahmia M, Aichouni M, Ait Messaoudene N (2018) Nanotechnology phenomena in the light of the solar energy. J Energ Environ Chem Eng 3:1–8
Gillett SL (2002) Nanotechnolgy: clean energy and resources for the future. Foresight, Reno, NV
Goddard WA III, Brenner D, Lyshevski SE, Iafrate GJ (2012) Handbook of nanoscience, engineering, and technology. CRC press. Taylor & Francis Group, Boca Raton, FL
Goh WJ, Makam VS, Hu J, Kang L, Zheng M, Yoong SL, Udalagama CN, Pastorin G (2012) Iron oxide filled magnetic carbon nanotube–enzyme conjugates for recycling of amyloglucosidase: toward useful applications in biofuel production process. Langmuir 28:16864–16873
Green MA, Emery K, Hishikawa Y, Warta W (2009) Solar cell efficiency tables (version 33). Prog Photovolt Res Appl 17:85–94
Guardian T (2014) Nanotechnology. Small world: nanotech. The nanotechnology in your clothes. theguardian.com/science/small-world/2014/feb/14/nanotechnology-clothes-nanoparticles
Guerra F, Attia M, Whitehead D, Alexis F (2018) Nanotechnology for environmental remediation: materials and applications. Molecules 23:1760
Hashim AA (2011) Advances in nanocomposite technology. BoD–Books on Demand. Intech. Janeza Trdine, Croatia

Hendriks A, Zeeman G (2009) Pretreatments to enhance the digestibility of lignocellulosic biomass. Bioresour Technol 100:10–18
Hu S, Guan Y, Wang Y, Han H (2011) Nano-magnetic catalyst KF/CaO–Fe3O4 for biodiesel production. Appl Energy 88:2685–2690
Hussain A, Arif SM, Aslam M (2017) Emerging renewable and sustainable energy technologies: state of the art. Renew Sust Energ Rev 71:12–28
Hussein AK (2015) Applications of nanotechnology in renewable energies—a comprehensive overview and understanding. Renew Sust Energ Rev 42:460–476
Hussein AK (2016) Applications of nanotechnology to improve the performance of solar collectors–recent advances and overview. Renew Sust Energ Rev 62:767–792
Hussein AKJR, Reviews SE (2015) Applications of nanotechnology in renewable energies—a comprehensive overview and understanding. Renew Sust Energ Rev 42:460–476
International Energy Agency (IEA) (2007) Renewables in global energy supply. An IEA fact sheet. IEA Publication 9, rue de la Federation. Paris, France
Jang JS, Kim HG, Joshi UA, Jang JW, Lee JS (2008) Fabrication of CdS nanowires decorated with TiO2 nanoparticles for photocatalytic hydrogen production under visible light irradiation. Int J Hydrog Energy 33:5975–5980
Jassby D, Cath TY, Buisson H (2018) The role of nanotechnology in industrial water treatment. Nat Nanotechnol 13:670
Kadırgan F (2006) Electrochemical nano-coating processes in solar energy systems. Int J Photoenerg 2006:1–5
Kalashnikova I, Das S, Seal S (2015) Nanomaterials for wound healing: scope and advancement. Nanomedicine 10:2593–2612
Kalogirou SA (2005) Seawater desalination using renewable energy sources. Prog Energy Combust Sci 31:242–281
Kamat PV (2007) Meeting the clean energy demand: nanostructure architectures for solar energy conversion. J Phys Chem C 111:2834–2860
Kandil MM (2016) The role of nanotechnology in electronic properties of materials. Technical Report
Khan I, Saeed K, Khan I (2017) Nanoparticles: properties, applications and toxicities. Arab J Chem
Khezri K, Saeedi M, Dizaj SM (2018) Application of nanoparticles in percutaneous delivery of active ingredients in cosmetic preparations. Biomed Pharmacother 106:1499–1505
Kim JS, Kuk E, Yu KN, Kim J-H, Park SJ, Lee HJ, Kim SH, Park YK, Park YH, Hwang C-Y (2007) Antimicrobial effects of silver nanoparticles. Nanomedicine 3:95–101
King T, Osmond-McLeod MJ, Duffy LL (2018) Nanotechnology in the food sector and potential applications for the poultry industry. Trends Food Sci Technol 72:62–73
Krishna VD, Wu K, Su D, Cheeran MC, Wang J-P, Perez A (2018) Nanotechnology: review of concepts and potential application of sensing platforms in food safety. Food Microbiol 75:47–54
Lewis NS (2007) Toward cost-effective solar energy use. Science 315:798–801
Lewis NS, Nocera DG (2006) Powering the planet: chemical challenges in solar energy utilization. Proc Natl Acad Sci 103:15729–15735
Lohani A, Verma A, Joshi H, Yadav N, Karki N (2014) Nanotechnology-based cosmeceuticals. ISRN Dermat 2014:843687
Lu Y, Biswas MC, Guo Z, Jeon J-W, Wujcik EK (2018) Recent developments in bio-monitoring via advanced polymer nanocomposite-based wearable strain sensors. Biosens Bioelectron 123:167–177
Luther W, Eickenbusch H, Kaiser OS, Brand L (2015) Application of nanotechnologies in the energy sector. Hessen Trade & Invest GmbH. Wiesbaden, Germany
Mathew J, Joy J, George SC (2018) Potential applications of nanotechnology in transportation: a review. J King Saud Univ Sci 31(4):586–594
Matteo C, Candido P, Vera R, Francesca V (2012) Current and future nanotech applications in the oil industry. Am J Appl Sci 9(6):784–793

McWilliams A (2010) Nanotechnology: a realistic market assessment. BCC Res, Wellesley, Mass, USA
Morones JR, Elechiguerra JL, Camacho A, Holt K, Kouri JB, RamÍrez JT, Yacaman MJ (2005) The bactericidal effect of silver nanoparticles. Nanotechnology 16:2346
Nagar MDS, Agrwal G (2013) Nanofluid and its application. Int J Pharm Res Sci 1:297–327
Nano Werk (2013) Nanotechnology in sports equipment: the game changer [Online]. Available: https://www.nanowerk.com/spotlight/spotid=30661.php. Accessed
NanoConnectScandinavia (2018) Nanotechnology for energy applications brochure
National Cancer Instite (2017) Cancer and nanotechnology [Online]. Available: https://www.cancer.gov/sites/nano/cancer-nanotechnology. Accessed
National Nanotechnology Initiative (2012) Benefits and applications [Online]. Available: https://www.nano.gov/you/nanotechnology-benefits. Accessed
Nikitin A, Li X, Zhang Z, Ogasawara H, Dai H, Nilsson A (2008) Hydrogen storage in carbon nanotubes through the formation of stable C– H bonds. Nano Lett 8:162–167
Nizami A-S, Rehan M (2018) Towards nanotechnology-based biofuel industry. Biofuel Res J 5:798–799
Oke AE, Aigbavboa CO, Semenya K (2017) Energy savings and sustainable construction: examining the advantages of nanotechnology. Energy Procedia 142:3839–3843
Pandey G (2018) Nanotechnology for achieving green-economy through sustainable energy. Thought 11:942–950
Pandimurugan R, Thambidurai S (2017) UV protection and antibacterial properties of seaweed capped ZnO nanoparticles coated cotton fabrics. Int J Biol Macromol 105:788–795
Panwar N, Kaushik S, Kothari S (2011) Role of renewable energy sources in environmental protection: a review. Renew Sust Energ Rev 15:1513–1524
Peters RJ, Bouwmeester H, Gottardo S, Amenta V, Arena M, Brandhoff P, Marvin HJ, Mech A, Moniz FB, Pesudo LQ (2016) Nanomaterials for products and application in agriculture, feed and food. Trends Food Sci Technol 54:155–164
Poinern GEJ (2014) A laboratory course in nanoscience and nanotechnology. CRC Press, Boca Raton
Pulizzi F, Sun W (2018) Treating water with nano. Nature Publishing Group. UK
Pushparaj VL, Shaijumon MM, Kumar A, Murugesan S, Ci L, Vajtai R, Linhardt RJ, Nalamasu O, Ajayan PM (2007) Flexible energy storage devices based on nanocomposite paper. Proc Natl Acad Sci 104:13574–13577
Qu X, Brame J, Li Q, Alvarez PJ (2012) Nanotechnology for a safe and sustainable water supply: enabling integrated water treatment and reuse. Acc Chem Res 46:834–843
Quandt A, Özdoğan C (2010) Feynman, biominerals and graphene–basic aspects of nanoscience. Commun Nonlinear Sci Numer Simul 15:1575–1582
Raghav SB, Dinesh V (2016) Recent developments on nanotechnology in solar energy. Int J Eng Comp Sci 5(2):15829–15834
Ralph Seitz BPM, Thielmann A, Sauer A, Meister M, Pero M, Kleine O, Rohde C, Bierwisch A, De Vries M, Kayser V (2013) Nanotechnology in the sectors of solar energy and energy storage. International Electrotechnical Commission (IEC), Geneva, Switzerland
Ramsden J (2018) Applied nanotechnology: the conversion of research results to products. William Andrew, Cambridge, MA
Rivero PJ, Urrutia A, Goicoechea J, Arregui FJ (2015) Nanomaterials for functional textiles and fibers. Nanoscale Res Lett 10:501
Sahaym U, Norton MG (2008) Advances in the application of nanotechnology in enabling a 'hydrogen economy'. J Mater Sci 43:5395–5429
Sajith V, Sobhan C, Peterson G (2010) Experimental investigations on the effects of cerium oxide nanoparticle fuel additives on biodiesel. Adv Mech Eng 2:581407
Santos CS, Gabriel B, Blanchy M, Menes O, GarcÍA D, Blanco M, Arconada N, Neto V (2015) Industrial applications of nanoparticles–a prospective overview. Mater Today 2:456–465
Satyanarayana K, Mariano A, Vargas J (2011) A review on microalgae, a versatile source for sustainable energy and materials. Int J Energy Res 35:291–311

Schueler A (2011) Nanocomposite coatings for solar energy conversion: large opportunities for small structures. Lecture series at ETHZ, Zurich

Serrano E, Rus G, Garcia-Martinez J (2009) Nanotechnology for sustainable energy. Renew Sust Energ Rev 13:2373–2384

Sharma K, Sharma V, Sharma S (2018) Dye-sensitized solar cells: fundamentals and current status. Nanoscale Res Lett 13:381

Simonazzi A, Cid AG, Villegas M, Romero AI, Palma SD, BermÚdez JM (2018) Nanotechnology applications in drug controlled release. In: Drug targeting and stimuli sensitive drug delivery systems. Elsevier. William Andrew Publishing, New York, United States

Sun S-P, Zeng X, Lemley AT (2013) Nano-magnetite catalyzed heterogeneous Fenton-like degradation of emerging contaminants carbamazepine and ibuprofen in aqueous suspensions and montmorillonite clay slurries at neutral pH. J Mol Catal A Chem 371:94–103

Suthar RG, Gao B (2017) Nanotechnology for drinking water purification. In: Water purification. Elsevier. Academic Press, San Diego, United States

Tan A, Chawla R, Natasha G, Mahdibeiraghdar S, Jeyaraj R, Rajadas J, Hamblin MR, Seifalian AM (2016) Nanotechnology and regenerative therapeutics in plastic surgery: the next frontier. J Plast Reconstr Aesthet Surg 69:1–13

Taylor RA, Phelan PE, Otanicar TP, Walker CA, Nguyen M, Trimble S, Prasher R (2011) Applicability of nanofluids in high flux solar collectors. J Renew Sustain Ener 3:023104

Teng T-J, Arip MNM, Sudesh K, Nemoikina A, Jalaludin Z, Ng E-P, Lee H-L (2018) Conventional technology and nanotechnology in wood preservation: a review. Bioresources 13:9220–9252

Tiwari JN, Tiwari RN, Kim KS (2012) Zero-dimensional, one-dimensional, two-dimensional and three-dimensional nanostructured materials for advanced electrochemical energy devices. Prog Mater Sci 57:724–803

Understandingnano.Com (2013) Computer memory improvements with nanotechnology,computer memory: applications under development [Online]. Available: http://www.understandingnano.com/nanotechnology-computer-memory.html. Accessed

Wong KV, Perilla N, Paddon A (2014) Nanoscience and nanotechnology in solar cells. J Energy Resour Technol 136:014001

Ye Y, Ahn C, Witham C, Fultz B, Liu J, Rinzler A, Colbert D, Smith K, Smalley R (1999) Hydrogen adsorption and cohesive energy of single-walled carbon nanotubes. Appl Phys Lett 74:2307–2309

Yu K, Chen J (2009) Enhancing solar cell efficiencies through 1-D nanostructures. Nanoscale Res Lett 4:1

Yu X, Yang H (2017) Pyrethroid residue determination in organic and conventional vegetables using liquid-solid extraction coupled with magnetic solid phase extraction based on polystyrene-coated magnetic nanoparticles. Food Chem 217:303–310

Yu X, Ang HC, Yang H, Zheng C, Zhang Y (2017) Low temperature cleanup combined with magnetic nanoparticle extraction to determine pyrethroids residue in vegetables oils. Food Control 74:112–120

Yunus IS, Harwin, Kurniawan A, Adityawarman D, Indarto A (2012) Nanotechnologies in water and air pollution treatment. Environ Technol Rev 1:136–148

Zekić E, VukoviĆ Ž, HalkijeviĆ I (2018) Application of nanotechnology in wastewater treatment. Građevinar: časopis Hrvatskog saveza građevinskih inženjera 70:315–323

Zhang W, Ravi S, Silva P (2011) Application of carbon nanotubes in polymer electrolyte based fuel cells. Rev Adv Mater Sci 29:1–14

Zhou S, Yang X, Lin Y, Xie J, Wang D (2011) A nanonet-enabled Li ion battery cathode material with high power rate, high capacity, and long cycle lifetime. ACS Nano 6:919–924

Zhu J, Sadu R, Wei S, Chen DH, Haldolaarachchige N, Luo Z, Gomes J, Young DP, Guo Z (2012) Magnetic graphene nanoplatelet composites toward arsenic removal. ECS J Solid State Sci Technol 1:M1–M5

Züttel A, Sudan P, Mauron P, Kiyobayashi T, Emmenegger C, Schlapbach L (2002) Hydrogen storage in carbon nanostructures. Int J Hydrog Energy 27:203–212

Chapter 16
Wastewater and Industrial Effluent Treatment by Using Nanotechnology

Sourav Maity, Debopriya Sinha, and Angana Sarkar

16.1 Introduction

Human activities are known to affect the quality of potable water. Any water whose quality has been degraded due to human use is wastewater. Wastewater is basically the water that has been already used for domestic purposes, industrial use and commercial or agricultural activities. It can also be a result of run off from surface or storm water or may be inflow from sewer and/or sewer infiltration. Wastewater characteristics depend on its source. Wastewater can be classified as – domestic, municipal and industrial wastewater having its source from households, communities and industrial activities, respectively. The pollutants present in wastewater can be physical, chemical (both organic and inorganic) or biological in nature.

The physical pollutants will mainly include the suspended solids, whereas the chemical pollutants will include heavy metals such as lead; mercury; chromium; soluble organic materials like drugs, fruit sugars, soluble proteins, etc.; insoluble organic materials (ammonia, hydrogen sulphide); and emulsions like paints, adhesives and emulsified oils. Biological pollutants may include yeast surplus (from breweries), tissues, skin, hair (from tanneries), etc.

The rapidly increasing industrial growth has led to the formation of huge amount of industrial water loaded with toxic pollutants that are hazardous for the environment as well as human health. Industrial wastewater is one of the most important sources of water pollution (Sethi 2018). In this book chapter, we will primarily focus on industrial wastewater. There are a number of industrial activities that are

S. Maity · D. Sinha
National Institute of Technology, Rourkela, Odisha, India

A. Sarkar (✉)
Department of Biotechnology & Medical engineering, National Institute of Technology, Rourkela, Odisha, India
e-mail: sarkara@nitrkl.ac.in

I. Bhushan et al. (eds.), *Nanomaterials and Environmental Biotechnology*, Nanotechnology in the Life Sciences,
https://doi.org/10.1007/978-3-030-34544-0_16

responsible for generating industrial wastewater. Some of them include industrial site drainage, cooling waters used in industries such as heat slime, industrial processing waters, water used in hydraulic fracturing, emulsions from paper mills, etc. Battery manufacturing units release heavy metals such as cadmium, chromium, mercury and lead along with oil, grease, silver and zinc. The pollutants discharged by organic chemical manufacturers differ from plant to plant depending upon the variety of products produced. The most common organic compounds released from these units include benzene, chloroform, naphthalene, phenols, toluene and vinyl chloride (Ebrahiem et al. 2017). They may also release metal pollutants like chromium, copper, lead, nickel and zinc. Electric power plants particularly coal-based power stations release metals like mercury, cadmium and lead as well as compounds of nitrogen, arsenic and selenium of significant levels. At times, the wastewater stream may contain fly ash, bottom ash and flue gas (which are transferred from the polluted air to water via means of wet scrubbers). Food processing units may contain certain dissolved organic matter along with high loads of particulate matter and some surfactants (Britz et al. 2006). The cooling waters of iron and steel industries are contaminated with ammonia, cyanide, benzene, naphthalene, anthracene, phenols, cresols and complex organic compounds collectively known as polycyclic aromatic hydrocarbons (PAH). Pollutants in the wastewater may also include hydraulic oils, tallow and even particulate solids. Discharges from paper and pulp manufacturing units include dioxins, furans and phenols. Apart from these, the discharged water also contained lots of sugars and lignocelluloses. Textile dyeing industries effluents contain synthetic and natural dyestuff, guar, wetting agents and dyes (Robinson et al. 2001). Effluents from soap and detergent industries contain complex chemicals, and they also had high levels of zinc and iron in them along with oil and grease. Thus, the industrial wastewater was treated using methods such as dissolved air floatation, chemical coagulation and finally activated sludge process.

The characteristics of effluents generated from different industries and manufacturing units differ and so do their treatment technologies. Thus, many treatment technologies had been designed and developed in the past for removal of pollutants from industrial wastewater such as distillation, adsorption, chemical immobilization, aerated lagoons and so on. The characteristics of oil (droplet size, suspension state) are very important to effectively remove it from wastewater. API separators and hydrocyclones can be used in many industries to remove oil. For removal of biodegradable organic materials, methods such as trickling bed filter or activated sludge process were used. If wastewater contains synthetic organic compounds (from paint factories or pharmaceuticals), the advanced oxidation processing, distillation and vitrification can be applied. For wastewaters rich in acids and alkalis, neutralization is the preferred method of treatment. With modernization the type of waste being generated is changing day by day, and so the treatment technologies need to be advanced and efficient enough to effectively remove contaminants from wastewater. Advanced wastewater treatment is comprised of several processes such as reverse osmosis, membrane filtration, chemical coagulation and chemical precipitation depending upon the particular type of effluent produced by the respective industry. Earlier the conventional methods mainly aimed at removing the suspended materials present in wastewater and treatment of organic pollutants in it. Recently the focus has shifted to other important contaminants in wastewater such as dissolved solids, heavy metals, toxicants, volatile compounds, solvents, radioactive wastes, etc. so that finally the discharged water after proper treatment can be reused for some other purpose and should not have any negative effect on human health or on the ecosystem.

Wastewater treatment involving traditional methods has become increasingly challenged due to the identification of various other contaminants, rapid population growth and increasing industrial activities. Thus the need for switching toward advanced treatment technologies is the need of the hour, which can remove various probable harmful organic or inorganic compounds or microbial contaminants that cannot be removed effectively and efficiently by using conventional treatment processes. Advanced treatment processes include membrane filtration, nanotechnology, advanced photo-oxidation process, automatic variable filtration (AVF) technology, microbial fuel cells, natural treatment systems, coke oven by-product treatment and many more.

16.2 Existing Pollutants and Their Traditional Treatment Technologies

The nature and characteristics of pollutants released in industrial wastewater depend on the type of products manufactured by the industries. Most of the industries and manufacturing units produce wastewater which has to be treated to reduce the pollutant load in water. This water after treatment can be reused for some other purposes or if released into the environment will have less chances of pollution. Industrial wastewater treatment includes numerous processes that render the wastewater (an undesirable by-product of industries) less polluting. The treatment technology to be used depends on a lot of factors, such as – composition of the wastewater, type of pollutants in the wastewater, operation costs, etc. The wastewater from textile industries contains impurities that are present in suspended form. They are first coagulated and precipitated in order to generate micro-flocs, and then these micro-flocs were removed by filtration. For removal of COD and colour, simple aerobic treatment is carried out. The dyes present in the wastewater are very difficult to treat by this method because they are organic molecules which are recalcitrant. Hence, they show resistance to aerobic treatment. For colour removal, the technologies developed are classified as physical, chemical and biological (Robinson et al. 2001). The most preferred wastewater treatment technology is biological treatment because it is economical when compared to other methods. Biological treatment methods include fungal decolouration, microbial degradation and adsorption by microbial biomass. Chemical methods involve coagulation, flocculation, floatation, filtration, electro-floatation, electro-kinetic coagulation, conventional oxidation and electrochemical processes. Out of all the existing physical methods, adsorption is the most preferred method because it does not form any harmful substance. The spent grains released by the breweries usually undergo membrane filtration. Highly suspended solids, bleaching agents and solvents released from paper and pulp industries are subjected to adsorption, sedimentation chemical coagulation, reverse osmosis and advanced coagulation.

The wastewater from food industries contains surfactants, dissolved organic matter and particulate load which are treated by dissolved air floatation and gravity oil separators (Britz et al. 2006). Activated sludge treatment method is used to remove oestrogenic hormones that are given out in wastewater from pharmaceutical industries. For most of the biodegradable organic wastes present in wastewater, either activated sludge treatment or trickling bed filter method is used. Heavy metals like iron, lead and zinc, paints and cyanide can be removed by ion exchange or by chemical precipitation. Advanced oxidation process is usually used to treat industrial effluents where they transform toxic organic compounds into biodegradable substances. It is used for treatment of synthetic organic materials like solvents, paints, etc. It can also be used to remove micro-pollutants from wastewater. At times sonolysis has been used to remove chemical contaminants from wastewater which involves acoustic cavitation (Shah 2016). The most widely used adsorbent is activated carbon due to its extensive micro-porous structure and high surface reactivity and adsorption capacity. Membrane separation technologies are used to remove dyes and paints from wastewaters. Simple sedimentation techniques are used to remove most of the solids present in industrial wastewater with the solids obtained as slurry or may be sludge. Oils that float in the industrial effluent can be removed via skimming devices, which is a cheap and dependable method to remove oil, grease and hydrocarbon from wastewater. The effluents from the dairy industry include dissolved sugars and proteins, fats and other additive residues, and it may also contain pathogens from production processes (Britz et al. 2006). The colour and odour from dairy wastewater can be removed by activated carbon treatment. Fats, oils and grease can be removed using enzymatic hydrolysis (Cammarota et al. 2001).

The application of traditional and conventional methods of treatment to industrial wastewater is becoming a challenging task since the type and nature of contaminants in the wastewater are changing day by day, and thus their effectiveness and efficiency have become restricted. Thus, new technologies for treatment of industrial effluents must be developed that will have satisfactory efficiency in removing pollutants from wastewater.

16.3 Advanced Technologies for Wastewater Treatment

Wastewater treatment involving traditional methods has become more and more challenging with the identification of increasing variety of contaminants (let it be toxic or non-toxic), increasing industrial activities and rapid population growth. So, advanced wastewater treatment technologies are the need of the hour to remove various probable toxic compounds that could not be removed effectively and efficiently by using traditional (or conventional) wastewater treatment processes.

Some of the new advanced technologies that have been introduced for wastewater treatment are as follows (Shah 2018; Sethi 2018).

16.3.1 Membrane Filtration

This technology is of very much importance for the development and designing of advanced, new and improved water reclamation systems. A wide variety of dissolved contaminants could be removed by using micro and ultrafiltration membranes, which provide an excellent wastewater pretreatment process. Membrane bioreactor filtration technique is being used extensively for advance treatment followed by RO and UV treatment to produce non-potable water.

16.3.2 Nanotechnology

The evolution of nanotechnology has remodelled the wastewater treatment process. The most important feature of nanotechnology is that it can be incorporated with any other existing technologies to modify and/or clarify the concept with ease. Ongoing researches focus for production of fabricated membranes from nanomaterials to break down the toxic compounds from the wastewater during the treatment itself.

16.3.3 Automatic Variable Filtration (AVF) Technology

AVF technology used for wastewater treatment works on the principle in which the upward flow of the influent is rinsed by the downward flow of the filter media. The main advantage of this process is that there is no necessity for any further filter media cleaning or fresh water since during the treatment process itself, the filtered influent clears the filter media.

16.3.4 Advanced Photo-Oxidation Process (APOP)

APOP is a simple system used for wastewater treatment for disinfection and for removal of endocrine disruptors and other hazardous compounds.

16.3.5 Microbial Fuel Cells

It's the technology which involves the extraction of electrical energy from the organic matters existing in the wastewater stream directly by electron transfer to acquire the energy produced by microorganisms.

16.3.6 New Urban Sanitation Technology

This is the treatment which aims at reusing of energy and minerals from wastewater with unification of anaerobic digestion and electroflocculation (Elflox) technologies.

16.3.7 Natural Treatment Systems (NTSs)

NTSs incorporate a variety of biological, chemical and physical techniques which work concurrently to remove variety range of contaminants including nutrients, micro-constituents inclusive of endocrine disrupting chemicals and pathogens from wastewater.

16.3.8 Coke Oven (CO) By-Product Wastewater Treatment

In the CO effluent, the predominant pollutant is ammonia which would worsen the water pollution scenario if discharged untreated, so its recovery is of very much importance. Trickling filter and activated sludge process are the two most common techniques for the treatment of CO effluents. This treatment process is used by the steel-producing industries in India.

16.3.9 Urine Separating Process

Urine is part of domestic wastewater which comprises 90% and 50% of the nitrogen and phosphorus, respectively. Urine separating toilets have been developed and technologies designed for treating this for the production of fertilizer products with minimal requirement from outside sources.

Figure 16.1 illustrates the overview of various above advanced wastewater treatment processes.

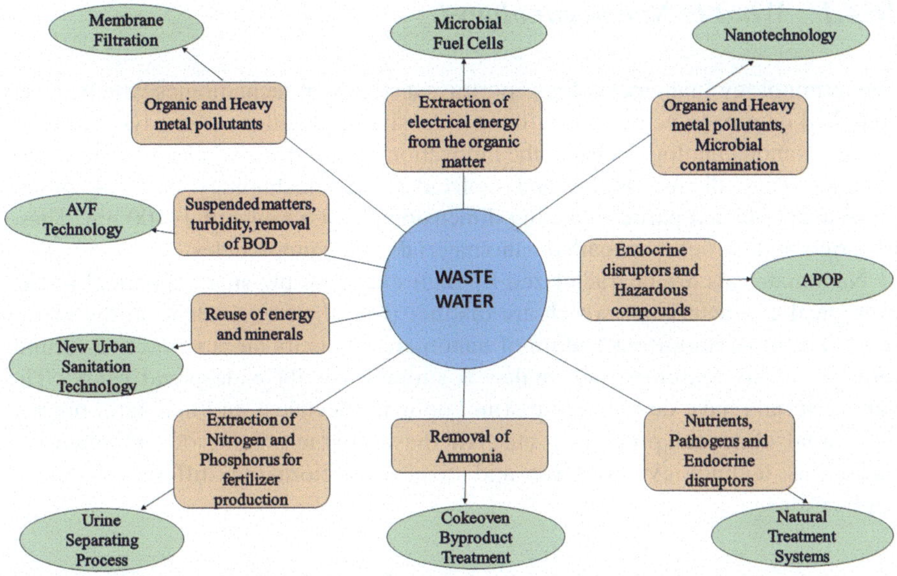

Fig. 16.1 Overview of advanced wastewater treatment processes

16.4 Nanotechnology

The genealogy of the term 'nano' is Greek, meaning of which is 'dwarf'. In scientific terms, a nanometre is defined as one billionth of a matter (10^{-9}) and can also be represented by the length of 10 hydrogen atoms lined up in a queue. The process of nanotechnology started occurring in nature, billions of years ago, when molecules began organizing themselves into complex forms and structures that launched life on Earth. By gradual evolution, mutation and adaptations, plants acquired the power to convert carbon dioxide to oxygen with the help of sunlight by the process which is known as 'photosynthesis'. This photosynthesis process still takes place within tiny structures in plants and is known as 'chloroplasts'. Chloroplast comprises several nanoscale 'thylakoid discs', which consists of a green pigment called 'chlorophyll'. Another similar example of natural nanotechnology can be 'chemical catalysts', through the process 'catalysis', or in bioscience it is known as 'enzymes'. The biomolecules that help in accelerating the rate of a chemical reaction are called enzymes.

The concept that seeded the idea of nanotechnology was first brought into the discourse by a well-known physicist named Richard Feynman in the year 1959, in his talk 'There's Plenty of Room at the Bottom'. But Norio Taniguchi, a Tokyo Science University Professor, first coined the term 'nanotechnology' in the year 1974.

16.4.1 What Is Nanotechnology?

The terminology ‘nanotechnology’ portrays a stretch of technologies which can be complied on a nanometre scale with an extensive application as an advanced technology. Nanotechnology helps in the innovation of new devices, materials and technologies using distinct atoms and molecules. Nanotechnology involves around those materials and particles having dimension about a range of 1–100 nm. These materials and particles are called nanomaterials and nanoparticles.

Nanomaterials are characterized by their exquisite physical, chemical and/or biological characteristics which are distinct from their equivalent on the macro level. The most important feature of nanomaterials is its large surface to volume ratio; it is this feature which makes it suitable for its widespread usage. The nanomaterials have completely distinct optical, electrical and magnetic properties, rapid chemical processes, etc. All these characteristics of nanomaterials make this technology attractive and also revolutionizing different fields of applications.

16.4.2 Nanotechnology in Wastewater Treatment

The unique properties of various nanomaterials enable to innovate novel technologies in terms of wastewater treatment. Nanotechnology is applied in sensing, monitoring and removal of wide range of pollutants. Organic pollutants like benzene, chloroform, polycyclic aromatic hydrocarbons (PAHs), etc. and heavy metal pollution like mercury, lead, cadmium, cobalt, copper, chromium, arsenic, etc. pose a serious threat to the environment due to its toxicity to living organisms including humans. Nanomaterials are expected to enhance the kinetics and efficiency of various chemical and physiochemical processes owing to its large surface area, catalytic activities and high reactivity, hence reducing chemical and energy consumption which is of uttermost importance in wastewater treatment. Various nanotechnology methods such as adsorption, nanofiltration, photocatalysis and membrane processes involve the use of TiO_2, ZnO, carbon nanotubes, ceramic membranes, metal (oxides) and magnetic nanoparticles are used to resolve or reduce pollutant toxicity problems in wastewater.

Nanomaterials used for water treatment and/or purification can be defined in four classes (Chorawala and Mehta 2015):

- Dendrimers
- Metal-containing nanoparticles
- Zeolites
- Carbonaceous nanomaterials

16.4.2.1 Adsorption

The adherence of atoms, ions or molecules from a dissolved solid, liquid or gas to a surface is called adsorption. Efficiency of conventional adsorbent in wastewater treatment is usually limited due to its surface area or the active sites and the adsorption kinetics. Where on the other hand in wastewater treatment process the sorption capacity of the nanosorbents is very high and specific? Nanoparticles exhibit two important characteristics that make them an excellent adsorbent. That is the large surface area of nanoparticles and surface multifunctionality, i.e. the ability to chemically react and bind with different contiguous atoms and molecules with ease. These properties make a nanoparticle more effective adsorbent for various contaminants in wastewater. The most important nanosorbent for wastewater treatment is carbon nanotubes (CNTs) which is used to treat heavy metals like lead, copper, zinc and cadmium and also organic compounds like polycyclic aromatic hydrocarbons (PAHs) (Chen et al. 2007; Li et al. 2003; Lu et al. 2006). Metal-based nanosorbent mainly metal oxides like titanium dioxide, alumina and iron oxide are used for the treatment of heavy metals like lead, mercury, chromium, arsenic, etc. (Sharma et al. 2009). Polymeric nanosorbents such as dendrimers are also used to treat heavy metals in wastewater (Diallo et al. 2005).

16.4.2.2 Nanofiltration

Membrane filtration plays a significant role in providing high level of water purification since it removes various types of contaminants present in wastewater. Nanofiltration (with reverse osmosis [RO]) is a high-pressure membrane treatment process for wastewater. But with respect to RO, nanofiltration process requires much lower drive pressure so energy consumption is also low. This process is highly efficient in the removal of organic and inorganic contaminants, bacteria and viruses (Zekic et al. 2018). Types of membranes used in nanofiltration technique can be classified in three categories (Pendergast and Hoek 2011):

- Nanostructured ceramic
- Biologically inspired membranes
- Organic-inorganic membranes

Some benefits of using nanofiltration for wastewater treatment are (Tambe Patil 2015):

- Lower operating costs
- Lower energy costs
- Reduction of total dissolved solids (TDS)
- Reduction of heavy metals like arsenic, etc.
- Reduction of volatile organic compounds (VOCs) and pesticides
- Reduction of nitrates and sulphates
- Hard water softening

16.4.2.3 Nanofiber

Electrospinning is an economical and efficient process to produce ultra-fine fibres from a wide range of materials like ceramics, polymers or even metals; the resultant nanofibers possess exceptional properties from minute pore size, large surface area to high porosity which makes it ideal for its usage in wastewater treatment. For example, for removal of heavy metal ion, electrospun polyacrylonitrile nanofiber mats are used (Li and Xia 2004). In industrial wastewater treatment process, an advanced method is used where nanofiber technology in combination with biological process helps in the removal of toxic xenobiotics efficiently. The most important advantage of using this technology is that the biofilm formed from the bacteria forms both on the carrier surface and closer to its epicentre (within the carrier) so that the bacteria can be protected against the harmful effects from the adjacent environment (Prachi et al. 2013).

16.4.2.4 Photocatalysis

Photocatalytic oxidation in wastewater treatment is an advanced oxidation process which is used for the removal of trace elements and also microbial pathogens. It is a convenient pretreatment to enhance the biodegradability of non-biodegradable and hazardous contaminants in wastewater (Murakami et al. 2009). Organic compounds which are recalcitrant are also treated by photocatalysis (Han et al. 2009). The major drawback for wide application of this process is its slow kinetics since the photocatalytic activity is limited. In wastewater treatment, photocatalyst which is most widely used is titanium dioxide (TiO_2) due to its chemical stability, low toxicity, raw material abundancy and low cost. Other photocatalysts include fullerene and aminofullerene, but they are expensive and availability is not like TiO_2.

16.4.2.5 Nanocatalysts

In wastewater treatment processes, nanocatalysts are widely used as the catalytic activities are increased by them due to its unique properties like having a large surface area with compatible shape attributes. Nanocatalysts increase reactivity and degeneration of the contaminators. Catalytic nanoparticles which are widely used are zero valence metal, semiconductor materials and bimetallic elements, and these are used for the degeneration of contaminators like polychlorinated biphenyls (PCBs), halogenated aliphatic, organochlorine pesticides, azo dyes, nitro aromatics and halogenated herbicides (Bhattacharya et al. 2013). Development of highly efficient nanocatalyst particles has been done for degrading microbial contaminants in wastewater like silver (Ag) nanocatalyst, n-doped TiO_2 and ZrO_2 nanocatalyst (Xiu et al. 2011, 2012; Liau et al. 1997; Li et al. 2008).

16.4.2.6 Sensing and Monitoring

In wastewater treatment, the prime challenge is monitoring water quality parameters due to concentration being low of certain contaminators and deficit of rapid detection of pathogens. Sensors usually has three major constituents: recognition element, a signal transducer and a detector.

Detection of pathogens is of utmost importance as it is related directly to public health. Coliform test which is a conventional indicator system is slow and fails to sense or monitor the existence of some major or emerging pathogens, which include bacteria (*Helicobacter* and *Legionella*), viruses (hepatitis A and E) and protozoan (*Cryptosporidium* and *Giardia*). Nanomaterials that are most commonly used for detection of pathogens are quantum dots (QDs), carbon nanotubes (CNTs), magnetic nanoparticles, dye-doped nanomaterials and noble metals (Vikesland and Wigginton 2010; Theron et al. 2010; Yan et al. 2007).

For both detection and concentration check for trace metals or organic contaminants, carbon nanotubes (CNTs) are being used since their adsorption capacity is high as well as the kinetics is quite fast (Duran et al. 2009; Cai et al. 2003). For detecting pesticides at ppb levels nano–Au is used (Lisha and Pradeep 2009) and transformed nano–Au detects Hg^{2+} and CH_3Hg^+ (Lin and Tseng 2010).

Table 16.1 represents a brief analysis of application of nanotechnology in treatment of wastewater.

16.5 Pros and Cons of Nanotechnology

As we know there are few things in this world that do not have two-faced implications. Nanotechnology also has both pros and cons (Sharma and Sharma 2012).

Now coming to the pros:

- Efficient – Use of raw materials will become more efficient with the use of nanoscale reagents or catalysts which increases the rate of reaction and other chemical reaction efficiencies.
- Monitoring and protection of the environment – Designing biosensors utilizing nanomaterials helps to detect environmental pollutions. And designing suitable technologies using nanotechnology for targeted pollutant(s) above (or at the borderline of) the permissible limits can be eliminated.
- Biological applications – Using advanced nanotechnology, ultrasmall probes can be developed on planetary surfaces which would help agricultural practices and to control air, water and soil contamination.
- Biomedical applications – Advanced nanotechnology can be used to develop diagnostic tools and for treatment in medical field.
- High precision in reducing the amount of waste.
- Remediating environmental damages.

Table 16.1 Application of nanotechnology in treatment of wastewater

Sl. No.	Type of application	Nanomaterials used	Pollutants	References
1	Adsorption	Carbon nanotubes (CNTs)	Organic compounds such as PAHs (polycyclic aromatic hydrocarbons)	Chen et al. (2007)
			Heavy metals such as lead, copper, zinc and cadmium	Li et al. (2003), Lu et al. (2006)
		Metal-based nanosorbent such as titanium dioxide, alumina and iron oxide	Heavy metals such as arsenic, mercury, lead, copper, nickel, cadmium and chromium	Sharma et al. (2009)
		Polymeric nanosorbent such as dendrimers	Heavy metals such as copper	Diallo et al. (2005)
2	Nanofiltration	Nanostructured ceramic	Heavy metals, nitrates and sulphates, pesticides and VOC (organic chemical)	Kaufman et al. (2010), Lens et al. (2013), Pendergast and Hoek (2011)
		Nano-zeolites		
		Biological inspired membranes such as aquaporin-z from *E. coli*		
3	Nanofiber	Polymers, ceramics or even metals	Organic compounds	Cloete et al. (2010), Li and Xia (2004)
			Heavy metals	
4	Photocatalysis	Titanium dioxide (TiO_2)	Organic compounds	Han et al. (2009), Murakami et al. (2009)
			Pretreatment for hazardous, non-biodegradable and recalcitrant organic compounds	
5	Nanocatalysts	Nano–Ag	Microbial contamination	Xiu et al. (2011, 2012), Liau et al. (1997), Li et al. (2008), Vecitis et al. (2010)
		Nano–TiO_2		
		Nano–ZnO		
		Carbon nanotubes (CNTs)		
		Fullerol and amino fullerene		
		Semiconductor nanoparticles	Organic compounds such as PCBs	Bhattacharya et al. (2013)
		Zero-valence nanoparticles		
		Bimetallic nanoparticles		

(continued)

Table 16.1 (continued)

Sl. No.	Type of application	Nanomaterials used	Pollutants	References
6	Sensing and monitoring	Magnetic nanoparticles, noble metals, quantum dots (QDs), nanoparticles doped with dye and CNTs	Detection of pathogens	Vikesland and Wigginton (2010), Theron et al. (2010), Yan et al. (2007)
		Carbon nanotubes (CNTs)	Trace metals or organic pollutant detection and concentration	Duran et al. (2009), Cai et al. (2003)
		Nano–Au	Pesticide detection	Lisha and Pradeep (2009)
			Detection of Hg^{2+} and CH_3Hg^+	Lin and Tseng (2010)
		Quantum dot (QD)-modified TiO_2 nanotubes	Detection of polycyclic aromatic hydrocarbons (PAHs)	Yang et al. (2010)

Potential risks involved:

- Evaluation involving complete risk assessment of human health safety at all stages of nanotechnology is still not clear.
- Life cycle risk assessment of other environmental implications at all stages is also not clear.
- Synthesizing nanoparticles requires high energy which causes high energy demand.
- Recovery and recycling rates of nanoparticles are very low.
- Lack of experienced engineers and trained workers is of major concern.

16.6 Future Aspects

Nanotechnology is a promising and fresh technology which is being considered for wastewater treatment to provide eco-friendly, economical and convenient treatment methods. Nanoparticles of wide varieties have proved its effectiveness and efficiency in detection and removal of various water contaminants.

Future research aspects should target how to incorporate nanotechnology with the existing wastewater treatment technologies. Since its not practically possible to cease all the existing ongoing wastewater treatment facilities in the ongoing period. So, designing and developing nanoparticles which could be made compatible with minimum changes to the existing technology and infrastructure are of much importance.

The limitation of the use of nanotechnology in wastewater treatment is its economic cost-effectiveness in terms of infrastructure and operating costs and that it might be difficult to retrieve the nanoparticles from the treated solutions, which results in the loss of nanoparticles. The toxicity of the nanoparticles and its environmental impacts are still being investigated to study its impacts on human healthiness and the environment. So development of ecomical nanoparticles eith enhanced efficiency and prevail over the limitations of using nanotechnology in wastewater treatment is one of the most important future prospect of this technology.

It can be envisioned that this advanced technology, with careful management to avoid any undesirable consequences, can be a huge contribution and establish itself as the best wastewater treatment technology.

References

Bhattacharya S, Saha I, Mukhopadhya A, Chattopadhyay D, Ghosh UC, Chatterjee D (2013) Role of nanotechnology in water treatment and purification: potential applications and implications. Int J Chem Sci Technol 3(3):59–64. ISSN 2249-8532

Britz TJ, van Schalkwyk C, Hung YT (2006) Treatment of dairy processing wastewaters. pp. 1–28

Cai YQ, Jiang GB, Liu JF, Zhou QX (2003) Multiwalled carbon nanotubes as a solid-phase extraction adsorbent for the determination of bisphenol a, 4-n-nonylphenol, and 4-tertoctylphenol. Anal Chem 75(10):2517–2521

Cammarota MC, Teixeira GA, Freire DMG (2001) Enzymatic pre-hydrolysis and anaerobic degradation of wastewaters with high fat contents. Biotechnol Lett 23(19):1591–1595

Chen W, Duan L, Zhu DQ (2007) Adsorption of polar and nonpolar organic chemicals to carbon nanotubes. Environ Sci Technol 41(24):8295–8300

Chorawala KK, Mehta MJ (2015) Applications of nanotechnology in wastewater treatment. Int J Innov Emerg Res Eng 2(1):21–26. P-ISSN: 2394-5394

Cloete TE, de Kwaadsteniet M, Botes M, Lopez-Romero JM (2010) Nanotechnology in water treatment applications. Caister Academic Press, 2010

Diallo MS, Christie S, Swaminathan P, Johnson JH, Goddard WA (2005) Dendrimer enhanced ultrafiltration. 1. Recovery of Cu(II) from aqueous solutions using PAMAM dendrimers with ethylene diamine core and terminal NH2 groups. Environ Sci Technol 39(5):1366–1377

Duran A, Tuzen M, Soylak M (2009) Preconcentration of some trace elements via using multiwalled carbon nanotubes as solid phase extraction adsorbent. J Hazard Mater 169(1e3):466–471

Ebrahiem EE, Al-Maghrabi MN, Mobarki AR (2017) Removal of organic pollutants from industrial wastewater by applying photo-Fenton oxidation technology. Arab J Chem 10:S1674–S1679

Han XG, Kuang Q, Jin MS, Xie ZX, Zheng LS (2009) Synthesis of titania nanosheets with a high percentage of exposed (001) facets and related photocatalytic properties. J Am Chem Soc 131(9):3152

Kaufman Y, Berman A, Freger V (2010) Supported lipid bilayer membranes for water purification by reverse osmosis. Langmuir 26(10):7388–7395

Lens PNL, Virkutye J, Jegatheesan V, Kim SH, Al-Abed S (2013) Nanotechnology for water and wastewater treatment. IWA Publishing, 2013

Li D, Xia YN (2004) Electrospinning of nanofibers: reinventing the wheel? Adv Mater 16(14):1151–1170

Li YH, Ding J, Luan ZK, Di ZC, Zhu YF, Xu CL, Wu DH, Wei BQ (2003) Competitive adsorption of Pb2þ, Cu2þ and Cd2þ ions from aqueous solutions by multi-walled carbon nanotubes. Carbon 41(14):2787–2792

Li QL, Mahendra S, Lyon DY, Brunet L, Liga MV, Li D, Alvarez PJJ (2008) Antimicrobial nanomaterials for water disinfection and microbial control: potential applications and implications. Water Res 42(18):4591–4602

Liau SY, Read DC, Pugh WJ, Furr JR, Russell AD (1997) Interaction of silver nitrate with readily identifiable groups: relationship to the antibacterial action of silver ions. Lett Appl Microbiol 25(4):279–283

Lin YH, Tseng WL (2010) Ultrasensitive sensing of Hg(2þ) and CH(3)Hg(þ) based on the fluorescence quenching of lysozyme type VI-stabilized gold nanoclusters. Anal Chem 82(22):9194–9200

Lisha KP, Pradeep T (2009) Enhanced visual detection of pesticides using gold nanoparticles. J Environ Sci Health B 44(7):697–705

Lu CS, Chiu H, Liu CT (2006) Removal of zinc(II) from aqueous solution by purified carbon nanotubes: kinetics and equilibrium studies. Ind Eng Chem Res 45(8):2850–2855

Murakami N, Kurihara Y, Tsubota T, Ohno T (2009) Shape-controlled anatase titanium(IV) oxide particles prepared by hydrothermal treatment of peroxo titanic acid in the presence of polyvinyl alcohol. J Phys Chem C 113(8):3062–3069

Pendergast MTM, Hoek EMV (2011) A review of water treatment membrane nanotechnologies. J Energy Environ Sci 4(6):1946–1971

Prachi, Gautam P, Madathil D, Nair ANB (2013) Nanotechnology in waste water treatment: a review. Int J ChemTech Res. CODEN(USA): IJCRGG ISSN: 0974-4290 5(5):2303–2308

Robinson T, McMullan G, Marchant R, Nigam P (2001) Remediation of dyes in textile effluent: a critical review on current treatment technologies with a proposed alternative. Bioresour Technol 77(3):247–255

Sethi S (2018) Wastewater management: new technologies for treatment. Construction Times, Feb 2018

Shah MP (2016) Industrial wastewater treatment: a challenging task in the industrial. Adv Recycling Waste Manag 2:115

Shah S (2018) Wastewater management: new technologies for treatment. IN FOCUS

Sharma V, Sharma A (2012) Nanotechnology: an emerging future trend in wastewater treatment with its innovative products and processes. Int J Enhan Res Sci Technol Eng 1(2):1–8, ISSN No: 2319-7463

Sharma YC, Srivastava V, Singh VK, Kaul SN, Weng CH (2009) Nano-adsorbents for the removal of metallic pollutants from water and wastewater. Environ Technol 30(6):583–609

Tambe Patil BB (2015) Wastewater treatment using nanoparticles. J Adv Chem Eng 5:131. https://doi.org/10.4172/2090-4568.1000131

Theron J, Cloete TE, de Kwaadsteniet M (2010) Current molecular and emerging nano biotechnology approaches for the detection of microbial pathogens. Crit Rev Microbiol 36(4):318–339

Vecitis CD, Zodrow KR, Kang S, Elimelech M (2010) Electronic-structure-dependent bacterial cytotoxicity of single-walled carbon nanotubes. ACS Nano 4(9):5471–5479

Vikesland PJ, Wigginton KR (2010) Nanomaterial enabled biosensors for pathogen monitoring e a review. Environ Sci Technol 44(10):3656–3669

Xiu ZM, Ma J, Alvarez PJJ (2011) Differential effect of common ligands and molecular oxygen on antimicrobial activity of silver nanoparticles versus silver ions. Environ Sci Technol 45(20):9003–9008

Xiu ZM, Zhang QB, Puppala HL, Colvin VL, Alvarez JJP (2012) Negligible particle-specific antibacterial activity of silver nanoparticles. Nano Lett 12(8):4271–4275

Yan JL, Estevez MC, Smith JE, Wang KM, He XX, Wang L, Tan WH (2007) Dye-doped nanoparticles for bioanalysis. Nano Today 2(3):44–50

Yang LX, Chen BB, Luo SL, Li JX, Liu RH, Cai QY (2010) Sensitive detection of polycyclic aromatic hydrocarbons using CdTe quantum dot-modified TiO(2) nanotube array through fluorescence resonance energy transfer. Environ Sci Technol 44(20):7884–7889

Zekić E, Vuković Ž, Halkijević I (2018) Application of nanotechnology in wastewater treatment. Građevinar 70(04):315–323

Chapter 17
Biomolecular and Cellular Manipulation and Detection (Nanofluidics and Micro- and Nanotechnologies in Integrative Biology)

Shagun Gupta, Vijeshwar Verma, and Vipan Kakkar

17.1 General Introduction

Owing to their size compatibility, it has become possible to interface nanomaterials with biological samples, thus foreshadowing the inevitable era of nanotechnology and microfluidics resulting in an emerging interdisciplinary field known as bionano-electronics (Center for Synthetic Biology). Bionanoelectronics also appends improvements in molecular-scale electronics, a branch of nanotechnology that utilizes single molecule, or nanoscale collection of single molecules, as electronic components.

Molecular electronics, also termed as single molecule electronics, had made it feasible to fabricate hardware by synthesizing the devices and circuits in layers that have the capability of self-assembling, similar to the evolution of structures in living organisms. Devices having different functions are being manufactured using various techniques depending on chemical attractions and can be used for variety of applications rather than the expensive and complex processes previously used to etch electronic circuits (Rakshit et al. 2004; Venere 2005). Molecules might be able to complement or supplement silicon, and in today's world, instead of traditional electronic components, various devices such as computers, sensors, and so on can use molecules, such as DNA, RNA, and proteins. This has spawned new concept – "biochips," described in subsequent sections – which have the potential to utilize proteins in sensors for detecting pollutants and contaminants in sand, water, and air and for the analysis of blood, urine, and other biological samples (Venere 2005; Rakshit et al. 2004; Sørensen 2006; Hersam and Reifenberger 2004).

S. Gupta · V. Kakkar (✉)
School of Electronics & Communication Engineering, Shri Mata Vaishno Devi University, Katra, India
e-mail: vipan.kakar@smvdu.ac.in

V. Verma
School of Biotechnology, Shri Mata Vaishno Devi University, Katra, India

I. Bhushan et al. (eds.), *Nanomaterials and Environmental Biotechnology*, Nanotechnology in the Life Sciences,
https://doi.org/10.1007/978-3-030-34544-0_17

17.2 Buckyballs and Nanotubes

One of the major challenges in developing molecular electronics is the better understanding of how electricity is transferred between silicon contacts and molecules in a circuit (Rakshit et al. 2004; Venere 2005). On August 12, 2005, issue of Physical Review Letters, scientists at Purdue University (West Lafayette, IN) reported that they had designed a nanotech simulation tool that explains the flow of current between individual molecules and silicon atoms so that researchers will be able to fabricate "molecular electronic" devices for advanced sensors and future computers (Venere 2005). According to them, this will help the researchers to precisely analyze the changes in electrical conductivity by employing this new simulation tool. These changes in electrical conductivity depend on how molecules are linked to silicon, information that is highly critical for designing the devices properly. In order to show the flow of current, they have used the tool called buckminsterfullerenes, or "buckyballs," named after architect R. Buckminster Fuller. A buckyball molecule, shown in Fig. 17.2, consists of 60 carbon atoms and is a soccer-ball-shaped molecule with a width of about 1 nm, which is approximately 10 atoms wide. They have predicted the flow of electricity for the three ways in which buckyballs and silicon can be connected. Figure 17.1 shows the geometrical structures with the experimental and theoretical results of all the three scenarios. In the first scenario, there is no chemical bond, and the buckyball is simply sitting on top of the silicon as shown in Fig. 17.1a. In the next scenario (Fig. 17.1b), heating or annealing of silicon has been done in order to connect the molecule to the silicon. And in the third scenario, a natural defect subsisting in silicon has been considered in which the buckyball is resting inside a tiny pit, as shown in Fig. 17.1c. This model precisely plotted the conduction and voltage changes in the three types of connections, and it was seen that the predictions agreed with actual data obtained by the experiments performed by different researchers in laboratories for measuring the current flow alterations (Liang and Ghosh 2005).

Buckyballs are also being explored for improving the (magnetic resonance imaging) MRI's ability for the detection of brain tumors. They can also be employed for destroying these tumor cells by delivering the payloads of radiation. Experiments on rats have been performed and have shown the MRI detection sensitivity has been increased by at least 40-fold with the buckyballs being packed with MRI contrast metal gadolinium. A modified version of buckyballs utilizing a fluorescent metal atom called terbium has also been created by the scientists for guiding the surgeons in removing the tumors with high precision and for those cancer cells that have been missed by surgeons; another metal named lutetium can be added that can provide a lethal dose of radiation to these cells. For the successful testing in humans, this research still requires few more years (Amiji 2008). The "buckyball" concept is now almost over a decade old and has created much excitement among the researchers because of its potential in different applications, but as a commercial product, it is still in its infancy. Another geometry similar to buckyballs but tubular in shape came into existence in 1991 by a Japanese physicist "Sumio Iijima" and has more practical implementations than buckyballs (ETHW 2017). Such tubular

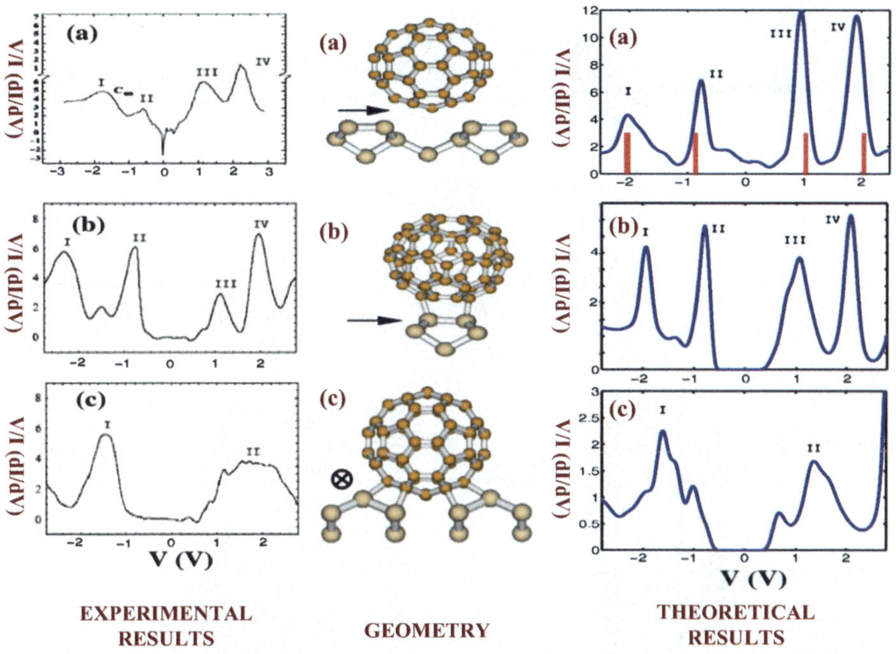

Fig. 17.1 Comparison of experimental conductance–voltage characteristics (G-V) with the theoretical results of G-V available in the literature for the different cases of bonding geometries (Liang and Ghosh 2005)

structures, known as nanotubes, consist of carbon atoms or molecules in the form of tubes or even tubes-within-tubes. Due to the hollow-shaped geometries of both the balls and tubes, as shown in Fig. 17.2, they can be used for drug-delivery application, as carriers for drugs or other chemicals. These structures made of carbon atoms have also found applications in the field of electronics for demonstrating the various electronics components such as transistors, tiny wires, etc. (ETHW 2015). Nanotubes have been added to the batteries in order to increase their surface area, and light-emitting nanotubes are being used in the field of telecommunication. Further, nanotubes are also being used for interacting with nerve cells by sending signals to them.

17.2.1 Application of Nanotubes in Integrative Biology

17.2.1.1 Sending Signals to Nerve Cells via Nanotubes/Neuron-Nanotube Electric Interface

Nanotubes, as defined above, are tiny hollow filaments of carbon with the size of about one- ten thousandth of the diameter of a human hair. Nanotubes have gained much fame due to their versatile properties and have considered as one among the

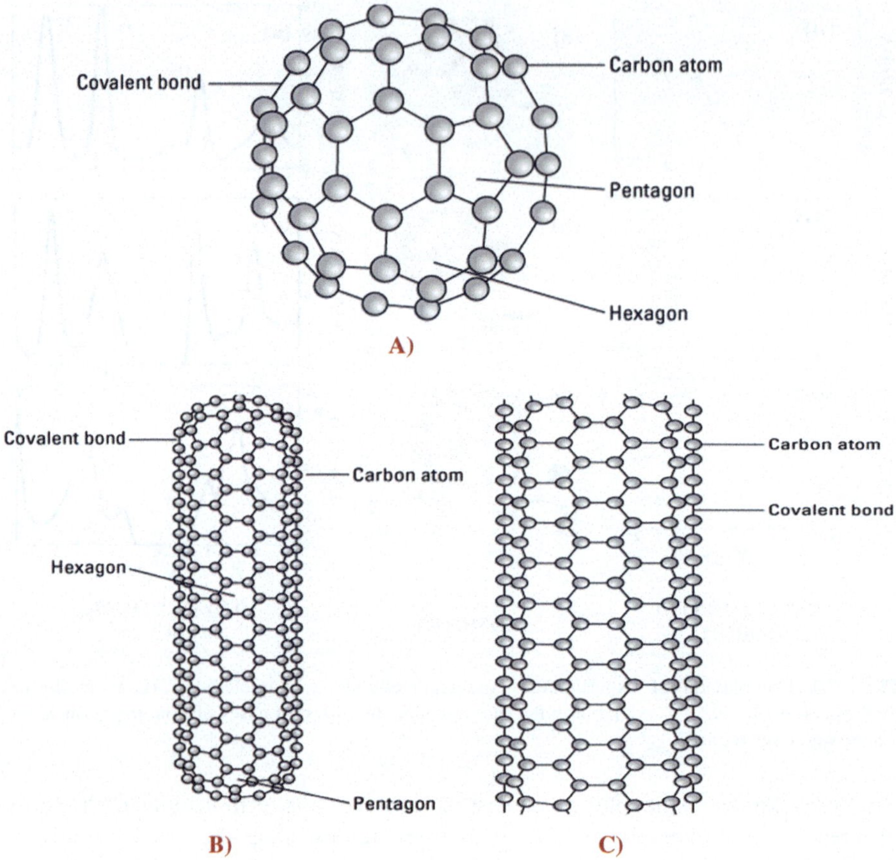

Fig. 17.2 Geometrical structures of (**a**) buckyball molecule, (**b**) a carbon nanotube with close ends, and (**c**) a carbon nanotube with open ends. (Earlboysen 2011a, b. Copyright ©2011, John Wiley and Sons)

most versatile materials ever discovered. They can be cylindrical in shape with open ends or can be a close-ended geometry as shown in Fig. 17.2b, c, respectively. Carbon nanotubes (CNTs) have high strength and density features when compared to steel, are better conductors than copper, can be used as a replacement of silicon in semiconductor chip industry, and can serve as a surface for the growth of cells, if deposited on transparent plastic in the form of thin films. Due to these remarkable characteristics, nanotubes have found applications in almost every field ranging from designing the elevator cables to lift the payloads into Earth orbit to the miniaturized computers or chips that are even smaller than human cells. In May 2006, the researchers at Rice University and University of Texas Medical Branch at Galveston (UTMB) highlighted the electrical interface capability of nanotube films between various biomedical instruments or prosthetic devices and living tissues, in a paper published in *Journal of Nanoscience and Nanotechnology* (Liopo et al. 2006). They have reported that the electrical communication between these two entities has been made possible through a transparent conductive layer designed by them by stimulating the cells. Neurons that had been cultured from experimental rats and neuroblastoma cells mainly utilized in test-tube experiments are the two different cell types employed by the researchers' group in their experiments. Both the types of cells have been placed on single-walled carbon nanotubes' (SWNTs) mats seeded on flexible transparent plastic sheets and having the thickness equivalent to ten layers. An important neural protein has been absorbed by the nanotubes forming a roughly textured carpet for the readily growth of nerves on them. Through these SWNTs, electrical pulses were transmitted and individual cells' responses were recorded by a miniaturized electrode located next to these cells using a microscope. The researchers have also studied the effect of various types of SWNTs on the development and growth of neuroblastoma cells. They have compared the cells that were cultured on unaltered native CNTs and using traditional tissue culture plastic to the cells that were placed on functionalized SWNTs' mats. Functionalization of SWNTs indicates adhering of the surfaces of CNTs with additional molecules in order to improve the CNTs' electrical properties for proper growth of cells. It was seen that native CNTs resulted in better attachment of neuron and supports the growth well. Next step in the research is to make the attachment of neurons better so that they can communicate in an improved manner leading to superior biocompatibility of these surfaces. This can be done by finding an updated way of functionalizing the nanotubes (Liopo et al. 2006; Bourzac 2006).

If CNTs are able to record the continuing electrical activity in cells with high sensitivity, then a device for performing the things like prosthetic control can be designed using them to both deliver and sense stimuli to the cells (Jain 2013). Thus, nanotubes are a good hope for stimulating or replacing the damaged or missing nerve cells in the spinal cord, brain, and eye. Cochlear implant in the people with severe hearing disability is one of the practically implemented applications of neural prosthetics. In this, the electrodes are utilized that can respond to the sound and can transmit the electrical signals directly to the disabled person's brain. Retinal prosthetic concept has also gained much popularity which will help restore the vision of blind people (Bourzac 2006).

In 2016, Usmani et al. in their study have shown that CNTs have the potential as a nervous system prosthesis and can support nerve fiber growth (Usmani et al. 2016). Thus, they can provide a functional reconnection and help in bridging the segregated neural explants. Biocompatible material has been used for in vivo applications, and its implantation in the small rodents' brain has not caused large scars and, thus, did not result in a noticeable immune response unlike the traditional electrodes that caused scarring and inflammation (Amiji 2008).

17.2.1.2 Cell Membrane Interaction with Nanotube Transistor

Advancements in nanotechnology and microfluidics have made it feasible to integrate nanoscale-fabricated structures with various biological molecules and processes resulting in better sensing and electronic control over biological systems.

Various interdisciplinary state-of-the-art technologies like lab-on-chip for early detection of diseases, organ-on-chip, artificial sensory systems, etc. are the outcomes of such integrations and are critically dependent on the nature of interfaces between diversified fields. In March 2005, researchers from University of California, Los Angeles, USA, and Nanomix, Inc., USA, reported the demonstration of an intact biological system interaction with a nanoelectronic device for the first time claiming that this achievement could ultimately be helpful in an artificial eye creation (Artukovic et al. 2005). They tried to overcome the issue of complex hierarchy in biosystems, as organisms are made up of multiple cells and cell membranes comprise tiny objects such as individual proteins located at different distances, by maintaining the interaction of individual nanotubes with individual proteins and interaction of nanotube networks with multiple cell membranes. They had utilized a field-effect transistor (FET) made up of nanotube network with the cell membrane of a bacterium named *Halobacterium salinarum* (Liz Kalaugher 2005). Bacteriorhodopsin, a protein having a light-sensitive membrane, was present in the cell membrane, and metal electrodes' connected dense network of CNTs was at the heart of the transistor. Chemical vapor deposition (CVD) technique had been employed for growing the CNT network FET onto a substrate, and the cell membrane was deposited on this substrate for connecting the components. A silicon chip was then placed on top of a 5-nm-thick cell membrane layer, and rhodopsin dipoles' orientation in the membrane layer was altered by applying an electrical potential to the upper chip. The electrical properties of FET were tested with the different orientations of neighboring dipoles, and it was seen that bacteriorhodopsin's electric dipole remained intact and FET operation was retained. Transfer characteristics of FET had shown that bacteriorhodopsin's dipole lies about two-thirds of the way from the exterior of the cell to the inside of the membrane. In order to calculate the effect of this dipole on nanotubes, a simple electrostatic model as shown in Fig. 17.3a was used by the researchers. In this model, rhodopsin molecules form a constant dipole density line above a nanotube and with density equivalent to "π"; a charge density denoted by "$\lambda = -r\pi/d^2$" has been induced by the line of dipoles as mentioned in Fig. 17.3b. Thus, with the knowledge of both the dipole-moment of bacteriorhodopsin and the induced charge, one can calculate the distance of dipoles from the nanotubes. Conclusively, the researchers have successfully demonstrated that it is possible to directly connect the living cells with the nanoelectronic devices and such nanodevices can be used as an investigative tool in various applications (Bradley et al. 2005).

Further, the response of such electronic device can be utilized to understand various crucial aspects of the biological system like the distribution of charge within the proteins and so on. The scientists have justified their vision of an artificial eye in which the photon-to-electron conversion is required and which can be achieved by using a light-absorbing molecular layer coupled to neurons attached to an electronic device. These nanoelectronic devices also have the potential to directly connect and manipulate living cells leading to the possibility of detection and modification of different bioprocesses at the cellular level. Thus, nanobioelectronics can achieve the breakthroughs in the field of "cellectronics" (Liz Kalaugher 2005).

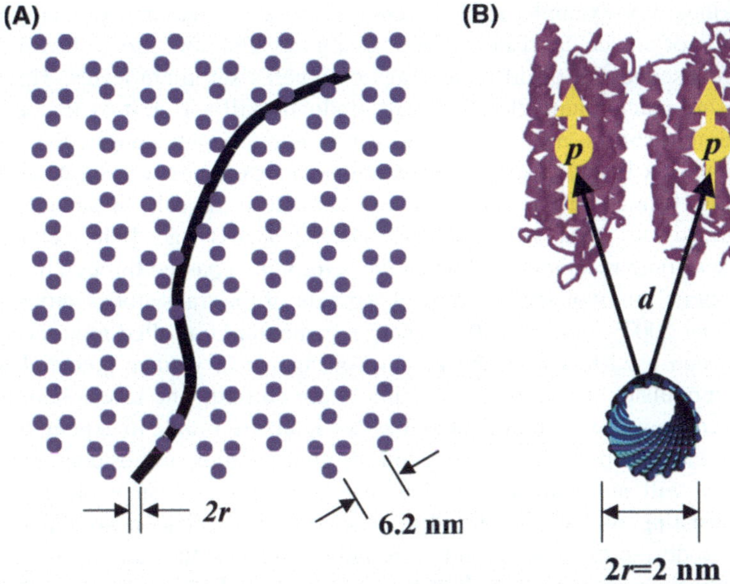

Fig. 17.3 (**a**) Electrostatic model illustrating the cell membrane geometry with respect to nanotubes. Trimers are arranged on a hexagonal lattice, and the purple dots resembling rhodopsin assemble into these trimers. (**b**) Model with the dimension specifications as used in the calculations (Bradley et al. 2005)

To enhance the achievements in nanobioelectronics, scientists have also engulfed themselves in nanomaterial research. Developments in nanotechnology have resulted in the discoveries of various nanomaterials having the capabilities to be used in biological environments and are termed as nanoporous biointerfaces. One such biointerface that has gained popularity in biomedical applications is nanoporous anodic aluminum oxide (AAO) owing to its exclusive features such as increased surface area, highly biocompatible nature, etc. Due to these unique properties, AAO membranes can act as cell interfaces and have found applications in immunoisolation devices, coculture substrates, implant modifications, etc. (Brüggemann 2013).

In order to manipulate and to study the behavior of cells without crushing them in the process, a cell-transistor interface has been developed by Max Planck Institute (Munich, Germany) researchers in June 2007 with the notion that the development of such an interface will usher in a new era of bioelectronics (John Wiley & Sons 2007). To demonstrate this cell-transistor interface, firstly an array of transistors have been employed by the researchers of institute so as to grow the living cells atop and thereby enabling a silicon chip to directly monitor the activity of cells and to detect the efficacy of new drugs on these grown living cells. The results can be read and interpreted from the chip instantly, thus will help in expediting the drug development.

The technique of growing the living cells on a transistor array is better than the traditional methods in which the cells need to be destroyed in order to analyze their inner workings. For example, a patch-clamp electrode is required to facilitate the interpretation of movement results of ion stream into and out of the cell while analyzing the consequences of different drugs on a cell's serotonin levels. This technique of using patch-clamp electrode will gradually kill and destroy the cell, and thus the sample can be used only up to a few hours. Contrastingly, the method employed by Max Planck researchers maintains the usefulness of sample and cell-transistor interface as long as nutrients are provided to the cell. In this technique, firstly the transistor gate is located which has been placed on each living cell directly below the serotonin receptors, and when the serotonin is passed through the receptor's ion channel, the voltage potential at the gate of the transistor is varied (John Wiley & Sons 2007; Johnson 2007). Then, a computer-controlled monitoring circuitry has been used to record the activity of change in voltage potential which ultimately modulates the transistor's source-drain current. The researchers finally proved the concept and have evaluated that this interface is able to explain the dosage/effect relationship of various drugs on serotonin and the use of computer control circuitry will automatically lead to the automation of such bioelectronics devices consisting of batteries for drug discovery tests. The next step in their research is to design the general and more-universal test setups and techniques for various applications requiring the harnessing of living cells like detection of different toxins (John Wiley & Sons 2007).

According to scientists, this research, involving studying the electrical interfacing between living cells and semiconductors, will provide a platform for different applications and will be used as a basis for the development and acceleration in the fields of medical prosthetics, neurocomputation, biosensors, and brain research (Johnson 2007).

17.2.1.3 Artificial Retina

Current research scenario has entered into an era where retinal implants now seem to be a practical concept owing to the developments in nanotechnology and nanomaterials. Due to the thin structure, flexible characteristics of SWNTs and greater efficiency and versatility of other nanoengineered materials, they are compatible with eye mechanics such as eye muscle contractions (Amiji 2008). In order to treat macular degeneration, the most common cause of vision loss in elderly people especially in the age group above 65 years, the researchers at University of Texas Medical Branch suggested the replacement of light-sensing nerves with the combination of CNTs and other nanoparticles that have the capability of sensing the light, converting it into an electrical signal and then transmitting the message to the nerves so as to establish a communication between the brain and the eye. In the patients suffering from macular degeneration, these nerves remain intact and the light-sensing cells in the center of the retina tear apart. Nanoparticles used in artificial retina can even help and provide color resolution (Bourzac 2006; Jain 2013; Amiji 2008).

Apart from nanotubes and buckyballs, advancements in nanotechnology have led to the emergence of certain new fields like nanorobotics, and some nanostructures such as nanobombs, nanoactuators, nanosensors, nanowires, etc. have evolved in recent years which have been explained in subsequent sections with their functionalities. Various disciplines ranging from nanofabrication processes required for manufacturing nanoactuators, nanosensors, and nanoscale robots to physical modeling at nanoscales have been brought together by the emanating field of nanorobotics (Berger 2009).

17.3 Nanobots

Nanorobotics is an emerging technology focusing on designing the machines or robots with the components' size in nanometers (10^{-9} m) (Ghosh and Fischer 2009). Such devices are popularly termed as nanorobots (nanobots), nanites, nanoids, nanomites, or nanomachines and are in the research and development phase (Ignatyev 2010; Yarin 2010). The possibility of integration of photolithography, nanoelectronics, and new biomaterials have opened up the new avenue for manufacturing nanorobots for general medical applications, such as drug delivery, diagnosis, surgical instrumentation, and so on (Cerofolini et al. 2010; Fisher 2008; Yarin 2010; Scudellari 2018). The field of medicine can be improved to a greater extent due to the advantage of nanotechnology to transport the numerous amounts of nanobots in a single injection. With the advent of artificial intelligence, it is possible to program these nanobots to adapt the environment and perform the functions accordingly prior injecting them into the body, thus will boost the proper monitoring and treatment of a disease (Saadeh and Vyas 2014; AZoNano 2017). Nanobots have been used in the Blue Brain Project of IBM which aims at developing the world's first virtual brain that will enable human beings to scan themselves into the computers and will eventually help to preserve the knowledge, skill, and intelligence of a person forever (Markram 2006). The nanobots will monitor the activities and structure of the central nervous system and will help in uploading the natural human brain into computers by traveling throughout the circulatory system including the spine and brain. Thus, the nanobots can act as an interface between the natural brain and computer and, thereby, will be useful for patients with Alzheimer's disease and for persons with hearing disability, allowing them to hear via direct nerve stimulation (Markram 2006; Markram 2012).

Nanorobotics have found applications in the field of hematology which involves the various functions ranging from the development of artificial methods for oxygen transportation inside the body after an extensive attack to the formation of clots with the better clotting capabilities in life-threatening hemorrhage situations. Hypothetical nanobots termed as "respirocytes" can act as artificial red blood cells when injected and perform their functions in emergency cases. Another type of nanobots knows as "clottocytes" have the capability of mimicking the natural platelets and work as artificial platelets for halting the bleeding (AZoNano 2017; Saha 2009).

In the field of cancer theranostics, nanobots designed with embedded chemical sensors and having enhanced detection abilities can contribute in overall management of patients as the chances of survival increases and mortality rate decreases if the cancer is detected in early stages. In addition to accelerating the prognosis of cancer, nanobots with chemotactic sensors have also the potential of providing the therapy and targeted treatment according to the different stages and phases of cancer through specific programming (AZoNano 2017; Saxena et al. 2015).

Nanobots can also be used to prevent the biohazards and to improve the response to epidemics. In resource-constrained situations with limited public infrastructure and where laboratory analysis is not feasible, it is possible to transmit and analyze real-time information by using nanobots designed with protein-based biosensors which is much needed in especially war-torn areas and in regions devastated by epidemic disease (Cavalcanti et al. 2008; AZoNano 2017).

17.4 Nanoactuators

Nanobiotechnology has enabled us to control the substances not only at the molecular level, but now with the enhancement in research and applied science, we can manipulate these substances even at the atomic level, having the dimensions in nanoscales. Cells in our bodies, microorganisms, and all the biological molecules including enzymes, nucleic acid (DNA, RNA), receptors, proteins, etc. are all nanosized; thus the joint use of nanotechnology and biotechnology has various potential applications ranging from early detection of several lethal diseases to providing accurate treatment so as to cure the root cause of these diseases (Shoseyov and Levy 2008). Nanoactuators, one of the significant nanotoolboxes, are inevitable outcomes of nanobiotechnology. A nanoactuator or a molecular dynamo has been developed and patented by the researchers of University of Portsmouth, UK, in October 2006 to function as an interface between the modern computer era and the living organisms. The size of this actuator working as a DNA-based electronic switch is comparable to almost one-thousandth of a strand of human hair, and so is invisible to human naked eye (NanowerkNews 2006). This molecular nanoactuator comprises of a biological motor powered by adenosine triphosphate (ATP, a naturally occurring energy source found in living cells), a microchip's miniscule channel in which a DNA strand is anchored and a magnetic bead. In order to generate the electricity, these elements work together to create a dynamo effect and ultimately, a device is developed that can emit the electric signals to communicate with the computer. This DNA switch can be used as a biosensor for the detection of airborne pathogens and is also useful for detection of toxins. With further modifications, this nanoactuator technology-based molecular switch can act as an interface between human muscles and external devices and thus will be useful for designing artificial limbs controlled by computers for disabled persons. It has the capability of actually acting as a communicator or mediator between the silicon (electronic) world and biological world (NanowerkNews 2006; Phys.org 2006).

17.5 Nanobombs

Nanotechnology has also been applied to improve drug delivery by overcoming the major issues associated with delivery of drugs for the treatment of various diseases including cancer. In the recent years, nanomedicine (an application of nanobiotechnology) has evolved much by focusing on the development of different types of nanoparticles for drug delivery (Hood 2004; Freitas 2005). Cancer nanomedicine is aiming to cover a wide range of applications from imaging and diagnosing to treating cancers with targeted therapies. Cantilevers, nanopores, nanotubes, quantum dots, nanoshells, dendrimers, nanobombs, and nanowires are some of the tools of nanotechnology having applications in the treatment of different cancers (Pathak and Katiyar 2007; Misra et al. 2010; Azad 2015; Sahoo et al. 2017). Among these, the nanobombs hold great promise as a therapeutic agent for killing cancer cells. An advantage over other carbon nanotube treatments being considered by scientists is that with nanobombs, the nanoparticles injected into the body are also destroyed along with the cancer cells (ScienceDaily 2005; OSUCCC-James 2015). In 2015, researchers at Ohio State University Wexner Medical Center, USA, have reported the development of nanobombs that can burst after swelling when exposed to near-infrared (NIR) laser light. These bomb-like nanoparticles might help in overcoming a biological barrier that has blocked the development of agents such as microRNAs required for the outright killing of cancer cells or for stalling their growth. These agents perform these operations by varying the activity, i.e., the expression, of genes in cancer cells (Wang et al. 2016; Zhang et al. 2015).

A novel treatment protocol employing CO_2 bubbling-based "nanobomb" system (shown in Fig. 17.4) has been developed recently for targetedly suppressing Panc-1 pancreatic tumor via low-intensity ultrasound-activated inertial cavitation. This exogenous therapeutic low-intensity ultrasound (US) irradiation causes the elevation in temperature, and the endogenous acidic environment around/within tumor results in low pH value. In response to variation in temperature and pH, this dual-responsive system produces and adsorbs the CO_2 bubbles due to the presence of L-arginine (LA) and hollow mesoporous silica nanoparticle (HMSN) carriers. Depending upon the increase in temperature and reduction in pH value, CO_2 bubbles will be released quickly by a dual-responsive CO_2 release system. Then, before dissolution, these bubbles will explode timely as a result of triggering action performed by the therapeutic ultrasound waves. Destruction of blood vessels within panc-1 tumor and instant necrosis of panc-1 cells have been induced by the related bio-effects, such as mechanical effects, cavitation, shock waves, and so on, as a result of the explosion of CO_2 bubbles. Consequently, the growth of panc-1 solid tumor will be inhibited and also the normal organs will not be affected much, thereby, minimizing the side effects to these organs. This novel treatment approach is a localized physiotherapy, which is expected to greatly reduce damages to normal organs during cancer therapy unlike the conventional techniques used for treatments like chemotherapy (Zhang et al. 2015; Wang et al. 2016).

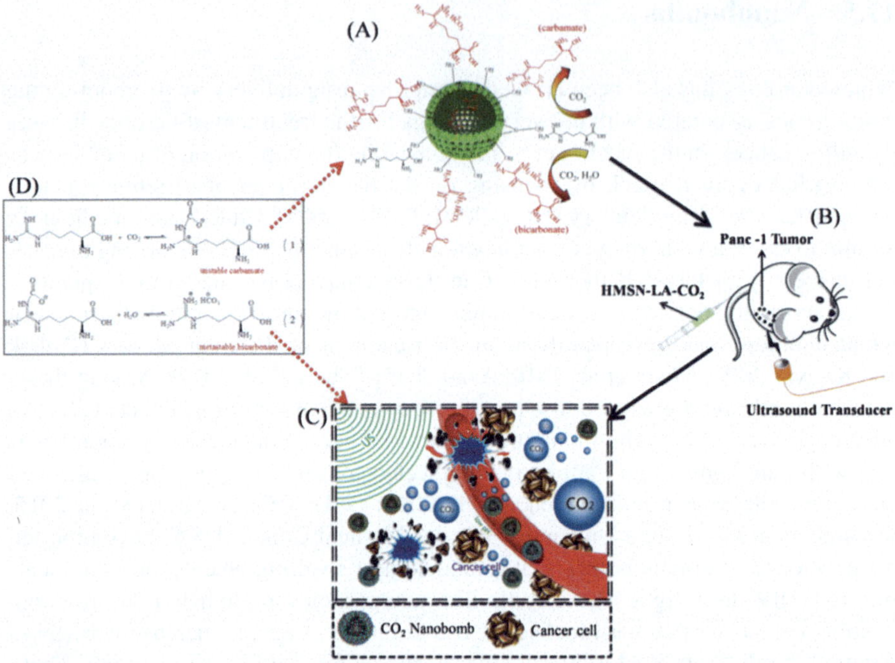

Fig. 17.4 CO_2 nanobomb system (US+HMSN-LA-CO2) explosion technical route diagram for inhibiting the panc-1 tumor growth. (**a**) Preparation procedure illustration of HMSN-LA-CO_2 (dual-responsive CO_2 nanobomb), (**b**) schematic explanation of therapeutic procedure of HMSN-LA-CO_2, (**c**) illustration of CO_2 nanobomb system's therapeutic mechanism, and (**d**) reaction formulae used for the chemical adsorption and desorption of CO_2 by HMSN-LA (Zhang et al. 2015)

17.6 Nanowires

Nanowires (NWs) are the nanostructures in the form of wires having the dimensions in the range of 10^{-9} m. Quantum mechanical effects are considerably important at nanoscales; thus these nanostructures at such a scale are also known as "quantum wires" (ScienceDaily). The remarkable properties of these nanowires have enabled them to be used in various applications (UnderstandingNano). One such nanowire is molecular nanowire which is designed by repeating the organic or inorganic molecular units. In 2007, researchers at Yale Institute for Nanoscience and Quantum Engineering, USA, published a novel approach for synthesizing the nanowires (YaleUniversity 2007). The integration of these nanostructures directly with the MEMS (microelectronic systems) has been made possible for the first time by this approach. The technique used by the researchers involved the plugging into the biochemical system of cells electronically. A tried-and-true process of wet-etch lithography on silicon-on-insulator wafers available commercially has been used in order to overcome the various hurdles associated with the synthesis of NWs. The nanowires synthesized with this novel technique are structurally stable and have the

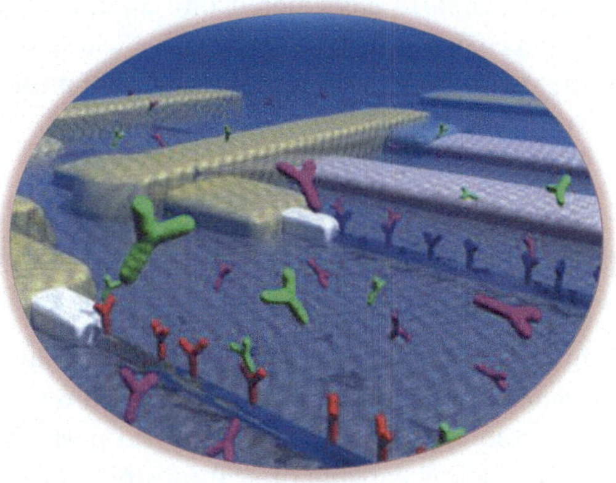

Fig. 17.5 Schematic illustration of nanowire sensors in a solution (YaleUniversity 2007)

capability of detecting the antibodies or other biomolecules with high sensitivity at affordable costs, thus possessing the potential of revolutionizing the medical diagnostic and therapeutic fields. These NW sensors (Fig. 17.5) with label-free reagents can also sense the live cellular immune response on real-time basis unlike traditional assays which usually takes hours to detect the response of cells of immune system (YaleUniversity 2007). This advancement in nanotechnology has speeded up the diagnostic field as a fully automatic sensor platform for the detection of cellular and molecular arrays can be thought of due to the possibility of seamless integration of this approach with CMOS technology (Stern et al. 2007).

Enhancements in the development of nanomaterials have made it feasible to directly interconnect the living cells with the external world, thus allowing the probing and manipulation of various biological processes occurring across the cell membranes, inside the whole cells, or between the neighboring cells. These nanostructures can act as interfaces between the outer world and inside body mechanism apart from functioning either at the vicinity of a biomolecule's surface or inside a cell. In 2007, a report was published in *Journal of the American Chemical Society* in which the authors have highlighted the signal-carrying ability of silicon nanowires (Si NWs) to the living cells (Kim et al. 2007). They have demonstrated that as the cells grow, Si NWs having dimensions in the range of 1–100 nm can directly penetrate into them without causing any damage. The interface capability of Si NWs for carrying out the transmission of signals will help in connecting the prosthetics to the living tissues electronically and also for developing the biosensors (Kim et al. 2007).

Rapid progression in various fields like nanotechnology, molecular biology, microfluidics, biosensors, VLSI, etc. has empowered the integration of all such processes on a single platform for POC (point-of-care) detection and management of various diseases so as to pacify their consequences and to prevent the situations of epidemics and pandemics. Lab-on-chip is one of the state-of-the-art technologies that have been developed owing to the possibility of easy integration of all the above mentioned interdisciplinary fields (Gupta et al. 2016). Further, current research scenario has entered into an era where collaboration of these diversified fields has made it feasible to mimic a whole organ on a chip. In the subsequent sections, the features and applications of lab-on-chip and organ-on-chip have been highlighted.

17.7 Lab-on-Chip

Lab-on-chip technology implies those techniques that perform various laboratory operations on a miniaturized scale such as chemical synthesis and analysis on a single chip leading to a handheld and portable device. In other words, a LoC is a device which is capable of scaling single or multiple laboratory functions down to chip format. The size of this chip ranges from millimeters to a few square centimeters (Volpatti and Yetisen 2014).

LoC technology has several advantages compared with conventional techniques such as minimal sample requirement, rapid analysis time, ease of use, minimized exposure to hazardous materials, and reduced waste generation (Soutter 2019). LoCs are capable of analyzing metabolites, molecules, proteins, nucleic acids, cells, and viruses. It can perform fluidic processes, sensing, analysis, and separation of biochemical samples. LoC is basically the integration of fluidics, electronics, optics, and biosensors (Giannitsis and Min 2010). Due to the advent of advanced technologies such as MEMS, NEMS, the integration of large number of interdisciplinary modules, as shown in Fig. 17.6, on a single chip (Gene-Quantification) is possible making the LoC designing much easier (Kakkar 2018, Majlis 2013).

LoCs have found applications in various fields including pharmacy, chemistry, and biomedical area (Gupta et al. 2016), such as immunoassay LoCs used for bacteria detection, flow cytometer LoC (for HIV), DNA chips, gene chips, LoC for cellular analysis, real-time PCR detection chips, etc. (Chin et al. 2012; Grenier et al. 2011). Today, the research focus is shifting from cells and molecules to organs and tissues on a chip as explained in the next section.

17.8 Organs-on-Chip

The spectacular growth in the entrenched fields of microfluidics and labs-on-chip has led to the possibility of fabricating organ-on-chip devices. This development of organs and tissues on a chip will help researchers to fill the void between the ongoing in vivo experimentation and the available standard in vitro tests. Organs-on-chips, a type of microengineered in vitro tissue models, have the potential of achieving better throughput through parallelization by combining realistic physiological cell culture microenvironment with artificial engineering approaches (Van Den Berg 2013).

One such organ-on-chip termed as blood-brain barrier (BBB) on chip has been described in an article published in 2013 IEEE 26th International Conference on

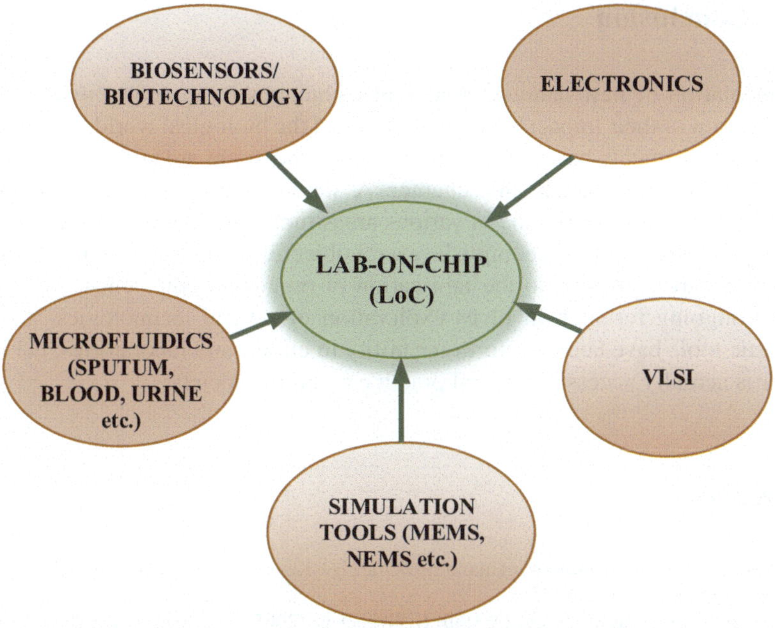

Fig. 17.6 Interdisciplinary fields of LoC (Gupta et al. 2016; Gupta and Kakkar 2018)

MEMS (Van Den Berg 2013). Human bodies have unique feature of preserving the homeostasis of the brain and preventing toxic elements from entering into the brain. This feature is termed as blood-brain barrier (BBB) which doesn't function properly in patients suffering from neurodegenerative disorders. The authors in this article have presented a miniaturized BBB model by immortalizing the endothelial cell line (hCMEC/D3) of human brain on a microfluidic chip consisting of integrated electrodes used for analyzing the tightness of barrier by calculating the TEER (transendothelial electrical resistance) (Van Den Berg 2013; Griep et al. 2013). Then, on a single nominal device, both the biochemical and mechanical modulation of barrier function is done by utilizing tumor necrosis factor alpha (TNF-α) for stimulation and by exposing it to fluid shear stress respectively. This BBB microfluidic platform will help in evaluating the passage of drugs in order to treat various neurodegenerative disorders (Griep et al. 2013).

In December 2017, scientists at Lawrence Livermore National Laboratory (LLNL) have claimed the development of a brain-on-chip device in order to predict and test the effects of various chemical and biological agents on the brain. This device has the potential of monitoring various brain disorders and also analyzing the treatment provided by different pharmaceutical drugs without any human or animal intervention (ScienceDaily 2017).

Thus, Nano-neuroelectronic systems designed by integrating multidisciplinary will enable brain researchers to enter an exciting era that promises to bring new insights into neuronal communication.

17.9 Conclusion

Implementation of new nanotechnological techniques in an economical way and magnification of their impact have revolutionized the biological world. Many breakthrough achievements have already been made, and some are still in the development phase as described in this chapter. A state-of-the-art LoC technology has proven to be a useful technique in various areas including theranostics and is serving as a promising tool to accomplish several objectives ranging from the designing of point-of-care platforms to the fabrication of organs-on-chip. Thus, we can say that the ongoing research involves exploration of several technologies, and new diagnostic tools have been put forth for utility in clinics and laboratories, but as yet meager is actually accessible in real practice to aid the society.

References

Amiji MM (2008) The handbook of nanomedicine. By Kewal K. Jain. ChemMedChem: Chem Enabling Drug Discov 3:1977

Artukovic E, Kaempgen M, Hecht D, Roth S, Grüner G (2005) Transparent and flexible carbon nanotube transistors. Nano Lett 5:757–760

Azad LCAK (2015) Nanomedicine and prospect of medical science. J Anesth Crit Care Open Access 2:00053

AZoNano (2017) Application of nanorobotics (nanobots). Available: https://www.azonano.com/article.aspx?ArticleID=4679

Berger M (2009) Nano-Society: pushing the boundaries of technology. Royal Society Of Chemistry, Cambridge, UK

Bourzac K (2006) This is your brain on nanotubes. Available: https://www.technologyreview.com/s/405853/this-is-your-brain-on-nanotubes/

Bradley K, Davis A, Gabriel J-CP, Grüner G (2005) Integration of cell membranes and nanotube transistors. Nano Lett 5:841–845

Brüggemann D (2013) Nanoporous aluminium oxide membranes as cell interfaces. J Nanomater 2013:32

Cavalcanti A, Shirinzadeh B, Zhang M, Kretly L (2008) Nanorobot hardware architecture for medical Defense. Sensors 8:2932–2958

Center for Synthetic Biology, U. Bionanoelectronics [Online]. Available: https://synbio.ku.dk/research/unik/research/bionanoelectronics/

Cerofolini G, Amato P, Masserini M, Mauri G (2010) A surveillance system for early-stage diagnosis of endogenous diseases by swarms of nanobots. Adv Sci Lett 3:345–352

Chin CD, Linder V, Sia SK (2012) Commercialization of microfluidic point-of-care diagnostic devices. Lab Chip 12:2118–2134

Earlboysen NB (2011a) What are carbon nanotubes? In: Nanotechnology for dummies, 2nd edn. Wiley, Hoboken

Earlboysen NB (2011b) What is a Buckyball (C60). In: Nanotechnology for dummies, 2nd edn. Wiley, Hoboken

ETHW (2015) Discovering the buckyball. Available: https://ethw.org/Discovering_the_Buckyball

ETHW (2017) Buckyballs and nanotubes. Available: https://ethw.org/Buckyballs_and_Nanotubes

Fisher B (2008) Biological research in the evolution of cancer surgery: a personal perspective. Cancer Res 68:10007–10020

Freitas RA (2005) Current status of nanomedicine and medical nanorobotics. J Comput Theor Nanosci 2:1–25

Gene-Quantification. Lab on a chip. Available: http://lab-on-chip.gene-quantification.info

Ghosh A, Fischer P (2009) Controlled propulsion of artificial magnetic nanostructured propellers. Nano Lett 9:2243–2245

Giannitsis A, Min M (2010) Usage of microfluidic lab-on-chips in biomedicine. In: 2010 12th Biennial Baltic Electronics Conference. IEEE, pp 249–252

Grenier K, Dubuc D, Chen T, Chrétiennot T, Poupot M, Fournié J (2011) Microfluidic on-chip for biomedical applications. In: 2011 IEEE bipolar/BiCMOS circuits and technology meeting. IEEE, pp 129–132

Griep L, Wolbers F, De Wagenaar B, Ter Braak PM, Weksler B, Romero IA, Couraud P, Vermes I, Van Der Meer AD, Van Den Berg A (2013) BBB on chip: microfluidic platform to mechanically and biochemically modulate blood-brain barrier function. Biomed Microdevices 15:145–150

Gupta S, Kakkar V (2018) Recent technological advancements in tuberculosis diagnostics-a review. Biosens Bioelectron 115:14–29

Gupta S, Ramesh K, Ahmed S, Kakkar V (2016) Lab-on-chip technology: a review on design trends and future scope in biomedical applications. Int J Bio-Sci Bio-Technol 8:311–322

Hersam MC, Reifenberger RG (2004) Charge transport through molecular junctions. MRS Bull 29:385–390

Hood E (2004) Nanotechnology: looking as we leap. Environ Health Perspect 112(13):A740–A749. doi:10.1289/ehp.112-a740

Ignatyev MB (2010) Necessary and sufficient conditions of nanorobot synthesis. Dokl Math. Springer 82:671–675

Jain KK (2013) Applications of biotechnology in neurology. Springer Science & Business Media, New York

John Wiley & Sons, I (2007) Bioelectronics: progress toward drug screening with a cell–transistor biosensor. Available: https://phys.org/news/2007-06-bioelectronics-drug-screening-celltransistor-biosensor.html

Johnson RC (2007) Cell-transistor interface clears biolectronics hurdle. Available: https://www.eetimes.com/document.asp?doc_id=1268437

Kakkar V (2018) An ultra low power system architecture for implantable medical devices. IEEE Access 7:111160–111167

Kim W, Ng JK, Kunitake ME, Conklin BR, Yang P (2007) Interfacing silicon nanowires with mammalian cells. J Am Chem Soc 129:7228–7229

Liang G-C, Ghosh A (2005) Identifying contact effects in electronic conduction through C60 on silicon. Phys Rev Lett 95:076403

Liopo AV, Stewart MP, Hudson J, Tour JM, Pappas TC (2006) Biocompatibility of native and functionalized single-walled carbon nanotubes for neuronal interface. J Nanosci Nanotechnol 6:1365–1374

Liz Kalaugher (2005) Nanotube transistor interacts with cell membrane. Nanotechweb. [ONLINE]. Available: http://www.physics.ucla.edu/research/biophysics/news/pdf/nanotube_transistor.pdf

Majlis BY (2013) Mems and lab on chip: interfacing macro to nano world. In: RSM 2013 IEEE regional symposium on micro and nanoelectronics. IEEE, p Vii

Markram H (2006) The blue brain project. Nat Rev Neurosci 7:153

Markram H (2012) The human brain project. Sci Am 306:50–55

Misra R, Acharya S, Sahoo SK (2010) Cancer nanotechnology: application of nanotechnology in Cancer therapy. Drug Discov Today 15:842–850

NanowerkNews (2006) DNA switch developed to interface living organisms with computers. Available: https://www.nanowerk.com/news/newsid=958.php

OSUCCC-James (2015) Nanobombs might deliver agents that alter gene activity in cancer stem cells. Available: http://cancer.osu.edu/news-and-media/news/nanobombs-might-deliver-agents-that-alter-gene-activity-in-cancer-stem-cells

Pathak P, Katiyar V (2007) Multi-functional nanoparticles and their role in cancer drug delivery–a review. Virus 30:100 nm

Phys.Org (2006) am develops DNA switch to interface living organisms with computers. [ONLINE]. Available: https://phys.org/news/2006-10-team-dnainterface.html

Rakshit T, Liang G-C, Ghosh AW, Datta S (2004) Silicon-based molecular electronics. Nano Lett 4:1803–1807

Saadeh Y, Vyas D (2014) Nanorobotic applications in medicine: current proposals and designs. Am J Robot Surg 1:4–11

Saha M (2009) Nanomedicine: promising tiny machine for the healthcare in future-a review. Oman Med J 24:242

Sahoo SK, Misra R, Parveen S (2017) Nanoparticles: a boon to drug delivery, therapeutics, diagnostics and imaging. In: Nanomedicine in cancer. Pan Stanford, Singapore

Saxena S, Pramod B, Dayananda B, Nagaraju K (2015) Design, architecture and application of nanorobotics in oncology. Indian J Cancer 52:236

ScienceDaily. Nanowire [Online]. Available: https://www.sciencedaily.com/terms/nanowire.htm

ScienceDaily (2005) University of Delaware researchers develop cancer 'nanobomb'. Available: https://www.sciencedaily.com/releases/2005/10/051013222044.htm

ScienceDaily (2017) 'Brain-on-A-chip' to test effects of biological and chemical agents, develop countermeasures. Available: https://www.sciencedaily.com/releases/2017/12/171218092556.htm

Scudellari M (2018) Nanobots glide through living cells. Available: https://spectrum.ieee.org/the-human-os/biomedical/devices/nanobots-glide-through-living-cells

Shoseyov O, Levy I (2008) Nanobiotechnology: bioinspired devices and materials of the future. Springer Science & Business Media, New York

Sørensen JK (2006) Synthesis of new components, functionalized with (60) fullerene, for molecular electronics. In: 4th annual meeting-CONT

Soutter W (2019) Lab-on-a-chip technology – applications for life science. Available: http://www.agilent.com/cs/library/articlereprints/public/59883035.pdf, https://www.azonano.com/article.aspx?ArticleID=3081

Stern E, Klemic JF, Routenberg DA, Wyrembak PN, Turner-Evans DB, Hamilton AD, Lavan DA, Fahmy TM, Reed MA (2007) Label-free immunodetection with CMOS-compatible semiconducting nanowires. Nature 445:519

UnderstandingNano. Nanowires: uses and applications of nanowires. Available: http://www.understandingnano.com/nanowires-applications.html

Usmani S, Aurand ER, Medelin M, Fabbro A, Scaini D, Laishram J, Rosselli FB, Ansuini A, Zoccolan D, Scarselli M (2016) 3D meshes of carbon nanotubes guide functional reconnection of segregated spinal explants. Sci Adv 2:E1600087

Van Den Berg A (2013) Labs on chips for biomedical applications. In: 2013 IEEE 26th international conference on Micro Electro Mechanical Systems (MEMS). IEEE, pp 149–152

Venere E (2005) Purdue simulation to help merge molecules with silicon electronics. Available: https://news.uns.purdue.edu/html4ever/2005/050816.Liang.buckyballs.html

Volpatti LR, Yetisen AK (2014) Commercialization of microfluidic devices. Trends Biotechnol 32:347–350

Wang H, Agarwal P, Zhao S, Yu J, Lu X, He X (2016) A near-infrared laser-activated "nanobomb" for breaking the barriers to microRNA delivery. Adv Mater 28:347–355

YaleUniversity (2007) Breakthrough in nanodevice synthesis revolutionizes biological sensors. Available: https://www.eurekalert.org/pub_releases/2007-01/yu-bin013007.php

Yarin A (2010) Nanofibers, nanofluidics, nanoparticles and nanobots for drug and protein delivery systems. Sci Pharm 78:542

Zhang K, Xu H, Chen H, Jia X, Zheng S, Cai X, Wang R, Mou J, Zheng Y, Shi J (2015) CO2 bubbling-based 'nanobomb' system for targetedly suppressing panc-1 pancreatic tumor via low intensity ultrasound-activated inertial cavitation. Theranostics 5:1291

Chapter 18
Bio-Based Nano-Lubricants for Sustainable Manufacturing

Rahul Anand, Mir Irfan Ul Haq, and Ankush Raina

18.1 Introduction

In today's contemporary manufacturing industry, there is rising demand of high productivity with low cost along with improved quality. In order to achieve the present production rate demand and challenge of cost competitiveness across the global markets, material removal rate should be enhanced. Increased material removal rate demands higher values of cutting parameters, i.e. speed, depth of cut and feed rate, which carry a great risk of tool wear and failure due to enormous heat generation and friction at cutting zone between chip and tool (Bruni et al. 2006). Plastic deformation also tends to deteriorate the surface integrity which is an important property in evaluation of productivity of machining processes (Davim et al. 2008). Chip flowing over the tool surface tends to carry the major portion of heat, and some portion is conducted by tool itself; thus reducing the temperature at cutting zone will largely enhance tool life and surface integrity and reduce tool deterioration (Dudzinski et al. 2004). Thus cutting fluids in modern industry act as an accessory in reducing the shear zone temperature by reducing chip tool contact length and improving the thermal conductivity at interface which greatly optimizes the cutting parameters and improved tool life (Vieira et al. 2001). Primary function of cutting fluid used in conventional machining is to simultaneously cool and lubricate the cutting zone thereby decreasing the coefficient of friction which in turn reduces the cutting forces required for conventional machining operations like turning, drilling, milling and grinding (Adler et al. 2006).

However issues related to excessive and conventional use of mineral oil-based cutting fluids have been debated by the researchers lately which involves disposal of toxic mineral oils in soil and water, and improper handling of mineral oil-based

R. Anand · M. I. U. Haq (✉) · A. Raina
School of Mechanical Engineering, Shri Mata Vaishno Devi University, Katra, Jammu & Kashmir, India

I. Bhushan et al. (eds.), *Nanomaterials and Environmental Biotechnology*, Nanotechnology in the Life Sciences,
https://doi.org/10.1007/978-3-030-34544-0_18

cutting by machine operators may cause various health hazards and results in degradation of health of the operator (Soković and Mijanović 2001). Shashidhara and Jayaram (2010) estimated 80% of total health problems are caused due to mineral oil-based cutting fluids. Purchasing cost of cutting fluids includes 7–17% of total machining cost (Klocke and Eisenblätter 1997). European Union almost consumes annually 320 Mt of cutting fluids during machining processes out of which two thirds of used fluid has to be disposed (Lawal et al. 2013). Shokrani et al. (2012) estimated non-biodegradable mineral oil-based cutting fluid increases disposal cost fourfold the purchasing cost which also includes physical and chemical treatment cost required before disposing them.

Due to increased disposal, treatment cost and health and environmental problems, researchers are now investing their interest discovering biodegradable vegetable and esoteric oil (Tschätsch and Reichelt 2009). These vegetable and ester oils have biodegradability of 90–95% and have environmental compatibility in terms of biodegradability, toxicity, safe disposal and additives (Kodali and Nivens 2001; Pop et al. 2008). These vegetable oils provide a vast range of rheological properties in terms of flash point, viscosity index and low volatility. Debnath et al. (2014) listed out detailed developments in biodegradable vegetable-based cutting fluids and their influence on cutting parameters along with reduction in ecological and economical complications which were found in mineral oils. Simultaneously researchers are more focused on achieving more clean production by improving the methods cutting fluid applications at cutting zone (Fratila 2009). Dry machining can never be one solution to the problem as it affects tool life badly (Diniz and Micaroni 2002). So more convenient ways of near-dry machining (MQL) are now being adapted to minimize the use of cutting fluids. Sharma et al. (2009) talk about more such lubrication techniques (cryogenic cooling, MQL, flood and high-pressure cooling) during conventional machining processes which lead to clean and sustainable manufacturing. Lawal et al. (2013) in his research compared vegetable oil-based MQL lubrication technique and conventional flood lubrication and concluded that MQL with vegetable oil-based lubrication has better results than other conventional techniques in terms of process parameters and cost.

With technological advancements in manufacturing industry, nano-cutting fluids are now being imposed which profitably meet the challenge of heat dissipation during machining by providing stability and improved thermal conductivity to base oil at wide range of temperatures. These nano additives can be classified into metallic, non-metallic and nanofibres with diameter range up to 100 nm which are dispersed in base cutting oil (Chol and Estman 1995). Experimental research (Sharma et al. 2015b, 2016; Su et al. 2016) reveals these nanoparticles dispersed in base oil easily penetrate into cutting zones, and due to their heat extraction and friction reduction properties, they reduce too wear and cutting forces. MWCNT (multiwall carbon nanotubes) (Sharma et al. 2019), Al_2O_3 (aluminium oxide) (Arumugam et al. 2018), ZnO (zinc oxide) (Battez et al. 2008) and MoS_2 (molybdenum disulphide) (Rajendhran et al. 2018) are some of the nanoparticles which have improved thermal and tribological properties. Due to enhanced machining performance, addition of nanoparticles has minimized the use of cutting fluids, saving the environment and

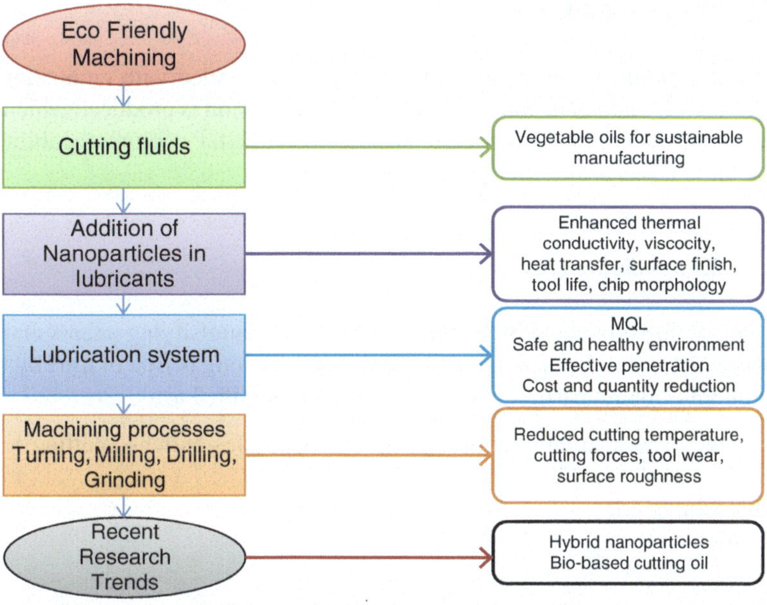

Fig. 18.1 Flow chart representation of the machining processes involved

decreasing the cost of production. This paper widely discuss about vegetable oils and heat extraction and friction reduction mechanisms of nanofluids with MQL technique in conventional machining operations turning, drilling, milling and grinding (Fig. 18.1).

18.1.1 Types of Cutting Fluids

El Baradie (1996) in his research classified cutting fluids along with their composition and also addressed issues like green machining, recycling of cutting fluids and their disposal.

18.1.1.1 Neat Cutting Oils

It mainly includes pure mineral oils or mineral oils with some load-carrying additives such as:

(a) Blends of mineral oil and fatty oil
(b) Blends of mineral oil and sulphurized fatty oil
(c) Blends of mineral oil and chlorofatty oil
(d) Blends of mineral oil and sulpho-chlorinated fatty oil
(e) Extreme pressure additives

18.1.1.2 Water-Soluble Fluids

Water-soluble fluids are predominantly mineral oils blended with emulsifiers, and when this blend is added to water, an oil-in-water emulsion is produced which facilitates the high-speed cutting process because of its better cooling capabilities (). Water-soluble fluids are further classified as.

18.1.1.3 Emulsifiable Oils

These are soluble oils made by blending the oil with emulsifying agents and other materials which form the dispersion phase in water for long period of time by breaking the oil into minute particles. These are further classified as:

(a) General purpose soluble oil (it includes nanoparticles-enriched cutting.)
(b) Clear type soluble oils
(c) Fatty soluble oils
(d) EP soluble oils

18.1.1.4 Chemical (Synthetic Fluids)

These chemical coolants and lubricants include inorganic chemical agents dissolved in water with no mineral oil. It includes soaps and wetting agents for reduction in surface tension, nitrides for corrosion resistance, chlorine and sulphur for lubrication, glycols as blending agents and humectants and germicides to control bacterial growth.

18.1.1.5 Semisynthetic Fluids

This type includes combination of chemical fluids and emulsifiable oils in water which forms stabilized emulsion with small droplet size that can facilitate heavy-duty machining.

18.1.2 Methods of Application of Cutting Fluids in Conventional Machining

During machining of metals, conversion of energy to heat takes place in the regions of plastic deformation due to work involved in shearing the workpiece (da Silva and Wallbank 1999). High temperatures are thus produced on the cutting edges and chip tool interfaces which control the tool wear rate, tool life, MRR and surface finish of the material. Large part of this heat is conducted away by the chip, and the rest is conducted by the tool which acts as a sink. Researches have shown that on reducing the cutting, temperature will increase tool life and reduce the tool wear (Khan and Ahmed 2008). Cutting fluids are thus applied to carry away the heat from the cutting tool and workpiece interface thereby reducing the temperature which is critical for tool wear. 7–17% of the total cost of machining is acquired by the cutting fluid alone (Klocke and Eisenblätter 1997). Several techniques have been developed to increase the efficiency of cutting fluid and enhance the machining process by environment-friendly means.

Along with the improvement in the conventional cooling techniques and reduction in temperature rise, researchers have focused on environmental problems like disposability, toxicity, misting, staining and self-cleanliness involved using cutting fluids.

18.1.2.1 Cryogenic Cooling

In cryogenic cooling liquid nitrogen (LN) is applied at cutting tool, workpiece or at the cutting zone in order to reduce the cutting temperature and enhance low-temperature machining (Evans and Bryan 1991). The nitrogen absorbs heat and evaporates to atmosphere thereby cooling the cutting edges of the tool and by forming a fluid gas layer between the chip and tool interface. Cryogenic cooling is an environmentally friendly and efficient way of maintaining a safe working temperature well below the critical or softening temperature of tool (Dhar et al. 2002). Thus reduction in temperature results in reduced flank and crater wear thereby improving the product quality and productivity. Cryogenic cooling is more predominant at lower cutting speed due to more effective penetration of cryogen between chip and tool interface (Khan and Ahmed 2008). Its cooling capability decreases with increase in cutting speed due to obstruction in penetration of cryogen in hot chip tool interface. Turning with carbide inserts with liquid nitrogen cryogenic cooling, it was observed that tool life improves which attributes to reduction in notching, abrasion, adhesion and diffusion type of wear.

18.1.2.2 Solid Lubricant/Coolant

Use of conventional cutting fluids cause environmental pollution, ecological imbalance and serious biological hazards to workers along with increased cutting cost (Sreejith and Ngoi 2000; Tan et al. 2002). Now all these factors instigate researchers to avoid use of cutting fluids with some other cooling methods. Solid lubrication is one of the techniques which strive to minimize the use of cutting fluid (Shaji and Radhakrishnan 2003). Molybdenum disulphide (MoS_2) and graphite are most commonly used solid lubricants due to their lamellar structure; these lamellas incline between loaded surfaces and glide over each other in order to reduce friction and contact. Boron nitride, calcium fluoride and tungsten disulphide are other materials which can be used. During turning of AISI 52100 steel at high speed of 125 m/s, it was observed that due to layered lattice structure, cutting forces were minimized, and due to good adhesion property of molybdenum disulphide, lower value of surface roughness (13–15% reduction) than graphite (7–10% reduction) was observed (Dilbag and Rao 2008; Lathkar and Bas 2000).

18.1.2.3 High-Pressure Cooling Technique

High-pressure coolant delivery technique allows the coolant to penetrate in tool-chip and tool-workpiece interface thereby enhancing the cooling effect and lowering the tool wear rate by effectively lubricating the contact regions (Diniz and Micaroni 2007; Wertheim et al. 1992). Pressurized coolant thus forms a hydraulic layer of wedge shape between tool-workpiece interface which facilitates high-speed machining (Mazurkiewicz et al. 1989). Researchers have found that pressurized coolant not only increase the effectiveness of cooling but also reduce the chip tool contact time by taking the chip off the rake surface thereby significantly improving the tool life (Ezugwu and Bonney 2005). Turning titanium under high-pressure conditions reduces chip tool interface temperature and cutting forces and improves surface roughness as chip tool contact time is reduced (De Lacalle et al. 2000). Cutting forces significantly reduce cutting forces while machining Inconel 718 alloy due to improved lubrication at tool-workpiece interface (Ezugwu et al. 2005).

18.1.2.4 Air/Vapour/Gas Cooling

For protection of ecology and environment, the use of green cutting fluids has become a major concern in machining. Water vapour and gas emerge as best alternatives for cooling and lubrication than conventional synthetic cutting fluids. Liquid Nitrogen and Nitrogen gas at high pressure serve best operating conditions for machining. Application of nitrogen, oxygen and CO_2 provides lower cutting forces than dry and wet cooling operations with reduced surface roughness (Altan et al. 2002; Çakıra et al. 2004; Godlevski et al. 1998). Mist application of air-oil at 0 °C provides simultaneous effect of cooling and lubrication which reduces wear by 13% as compared to air-jet cooling (Ko et al. 1999). During machining of the AISI 1045 steel (in presence of nitrogen gas, water vapour, mixture of gas and water vapour), water vapour alleviate friction by filling up the cavities in the cutting zone by capillary action which thus improves tool life twice than dry machining (Liu et al. 2007). During turning of En32b plain carbon steel, the nitrogen gas provided the best cutting conditions with reduced crater (55%) and flank wear (30%) than dry machining by reducing the chip tool contact length (Stanford et al. 2009).

18.1.2.5 Minimum Quantity Lubrication

MQL (minimum quantity lubrication) also referred as near-dry machining or micro-machining is an environmental-friendly alternative to flood and dry machining. Lubricant is mixed with air and is injected on the cutting surfaces through different techniques (Varadarajan et al. 2002). MQL is the best lubrication technique that leads to clean and sustainable production along with better heat convection, tool life and surface finish and improved chip morphology (Philip et al. 2001; Wertheim et al. 1992). Lubricant is either mixed inside or outside the nozzle with compressed air to form an aerosol which is applied to the cutting interface with the help of the nozzle with industrial range of 10–100 ml/h (Kamata and Obikawa 2007). Biodegradable vegetable oils as cutting fluids are preferred because of their superior lubrication performance at high pressure. MQL helps in reducing wear that are sensitive to temperature (abrasion wear and diffusion wear) thereby maintaining the hardness and reducing the notch growth in tool. In investigations it was observed that chip tool contact length and cutting forces reduced with MQL technique than conventional dry and wet turning. MQL of air-oil mixture when sprayed at 0.276 Mpa through 0.762 mm via nozzle at low cutting speed reduces tangential cutting force by 26.6% than in dry machining and 32.2% of that in flood cooling (Li and Liang 2007).

18.1.2.6 Nano-Enriched Cutting Fluids

Researchers have investigated that addition of nanoparticles enhance the machining and tribological properties of the cutting fluids. Nanoparticles with properties like rolling, sliding and filming action ensure better lubrication thereby reducing power consumption and cutting forces. Observations also tell that addition of nanoparticles not only penetrates inside shearing zone but also increases heat transfer capability of the cutting fluid as nanoparticles have much higher thermal conductivity than liquid alone as shown in the table. Nanoparticles with MQL lubrication technique are now being investigated as new research trend in the field of lubrication.

18.1.3 MQL (Minimum Quantity Lubrication) Application Technique

MQL technique involves spraying mixture of lubricant with air at a high pressure at machining zone with help of nozzle (Duchosal et al. 2013). MQL system consists of nozzle jet to spray the air-oil mixture, pressure gauges and oil reservoir. Due to the venture effect, a vacuum is created inside the mixing chamber which sucks the lubricating oil from the oil sump (Khan et al. 2009). The compressed air passing through the chamber atomizes the lubricant stream into fine aerosol micron-sized particles which is sprayed in cutting zones as mist by the nozzle. It effectively acts as a coolant and penetrates deep into the tool-workpiece interface.

There are two main methods of cutting fluid application to the machining zones in MQL.

18.1.3.1 Internal Application

In this application the oil and compressed air are applied to the cutting zone interface through tool itself. It's also called as through-tool application. Further it is also divided into two types.

(i) *Single channel*: In this system of MQL technique, compressed air and oil are mixed in mixing chamber before being supplied through cutting tool to machining zone.
(ii) *Dual channel*: Compressed air with cutting fluid in this delivery system are differently supplied and are mixed just before the machining zone.

18.1.3.2 External Application

In this application the oil and compressed gas are applied to the cutting zone through external nozzle arrangement. Furthermore there are two more methods of external application.

(i) *Ejector nozzle*: The oil and compressed air are supplied differently, and atomization takes place just outside the nozzle.
(ii) *Conventional nozzle*: Micro-sized fine aerosol mist is made in external atomizer and is then sprayed through the conventional nozzle.

Zeilmann and Weingaertner (2006) investigated the effect of external and internal delivery applications on drilling temperature by drilling titanium alloy (Ti-6AL-4V). It was observed that temperature rise in internal delivery system was 50% less than external delivery system due to improper penetration of externally supplied aerosol mixture. Tsao (2007) on milling aluminium alloy (A6061P-T651) investigated that using sulphurous boric acid cutting fluid in MQL reduces average flank wear by 12.5% compared to dry milling. Along with flank wear, surface finish and tool life also improved using MQL tan wet and dry machining using different types of carbide tools (Kamata and Obikawa 2007). Kishawy et al. (2005) investigated cutting forces and wear and surface roughness conditions in both MQL flood and dry lubrication. MQL showed preferable results over dry machining due to its economic and environmental advantages.

18.2 Vegetable Oil-Based Lubricants

18.2.1 Physicochemical Properties of Vegetable Oil-Based Lubricants

Bio-based lubricants which are usually vegetable oils and other natural sources possess less toxicity and eco-friendly biodegradability. Vegetable oils which can be edible and nonedible and which differ in geographical and tropical regions of the world can be used as bio-lubricants. Main emphasis nowadays is laid on nonedible vegetable oils which are considered as unused crops (Atabani et al. 2013). Principal component of these vegetable oils is triacylglycerols which constitute 98% composition followed by diglycerols and fatty acids (Rudnick 2005). This triglyceride structure is responsible for structural stability and high viscosity at high machining temperatures.

18.2.1.1 Viscosity

Viscosity plays an important role in reducing wear and metal-to-metal contact (Salimon et al. 2010). High viscosity measures more resistance to flow, and less viscous lubricant has less resistance to flow (Mobarak et al. 2014). Long carbon chain length, compact branched and saturated structure enhance the viscosity of the oil (Knothe and Steidley 2005; Rudnick 2005). There are various modification techniques through which viscosity of the vegetable oils can be improved to sustain viscous properties at higher machining temperatures. Presence of one double bond will increase the viscosity of oil, whereas presence of two or more double bonds will decrease the viscosity (Rodrigues et al. 2006). Epoxidized soybean oil possesses high viscosity at extreme cutting temperatures than other vegetable oils (Ting and Chen 2011).

18.2.1.2 Viscosity Index

It is the measure of variation in viscosity with temperature. The higher the viscosity index of oil, the lower is the variation with increase temperature. Oil having high viscosity index will maintain the thickness of the film at higher temperatures which reduces wear and enhances cutting conditions (Masjuki et al. 1999). Generally vegetable oils possess high viscosity index than mineral oils. Canola, palm oil-based lubricants, showed higher viscosity index than mineral-based cutting oils due to their enhanced intermolecular interactions and polyunsaturated structure (Sripada et al. 2013).

18.2.1.3 Flash Point

Flash point is an important property of a lubricant which refers to the lowest temperature at which a particular lubricant will turn to form vapours. Higher flash point thus minimizes the risk of leakage and thus provides safe machining operation at various range of temperatures (Gnanasekaran and Chavidi 2018).

18.2.1.4 Pour Point

Pour point is another important characteristic of lubricating oil which is defined as lowest temperature at which lubricant will become semi-solid and loses its ability to flow. Thus lubricants with low flash point are more suitable for machining at low temperatures (Benchaita and Lockwood 1993). If lubricant doesn't possess low pour point property, the lubricant will cease to flow at low temperature and will increase friction and wear during machining. For most of the vegetable oil pour point is lower than −10 °C which results in poor flow of oil at low temperatures due to bending and continuous stacking of triglyceride structure and formation of macro-crystals (Asadauskas and Erhan 1999; Quinchia et al. 2012; Rhee et al. 1995). These crystal structures and strong intermolecular bonding sometimes cease the ability of oil to flow at low temperatures which attributes to loss in kinetic energy due to self-stacking (Erhan et al. 2006). Oils that have more unsaturated fatty chains will have lower pour points; rapeseed oil (−21 °C) has lower pour point than olive oil (−10 °C) due to increased unsaturated fatty chains which restricts the formation of packed crystals during cooling (Gryglewicz et al. 2003).

18.2.1.5 Oxidation Stability

Oxidation stability of a lubricant refers to the capability of the vegetable oil to resist reaction with oxygen, the rate of which is also influenced by pressure, temperature and presence of impurities (Mobarak et al. 2014). A lubricant should possess high oxidative stability as low oxidation stability may lead to polymerization which increases viscosity and degrades the functional properties of lubricant on machining (Asadauskas et al. 1996; Lal and Carrick 1994; Suda et al. 2002). Critical factor that influences the oxidation stability is the presence of polyunsaturated compounds in the molecular chains which can easily react with oxygen from oxides, peroxides and acids (Wu et al. 2000). Thus the higher is the unsaturation, the more is the chance of oxidation (Fox and Stachowiak 2007). Removal of β-CH group allows the formation of carboxylic acid which deteriorates the quality of lubricating oil (Zulkifli et al. 2014).

Table 18.1 Physicochemical properties of different vegetable oils

Vegetable oil	Viscosity 40 °C (mm²/s)	Density × 10^3 kg/m³	Flash point (°C)	Pour point (°C)	Viscosity index
Sunflower oil (Rac and Vencl 2009)	47.6	0.917	245	−20	218
Rapeseed oil (Arumugam and Sriram 2013)	35.0	0.930	320	−10	220
Palm oil (Barnwal and Sharma 2005)	39.6	0.918	267	–	–
Soybean oil (Honary 1996)	29.0	0.913	328	−10	246
Jatropha oil (Mofijur et al. 2012)	35.4	0.918	186		–
Karanja oil (Bobade and Khyade 2012)	40.2	0.924	225	−3	–
Coconut oil (Nakpong and Wootthikanokkhan 2010)	28.1	0.920	–	–	–
Safflower oil (Singh and Singh 2010)	31.3	0.914	260	−06.7	–
Calophyllum inophyllum oil (Atabani et al. 2013; Habibullah et al. 2015)	71.9	0.896	221	4.3	

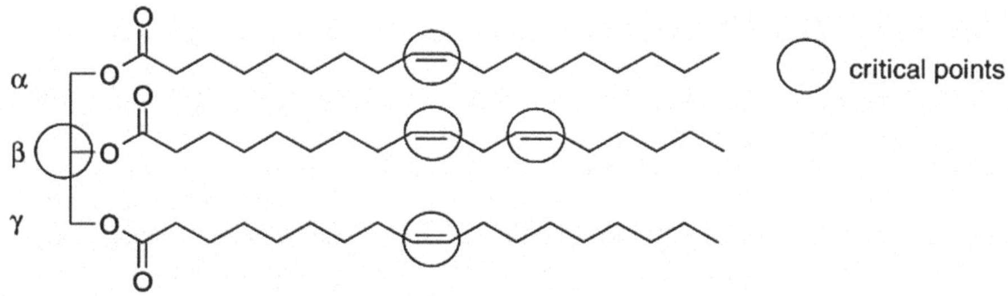

Plant oil-glycerine ester consists of different fatty acids with critical points and β-CH group (Wagner et al. 2001) (Table 18.1).

18.3 Role of Nanoparticles in Cutting Fluids

In recent approaches in order to increase the lubricity of base oil in MQL, scientists have used nanoparticles along with base oil which significantly improve thermal and heat transfer capabilities, reduce frictional coefficient and increase load-bearing capabilities of the mechanical systems. Adhesion and stress concentration were reduced to a large extent with the help of nano-cutting fluid thereby reducing mechanical failure and wear (Chang and Friedrich 2010). Sharma et al. studied improved thermophysical properties and reduced cutting forces, surface roughness and tool wear relative to base fluid used alone which improved machine performance (Sharma et al. 2015a). Kumar et al. (2012) recorded that increase in concentration of nanoparticles increases thermal conductivity which reduces the cutting zone temperature by easy extraction of heat from cutting zone. A variety of mechanisms that are involved in the enhancement of lubrication properties of the cutting fluid have been explored.

18.3.1 Mechanism of Nanolubrication

18.3.1.1 Ball Bearing/Rolling/Sliding Effect

Rapoport et al. (2002) investigated layered materials like graphite; MoS_2 when used as nano additives in cutting fluids eliminate the metal-to-metal contact between asperities due to their slippery nature that leads to fast displacement from contact area. Such effect facilitates the shearing nature of the cutting fluids. Wu et al. (2007) studied the transition from sliding to rolling friction and deposition of nanoparticles (CuO) in the worn surfaces which decreases the shear stress and improves the tribological property of cutting fluids.

18.3.1.2 Polishing Mechanism

It is a surface enhancement phenomenon in which nanoparticles penetrate into rubbing surfaces and polish the surface which decrease the contact wear of mating surfaces. Tao et al. (1996) studied the polishing effect of diamond nanoparticles dispersed in paraffin oil which increased the hardness of the rubbing surface than the test material. Lee et al. (2009b) studied the surface enhancement property of the nano oil by comparing the friction coefficient of disc type specimen dipped in mineral oil and nano oil.

18.3.1.3 Mending Mechanism

The nanoparticles accumulate over the contact surface and compensate for the lost mass on the contact surface thereby reducing the wear scars resulting in smooth surface (Lee et al. 2009a). Liu et al. (2004) investigated the mending effect and mechanism of copper nanoparticles on pin-on-desk setup. SEM and STM observations exhibit the presence of copper in the worn-out areas thereby reflecting the contribution of mending effect.

18.3.1.4 Formation of Tribofilm

This process produces a protective film over the mating surfaces. Nanoparticles on reacting with the surface layer of the material produce some tiny secondary particles which help in smoothening of nano bumps on the mating surfaces (Gupta and Harsha 2018). Hu et al. (2002) demonstrated the tribofilm formation by adding magnesium borate (10 nm) in SN 500 oil which reduced the coefficient of friction and surface wear by forming the protective wear film over the mating surfaces.

18.3.2 Preparation of Nanofluids

18.3.2.1 Two-Step Method

Most common method for the preparation of nanofluids is two-step method in which powder form with specific size of nanotubes and nanoparticles are produced as first step. During second step mixing of these nanoparticles is carried out in the base fluid directly with the help of magnetic stirrer, ultrasonic devices or homogenizer (Drzazga Michałand Dzido et al. 2012; Duangthongsuk and Wongwises 2009; Hwang et al. 2007; Khairul et al. 2014; Manimaran et al. 2014). Two-step method works well with oxide nanoparticles and metallic nanoparticles as well (Manna 2012). Al_2O_3/water and TiO_2/water nanofluids were prepared using two-step method of nanofluid preparation (Chon et al. 2005; Haghighi et al. 2013; Li et al. 2008; Murshed et al. 2005; Wang et al. 2009). Li et al. (2008) using two-step method prepared Cu/water fluid to check the influence of pH value and surfactant addition on thermal conductivity. By two-step method, preparation of nanofluids on mass scale is possible which has already increased the rate production in industries; however problem of transportation and storage cannot be avoided in this process (Yu et al. 2012). Formation of agglomerates due to high surface energy of nanoparticles is also a major disadvantage of two-step method which is further improved by one-step method in preparation of nanofluids (Mohammed et al. 2011). However two-step method can be used to form nanofluid solution with any base fluid (Wang and Mujumdar 2007).

18.3.2.2 One-Step Method

One-step method combines both production and deposition of nanoparticles as nanofluids. In this method of preparation, problem of storage and transportation is ward off, and agglomeration of nanoparticles is minimized which enhances the stability of nanoparticles (Li et al. 2009). In one-step process, nanofluid is directly prepared by physical vapour deposition (PVD) technique or by condensation technique known as VEROS (vacuum evaporation onto a running oil substrate) (Akoh et al. 1978). Eastman et al. (1996) directly dispersed the vapours of condensed metal to nanoparticles in the base fluid thereby modifying the VEROS technique of one-step method. Direct evaporation and condensation of one-step method developed by Choi et al. (2002) provides enhanced stability and largely controls the size of nanoparticles. Lo et al. (2005) introduced a submerged arc nanoparticle synthesis system (SANSS) using two different dielectric liquids which enhances stability and minimizes agglomeration effectively. Nanoparticles produced are of polygonal, square, circular and needle-like shape. Later on Chang and Kao (Chang and Kao 2007) modified this technique and improved the nanoparticle synthesis both quantitatively and qualitatively. Disadvantage of this method is residual reactants that are left after incomplete reaction later arise as impurity in nanofluid solution, and other disadvantage is its compatibility with low vapour pressure base fluids only (Wang and Mujumdar 2007).

18.3.3 Importance of Nanofluid Stability

Stability of nanoparticles means nanoparticles do not tend to agglomerate with passage of time at a significant rate (Yu et al. 2012). Stability is hence susceptibility of nanoparticles to agglomerate due to high surface energy which not only blocks microchannels but also influence thermophysical properties like thermal conductivity, density and viscosity with time (Ghadimi et al. 2011). Rate of agglomeration depends on the frequency of collision and probability of cohesion during collision due to *Brownian motion*. Stability of nanoparticles is the basic requirement for its wide application and utilization. Stability of nanofluids depends on preparation method, type of base fluids, sonication time and added surfactants. Stability is achieved when electrostatic repulsive forces between the nanoparticles are higher than attractive forces (Missana and Adell 2000; Popa et al. 2010). Agglomeration is generally reduced by steric and electrostatic mechanisms. In steric stabilization mechanism, additives like surfactants are added which form a protective layer over the nanoparticle and cease its ability to agglomerate or collide with other nanoparticle (Zhu et al. 2007). In electrostatic stabilization adsorption of ions create an ionic layer over the nanoparticle surface which enhances the repulsive forces between nanoparticles thereby increasing the stability of nanofluid (Mukherjee and Paria 2013) (Fig. 18.2).

18.4 Nanoparticle-Enriched Cutting Using MQL

18.4.1 MQL-Assisted Drilling with Nanoparticles

Minimum quantity of lubrication has largely reduced cost of machining and has enhanced machining parameters on large scale. Attempts to reduce the use of cutting fluid in machining are turning out to be a good alternative than flood and dry machining. MQL drilling is found out more promising than dry and wet conditions for low aspect ratio, and with further modifications in supply and cutting fluid, high aspect ratios can be achieved by Mathew and Vijayaraghavan (2017). Nam et al. (2011) studied the characteristics of nanofluid micro-drilling using MQL on aluminium 6061. Varying the concentration of nanodiamonds and type of base fluid, it

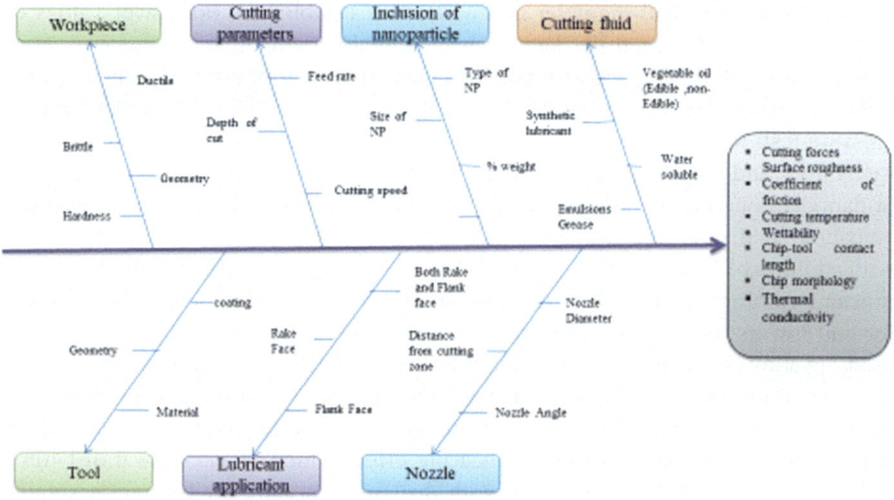

Fig. 18.2 Fishbone diagram representation of various parameters involved in MQL-assisted machining

was observed that nanodiamond-enriched MQL improves the quality of holes and lubrication which attributes to rolling, ball bearing and cooling characteristics of nanoparticles. Furthermore 1% of nanodiamond was more effective in reduction of torque and thrust forces than 2% of nanodiamond. Huang et al. (2016) analysed the micro-drilling operation with different drilling conditions. It was concluded that applying nanodiamonds in vegetable oil with MQL reduces the machining zone temperature due to improved heat transfer rate by nanoparticles. Application of nanoparticles also reduces the burr around the holes improving the quality of holes. Addition of nanoparticles reduces forces and torque required for drilling thereby improving the tool life of drill. Nam et al. (2015) conducted 25 experiments by multi-optimization techniques (RSM and GA) and deduced that minimum thrust, and torque forces at high spindle speed were obtained at 2 vol% of nanodiamonds in conventional oil. Considering same parameters (drill diameter, spindle speed, feed direction and concentration), Chai et al. (2016) dispersed MWCNT in hydrogenated oil using surfactants. TEM and FTR results analysis shows increase in thermal conductivity with some abrupt changes at different temperatures. 9.8% increase in thermal conductivity by adding 100 ppm nanoparticles at 50 °C was the highest observed value. Chatha et al. (2016) investigated influence of Al_2O_3 mixed nanofluid (1.5 vol%) on drilling 6063 aluminium with HSS tool. Addition of nanoparticles showed better machining results in terms of surface roughness, cutting speed and thrust forces as compared to dry, flood and pure MQL conditions due to their enhanced lubricating and cooling characteristics. Furthermore increase in number of holes drilled and reduction in chip contact length were observed with nanofluid MQL machining than dry and pure MQL machining. Garg et al. (2016) performed computational modelling using different techniques (ANFIS and GP), and relation between feed rate, drill diameter and MRR was established. For lower values of torque and thrust forces, drill diameter was critical parameter.

Mosleh et al. (2017) experimentally compared performance of nanodiamond (ND) and molybdenum disulphide (MoS_2) in alcohol-based lubricant under extreme pressure test. Results showed that addition of MoS_2 increases load-carrying capacity by 16% and increases service life of tool by 25%; on the other hand, dispersion of nanodiamond in base oil reduced the service life and has unfavourable response on tribological and lubrication properties of nanofluid. There is increase in tool life with addition of MoS_2 by 25%, while ND mixed nanofluid deteriorates both lubrication and tribological properties. Addition of TiO_2 nanoparticles increases heat transfer and lowers the drilling temperature at interface as compared to dry and base fluids (Salimi-Yasar et al. 2017). Nam and Lee (2018) investigated influence of nanodiamonds in vegetable oil on machinability of titanium alloy (Ti-6Al-4V). Results showed that nanodiamonds (35 nm) at low feed rate (10 mm) penetrate deep into drilling zones and provide ball-bearing effect which reduces the thrust force and torque required. Chip adhesion with drill surface and burr formation significantly reduced (Table. 18.2).

18.4.2 MQL-Assisted Grinding with Nanoparticles

Shen et al. (2008b) studied the influence of Al_2O_3-enriched nanofluid in grinding cast iron with MQL. Different modes of lubrication, dry, flooded, pure water, synthetic oil (5 vol%) with MQL, water-based MQL with Al_2O_3 and ND nanofluid of 1.5 vol%, were experimentally used in this study. It was observed that grinding ratio increases with larger fractions of Al_2O_3 (4 vol%) due to formation of tribofilm between grinding wheel and workpiece. MQL with Cimtech 500 outperformed other modes of lubrication in reducing surface roughness. 2.5% Al_2O_3 provided better surface finish than any other concentration in base fluid. Furthermore nanodiamond-rich nanofluid provided better results than Al_2O_3-rich nanofluid. On increasing flow rate in MQL, cooling rates at grinding interface can be enhanced.

Using same setup as Shen et al. (2008a, b), investigated effect of MoS_2 nanoparticles in grinding using paraffin oil, soybean oil and CANMIST as base oil with MQL lubrication technique. It was recorded that MoS_2 with CANMIST oil reduces grinding force by 27% thereby improving G-ratio by 46%, superior of all other base fluids.

Alberts et al. (2009) investigated experimentally the effect of nanofluid with varying size of graphite nanoparticles (xGnP) on machining D2 steel with MQL mode of lubrication. It was observed that large-size nanoparticles (15 μm) effectively reduce grinding forces, surface roughness and grinding energy than small-size nanoparticles (1 μm).

Kalita et al. (2010) studied the grinding performance of soybean mixed with MoS_2 nanofluid. 25% reduction in coefficient of friction was recorded with nano-mixed MQL as compared to flood lubrication and 10% reduction as compared to

Table 18.2 Literature review of MQL-assisted drilling with nanoparticles

Author	Base fluid	Nanoparticle (size)	Workpiece	Parameters			Mode of nanofluid supply	Process findings
				Cutting speed	Depth of Cut	Feed rate		
Nam et al. (2011)	Paraffin and vegetable oil	ND (30 nm)	Aluminium 6061	60,000 RPM	0.4 mm	50 mm/min	MQL	No. of holes drilled increased with use of nanofluid due to enhanced cooling and lubrication rate In order to reduce drilling torques and thrust forces, 1% of ND was effective than 2% of ND
Huang et al. (2016)	Vegetable oil	ND (<100 nm)	Aluminium alloy 7075-T6	48,000 RPM	0.52 mm	8 μm/rev	MQL	Addition of (2 wt%) of nanofluid reduces the drilling torque and improves hole quality by reducing the burrs
Nam et al. (2015)	Paraffin and vegetable oil	ND (30 nm)	Aluminium	60,000 RPM 45,000 RPM 30,000 RPM	–	15 mm/min	MQL	
Chai et al. (2016)	Hydrogenated oil	MWCNT (10–12 nm)	Steel	1800 RPM	20 mm	0.02 m/min	–	A 10% increase in thermal conductivity was observed with a very less addition of nanoparticles in the base fluid
Chatha et al. (2016)	Soybean oil	Al_2O_3 (20 nm)	Aluminium 6063	30, 53.7 m/min	20 mm	60 mm/min	MQL	HSS drills drilled nearly 200 holes with less burr formation in MQL and flooded lubrication in response to 27 in dry drilling Nanoparticles along with vegetable oil enhance the lubrication properties and reduce friction and wear thereby forming as protective coating on the surface

(continued)

Table 18.2 (continued)

Author	Base fluid	Nanoparticle (size)	Workpiece	Parameters			Mode of nanofluid supply	Process findings
				Cutting speed	Depth of Cut	Feed rate		
Mosleh et al. (2017)	Boelube oil 70,104	MoS_2 (70–100 nm) ND (3–7 nm)	4340 steel	9.144 m/min	–	0.63 mm/min	–	2–4% by weight concentration of MoS_2 increased the load-carrying capacity up to 16% of base fluid 0.5–1% by weight concentration of ND enhanced the tribological and lubricity of base fluid thereby increasing tool life by 25%
Garg et al. (2016)	Vegetable oil	ND (nanodiamond)	Aluminium	60,000 RPM 45,000 RPM 30,000 RPM		15 mm/min 12.5 mm/min 10 mm/min	MQL	Drill diameter and feed rate were optimized as critical parameters for higher MRR
Salimi-Yasar et al. (2017)	Soluble oil	TiO_2 (20 nm)	Steel	1800 RPM	20 mm	0.02 m/min	–	Addition of TiO_2 enhances the thermal and lubricating properties with increased heat transfer rate than base fluid alone
Nam and Lee (2018)	Vegetable oil	ND (35, 80)	Ti-6Al-4V	60,000 RPM	0.4 mm	10 mm/min	MQL	Small size (35 nm), high weight concentration 0.4% and low feed rate 10 mm/min enhanced machinability Burr formation and chip adhesion were minimized due to bearing characteristic of nanoparticles at drilling zone

pure MQL using soybean. Furthermore SEM and EDS analysis confirmed presence of Mo-S-P multilayer and tribolayer that reduces frictional wear between machining surfaces and enhances G-ratio without reducing material removal rate.

Kalita et al. (2012a) using EN24 and cast iron as working material studied the influence of MoS_2-rich paraffin oil and soybean oil under different lubricating conditions. 50% increase in G-ratio along with reduction of frictional coefficient (min 0.22) and 53% reduction in energy consumption was observed with both soybean and paraffin oil-based MQL lubrication. Changing the abrasive size on grinding wheel on same setup as 508–356 μm (Kalita et al. 2012b) recorded 48–52% reduction in force ratio along with reduction of peak temperature of cutting zone by 8.5% on addition of 2 wt% of nanoparticles. X-ray and SEM microanalysis showed the presence of plate-like lamella of Mo-S-P layers (molybdenum sulphur phosphorus tribofilm) thereby providing enhance lubricating properties by shearing and rolling mechanisms.

Prabhu et al. (2015) using fuzzy logic neutral technique analysed the surface roughness and microcracks on machining D3 tool steel. Using ANOVA analysis it was observed that cutting speed was the most important factor (65.1% with CNT and 58.38% with/without CNT) that influence the surface roughness of steel workpiece. Addition of MWCNT improved the surface roughness and microcracks. ANN model developed produces surface roughness close to actual value with minimum error of 0.88–9.3% and hence can be used to record surface finish values before machining.

Vasu and Kumar (2011) experimentally studied the influence of adding Al_2O_3 nanoparticles with emulsifier as base fluid on machining steel EN-31 workpiece. 20–30% reduction in machining temperature was observed at wheel workpiece interface. FEM model used records that 1% addition of Al_2O_3 reduces surface roughness and energy partition.

Prabhu and Vinayagam (2012) using Taguchi analysis investigated the influence of CNT nanoparticles in grinding AISI D3 tool steel. Depth of cut, spindle speed and feed rate were considered as design factors, and eight experiments were performed using L_8 orthogonal array. Minimum surface roughness (0.4478 μm) was recorded using CNT nanofluid, and crack width recorded with CNT also reduced from 2.22 to 1.93 mm. Using same setup as Vasu and Kumar (2011), Lee et al. (2012) considering antitoxic and super tribological properties of nanodiamond and Al_2O_3 mixed them with 150 and 20 nm each at 4% and 3% vol each in paraffin oil. Results of machining were compared with dry and pure oil-based MQL lubrication conditions. MQL with nanofluid was more effective in reducing cutting forces than dry and pure MQL. High concentration along with small diameter of nanoparticle reduces cutting forces significantly. Furthermore large size of nanoparticle effectively reduces surface roughness.

Mao et al. (2012) experimentally studied the performance of Al_2O_3 nanofluid with MQL technique on machining AISI 52100 hardened steel with aluminium oxide abrasive grinding wheel. Reduction in cutting zone temperature (40%), decrease in grinding forces and improved surface morphology was recorded with Al_2O_3 nanofluid MQL than pure water MQL grinding. Furthermore ball bearing and rolling mechanisms of nanoparticles effectively enhance lubricating and cooling conditions at cutting zone.

Mao et al. (2013b) also investigated influence of air pressure and nozzle angle during machining with Al_2O_3 nanofluids. Spraying parameter (0.2, 0.4, 0.6, 0.8 MPa) pressure with nozzle direction and spraying distance 20, 40 and 60 mm were used experimentally. It was observed that optimized parameters can be obtained with nozzle set at 15° with horizontal than tangential or directly towards wheel.

Prabhu and Vinayagam (2013) investigated effect of grinding glass with SWCNT using water as base fluid in MQL system of lubrication. Electrolytic process technique (ELID) was used with metal-bonded diamond grinding wheel for machining. Reduction in surface roughness and microcracks from 0.3758 to 0.1791 μm and 2.88 to 1.58 was calculated with atomic force microscopy.

Mao et al. (2013a) concluded that higher concentration (1, 3, 5 vol%) of Al_2O_3 nanoparticles reduces grinding forces (2.32, 2.19, 2.12 N/mm). It was further investigated that oil-based MoS_2 require lower grinding force than water-based nanofluid in MQL-based nanofluid machining.

Li et al. (2013) studied surface heat generation on the surface of three workpiece (45 steel, 2Cr13, zirconia ceramics) through numerical simulation. Grinding operation was performed under three lubricating conditions: dry, flood cooling and oil-based pure MQL and MQL with CNT nanoparticles. It was observed that CNT nanofluid with MQL effectively reduces the surface temperature than any other mode of lubrication thereby enhancing the surface quality and minimizing the burns on the surface of grinding wheel. Addition of surfactant (SDBS) reduced the chance of agglomeration and enhanced cooling and lubrication properties of the nanofluid.

Jia et al. (2014) experimentally investigated influence of machining with nanofluid of MoS_2, PCD (polycrystalline diamond) and ZrO_2 nanoparticles using MQL technique. Results were compared with dry, water-based flood cooling, pure MQL using soybean oil and nanoparticle with mixed MQL. It was observed that tangential forces decreased by 45.88% in pure MQL grinding, 62.34% in nanofluid-based MQL and 69.33% in flood grinding. Furthermore it was observed that MoS_2 forms a physical layer of molybdenum trioxide on the surface which reduces the load-carrying capacity at the grinding zone.

Wang et al. (2014) studied the relationship between crack formation and surface morphology of the workpiece (quenched 45 steel) using MoS_2 nanofluid-based MQL technique through mathematical model. Bump height reduces with increase in speed and with increase in number of cracks, whereas when a number of cracks reduce with increase in peripheral speed, the bump height increases with increasing crack length. Experimental values were smaller than simulated values which attributes to cooling and lubrication effects of nanoparticles.

ManojKumar and Ghosh (2015) investigated performance of MWCNT mixed nanofluid in machining of AISI 52100 steel using small amount of cutting fluid lubrication (SQL). Results of dry, oil-based flood lubrication, oil-based SQL and nanofluid mixed SQL were compared. It was observed that enhanced thermal conductivity which improved chip morphology and minimum coefficient of friction (0.07) was recorded with 1 vol% of MWCNT. G-ratio improved with MWCNT nanofluid, and results of surface roughness were comparable with flood lubrication using SQL.

Mao et al. (2014a) performed grinding of AISI 52100 steel with nanofluid using MQL technique. Canola oil as cutting fluid, dry, deionized water, Al_2O_3 with MQL and MoS_2 with MQL were used for experiments. It was observed that MQL with nanofluid due to their rolling and sliding properties at grinding zone reduce the frictional coefficient and grinding forces thereby improving the surface finish of the workpiece than any other mode of lubrication.

Zhang et al. (2015c) compared performance of different lubricating oils (soybean oil, rapeseed oil, palm oil) in grinding 45 steel workpiece using MoS_2 (2 wt%) nanofluid-based MQL. Addition of MoS_2 in soybean oil enhanced viscosity which further improved lubrication properties of base fluid reducing grinding forces and grinding energy. Adding CNT, MoS_2 and ZrO_2 nanoparticles at different concentrations, each (1, 2, 3 wt%) (Zhang et al. 2015b) concluded that 2 vol% of CNT showed best cooling effect due to its higher conductivity than any other concentration of nanoparticles. Using same experimental conditions (Zhang et al. 2015a), compared grinding energy of MoS_2, CNT and ZrO_2 in colza oil as base fluid. MoS_2 with jet MQL recorded 32.7% J/mm^3 less specific grinding energy than that of CNT and ZrO_2.

Zhang et al. (2016) experimentally investigated the influence of nanofluids (MoS_2, CNT, MoS_2-CNT) with different concentrations (2, 4, 6, 8, 10, 12 vol%). Increasing the surface roughness resulted in increase in surface roughness value from 2% to 12%. Furthermore lowest G-ratio was observed with MoS_2-CNT which attributes to physical collaboration of both nanoparticles in lubrication mechanism.

Setti et al. (2015) analysed the influence of Al_2O_3 and CuO nanofluids on machining titanium alloy with MQL technique. Considerable reduction in coefficient of friction and cutting temperature was observed with water-based Al_2O_3 nanofluid than with CuO-based cutting fluid. Chip morphology also revealed about cooling and antiwear and self-cleaning lotus effect of nanoparticles.

Sinha et al. (2017) experimentally analysed the performance of nanofluids (Ag, ZnO) while machining Inconel 718 using considerably small amount of cutting fluid using SQL. Significant reduction in grinding forces, wear and small frictional coefficient was recorded with use of ZnO nanoparticle with improved surface finish on machined workpiece. This attributes to better lubricity and spreading ability of ZnO nanoparticles.

Wang et al. (2017b) compared the performance of Al_2O_3 at different concentrations (0.5–4 vol%) using MQL technique in grinding nickel-based alloy. Minimum contact angle (45°) for maximum lubricity and spreadability was observed at 2 vol%. Best tribological properties of nanofluid were obtained with 2 vol% of Al_2O_3; increase in concentration reduces the surface quality of workpiece. 24.3% reduction in force ratio and 34.1% reduction in grinding energy than pure MQL has been observed. Wang et al. (2017a) further added different nanoparticles (MoS_2, SiO_2, ND, CNT, Al_2O_3, ZrO_2) in palm oil as base fluid. Al_2O_3 provided best antifriction property by reducing frictional coefficient by 20% and w-ratio by 65%. Improved lubrication properties on addition of nanoparticles attribute to formation of tribofilm and polishing effect at wear zones and sliding surfaces. Li et al. (2017) using same setup and conditions as Wang et al. (2017a) experimentally analysed physical properties of different nanoparticles in grinding nickel alloy. Increase in concentration by volume of nanoparticles (up to 2 vol%) increases thermal heat transfer and viscosity of nanofluid. CNT with 2 vol% concentration records the lowest temperature (108°), and SiO_2 records lowest grinding force ratio. Furthermore lower contact angle will improve lubrication effect (Table 18.3).

18.4.3 MQL-Assisted Turning with Nanoparticles

Krishna et al. (2010) experimentally studied the influence of mixing nanosolid suspensions of nanoboric acid in coconut oil and SAE-40 oil in machining AISI 1040. Turning operation was conducted on lathe with carbide inserts. It was recorded that flow rate of 10 ml/min and 0.5% concentration of nanoboric acid with coconut oil shows better performance in terms of heat transfer, tool wear and surface finish than nanoboric acid suspension in SAE-40 oil. With increase in concentration (up to 0.5 wt%) of nanoparticles, thermal conductivity increases, and due to reduced friction, there is a reduction in cutting temperature as well.

Yan et al. (2011) on mixing different concentrations of nanosolids MoS_2, Cu, CuO (1, 3, 5, 10, 20 wt%) and graphite fibre (0.2, 0.5, 1, 3, 5 wt%) in calcium-based greases in turning of RB-SiC ceramic, established that due to formation of thin tribolayer direct contact between tool and workpiece were significantly reduced. 10 wt% of copper nanoparticles in grease produces best surface finish and minimum tool wear which attributes to higher microplasticity of cu nanoparticles than others.

Khandekar et al. (2012) experimentally studied the effect of adding Al_2O_3 nanoparticles in turning AISI 4340 steel under dry, flood and nano-enriched servo-cut base fluid. Addition of 1% of Al_2O_3 enhances wettability of cutting fluid which enhances heat transfer property of cutting fluid. Enhanced heat dissipation and cooling characteristics of nano-enriched cutting fluid reduce cutting forces and tool wear during machining.

Amrita et al. (2013) experimentally investigated the influence of adding nanographite at different concentrations (0.1, 0.3, 0.5 wt%) in turning AISI 1040. Cutting fluid was supplied under different conditions of dry, flood and mist (5, 10, 15 ml/min) at different air pressures. With increase in pressure, droplet size of aerosol decreases which enhance lubrication and cooling property of cutting fluid. 15 ml/min and 0.5 wt% of Al_2O_3 decrease tool wear by 94.04% and 75.24% than dry and flood lubrication, respectively, for carbide tool. 71.92% reduction in tool wear was observed in case of HSS steel tool as compared to flood lubrication. Furthermore 76.25% and 56.98% reduction in cutting forces in case of carbide tool and 84.02% and 77.76% reduction in HSS steel was observed as compared to dry and flood machining.

Srikiran et al. (2014) investigated the influence of nanographite mixed cutting fluid in machining AISI 1040 steel using carbide insert. Cutting forces increase by

Table 18.3 Literature review of MQL-assisted grinding with nanoparticles

Author	Base fluid	Nanoparticle	Workpiece material	Parameters			Mode of nanofluid supply	Process findings
				Cutting speed	Depth of cut	Feed rate		
Shen et al. (2008b)	Deionized water, Cimtech 500	Al_2O_3, ND (40, 100, 200 nm)	Dura-Bar 100-70-02 ductile iron	30 m/s	10 μm	2400 mm/min	MQL	2.5% of Al_2O_3 provided better surface finish than 1% or 4% concentration across and along grinding direction Higher G-ratio was obtained by adding nanoparticles without any considerable change in grinding temperature
Shen et al. (2008a)	Paraffin oil Soybean oil CANMIST oil	MoS_2 (<100 nm)	Dura-Bar 100-70-02 ductile iron	30 m/s	10 μm		MQL	27% reduction in grinding forces was observed for CANMIST OIL, 21 for paraffin oil and 9% for soybean oil Higher G-ratio for higher concentrations for MoS_2 was observed
Alberts et al. (2009)	IPA TRIM SC200	xGnP (5–10 nm)	D2 tool steel (62 HRC)	25 m/s	50 mm	0.75 m/min	MQL	Larger size (15 μm) nanoparticles of graphite dispersed in IPA solution records reduction in specific energy, surface roughness and cutting forces
Kalita et al. (2010)	Soybean oil	MoS_2	Dura-Bar 100-70-02 ductile iron	30,000 mm/s	2 mm	40 mm/s	MQL	G-ratio improved without affecting the MRR MQL accompanied with MoS_2 reduced the coefficient of friction thereby reducing tangential grinding forces
Prabhu and Vinayagam (2010)	SAE20W40 oil	MWCNT (10–20 nm)	AISI D2 tool steel	2000 RPM 2500 RPM	0.1 mm 0.2 mm	1.9 mm/min 2.5 mm/min	MQL	Surface tribology improved from micro level to nano level by adding 2% of multiwall carbon nanotube (MWCNT)
Vasu and Kumar (2011)	TRIM E709 oil	Al_2O_3 (<100 nm)	EN-31 steel	1400 RPM	25 μm 50 μm 75 μm	100 mm/s 150 mm/s 200 mm/s	MQL	Using TRIM E709 emulsifier, the cutting temperature reduced to 20–30% compared to plain emulsifier

(continued)

Table 18.3 (continued)

Author	Base fluid	Nanoparticle	Workpiece material	Parameters			Mode of nanofluid supply	Process findings
				Cutting speed	Depth of cut	Feed rate		
Prabhu and Vinayagam (2012)	SAE20W40 oil	CNT (10–20 nm)	AISI D3 tool steel	2000 RPM 2500 RPM	0.1 μm 0.2 μm	1.9 mm/min 2.5 mm/min	MQL	Addition of CNT increases the flash point and fire point of the base fluid by 6.3 and 5% Surface finish was improved with the use of nanoparticles, from 0.54045 μm without to 0.4478 μm with CNT
Lee et al. (2012)	Paraffin oil	Al_2O_3 ND	SK-41C tool steel	80,000 RPM	5.0 μm	120 mm/min	MQL	Size of nano-Al_2O_3 was recorded as critical parameter in reducing the surface roughness than ND Small size and higher concentration of ND were more effective in reducing the grinding forces
Mao et al. (2012)	Deionized water	Al_2O_3 (40 nm)	AISI 52100 steel	31.4 m/s	10 μm	0.05 m/s	MQL	Added nanoparticles of Al_2O_3 provide the ball-bearing effect which reduces the grinding forces, reduces the interface temperature and provides better lubricating conditions than water alone
Kalita et al. (2012a)	Paraffin oil Soybean oil	MoS_2 (100 nm)	Dura-Bar 100-70-03 EN24 steel alloy	30 m/s	10 μm 20 μm	0.06 m/s 0.1 m/s	MQL	Lowest frictional coefficient of 0.22 and 53% reduction in the energy consumption was observed in grinding with nanofluids 50% increase in G-ratio and increase in tool life were also observed with nMQL
Kalita et al. (2012b)	Soybean oil	MoS_2 (100 nm)	Dura-Bar 100-70-03	30 m/s	10 μm 20 μm	0.06 m/s	MQL	Formation of a tribo-chemical film on frictional surfaces reduces the friction, wheel wear and energy

Mao et al. (2013b)	Deionized water	Al_2O_3 (60 nm)	AISI 52100 steel	31.4 m/s	10 μm	0.05 m/s	MQL	Angular position of nozzle and increased air pressure results in enhanced grinding conditions with reduced cutting speed and surface roughness Shorter spraying distance improves the grinding performance
Prabhu and Vinayagam (2013)	Water-soluble oil	SWCNT	Glass	1200 m/min	30 μm	20 m/min	–	ELID technique with CNT nanoparticles tend to improve surface finish from (reduced from) 1.788 to 0.722 μm and microcracks
Mao et al. (2013a)	Deionized water	Al_2O_3 (40, 80, 70 nm)	AISI 52100 steel	31.4 m/s	10 μm	0.05 m/s	MQL	Oil-based nanofluids have better surface finish and reduced cutting forces as compared to water-based nanofluids Higher conc. of Al_2O_3 lowers the grinding temperature and grinding forces Increase in diameter reduces the grinding forces but adversely affects the surface finish of workpiece
Mao et al. (2014b)	Deionized water	Al_2O_3 (10 nm)	–	–	–	–	MQL	It was concluded that the best dispersion property of Al_2O_3 was found when conc. is (0.5 wt%), pH value is 7 and ultrasonic vibration time is 1 h
Jia et al. (2014)	Soybean oil	MoS_2 ZrO_2 Polycrystal diamond (PCD) MoS_2	Hardened 45 steel	30 m/s	10 μm	4 m/min	MQL	Anti-attrition effects of MoS_2 were observed the best Both PCD and ZrO_2 provided the roller bearing characteristic at the sliding interface Reduced grinding forces, friction coefficient and energy were achieved at 6% conc. by mass

(continued)

Table 18.3 (continued)

Author	Base fluid	Nanoparticle	Workpiece material	Parameters			Mode of nanofluid supply	Process findings
				Cutting speed	Depth of cut	Feed rate		
Wang et al. (2014)	LB1000	MoS_2 (50 nm)	Quenched 45 steel	5–60 m/s	10–45 μm	0.01–0.1 m/s	MQL	Grinding temperature gets reduced by addition of nanoparticles, and there is elastic recovery of workpiece material due to effective cooling and lubrication
ManojKumar and Ghosh (2015)	Deionized water	MWCNT	Hardened AISI 52100	25 m/s	10 μm 20 μm	6 m/min 10 m/min		0.6 vol% of MWCNT proved better than soluble oil in reducing grinding forces MWCNT nanofluid has potential to substitute synthetic oils in grinding hardened steel
Mao et al. (2014a)	Deionized water Canola oil	Al_2O_3 (10 nm) MoS_2 (70 nm)	AISI 52100	31.4 m/s	10 μm	0.05 m/s	MQL	1.2 wt% of Al_2O_3 reduces grinding forces due to reduced coefficient of friction Grinding temperature also reduces due to increase in convective heat transfer
Zhang et al. (2015c)	Liquid paraffin Palm oil Rapeseed oil Soybean oil	MoS_2 (50 nm)	45 steel	30 m/s	10 μm	3 m/min	MQL	Optimum lubricating properties were achieved at 6% of MoS_2 in soybean oil Soybean having low viscosity proved suitable for nanoparticles as base oil in jet MQL
Zhang et al. (2015b)	Soybean oil	MoS_2 (50 nm) CNT ZrO_2	Hardened 45 steel	35 m/s	20 μm	0.05 m/s	MQL	2% volume conc. of CNT gave ideal experimental results with excellent cooling effects MoS_2 with soybean as base fluid gave best lubrication results

Setti et al. (2015)	Water	Al_2O_3 CuO	Ti-6Al-4V	17 m/s	5 μm	9 m/min	MQL	Nanofluids enhance micron level grain fracture and reduce surface temperature, wear and friction between wheel and workpiece by forming a tribofilm
Zhang et al. (2015a)	Colza oil	MoS_2 (50 nm) CNT ZrO_2	Hardened 45 steel	35 m/s	20 μm	0.05 m/s	MQL	Lubrication performance and surface finish enhanced with 2 vol% of MoS_2 among all three
Prabhu et al. (2015)	SAE20W40	MWCNT	AISI D3 tool steel	2000 RPM 2500 RPM	0.1 mm 0.2 mm	1.9 m/s 2.5 m/s	MQL	Using ANOVA cutting speed was found significantly affecting surface roughness
Zhang et al. (2016)	Synthetic oil	MoS_2 (30 nm) CNT MoS_2-CNT	Ni-based alloy GH4169 440 C steel	30 m/s	10 μm	3000 mm/min	MQL	Blend of MoS_2-CNT records the lower grinding ratio and surface roughness over MoS_2 and CNT used alone
Li et al. (2013)	Vegetable oil	CNT (50 nm)	2Cr13 steel 45 steel ZrO_2	40 m/s	1 μm 5 μm 10 μm	3 m/s	MQL	Better grinding performance can be achieved by 2 vol% of nanofluid due to improved tribological properties of CNT
Wang et al. (2017b)	Palm oil	Al_2O_3 (50 nm)	Nickel-based alloy GH4169 440C steel	30 m/s	10 μm	3000 mm/min	MQL	The force ratio and specific grinding ratio lower down with use of nanofluids in MQL than pure MQL-based fluid 2.0 vol% of Al_2O_3 provides best tribological performance
Sinha et al. (2017)	Deionized water	Ag Zno	Inconel 718	18 m/s	10 μm	6 m/min		Nanofluid (Zno) girding minimizes the chance of redeposition of debris and oxide layer formation on ground machined surface Due to better wettability and lubricity of nanofluids, no wear land was observed on the surface

(continued)

Table 18.3 (continued)

Author	Base fluid	Nanoparticle	Workpiece material	Parameters			Mode of nanofluid supply	Process findings
				Cutting speed	Depth of cut	Feed rate		
Wang et al. (2017a)	Palm oil	MoS_2 CNTs ND SiO_2 Alpha-Al_2O_3	Nickel-based alloy GH4169 440C steel	30 m/s	10 μm	3000 mm/min	MQL	Al_2O_3 provide best surface morphology due to formation of thick tribofilm which reduces friction and wear Nanofluids MoS_2, SiO_2 and Al_2O_3 provide better performance capabilities than base oil alone, and they are also termed as environmental-friendly nanofluids
Li et al. (2017)	Palm oil	MoS_2 CNTs ND SiO_2 Alpha-Al_2O_3	Nickel-based alloy GH4169 440C steel	30 m/s	10 μm	3000 mm/min	MQL	SiO_2 have lowest grinding force ratio with CNT having high conductivity
Pashmforoush and Bagherinia (2018)	Water	Cu	Inconel 738	20 m/s	10 μm 20 μm 30 μm	50 mm/s 100 mm/s 150 mm/s		Surface roughness decreased by 59.19% using copper nanofluid, and adhesion of workpiece chips on wheel also reduced by 62.16%

1.08 times when diameter of the nanographite decreased from (70–90 nm) to (5–10 nm) which attributes to increase in frictional resistance due to sliding and sticking phenomenon that increases cutting forces. Smaller size of nanoparticles results in increased surface roughness due to continuous rubbing, increase in temperature and tool chatter which degrades the surface quality of workpiece.

Chan et al. (2013) experimentally studied that surface properties and waviness of the workpiece can be enhanced by 19% and 22% with use of nano-droplet-enriched cutting fluid (NDCF). NDCF reduces the thrust forces by 26% and increases cutting to thrust force ratio reducing viscosity which can further improve surface roughness.

Saravanakumar et al. (2014) investigated the performance of silver nanoparticles-enriched cutting fluid in turning. It was experimentally observed that with addition of silver nanoparticles, there is reduction in cutting forces by 8.8% and surface roughness by 7.5% as compared to conventional fluids used for machining. Formation of a tribofilm with use of silver nanoparticles reduces the coefficient of friction and tends to increase tool life by bringing down the cutting temperature.

Amrita et al. (2014a) investigated different parameters in machining AISI1040 steel with uncoated carbide tool on addition of $nMoS_2$ (nano molybdenum sulphide), NBA (nanoboric acid) and FNG (functionalized nanographite). Addition of surfactant (SDBS) enhanced the stability of nanoparticles in base fluid. MQL technique with and without application of nanofluids provided better cutting performance than dry machining. $nMoS_2$-enriched cutting fluid significantly reduces surface roughness, wear and cutting forces compared to other nanofluids. NBA considerably reduces cutting temperature at the machining zone.

Sayuti et al. (2014b) designed his experiments using Taguchi method with different process parameters during turning of AISI 4140 hardened steel using SiO_2 nanoparticles. Using fuzzy logic and Taguchi analysis minimum surface roughness and tool wear was achieved by 0.5 wt% of SiO_2 at 30° and 60°, respectively. Increase in concentration degrades surface finish and increases tool wear.

Padmini et al. (2015) mixed $nMoS_2$ and boric acid (H_3BO_3) in sesame oil and coconut oil separately. The samples were then analysed under different lubricating conditions. Using MQL-based lubrication system with nano enrichment of both MoS_2 and boric acid, cutting forces and surface roughness were reduced up to 82% and 37% as compared to dry machining. $nMoS_2$ effectively reduces interface temperature of 37% and 31% in coconut and sesame oil, respectively.

Padmini et al. (2016) investigated influence of adding $nMoS_2$ in coconut, sesame and canola oil at different concentrations in turning of AISI 1040 steel. It was recorded that 0.5 wt% of MoS_2 effectively reduces cutting forces, surface roughness and tool wear by 37%, 39% and 44% as compared to dry machining. Furthermore $nMoS_2$ forms a fullerene-like nanofilm interface which results in reducing coefficient of friction.

Amrita et al. (2014b) investigated the influence of adding nanographite 0.3 wt% in water-soluble oil. AISI 1040 steel was turned using different lubrication techniques: dry, flood, pure fluid with MQL and nano-enriched cutting fluid, and results were compared experimentally. It was observed that mist application of nanofluid due to formation of fine aerosols enhances the cutting condition by bringing down interface temperature and reducing the tool wear. 54% reduction in cutting temperature, 25% in tool wear and 71% in surface roughness were observed as compared to flood lubrication.

Gupta et al. (2016) investigated influence of nMQL during turning in CNC machine using cubic boron nitride inserts. Al_2O_3, MoS_2 and graphite were mixed in vegetable oil for experimentation. After performing 29 experiments and optimization through ANOVA, it was recorded that cutting speed, cooling condition and feed rate contribute 40.20%, 36.50% and 20.13% and are significant parameters. Furthermore it was observed that graphite effectively reduce the cutting temperature as compared to other nanoparticles due to formation of lamellar structure between machining interfaces.

Su et al. (2016) investigated experimentally the performance of graphite-LB2000 and graphite-PriEco6000 nanofluids under different lubricating conditions. Experiments were conducted on turning AISI 1045 annealed carbon steel with uncoated carbide insert under dry, MQL and MQL with graphite-oil mixture. It was observed that MQL with graphite-enriched LB2000 nanofluid shows better performance than graphite-enriched PriEco6000 in terms of reduction in machining temperature and cutting forces.

Ali et al. (2017) studied influence of surfactant (SDBS w) in nano-enriched mixture of Al_2O_3 and soluble oil at different concentrations (0.1, 0.4, 0.6 wt%). Increasing the concentration of nanoparticles surface roughness of the workpiece reduces. Feed rate was calculated as most significant factor using ANOVA analysis.

Raju et al. (2017) investigated the performance of MWCNT (2 vol%) with distilled water. Machining with MWCNT reduces tool wear by 40% and cutting forces by 5–8% in comparison with base fluid and dry machining, respectively. Furthermore presence of nanoparticles reduced the contact angle by 33.33% as compared to conventional fluid which enhances the wear resistance capability of cutting fluid.

Behera et al. (2017) experimentally compared the spreading coefficient of different surfactants (CTAB, SDS, Tween-20, PVP) on cutting tool surface. Reduction in tangential force (29.72%), coefficient of friction (7.25%) and feed force (19.39%) were observed with Tween-20-NF compared to dry and other surfactant-rich nanomachining. Reduction of about 33.35% in chip tool interface was observed with Tween-20 NF compared to dry environment.

Hegab et al. (2018b) experimentally observed the performance of MWCNT-rich nanofluid on turning titanium alloy by varying cutting speeds. Using ANOVA analysis it was recorded that MWCNT-rich nanofluid gave better results in power consumption and flank wear than base fluid alone. Significant improvement on cutting temperature, cooling and wetting characteristics of base fluid was observed by using MWCNT. Continuing their work Hegab et al. (2018a) studied the effect of turning with nano-enriched cutting fluid on chip morphology and tool wear on machining Inconel 718. MWCNT and Al_2O_3 have been used to enhance the lubrication properties of base fluid. Less deformation in chip and enhancement in interface bonding of tool and chip were observed which attributes to increase in shear angle and superior cooling characteristics of cutting fluid.

Sartori et al. (2018) observed the cooling and lubrication characteristics of solid lubricants (graphite and PTFE) using MQL technique on turning Ti-6Al-4V alloy. Reduction in crater and nose wear was observed due to improved dynamic viscosity of solid lubricant-assisted MQL-MQC (Table 18.4).

18.4.4 MQL-Assisted Milling with Nanoparticles

Park et al. (2011) experimentally analysed the effect of nano-enhanced lubrication in ball milling AISI 1045 steel workpiece using MQL technique. Nanographene (xGnP) particles were mixed with vegetable oil, and experiment was conducted in vertical milling machine under dry, flood, water-soluble mineral oil, units cooltube 200 (vegetable oil) and xGnP nanoparticle in vegetable oil with MQL. 0.1 wt% and 1 μm diameter are regarded as optimum concentration and size of nanoparticle for enhanced cutting performance as compared to 1 wt% and 15 μm. The layered structure of nanographene reduces coefficient of friction significantly.

Sarhan et al. (2012) investigated the effect of SiO_2 nano-enriched milling of AA6061-T6 alloy using MQL. It was observed that SiO_2 mixed with mineral oil improved milling performance in terms of specific power, cutting forces and coefficient of friction as compared to base fluid alone which attributes to sliding and rolling action of nano SiO_2. Using same setup (Sayuti et al. 2013a) studied the effect of concentration, nozzle angle and cutting temperature. Using ANOVA analysis minimum cutting force was observed with 0.2 wt% SiO_2 and 60° nozzle angle, and minimum cutting temperature was observed with 15° nozzle angle at 2 bar air pressure. Surface roughness can be minimized with high concentration of nanoparticles and with 30° nozzle angle.

Sayuti et al. (2013b) studied the influence of carbon anion-enriched nanofluid on machining aerospace aluminium alloy (duralumin Al-2017-T4) using vertical milling machine. 0%, 0.5%, 1% and 1.5 wt% concentrations of carbon anion were used during machining, and it was observed that 21.99% and 46.32% reduction in cutting forces and surface roughness can be obtained by nano-enriched cutting fluid as compared to conventional oil lubrication. Higher concentration of nanoparticles reduced cutting forces and improved surface quality. Sayuti et al. (2014a) concluded that addition of SiO_2 improves the surface morphology during milling and reduces the thermal deformation of workpiece which attributes to the rolling action and formation of thin tribofilm that improves surface quality and conductivity.

Rahmati et al. (2014) analysed the milling performance on aluminium alloy and recorded improvement in surface finish by 3.87% with 0.5 wt% MoS_2-enriched cutting fluid as compared to conventional oil. With further increase in concentration of nanoparticles, surface finish deteriorates. FESEM and XRD were used to study surface morphology.

Najiha et al. (2015) analysed the performance of water-mixed nanoparticles (TiO_2) and compared the results with that of conventional oil-mixed nanoparticles. It was observed that both techniques produce almost same results for tool wear,

Table 18.4 Literature review of MQL-assisted turning with nanoparticles

Author	Base fluid	Nanoparticles	Workpiece material	Parameters: Cutting speed	Depth of cut	Feed rate	Mode of lubrication	Process findings
Krishna et al. (2010)	Coconut oil, SAE-40	Nanoboric acid (50 nm)	AISI 1040 steel	60, 80, 100 m/min	0.1 mm	0.14, 0.1, 0.2 mm/rev	MQL	Percentage increase in conc. of nanoboric acid increases the thermal conductivity and decreases specific heat 0.5% of nanoboric acid reduces cutting temperature and improves tool life and surface finish
Yan et al. (2011)	Calcium-based grease	MoS_2 GF Cu CuO	RB-Sic	1000 RPM	2 μm	20 mm/min	MQL	Best surface quality and less tool wear were obtained at 10 wt% cu conc. in grease
Khandekar et al. (2012)	Servocut S	Al_2O_3	AISI 4340	350 m/min	0.1 mm	0.1 mm/rev	MQL	Addition of 1% of Al_2O_3 in base fluid enhances the wettability of base fluid 50% reduction in cutting forces by using nanofluids as compared to dry machining 54.4% reduction in surface roughness when nanofluid is used
Amrita et al. (2013)	–	Nanographite (<80 nm)	AISI 1040	40.7 m/min	1 mm	0.14 mm/rev	MQL	Min tool wear, tool chip interface temperature and cutting forces were observed for 15 ml/min flow rate and 0.5 wt% of nanoparticles
Amrita et al. (2014b)	Conventional water (nano-SO)	Nanographite (<80 mm) 10 ml/min	AISI1040	40 m/min	–	0.14 mm/rev	MQL	70%, 25% and 20% reduction in cutting temperature was observed with nanofluid cutting than dry, flood and mist fluid applications Surface roughness reduced by 42%, 32% and 28%, respectively

Srikiran et al. (2014)	SAE-40 oil	Nanographite (5–10 nm) (15–30 nm) (40–60 nm) (70–90 nm)	AISI 1040	51 m/min 78 m/min 126 m/min 192.6 m/min	0.25 mm 0.50 mm 0.75 mm 1.0 mm	0.05 mm/rev 0.08 mm/rev 0.10 mm/rev 0.125 mm/rev	MQL	Cutting forces, frictional forces and tool interface temperature increase due to decrease in size of nanoparticles from 70–90 nm to 5–10 nm
Chan et al. (2013)	JAEGER SW-105 oil	–	6061-T651 brass Pure copper	2000 RPM 8000 RPM	30 μm 20 μm	30 mm/min 20 mm/min	–	Reduced roughness and enhanced waviness were observed by using cutting fluid with nano-droplet-enriched cutting fluid (NDCF)
Saravana-kumar et al. (2014)	Distilled water	Ag	–	50.4 m/min 64.8 m/min	3 mm 5 mm	0.1 mm/rev 0.2 mm/rev	MQL	Surface roughness decreased by 7.5% with use of nanoparticles Rate of heat transfer at chip tool interface improved with decrease in cutting forces by 8.8%
Amrita et al. (2014a)	Water-based soluble oil	FNG (80 nm) NBA (100) nMoS_2 (100)	AISI 1040	65 m/min	0.75 mm	0.14 mm/rev	MQL	MoS_2 (0.3 wt%) recorded better surface properties with lower wear and reduced cutting forces
Sayuti et al. (2014b)	Mineral oil	SiO_2 (5–15 nm)	AISI 4140	0.15 mm/rev	0.5 mm	0.15 mm/rev	MQL	Minimum tool wear was obtained with 0.5% conc. of nanoparticles and 2 bar pressure in mineral oil with 60° nozzle angle and improved surface roughness at 30° angle of nozzle
Padmini et al. (2015)	Coconut oil (CC), sesame oil (SS)	Boric acid MoS_2 (<100 nm)	AISI 1040 (30 HRC)	60 m/min	0.5 mm	0.14 mm/rev	MQL	As compared to dry machining, nMoS_2 with coconut and sesame oil reduces surface roughness by 46% nMoS_2 effectively reduced tool wear by 28% and 38% in coconut and sesame oil

(continued)

Table 18.4 (continued)

Author	Base fluid	Nanoparticles	Workpiece material	Parameters			Mode of lubrication	Process findings
				Cutting speed	Depth of cut	Feed rate		
Padmini et al. (2016)	Coconut oil (CC) Sesame oil (SS) Canola oil (CAN)	Boric acid MoS_2 (<100 nm)	AISI 1040	40 m/min 60 m/min 100 m/min	0.5 mm	0.14 mm/rev 0.17 mm/rev 0.20 mm/rev	MQL	5% coconut oil and $nMoS_2$ reduces cutting forces by 37%, cutting temperature by 24% and tool wear by 44% than any other nanofluid
Gupta et al. (2016)	Vegetable oil	Al_2O_3 (<50 nm) MoS_2 Graphite	Titanium (grade 2)	215 m/min 250 m/min	1 mm	0.10 mm/rev 0.15 mm/rev	MQL	Lower cutting temperature and cutting forces were observed using graphite-based nanofluids
Su et al. (2016)	LB2000 PriEco6000	Graphite	AISI 1045	55 m/min 96 mm/min	1 mm	0.10 mm/rev	MQL	Using nanofluids cutting forces and temperature reduced 11.9% and 21% with respect to dry machining
Ali et al. (2017)	SolCut	Al_2O_3 (<50 nm)	Ti-6Al-4V	75 m/min 85 m/min 95 m/min	1 mm	0.10 mm/rev 0.15 mm/rev 0.20 mm/rev	MQL	Taguchi method was applied to find set of combinations of parameters for producing optimum surface roughness, tool wear and power consumption
Raju et al. (2017)	Distilled water	MWCNT	EN31	1500 RPM	1 mm	0.1 mm/rev	–	Tool wear improved by 49% using MWCNT nanofluid as compared to dry machining, and 30% reduction was observed than conventional fluid 9–12% reduction in surface roughness and cutting forces reduce by 5–8% with use of MWCNT-enriched nanofluid over conventional fluid

Behera et al. (2017)	Deionized water	Al_2O_3	Inconel 718	60 m/min 80 m/min	0.5 mm	0.2 mm/rev	MQL	Reduction in chip curling, cutting forces and coefficient of friction was observed as a result of good spreading behaviour of nanofluids
Hegab et al. (2018b)	–	MWCNT Al_2O_3	INCONEL 718	40 m/min 50 m/min 60 m/min	–	0.2 mm/rev 0.3 mm/rev 0.4 mm/rev	MQL	Addition of MWCNT and Al_2O_3 reduces cutting forces due to increase in shear angle and adequate dissipation of heat
Hegab et al. (2018a)	ECOLUBRIC E200	MWCNT	Ti-6Al-4V	120 m/min 170 m/min 220 m/min	–	0.1 mm/rev 0.15 mm/rev 0.2 mm/rev	MQL	2 wt% of nanoparticles in base solution provides better cutting performance at less power consumption than 4 wt% of nanofluid
Sartori et al. (2018)	AstroCut HD XBP	PTFE (polytetra-fluoroethylene) (5 μm) Graphite	Ti-6Al-4V	80 m/min	0.25 mm	0.2 mm/rev	MQL MQC	Addition of PTFE assisted the cratering phenomenon, and tool wear reduced with less abrasion of cutting edge
Sharma et al. (2018)	Alumina	Graphene	AISI 304 steel	60 m/min 90 m/min 120 m/min	0.6 mm 0.9 mm 1.2 mm	0.08 mm/rev 0.12 mm/rev 0.16 mm/rev	MQL	Addition of graphene in alumina increases its wettability 1.25 vol% of alumina and Al-GnP hybrid nanofluid shows lowest coefficient of friction and wear

but edge chipping is reduced in water-based MQL due to higher cooling rate and higher concentration of aluminium in deposited layer on flank face. Using same setup (Najiha and Rahman 2016) compared the wear phenomenon for TiO_2-enriched water and conventional oil using MQL technique. Adhesion and abrasion wear were found to be major tool wear phenomenon in all three cases of lubrication (flood, MQL with vegetable oil, and MQL with water) which were comparable in all three modes of lubrication. Furthermore it was observed that due to cooling effect and high latent heat produced by water-based MQL, edge fracture and edge chipping were not seen. Varying concentrations of nanoparticles (0.5, 2.5, 4.5 wt%) of TiO_2 in same setup (Najiha et al. 2016) recorded minimum tool wear at 2.5 vol% compared to other lubricating conditions. Experimental results were then compared with those of three-level fuzzy logic system.

Kim et al. (2016) compared four different lubrication techniques (dry, flood, MQL with hBN in vegetable oil with and without chilly CO_2 gas in end milling of Ti-6Al-4V alloy). 0.1 wt% of hBN provided enhanced milling performance than other lubrication techniques. Cooling and lubrication effect of both chilly CO_2 gas and hBN effectively reduce surface roughness, chip adhesion of machined surface and tool wear.

Muthusamy et al. (2016) analysed the wear analysis on milling of AISI 304 with TiO_2 mixed ethylene glycol nanofluid with different concentrations (0.5, 1.0, 1.5 vol%). Milling AISI 304 steel with tin-coated carbide tool using TiO_2 mixed nanofluid increase tool life (54.9 min) as compared to water-soluble coolant (32.67 min). Enhanced tool life was obtained with 1 vol% of TiO_2 at all cutting speeds. It was observed from results shown by EDX and ESM that a layer formation by embedment of nanoparticles from nanofluid acts as a protective layer for cutting tool.

Lv et al. (2018) studied the effect of hybrid nanoparticles graphene dioxide and silicon dioxide (GO and SiO_2) in water-based MQL lubrication technique at varying concentrations of both. On end milling of AISI 304, it was observed that addition of 0.02:0.05% of GO:SiO_2 in water resulted in reduced tool wear, with improved surface finish and tool life. SiO_2 and GO nanoparticles penetrate into the pits of the friction region thereby forming a smooth surface which enhances the rolling effect of SiO_2 and slipping of GO sheets thereby forming a thin protective layer which distributes stress concentration and separates the sliding pairs minimizing wear (Table 18.5).

18.5 Future Scope

Previous researchers have focused more on lubricants containing single type nanoparticles in mineral oils preferably. In the above study, the authors have tried to experimentally investigate the blending of two different types of nanoparticles having two different properties and mechanisms they involve in. The blending of grapheme (GnP) with TiO_2 (titanium dioxide) in a fixed ratio (1:1) along with a biodegradable oil improved tribological and thermophysical properties of the oil for

Table 18.5 Literature review of MQL-assisted milling with nanoparticles

Author/year	Base fluid	Nanoparticles	Workpiece material	Parameters: Cutting speed/wheel speed	Parameters: Depth of cut	Parameters: Feed rate/ workpiece speed	Mode of lubrication	Process findings
Park et al. (2011)	Vegetable oil	xGnP (1 μm)	AISI 1045	3500 RPM 4000 RPM	1.0 mm 0.6 mm	2500 mm/ min	MQL	Addition of xGnP nanoparticles increases wettability and reduces friction 0.1 wt% of xGnP resulted in improved cutting performance without agglomeration
Sarhan et al. (2012)	ECOCUT SSN 322 mineral oil	SiO_2 (5–15 nm)	Aluminium alloy AA6061-T6	500 RPM	5 mm	100 mm/ min	MQL	SiO_2 nanoparticles reduce contact friction due to which there is less consumption of specific energy and power
Sayuti et al. (2013a)	ECOCUT SSN 322	SiO_2 (5–15 nm)	Aluminium alloy AA6061-T6	5000 RPM	5 mm	100 mm/ min	MQL	Minimum cutting force can be observed at 0.2 wt% of SiO_2, and minimum cutting temperature was observed at 15° nozzle orientation angle
Sayuti et al. (2013b)	Alumicut oil	Carbon anions (5–20 nm)	Duralumin AL-2017-T4	75.408 m/ min	1.0 mm	100 mm/ min	MQL	Best surface topography and reduced cutting forces were observed at 1.5 wt% conc. of carbon anions Rolling action of nanoparticles reduces cutting forces and surface finish by 21.995 and 46.32%
Sayuti et al. (2014a)	ECOCUT SSN 322	SiO_2	Aluminium alloy AA6061-T6	5000 RPM	5.0 mm	100 mm/ min	MQL	Addition of SiO_2 improved the surface morphology with formation of thin tribofilm thereby reducing frictional wear and thermal deformation during machining

(continued)

Table 18.5 (continued)

Author/year	Base fluid	Nanoparticles	Workpiece material	Parameters			Mode of lubrication	Process findings
				Cutting speed/wheel speed	Depth of cut	Feed rate/ workpiece speed		
Rahmati et al. (2014)	ECOCUT HSG 905S oil	MoS_2 (20–60 nm)	Aluminium alloy AA6061-T6	8000 RPM	5 mm	2100 mm/ min	MQL	0.5 wt% of MoS_2 improved the surface quality than 1 wt% of MoS_2
Najiha et al. (2015)	Deionized water	TiO_2 (40 nm)	Aluminium alloy AA6061-T6	5300 RPM 5500 RPM	3 mm	440 mm/ min	MQL	Higher cooling rates with reduced chipping of tool were observed with addition of TiO_2 in deionized water
Najiha and Rahman (2016)	Deionized water	TiO_2 (40 nm)	Aluminium alloy AA6061-T6	5300 RPM 5500 RPM	3 mm	440 mm/ min	MQL	2.5 vol% reduces BUE formation and reduces chipping of tool material with improved cooling conditions
Kim et al. (2016)	Vegetable oil	(hBN) hexagonal boron nitride	Titanium alloy Ti-6Al-4V (70 nm)	45,000 RPM	100 μm	5 μm/tooth	MQL	Chilling effect of CO_2 reduces cutting forces, tool wear and improved surface finish Micro-end-milling performance improved by adding 0.1 wt% of hBN
Muthusamy et al. (2016)	Ethylene glycol	TiO_2	AISI 304 steel	1500 RPM 2500 RPM	0.1 mm 0.3 mm	0.02 mm/ tooth 0.04 mm/ tooth	MQL	Tool life increased by 40.55% with addition of TiO_2 in base fluid
Park et al. (2017)	Vegetable oil	xGnPs	Ti-6Al-4V	46.5 m/min 76.4 m/min 100 m/min 120 m/min	2 mm	0.15 mm/ rev	MQL, cryogenic cooling	Addition of nanoparticles reduces tool wear and friction at higher cutting speed even when base fluid vaporize at high surface temperature
Lv et al. (2018)	Water PEG solution	GO (5–10 nm) SiO_2 (5–10 nm)	AISI-304	100 m/min	1 mm	0.12 mm/ tooth	MQL	0.02 wt% of GO and 0.50 wt% of SiO_2 in base fluid reduce tool wear and improve surface roughness due to their ball-bearing effect

manufacturing operations. The nanofluid was then experimentally tied for turning operation on a commercial lathe machine with minimum quantity lubrication technique to minimize the usage of the prepared nanofluid providing cost benefits.

1. The present work can further be extended to discover environment-friendly vegetable oils that have the potential to be used as cutting fluids in common manufacturing operations.
2. Suitable additives in the form of nanoparticles with different weight fractions, shapes and sizes can be used to enhance the thermophysical properties of cutting oils. This would be helpful in developing nano-lubricants with enhanced tribological properties for the machining of specific metals and alloys.
3. Moreover, the present work can further be extended to the optimization of nanoparticle volume fraction, their shape and their size. This would be helpful in developing nano-lubricants with improved tribological properties for the machining of specific metals and alloys.
4. Different tool and workpiece combinations can be checked for optimum tool life and economic analysis.

References

Adler DP, Hii W-S, Michalek DJ, Sutherland JW (2006) Examining the role of cutting fluids in machining and efforts to address associated environmental/health concerns. Mach Sci Technol 10(1):23–58

Akoh H, Tsukasaki Y, Yatsuya S, Tasaki A (1978) Magnetic properties of ferromagnetic ultrafine particles prepared by vacuum evaporation on running oil substrate. J Cryst Growth 45:495–500

Alberts M, Kalaitzidou K, Melkote S (2009) An investigation of graphite nanoplatelets as lubricant in grinding. Int J Mach Tools Manuf 49(12–13):966–970

Ali MAM, Azmi AI, Khalil ANM, Leong KW (2017) Experimental study on minimal nanolubrication with surfactant in the turning of titanium alloys. Int J Adv Manuf Technol 92(1–4):117–127

Altan E, Kiyak M, Cakir O (2002) The effect of oxygen gas application into cutting zone on machining. In: Proceedings of sixth biennial conference on engineering system design and analysis (ESDA2002), Istanbul, pp 1–5

Amrita M, Srikant RR, Sitaramaraju AV, Prasad MMS, Krishna PV (2013) Experimental investigations on influence of mist cooling using nanofluids on machining parameters in turning AISI 1040 steel. Proc Inst Mech Eng J J Eng Tribol 227(12):1334–1346. https://doi.org/10.1177/1350650113491934

Amrita M, Shariq SA, Manoj M, Gopal C (2014a) Experimental investigation on application of emulsifier oil based nano cutting fluids in metal cutting process. Procedia Eng 97:115–124

Amrita M, Srikant RR, Sitaramaraju AV (2014b) Performance evaluation of nanographite-based cutting fluid in machining process. Mater Manuf Process 29(5):600–605

Arumugam S, Sriram G (2013) Preliminary study of nano-and microscale TiO_2 additives on tribological behavior of chemically modified rapeseed oil. Tribol Trans 56(5):797–805

Arumugam S, Baskar S, Sankaranarayanan S, Athreya SH, Narayanan NL, Prasad SSD (2018) Influence of morphology of anti-wear nano additives on Tribological behavior of Chemically Modified Rapeseed Oil. IOP Conf Ser Mater Sci Eng 390:12017

Asadauskas S, Erhan SZ (1999) Depression of pour points of vegetable oils by blending with diluents used for biodegradable lubricants. J Am Oil Chem Soc 76(3):313–316

Asadauskas S, Perez JM, Duda JL (1996) Oxidative stability and antiwear properties of high oleic vegetable oils. Lubr Eng 52(12):877–882

Atabani AE, Silitonga AS, Ong HC, Mahlia TMI, Masjuki HH, Badruddin IA, Fayaz H (2013) Non-edible vegetable oils: a critical evaluation of oil extraction, fatty acid compositions, biodiesel production, characteristics, engine performance and emissions production. Renew Sust Energ Rev 18:211–245

Barnwal BK, Sharma MP (2005) Prospects of biodiesel production from vegetable oils in India. Renew Sust Energ Rev 9(4):363–378

Battez AH, González R, Viesca JL, Fernández JE, Fernández JMD, Machado A et al (2008) CuO, ZrO_2 and ZnO nanoparticles as antiwear additive in oil lubricants. Wear 265(3–4):422–428

Behera BC, Chetan, Setti D, Ghosh S, Rao PV (2017) Spreadability studies of metal working fluids on tool surface and its impact on minimum amount cooling and lubrication turning. J Mater Process Technol 244:1–16

Benchaita MT, Lockwood FE (1993) Reliable model of lubricant-related friction in internal combustion engines. Lubr Sci 5(4):259–281

Bobade S, Khyade V (2012) Detail study on the properties of Pongamia Pinnata (Karanja) for the production of biofuel. Res J Chem Sci 2(7):16–20

Bruni C, Forcellese A, Gabrielli F, Simoncini M (2006) Effect of the lubrication-cooling technique, insert technology and machine bed material on the workpart surface finish and tool wear in finish turning of AISI 420B. Int J Mach Tools Manuf 46(12–13):1547–1554

Çakıra O, Kıyak M, Altan E (2004) Comparison of gases applications to wet and dry cuttings in turning. J Mater Process Technol 153:35–41

Chai YH, Yusup S, Chok VS, Arpin MT, Irawan S (2016) Investigation of thermal conductivity of multi walled carbon nanotube dispersed in hydrogenated oil based drilling fluids. Appl Therm Eng 107:1019–1025

Chan CY, Lee WB, Wang H (2013) Enhancement of surface finish using water-miscible nano-cutting fluid in ultra-precision turning. Int J Mach Tools Manuf 73:62–70

Chang L, Friedrich K (2010) Enhancement effect of nanoparticles on the sliding wear of short fiber-reinforced polymer composites: a critical discussion of wear mechanisms. Tribol Int 43(12):2355–2364

Chang H, Kao M-J (2007) An innovative nanofluid manufacturing system. J Chin Soc Mech Eng 28(2):187–194

Chatha SS, Pal A, Singh T (2016) Performance evaluation of aluminium 6063 drilling under the influence of nanofluid minimum quantity lubrication. J Clean Prod 137:537–545

Choi SUS, Yu W, Hull JR, Zhang ZG, Lockwood FE (2002) Nanofluids for vehicle thermal management. SAE Trans 111:38–43

Chol SUS, Estman JA (1995) Enhancing thermal conductivity of fluids with nanoparticles. ASME 231:99–106

Chon CH, Kihm KD, Lee SP, Choi SUS (2005) Empirical correlation finding the role of temperature and particle size for nanofluid (Al_2O_3) thermal conductivity enhancement. Appl Phys Lett 87(15):153107

da Silva MB, Wallbank J (1999) Cutting temperature: prediction and measurement methods—a review. J Mater Process Technol 88(1–3):195–202

Davim JP, Gaitonde VN, Karnik SR (2008) Investigations into the effect of cutting conditions on surface roughness in turning of free machining steel by ANN models. J Mater Process Technol 205(1–3):16–23

De Lacalle LNL, Pérez-Bilbatua J, Sánchez JA, Llorente JI, Gutierrez A, Albóniga J (2000) Using high pressure coolant in the drilling and turning of low machinability alloys. Int J Adv Manuf Technol 16(2):85–91

Debnath S, Reddy MM, Yi QS (2014) Environmental friendly cutting fluids and cooling techniques in machining: a review. J Clean Prod 83:33–47

Dhar NR, Paul S, Chattopadhyay AB (2002) Role of cryogenic cooling on cutting temperature in turning steel. J Manuf Sci Eng 124(1):146–154

Dilbag S, Rao PV (2008) Performance improvement of hard turning with solid lubricants. Int J Adv Manuf Technol 38(5–6):529–535

Diniz AE, Micaroni R (2002) Cutting conditions for finish turning process aiming: the use of dry cutting. Int J Mach Tools Manuf 42(8):899–904

Diniz AE, Micaroni R (2007) Influence of the direction and flow rate of the cutting fluid on tool life in turning process of AISI 1045 steel. Int J Mach Tools Manuf 47(2):247–254

Drzazga Michałand Dzido G, Lemanowicz M, Gierczycki A (2012) Influence of nonionic surfactant on nanofluid properties. In: Proceedings of the 14th European conference on mixing, Warszawa, Poland, pp 10–13

Duangthongsuk W, Wongwises S (2009) Measurement of temperature-dependent thermal conductivity and viscosity of TiO_2-water nanofluids. Exp Thermal Fluid Sci 33(4):706–714

Duchosal A, Leroy R, Vecellio L, Louste C, Ranganathan N (2013) An experimental investigation on oil mist characterization used in MQL milling process. Int J Adv Manuf Technol 66(5–8):1003–1014

Dudzinski D, Devillez A, Moufki A, Larrouquere D, Zerrouki V, Vigneau J (2004) A review of developments towards dry and high speed machining of Inconel 718 alloy. Int J Mach Tools Manuf 44(4):439–456

Eastman JA, Choi US, Li S, Thompson LJ, Lee S (1996) Enhanced thermal conductivity through the development of nanofluids. MRS Online Proc Libr Arch 457:3

El Baradie MA (1996) Cutting fluids: Part I. Characterisation. J Mater Process Technol 56(1–4):786–797

Erhan SZ, Sharma BK, Perez JM (2006) Oxidation and low temperature stability of vegetable oil-based lubricants. Ind Crop Prod 24(3):292–299

Evans C, Bryan JB (1991) Cryogenic diamond turning of stainless steel. CIRP Ann Manuf Technol 40(1):571–575

Ezugwu EO, Bonney J (2005) Finish machining of nickel-base Inconel 718 alloy with coated carbide tool under conventional and high-pressure coolant supplies. Tribol Trans 48(1):76–81

Ezugwu EO, Bonney J, Fadare DA, Sales WF (2005) Machining of nickel-base, Inconel 718, alloy with ceramic tools under finishing conditions with various coolant supply pressures. J Mater Process Technol 162:609–614

Fox NJ, Stachowiak GW (2007) Vegetable oil-based lubricants—a review of oxidation. Tribol Int 40(7):1035–1046. https://doi.org/10.1016/j.triboint.2006.10.001

Fratila D (2009) Evaluation of near-dry machining effects on gear milling process efficiency. J Clean Prod 17(9):839–845

Garg A, Sarma S, Panda BN, Zhang J, Gao L (2016) Study of effect of nanofluid concentration on response characteristics of machining process for cleaner production. J Clean Prod 135:476–489

Ghadimi A, Saidur R, Metselaar HSC (2011) A review of nanofluid stability properties and characterization in stationary conditions. Int J Heat Mass Transf 54(17–18):4051–4068

Gnanasekaran D, Chavidi VP (2018) Properties of vegetable fluids: a green insulator for power sector. In: Vegetable oil based bio-lubricants and transformer fluids: applications in power plants. Springer Singapore, Singapore, pp 125–155. https://doi.org/10.1007/978-981-10-4870-8_7

Godlevski VA, Volkov AV, Latysher VN, Maurin LN (1998) Water steam lubrication during machining. Tribologia 162(6):890–901

Gryglewicz S, Piechocki W, Gryglewicz G (2003) Preparation of polyol esters based on vegetable and animal fats. Bioresour Technol 87(1):35–39

Gupta RN, Harsha AP (2018) Tribological study of castor oil with surface modified CuO nanoparticles in boundary lubrication. Ind Lubr Tribol 70(4). https://doi.org/10.1108/ilt-02-2017-0030

Gupta MK, Sood PK, Sharma VS (2016) Optimization of machining parameters and cutting fluids during nano-fluid based minimum quantity lubrication turning of titanium alloy by using evolutionary techniques. J Clean Prod 135:1276–1288. https://doi.org/10.1016/j.jclepro.2016.06.184

Habibullah M, Masjuki HH, Kalam MA, Gulzar M, Arslan A, Zahid R (2015) Tribological characteristics of Calophyllum inophyllum–based TMP (trimethylolpropane) ester as energy-saving and biodegradable lubricant. Tribol Trans 58(6):1002–1011

Haghighi EB, Nikkam N, Saleemi M, Behi M, Mirmohammadi SA, Poth H et al (2013) Shelf stability of nanofluids and its effect on thermal conductivity and viscosity. Meas Sci Technol 24(10):105301

Hegab H, Umer U, Deiab I, Kishawy H (2018a) Performance evaluation of Ti-6Al-4V machining using nano-cutting fluids under minimum quantity lubrication. Int J Adv Manuf Technol 95(9–12):4229–4241

Hegab H, Umer U, Soliman M, Kishawy HA (2018b) Effects of nano-cutting fluids on tool performance and chip morphology during machining Inconel 718. Int J Adv Manuf Technol 96(9–12):3449–3458

Honary LAT (1996) An investigation of the use of soybean oil in hydraulic systems. Bioresour Technol 56(1):41–47

Hu ZS, Lai R, Lou F, Wang L, Chen Z, Chen G, Dong JX (2002) Preparation and tribological properties of nanometer magnesium borate as lubricating oil additive. Wear 252(5–6):370–374

Huang W-T, Wu D-H, Chen J-T (2016) Robust design of using nanofluid/MQL in micro-drilling. Int J Adv Manuf Technol 85(9–12):2155–2161

Hwang Y, Lee JK, Lee CH, Jung YM, Cheong SI, Lee CG et al (2007) Stability and thermal conductivity characteristics of nanofluids. Thermochim Acta 455(1–2):70–74

Jia D, Li C, Zhang D, Zhang Y, Zhang X (2014) Experimental verification of nanoparticle jet minimum quantity lubrication effectiveness in grinding. J Nanopart Res 16(12):2758

Kalita P, Malshe AP, Jiang W, Shih AJ (2010) Tribological study of nano lubricant integrated soybean oil for minimum quantity lubrication (MQL) grinding. Trans NAMRI/SME 38:137–144

Kalita P, Malshe AP, Kumar SA, Yoganath VG, Gurumurthy T (2012a) Study of specific energy and friction coefficient in minimum quantity lubrication grinding using oil-based nanolubricants. J Manuf Process 14(2):160–166

Kalita P, Malshe AP, Rajurkar KP (2012b) Study of tribo-chemical lubricant film formation during application of nanolubricants in minimum quantity lubrication (MQL) grinding. CIRP Ann Manuf Technol 61(1):327–330

Kamata Y, Obikawa T (2007) High speed MQL finish-turning of Inconel 718 with different coated tools. J Mater Process Technol 192:281–286

Khairul MA, Saidur R, Hossain A, Alim MA, Mahbubul IM (2014) Heat transfer performance of different nanofluids flows in a helically coiled heat exchanger. Adv Mater Res 832:160–165

Khan AA, Ahmed MI (2008) Improving tool life using cryogenic cooling. J Mater Process Technol 196(1–3):149–154

Khan MMA, Mithu MAH, Dhar NR (2009) Effects of minimum quantity lubrication on turning AISI 9310 alloy steel using vegetable oil-based cutting fluid. J Mater Process Technol 209(15–16):5573–5583

Khandekar S, Sankar MR, Agnihotri V, Ramkumar J (2012) Nano-cutting fluid for enhancement of metal cutting performance. Mater Manuf Process 27(9):963–967

Kim JS, Kim JW, Kim YC, Lee SW (2016) Experimental study on environmentally-friendly micro end-milling process of Ti-6Al-4V using nanofluid minimum quantity lubrication with chilly gas. In: ASME 2016 11th international manufacturing science and engineering conference, p V002T05A006

Kishawy HA, Dumitrescu M, Ng E-G, Elbestawi MA (2005) Effect of coolant strategy on tool performance, chip morphology and surface quality during high-speed machining of A356 aluminum alloy. Int J Mach Tools Manuf 45(2):219–227

Klocke F, Eisenblätter G (1997) Dry cutting. CIRP Ann Manuf Technol 46(2):519–526

Knothe G, Steidley KR (2005) Kinematic viscosity of biodiesel fuel components and related compounds. Influence of compound structure and comparison to petrodiesel fuel components. Fuel 84(9):1059–1065

Ko TJ, Kim HS, Chung BG (1999) Air–oil cooling method for turning of hardened material. Int J Adv Manuf Technol 15(7):470–477

Kodali DR, Nivens S (2001) Biodegradable high performance lubricants derived from natural oils. In: 2nd World Tribology Congress, abstract of papers, Vienna, p 235

Krishna PV, Srikant RR, Rao DN (2010) Experimental investigation on the performance of nanoboric acid suspensions in SAE-40 and coconut oil during turning of AISI 1040 steel. Int J Mach Tools Manuf 50(10):911–916

Kumar TA, Pradyumna G, Jahar S (2012) Investigation of thermal conductivity and viscosity of nano fluids. J Environ Res Dev 7(2):768–777

Lal K, Carrick V (1994) Performance testing of lubricants based on high oleic vegetable oils. J Synth Lubr 11(3):189–206

Lathkar GS, Bas USK (2000) Clean metal cutting process using solid lubricants. In: Proceeding of the 19th AIMTDR conference, Narosa, Madras, pp 15–31

Lawal SA, Choudhury IA, Nukman Y (2013) A critical assessment of lubrication techniques in machining processes: a case for minimum quantity lubrication using vegetable oil-based lubricant. J Clean Prod 41:210–221

Lee C-G, Hwang Y-J, Choi Y-M, Lee J-K, Choi C, Oh J-M (2009a) A study on the tribological characteristics of graphite nano lubricants. Int J Precis Eng Manuf 10(1):85–90

Lee K, Hwang Y, Cheong S, Choi Y, Kwon L, Lee J, Kim SH (2009b) Understanding the role of nanoparticles in nano-oil lubrication. Tribol Lett 35(2):127–131

Lee P-H, Nam JS, Li C, Lee SW (2012) An experimental study on micro-grinding process with nanofluid minimum quantity lubrication (MQL). Int J Precis Eng Manuf 13(3):331–338

Li K-M, Liang SY (2007) Performance profiling of minimum quantity lubrication in machining. Int J Adv Manuf Technol 35(3–4):226–233

Li XF, Zhu DS, Wang XJ, Wang N, Gao JW, Li H (2008) Thermal conductivity enhancement dependent pH and chemical surfactant for Cu–H_2O nanofluids. Thermochim Acta 469(1–2):98–103

Li Y, Zhou J, Tung S, Schneider E, Xi S (2009) A review on development of nanofluid preparation and characterization. Powder Technol 196(2):89–101

Li CH, Li JY, Wang S, Zhang Q (2013) Modeling and numerical simulation of the grinding temperature field with nanoparticle jet of MQL. Adv Mech Eng 5:986984

Li B, Li C, Zhang Y, Wang Y, Yang M, Jia D et al (2017) Effect of the physical properties of different vegetable oil-based nanofluids on MQLC grinding temperature of Ni-based alloy. Int J Adv Manuf Technol 89(9–12):3459–3474

Liu G, Li X, Qin B, Xing D, Guo Y, Fan R (2004) Investigation of the mending effect and mechanism of copper nano-particles on a tribologically stressed surface. Tribol Lett 17(4):961–966

Liu J, Han R, Zhang L, Guo H (2007) Study on lubricating characteristic and tool wear with water vapor as coolant and lubricant in green cutting. Wear 262(3–4):442–452

Lo C-H, Tsung T-T, Chen L-C, Su C-H, Lin H-M (2005) Fabrication of copper oxide nanofluid using submerged arc nanoparticle synthesis system (SANSS). J Nanopart Res 7(2–3):313–320

Lv T, Huang S, Hu X, Ma Y, Xu X (2018) Tribological and machining characteristics of a minimum quantity lubrication (MQL) technology using GO/SiO_2 hybrid nanoparticle water-based lubricants as cutting fluids. Int J Adv Manuf Technol 96(5–8):2931–2942

Manimaran R, Palaniradja K, Alagumurthi N, Sendhilnathan S, Hussain J (2014) Preparation and characterization of copper oxide nanofluid for heat transfer applications. Appl Nanosci 4(2):163–167

Manna I (2012) Synthesis, characterization and application of nanofluid—an overview. J Indian Inst Sci 89(1):21–33

ManojKumar K, Ghosh A (2015) Synthesis of MWCNT nanofluid and evaluation of its potential besides soluble oil as micro cooling-lubrication medium in SQL grinding. Int J Adv Manuf Technol 77(9–12):1955–1964

Mao C, Tang X, Zou H, Huang X, Zhou Z (2012) Investigation of grinding characteristic using nanofluid minimum quantity lubrication. Int J Precis Eng Manuf 13(10):1745–1752

Mao C, Zhang J, Huang Y, Zou H, Huang X, Zhou Z (2013a) Investigation on the effect of nanofluid parameters on MQL grinding. Mater Manuf Process 28(4):436–442

Mao C, Zou H, Huang X, Zhang J, Zhou Z (2013b) The influence of spraying parameters on grinding performance for nanofluid minimum quantity lubrication. Int J Adv Manuf Technol 64(9–12):1791–1799

Mao C, Huang Y, Zhou X, Gan H, Zhang J, Zhou Z (2014a) The tribological properties of nanofluid used in minimum quantity lubrication grinding. Int J Adv Manuf Technol 71(5–8):1221–1228
Mao C, Zou H, Zhou X, Huang Y, Gan H, Zhou Z (2014b) Analysis of suspension stability for nanofluid applied in minimum quantity lubricant grinding. Int J Adv Manuf Technol 71(9–12):2073–2081
Masjuki HH, Maleque MA, Kubo A, Nonaka T (1999) Palm oil and mineral oil based lubricants—their tribological and emission performance. Tribol Int 32(6):305–314
Mathew NT, Vijayaraghavan L (2017) Environmentally friendly drilling of intermetallic titanium aluminide at different aspect ratio. J Clean Prod 141:439–452. https://doi.org/10.1016/j.jclepro.2016.09.125
Mazurkiewicz M, Kubala Z, Chow J (1989) Metal machining with high-pressure water-jet cooling assistance—a new possibility. J Eng Ind 111(1):7–12
Missana T, Adell A (2000) On the applicability of DLVO theory to the prediction of clay colloids stability. J Colloid Interface Sci 230(1):150–156
Mobarak HM, Mohamad EN, Masjuki HH, Kalam MA, Al Mahmud KAH, Habibullah M, Ashraful AM (2014) The prospects of biolubricants as alternatives in automotive applications. Renew Sust Energ Rev 33:34–43
Mofijur M, Masjuki HH, Kalam MA, Hazrat MA, Liaquat AM, Shahabuddin M, Varman M (2012) Prospects of biodiesel from Jatropha in Malaysia. Renew Sust Energ Rev 16(7):5007–5020
Mohammed HA, Al-Aswadi AA, Shuaib NH, Saidur R (2011) Convective heat transfer and fluid flow study over a step using nanofluids: a review. Renew Sust Energ Rev 15(6):2921–2939
Mosleh M, Ghaderi M, Shirvani KA, Belk J, Grzina DJ (2017) Performance of cutting nanofluids in tribological testing and conventional drilling. J Manuf Process 25:70–76
Mukherjee S, Paria S (2013) Preparation and stability of nanofluids-a review. IOSR J Mech Civ Eng 9(2):63–69
Murshed SMS, Leong KC, Yang C (2005) Enhanced thermal conductivity of TiO_2—water based nanofluids. Int J Therm Sci 44(4):367–373
Muthusamy Y, Kadirgama K, Rahman MM, Ramasamy D, Sharma KV (2016) Wear analysis when machining AISI 304 with ethylene glycol/TiO_2 nanoparticle-based coolant. Int J Adv Manuf Technol 82(1):327–340. https://doi.org/10.1007/s00170-015-7360-3
Najiha MS, Rahman MM (2016) Experimental investigation of flank wear in end milling of aluminum alloy with water-based TiO_2 nanofluid lubricant in minimum quantity lubrication technique. Int J Adv Manuf Technol 86(9–12):2527–2537
Najiha MS, Rahman MM, Yusoff AR (2015) Flank wear characterization in aluminum alloy (6061 T6) with nanofluid minimum quantity lubrication environment using an uncoated carbide tool. J Manuf Sci Eng 137(6):61004
Najiha MS, Rahman MM, Kadirgama K (2016) Performance of water-based TiO_2 nanofluid during the minimum quantity lubrication machining of aluminium alloy, AA6061-T6. J Clean Prod 135:1623–1636
Nakpong P, Wootthikanokkhan S (2010) High free fatty acid coconut oil as a potential feedstock for biodiesel production in Thailand. Renew Energy 35(8):1682–1687
Nam J, Lee SW (2018) Machinability of titanium alloy (Ti-6Al-4V) in environmentally-friendly micro-drilling process with nanofluid minimum quantity lubrication using nanodiamond particles. Int J Precis Eng Manuf Green Technol 5(1):29–35
Nam JS, Lee P-H, Lee SW (2011) Experimental characterization of micro-drilling process using nanofluid minimum quantity lubrication. Int J Mach Tools Manuf 51(7–8):649–652
Nam JS, Kim DH, Chung H, Lee SW (2015) Optimization of environmentally benign micro-drilling process with nanofluid minimum quantity lubrication using response surface methodology and genetic algorithm. J Clean Prod 102:428–436
Padmini R, Krishna PV, Mohana Rao GK (2015) Performance assessment of micro and nano solid lubricant suspensions in vegetable oils during machining. Proc Inst Mech Eng B J Eng Manuf 229(12):2196–2204
Padmini R, Krishna PV, Rao GKM (2016) Effectiveness of vegetable oil based nanofluids as potential cutting fluids in turning AISI 1040 steel. Tribol Int 94:490–501

Park K-H, Ewald B, Kwon PY (2011) Effect of nano-enhanced lubricant in minimum quantity lubrication balling milling. J Tribol 133(3):31803–31808. https://doi.org/10.1115/1.4004339

Park K-H, Suhaimi MA, Yang G-D, Lee D-Y, Lee S-W, Kwon P (2017) Milling of titanium alloy with cryogenic cooling and minimum quantity lubrication (MQL). Int J Precis Eng Manuf 18(1):5–14

Pashmforoush F, Bagherinia RD (2018) Influence of water-based copper nanofluid on wheel loading and surface roughness during grinding of Inconel 738 superalloy. J Clean Prod 178:363–372

Philip PK, Varadarajan AS, Ramamoorthy B (2001) Influence of cutting fluid composition and delivery variables on performance in hard turning using minimal fluid in pulsed jet form. J Inst Eng PR Prod Eng Div 82(1):12–19

Pop L, Puşcaş C, Bandur G, Vlase G, Nuţiu R (2008) Basestock oils for lubricants from mixtures of corn oil and synthetic diesters. J Am Oil Chem Soc 85(1):71–76

Popa I, Gillies G, Papastavrou G, Borkovec M (2010) Attractive and repulsive electrostatic forces between positively charged latex particles in the presence of anionic linear polyelectrolytes. J Phys Chem B 114(9):3170–3177

Prabhu S, Vinayagam BK (2010) Nano surface generation of grinding process using carbon nano tubes. Sadhana 35(6):747–760

Prabhu S, Vinayagam BK (2012) AFM investigation in grinding process with nanofluids using Taguchi analysis. Int J Adv Manuf Technol 60(1–4):149–160

Prabhu S, Vinayagam BK (2013) Analysis of surface characteristics by electrolytic in-process dressing (ELID) technique for grinding process using single wall carbon nano tube-based nanofluids. Arab J Sci Eng 38(5):1169–1178

Prabhu S, Uma M, Vinayagam BK (2015) Surface roughness prediction using Taguchi-fuzzy logic-neural network analysis for CNT nanofluids based grinding process. Neural Comput & Applic 26(1):41–55

Quinchia LA, Delgado MA, Franco JM, Spikes HA, Gallegos C (2012) Low-temperature flow behaviour of vegetable oil-based lubricants. Ind Crop Prod 37(1):383–388

Rac A, Vencl A (2009) Performance investigation of chain saw lubricants based on new sunflower oil (NSO). Tribol Schmier 56(3):51

Rahmati B, Sarhan AAD, Sayuti M (2014) Morphology of surface generated by end milling AL6061-T6 using molybdenum disulfide (MoS_2) nanolubrication in end milling machining. J Clean Prod 66:685–691

Rajendhran N, Palanisamy S, Periyasamy P, Venkatachalam R (2018) Enhancing of the tribological characteristics of the lubricant oils using Ni-promoted MoS_2 nanosheets as nano-additives. Tribol Int 118:314–328

Raju RA, Andhare A, Sahu NK (2017) Performance of multi-walled carbon nanotube-based nanofluid in turning operation. Mater Manuf Process 32(13):1490–1496

Rapoport L, Leshchinsky V, Lvovsky M, Nepomnyashchy O, Volovik Y, Tenne R (2002) Mechanism of friction of fullerenes. Ind Lubr Tribol 54(4):171–176

Rhee I-S, Velez C, Von Bernewitz K (1995) Evaluation of environmentally acceptable hydraulic fluids. Defense Technical Information Center, Fort Belvoir

Rodrigues JA, Cardoso FP, Lachter ER, Estevão LRM, Lima E, Nascimento RSV (2006) Correlating chemical structure and physical properties of vegetable oil esters. J Am Oil Chem Soc 83(4):353–357

Rudnick LR (2005) Synthetics, mineral oils, and bio-based lubricants: chemistry and technology. CRC Press, Boca Raton

Salimi-Yasar H, Heris SZ, Shanbedi M (2017) Influence of soluble oil-based TiO_2 nanofluid on heat transfer performance of cutting fluid. Tribol Int 112:147–154

Salimon J, Salih N, Yousif E (2010) Biolubricants: raw materials, chemical modifications and environmental benefits. Eur J Lipid Sci Technol 112(5):519–530

Saravanakumar N, Prabu L, Karthik M, Rajamanickam A (2014) Experimental analysis on cutting fluid dispersed with silver nano particles. J Mech Sci Technol 28(2):645–651. https://doi.org/10.1007/s12206-013-1192-6

Sarhan AAD, Sayuti M, Hamdi M (2012) Reduction of power and lubricant oil consumption in milling process using a new SiO_2 nanolubrication system. Int J Adv Manuf Technol 63(5–8):505–512

Sartori S, Ghiotti A, Bruschi S (2018) Solid lubricant-assisted minimum quantity lubrication and cooling strategies to improve Ti6Al4V machinability in finishing turning. Tribol Int 118:287–294

Sayuti M, Sarhan AAD, Hamdi M (2013a) An investigation of optimum SiO_2 nanolubrication parameters in end milling of aerospace Al6061-T6 alloy. Int J Adv Manuf Technol 67(1–4):833–849

Sayuti M, Sarhan AAD, Tanaka T, Hamdi M, Saito Y (2013b) Cutting force reduction and surface quality improvement in machining of aerospace duralumin AL-2017-T4 using carbon onion nanolubrication system. Int J Adv Manuf Technol 65(9–12):1493–1500

Sayuti M, Erh OM, Sarhan AAD, Hamdi M (2014a) Investigation on the morphology of the machined surface in end milling of aerospace AL6061-T6 for novel uses of SiO_2 nanolubrication system. J Clean Prod 66:655–663

Sayuti M, Sarhan AAD, Salem F (2014b) Novel uses of SiO_2 nano-lubrication system in hard turning process of hardened steel AISI4140 for less tool wear, surface roughness and oil consumption. J Clean Prod 67:265–276

Setti D, Sinha MK, Ghosh S, Rao PV (2015) Performance evaluation of Ti-6Al-4V grinding using chip formation and coefficient of friction under the influence of nanofluids. Int J Mach Tools Manuf 88:237–248

Shaji S, Radhakrishnan V (2003) Analysis of process parameters in surface grinding with graphite as lubricant based on the Taguchi method. J Mater Process Technol 141(1):51–59

Sharma VS, Dogra M, Suri NM (2009) Cooling techniques for improved productivity in turning. Int J Mach Tools Manuf 49(6):435–453

Sharma AK, Tiwari AK, Dixit AR (2015a) Improved machining performance with nanoparticle enriched cutting fluids under minimum quantity lubrication (MQL) technique: a review. Mater Today Proc 2(4–5):3545–3551

Sharma AK, Tiwari AK, Dixit AR (2015b) Progress of nanofluid application in machining: a review. Mater Manuf Process 30(7):813–828

Sharma AK, Tiwari AK, Dixit AR (2016) Effects of minimum quantity lubrication (MQL) in machining processes using conventional and nanofluid based cutting fluids: a comprehensive review. J Clean Prod 127:1–18

Sharma AK, Tiwari AK, Dixit AR, Singh RK, Singh M (2018) Novel uses of alumina/graphene hybrid nanoparticle additives for improved tribological properties of lubricant in turning operation. Tribol Int 119:99–111

Sharma AK, Katiyar JK, Bhaumik S, Roy S (2019) Influence of alumina/MWCNT hybrid nanoparticle additives on tribological properties of lubricants in turning operations. Friction 7(2):153–168

Shashidhara YM, Jayaram SR (2010) Vegetable oils as a potential cutting fluid—an evolution. Tribol Int 43(5–6):1073–1081

Shen B, Malshe AP, Kalita P, Shih AJ (2008a) Performance of novel MoS_2 nanoparticles based grinding fluids in minimum quantity lubrication grinding. Trans NAMRI/SME 36(357):e364

Shen B, Shih AJ, Tung SC (2008b) Application of nanofluids in minimum quantity lubrication grinding. Tribol Trans 51(6):730–737

Shokrani A, Dhokia V, Newman ST (2012) Environmentally conscious machining of difficult-to-machine materials with regard to cutting fluids. Int J Mach Tools Manuf 57:83–101

Singh SP, Singh D (2010) Biodiesel production through the use of different sources and characterization of oils and their esters as the substitute of diesel: a review. Renew Sust Energ Rev 14(1):200–216

Sinha MK, Madarkar R, Ghosh S, Rao PV (2017) Application of eco-friendly nanofluids during grinding of Inconel 718 through small quantity lubrication. J Clean Prod 141:1359–1375

Soković M, Mijanović K (2001) Ecological aspects of the cutting fluids and its influence on quantifiable parameters of the cutting processes. J Mater Process Technol 109(1–2):181–189

Sreejith PS, Ngoi BKA (2000) Dry machining: machining of the future. J Mater Process Technol 101(1–3):287–291
Srikiran S, Ramji K, Satyanarayana B, Ramana SV (2014) Investigation on turning of AISI 1040 steel with the application of nano-crystalline graphite powder as lubricant. Proc Inst Mech Eng C J Mech Eng Sci 228(9):1570–1580
Sripada PK, Sharma RV, Dalai AK (2013) Comparative study of tribological properties of trimethylolpropane-based biolubricants derived from methyl oleate and canola biodiesel. Ind Crop Prod 50:95–103
Stanford M, Lister PM, Morgan C, Kibble KA (2009) Investigation into the use of gaseous and liquid nitrogen as a cutting fluid when turning BS 970-80A15 (En32b) plain carbon steel using WC–Co uncoated tooling. J Mater Process Technol 209(2):961–972
Su Y, Gong L, Li B, Liu Z, Chen D (2016) Performance evaluation of nanofluid MQL with vegetable-based oil and ester oil as base fluids in turning. Int J Adv Manuf Technol 83(9–12):2083–2089
Suda S, Yokota H, Inasaki I, Wakabayashi T (2002) A synthetic ester as an optimal cutting fluid for minimal quantity lubrication machining. CIRP Ann Manuf Technol 51(1):95–98
Tan XC, Liu F, Cao HJ, Zhang H (2002) A decision-making framework model of cutting fluid selection for green manufacturing and a case study. J Mater Process Technol 129(1–3):467–470
Tao X, Jiazheng Z, Kang X (1996) The ball-bearing effect of diamond nanoparticles as an oil additive. J Phys D Appl Phys 29(11):2932
Ting C-C, Chen C-C (2011) Viscosity and working efficiency analysis of soybean oil based bio-lubricants. Measurement 44(8):1337–1341
Tsao CC (2007) An experiment study of hard coating and cutting fluid effect in milling aluminum alloy. Int J Adv Manuf Technol 32(9–10):885–891
Tschätsch H, Reichelt A (2009) Cutting fluids (coolants and lubricants). In: Applied machining technology. Springer, Dordrecht/New York, pp 349–352
Varadarajan AS, Philip PK, Ramamoorthy B (2002) Investigations on hard turning with minimal cutting fluid application (HTMF) and its comparison with dry and wet turning. Int J Mach Tools Manuf 42(2):193–200
Vasu V, Kumar KM (2011) Analysis of nanofluids as cutting fluid in grinding EN-31 steel. Nanomicro Lett 3(4):209–214
Vieira JM, Machado AR, Ezugwu EO (2001) Performance of cutting fluids during face milling of steels. J Mater Process Technol 116(2–3):244–251
Wagner H, Luther R, Mang T (2001) Lubricant base fluids based on renewable raw materials: their catalytic manufacture and modification. Appl Catal A Gen 221(1–2):429–442
Wang X-Q, Mujumdar AS (2007) Heat transfer characteristics of nanofluids: a review. Int J Therm Sci 46(1):1–19
Wang X, Li X, Yang S (2009) Influence of pH and SDBS on the stability and thermal conductivity of nanofluids. Energy Fuel 23(5):2684–2689
Wang S, Li C, Zhang D, Jia D, Zhang Y (2014) Modeling the operation of a common grinding wheel with nanoparticle jet flow minimal quantity lubrication. Int J Adv Manuf Technol 74(5–8):835–850
Wang Y, Li C, Zhang Y, Li B, Yang M, Zhang X et al (2017a) Comparative evaluation of the lubricating properties of vegetable-oil-based nanofluids between frictional test and grinding experiment. J Manuf Process 26:94–104
Wang Y, Li C, Zhang Y, Yang M, Zhang X, Zhang N, Dai J (2017b) Experimental evaluation on tribological performance of the wheel/workpiece interface in minimum quantity lubrication grinding with different concentrations of Al_2O_3 nanofluids. J Clean Prod 142:3571–3583
Wertheim R, Rotberg J, Ber A (1992) Influence of high-pressure flushing through the rake face of the cutting tool. CIRP Ann Manuf Technol 41(1):101–106
Wu X, Zhang X, Yang S, Chen H, Wang D (2000) The study of epoxidized rapeseed oil used as a potential biodegradable lubricant. J Am Oil Chem Soc 77(5):561–563
Wu YY, Tsui WC, Liu TC (2007) Experimental analysis of tribological properties of lubricating oils with nanoparticle additives. Wear 262(7–8):819–825

Yan J, Zhang Z, Kuriyagawa T (2011) Effect of nanoparticle lubrication in diamond turning of reaction-bonded SiC. IJAT 5(3):307–312

Yu H, Hermann S, Schulz SE, Gessner T, Dong Z, Li WJ (2012) Optimizing sonication parameters for dispersion of single-walled carbon nanotubes. Chem Phys 408:11–16

Zeilmann RP, Weingaertner WL (2006) Analysis of temperature during drilling of Ti6Al4V with minimal quantity of lubricant. J Mater Process Technol 179(1–3):124–127

Zhang D, Li C, Jia D, Zhang Y, Zhang X (2015a) Specific grinding energy and surface roughness of nanoparticle jet minimum quantity lubrication in grinding. Chin J Aeronaut 28(2):570–581

Zhang D, Li C, Zhang Y, Jia D, Zhang X (2015b) Experimental research on the energy ratio coefficient and specific grinding energy in nanoparticle jet MQL grinding. Int J Adv Manuf Technol 78(5–8):1275–1288

Zhang Y, Li C, Jia D, Zhang D, Zhang X (2015c) Experimental evaluation of MoS_2 nanoparticles in jet MQL grinding with different types of vegetable oil as base oil. J Clean Prod 87:930–940

Zhang Y, Li C, Jia D, Li B, Wang Y, Yang M et al (2016) Experimental study on the effect of nanoparticle concentration on the lubricating property of nanofluids for MQL grinding of Ni-based alloy. J Mater Process Technol 232:100–115

Zhu H, Zhang C, Tang Y, Wang J, Ren B, Yin Y (2007) Preparation and thermal conductivity of suspensions of graphite nanoparticles. Carbon 45(1):226–228

Zulkifli NWM, Masjuki HH, Kalam MA, Yunus R, Azman SSN (2014) Lubricity of bio-based lubricant derived from chemically modified jatropha methyl ester. J Tribol 1:18–39

Chapter 19
Nanomaterials Used for Delivery of Bioactives

Akhlesh Kumar Jain and Umesh Gupta

19.1 Introduction

Progression in genetic engineering has led to propagation of gigantic diversity of bioactives which demands effective means of carriers for intracellular delivery in order to achieve specific objectives such as selective tumor targeting, genetic vaccination, regenerative medicine, and treatment of functional loss. Generally, these biologics are prone to enzymatic degradation and deactivation. Hence, immense arrangement of efforts has been made to develop nanometric size vehicles which could not only deliver the medicaments to the desired site of action but also protect for unwanted degradation. In this regard, nanoparticles have been shown great promise as a delivery vehicle for smaller molecules, plus large bioactives, i.e., proteins, peptides, vaccines, or nucleotides by either restricted or tissue-specific delivery. In addition, formulation scientists are fascinated about nanocarriers as a delivery vehicles as proportion of quantity of surface atoms or molecules to the total count of atoms or molecules enhanced drastically hence effective surface area multiplied exponentially (Hadjipanayis et al. 2010). Further, nanoparticles are in great number and could access regions of poor access such as injured tissues, tumor cells, inflamed organs, etc. due to their tiny size (Jong and Borm 2008). Nanotechnology concentrates on encapsulating drugs in bio-friendly nanocomposites, i.e., polymeric nanoparticles, nanoliposomes, solid lipid nanoparticles, micellar systems, and bioconjugates. A schematic diagram of different varieties of nanocarriers used for delivery of bioactives is depicted in Fig. 19.1. These carriers are usually explored to enhance oral bioavailability, to sustain medicament release

---

A. K. Jain (✉)
School of Pharmaceutical Sciences, GGV Central University, Bilaspur, Chhattisgarh, India

U. Gupta
Department of Pharmacy, Central University of Rajasthan, Kishangarh, Rajasthan, India

I. Bhushan et al. (eds.), *Nanomaterials and Environmental Biotechnology*, Nanotechnology in the Life Sciences,
https://doi.org/10.1007/978-3-030-34544-0_19

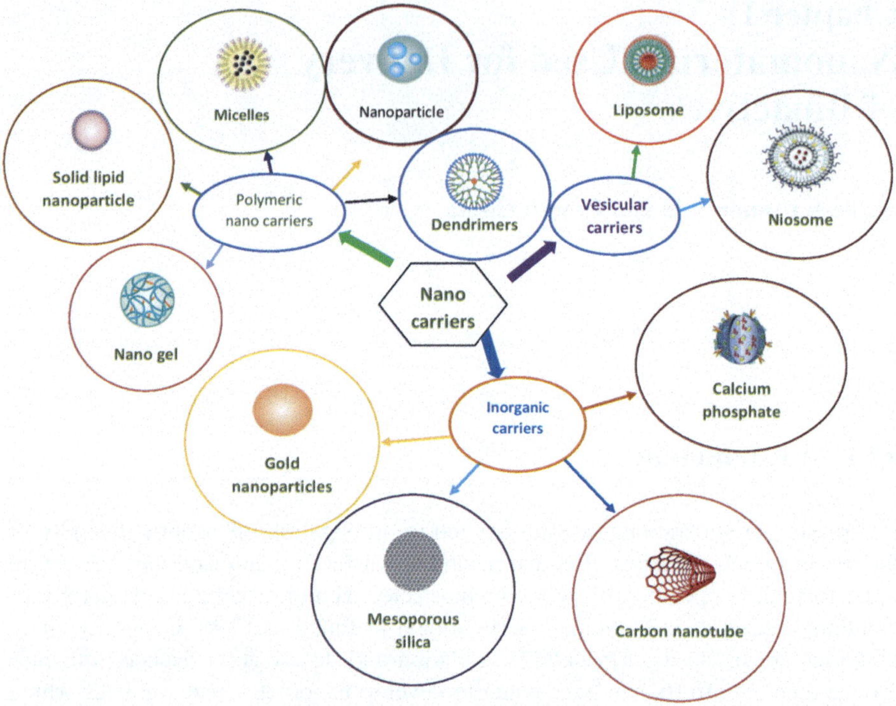

Fig. 19.1 Classification of most commonly used nanocarriers for drug delivery

in desired organ, to dissolve the therapeutics for intravascular administration, to enhance the drug stability against enzyme-mediated degradation, and to achieve targeted (cellular/organ) delivery of drugs. In addition, the release of encapsulated cargo from nanocarriers can be controlled in the organ of interest in order to produce desired therapeutic level for desired period of time to generate maximum therapeutic benefits. Usually, nanoparticles have shown greater cellular uptake compared to microparticles (Desai et al. 1996). Several important human disorder-related diseases have demonstrated significant improvement after treatment of protein-/peptide-loaded nanocarriers (Yu et al. 2016). Macromolecules have large size, high hydrophilicity and susceptibility to physical and chemical degradation, structural fragility, and complexity; hence, these characteristics strongly affect pharmacokinetic and pharmacodynamic behavior in vivo. Moreover, development processes such as higher temperature, exposure to organic solvents, etc. additionally compromise the stability of these macromolecules. To overcome all these challenges, extraordinary efforts are made to incorporate therapeutics in hydrogels, micellar systems, nanocapsules, nanoemulsions, nanoliposomes, and niosomes. Afterward, surface modification of these carriers by specific ligands was explored with an aim to deliver protein therapeutics into the organs of interest in active form. Nanomedicines approved by FDA are enlisted as per variety of carrier/material used in development of the formulation (Table 19.1). The key objective of

Table 19.1 List of nanomedicines approved by FDA classified by type of carrier/material used in the preparation of the formulation

Commercial name (Company)	Ingredient active	Carrier	Application	Advantage	Year approved
Doxil®/Caelyx™ (Janssen)	Doxorubicin	Liposomes	Kaposi's sarcoma; ovarian cancer; multiple myeloma	Increased site-specific delivery (tumor) and decreased systemic toxicity	1995; 2005; 2008
Abelcet® (Sigma Tau)	Amphotericin B lipid complex	Liposomes	Fungal infection	Decreased toxicity	1995
DaunoXome (Galen)	Daunorubicin	Liposomes	Kaposi's sarcoma	Increased site-specific delivery (tumor) and decreased toxicity	1996
DepoCyt© (Sigma Tau)	Cytarabine	Liposomes	Lymphomatous meningitis	Increased site-specific delivery (tumor) and decreased toxicity	1996
AmBisome® (Gilead Sciences)	Amphotericin B	Liposomes	Fungal and/or protozoal infections	Reduced nephrotoxicity	1997
Curosurf /poractant alfa (Chiesi Farmaceutici)	Proteins SP-B and SP-C	Liposomes	Lung activator for stress disorder; pulmonary surfactant for respiratory distress syndrome	Decreased toxicity and increased delivery for smaller volume	1999
Visudyne (Bausch and Lomb)	Verteporfin	Liposomes	Ocular histoplasmosis, myopia, decreased vision	Increased site-specific delivery, (lesion vessels) photosensitive release	2000
DepoDur® (Pacira Pharmaceuticals)	Morphine sulfate	Liposomes	Prolonged release	Loss of pain (postoperative)	2004
Marqibo® (Onco TCS)	Vincristine	Liposomes	Acute lymphoblastic leukemia	Increased site specific delivery (tumor) and decrease toxicity	2012
Onivyde® (Merrimack)	Irinotecan	Liposomes	Pancreatic cancer	Increased site specific delivery (tumor) and decrease toxicity	2015

(continued)

Table 19.1 (continued)

Commercial name (Company)	Ingredient active	Carrier	Application	Advantage	Year approved
Adagen (Sigma Tau Pharmaceuticals®)	Pegademase bovine	PEGylated adenosine deaminase enzyme	Immunodeficiency disease	Improved circulation time in body and decreased immunogenicity	1990
Oncaspar (Enzon Pharmaceuticals)	Asparaginase	PEGylated asparaginase	Acute lymphoblastic leukemia	Improved protein stability due to PEGylation	1994
Copaxone® (Teva)	Glatopa	Glutamate, alanine, lysine, and tyrosine random polymer	Multiple sclerosis	Regulation of clearance and polymer with controlled molecular weight	1996
Renagel® (Sanofi)	Sevelamer hydrochloride or sevelamer carbonate	Poly(allylamine hydrochloride)	Chronic renal diseases	Increased site-specific delivery and increase in circulation time in body	2000
PegIntron® (Merck)	Interferon-alpha (IFN-α2b)	PEGylated IFN-α2b protein	Hepatitis C	Improved protein stability due PEGylation	2001
Pegasys® (Genentech)	Interferon-alpha (IFN-α2a)	PEGylated IFN-α2a protein	Hepatitis B and C	Improved protein stability due PEGylation	2002
Eligard® (Tolmar)	Leuprolide acetate	Polymer (PLGH (poly(dl-lactide-co-glycolide)	Prostate cancer	Prolonged drug delivery and circulation time in body	2002
Neulasta® (Amgen)	PEG-filgrastim	PEGylated granulocyte colony- stimulating factor (GCSF) protein	Neutropenia induced by chemotherapy	Improved protein stability due to PEGylation	2002
Somavert® (Pfizer)	PEG-visomant	PEGylated HGH receptor antagonist	Acromegaly	Improved protein stability due to PEGylation	2003
Macugen® (Bausch & Lomb)	PEG-aptanib	PEGylated anti vascular endothe- lial growth factor aptamer	Macular degeneration; neovas cular age-related (decreased vision)	Improved stability due to PEGylation	2004

Mircera® (Hoffman-La Roche)	Methoxy polyethylene glycol-epoetin beta	Chemically synthesized erythropoiesis-stimulating agent	Anemia associated with renal failure due to diseases	Improved stability due to PEGylation	2007
Cimzia® (UCB)	Certolizumab pegol	PEGylated antibody fragment(Certolizumab)	Crohn's disease; rheumatoid arthritis; psoriatic arthritis and ankylosing spondylitis	Increased stability and circulation time in body	2008; 2009; 2013; 2013
Krystexxa® (Horizon)	Pegloticase	PEGylated porcine-like uricase	Chronic gout	Improved protein stability due to PEGylation	2010
Plegridy® (Biogen)	Interferon beta (IFNβ 1a)	PEGylated IFN-β1a protein	Multiple sclerosis	Improved protein stability due to PEGylation	2015
ADYNOVATE (Baxalta)	Factor VIII	PEGylated factor VIII	Hemophilia	Improved protein stability due to PEGylation	2015
Rapamune® (Wyeth Pharmaceuticals)	Sirolimus	Nanocrystals	Immunosuppressant	Increased bioavailability	2000
Megace ES (Par Pharmaceuticals)	Megestrol acetate	Nanocrystals	Anti-anorexic	Reduced posology	2001
Aviriza (Pfizer)	Morphine sulfate	Nanocrystals	Mental stimulant	Prolonged release and increased bioavailability	2002/2015
Ritalin LA® (Novartis)	Methylphenidate HCl	Nanocrystals	Mental stimulant	Increased drug loading and bioavailability	2002
Zanaflex® (Acorda)	Tizanidine HCl	Nanocrystals	Muscle relaxant	Increased bioavailability and decreased posology	2002
Emend® (Merck)	Aprepitant	Nanocrystals	Antiemetic drug	Increased absorption and bio- availability	2003
Vitoss® (Stryker)	Calcium phosphate	Nanocrystals	Bone substitute	Mimics bone structure by cell adhesion and growth	2003
OsSatura ® (IsoTis Orthobiologics)	Hydroxyapatite	Nanocrystals	Bone substitute	Mimics bone structure by cell adhesion and growth	2003

(continued)

Table 19.1 (continued)

Commercial name (Company)	Ingredient active	Carrier	Application	Advantage	Year approved
Ostim® (Heraseus Kulzer)	Hydroxyapatite	Nanocrystals	Bone substitute	Mimics bone structure by cell adhesion and growth	2004
Tricor® (Lupin Atlantis)	Fenofibrate	Nanocrystals	Hyperlipidemia	Increased bioavailability	2004
Focalin XR® (Novartis)	Dexmethylphenidate HCl	Nanocrystals	Mental stimulant	Increased bioavailability	2005
NanOss (Rti Surgical)	Hydroxyapatite	Nanocrystals	Bone substitute	Mimics bone structure by cell adhesion and growth	2005
EquivaBone® (Zimmer Biomet)	Hydroxyapatite	Nanocrystals	Bone substitute	Mimics bone structure	2009
Invega® Sustenna® (Janssen Pharms)	Paliperidone palmitate	Nanocrystals	Schizophrenia schizoaffective disorder	Decreased release of poor water-soluble drugs	2009/2014
Ryanodex® (Eagle Pharmaceuticals)	Dantrolene sodium	Nanocrystals	Malignant hypothermia	Allows higher administration at higher doses	2014
Estrasorb® (Novav)	Estradiol	Micelles	Menopause hormone therapy	Sustained release	2003
Abraxane® (Celgene)	Paclitaxel (ABI-007)	Albumin-bound paclitaxel nanoparticles	Breast cancer; non-small cell lung cancer and pancreatic cancer	Increase site-specific delivery (tumor) and solubility	2005, 2012, 2013
INFeD (Sanofi Aventis)	Iron	Iron dextran (low MW)	Chronic kidney failure with iron deficiency	Increased dose capacity	1957
DexIron/Dexferrum (Sanofi Aventis)	Iron	Iron dextran (high MW)	Chronic kidney failure with iron deficiency	Increased dose capacity	1957
Feridex/Endorem (AMAG Pharmaceuticals®)	Superparamagnetic iron oxide nanoparticles (SPION	SPION coated with dextran	Imaging material	Superparamagnetic character	1996, 2008
Ferrlecit® (Sanofi Aventis)	Sodium ferric	Sodium ferric gluconate	Chronic kidney failure with iron deficiency	Increased dose capacity	1999

Venofer (Luitpold Pharmaceuticals)	Iron oxide	Iron sucrose	Chronic kidney failure with iron deficiency	Increased dose capacity	2000
GastroMARK; umirem (AMAG Pharmaceuticals™)	Superparamagnetic iron oxide nanoparticles (SPION	SPION coated with silicone	Imaging material	Superparamagnetic character	2001/2009
Feraheme (AMAG pharmaceuticals)	Ferumoxytol ultrasmall superparamagnetic iron oxide nanoparticles (SPION)	Ferumoxytol SPION with polyglucose sorbitol carboxymethyl ether	Chronic kidney failure with iron deficiency	Prolonged steady release and decreased number of doses	2009
Nanotherm (MagForce)	Iron oxide	Aminosilane-coated Iron nanoparticles	Brain tumor	Thermotherapy for destroy tumor cells or sensitized for additional therapies	2010

preparation of nanocarriers is to organize particle size, surface characteristics, as well as efficient delivery in fully active forms. As a result, critical characterizations play a vital role in controlling in vitro as well as in vivo behavior of nanoparticles.

19.2 Classification of Nanocarriers

19.2.1 Liposomes

Liposomes are made of phospholipid bilayer enclosing aqueous cavity which could encapsulate small molecular weight drugs, peptides, proteins, and nucleotides (Yousefi et al. 2009; Patel et al. 2011). Usually, they have a particle size from 25 nm to several micrometers as per the number of bilayers present in the liposomes. Liposomes are proved to be versatile carriers because they offer flexibility in terms of vesicle size, surface charge, bilayer composition, and encapsulation ability which also make these carriers useful for delivery of bioactives. Lipid combination made of distearoyl phosphatidylcholine, dimyristoyl phosphatidylglycerol, and cholesterol (Anderson et al. 1999); a blend of phosphatidylcholine, castor oil, and polyethylene glycol attached to distearoyl phosphatidylethanolamine (Song et al. 1996); and a blend of distearoyl phosphatidylglycerol, distearoyl phosphatidylcholine, and cholesterol (Rezler et al. 2007) are usually used for the preparation of liposomes. Usually aqueous compartment contains hydrophilic agents, whereas lipophilic drugs are inserted into the lipid bilayer (Patel et al. 2009). Bioactives which are prone to degradation are prepared using reverse-phase evaporation process which bypasses unwanted exposure of/to harmful solvents. Liposomes having nano size range can also prolong the release of entrapped bioactives, warranting sustained effect at the site of interest. Pulmonary delivery of insulin is achieved via liposomes which has shown number of advantages as an alternative to insulin injection such as sustained blood glucose levels with a higher pharmacological bioavailability compared to dry powder inhalation (Bi et al. 2008). Further, Bak proteins and voltage-dependent anionic channel (VDAC) were loaded into the liposomes and produce apoptosis into mammalian cells through cytochrome C and caspase activation (Liguori et al. 2008).

Removal of liposomes from blood circulation takes place in a size-dependent manner through membrane fusion actions and interactions with specific serum proteins (opsonins). Despite the fact that cell surface receptors cannot detect liposome directly as a foreign particle, they identify liposomes via cellular and serum proteins that are adhered to liposomal surface (Ishida et al. 2002). Circulation time of liposomes can be enhanced, and also macrophage uptake can be reduced by means of PEGylation which provides anti-opsonizing properties to the liposomes, once it is decorated on the exterior surface by forming a hydrophilic layer of polymer, hence sheltering the charge associated with liposome surface. Usually, RES cells identify and react with liposomes through opsonins, but hydrophilicity of PEG layer helps in escaping the liposomes. Kedar et al. (1994) have significantly enhanced survival time of mice with metastatic carcinoma (earlier subjected to cyclophosphamide chemotherapy) was enhanced by two to six times followed by treatment with SSL-IL-2 compared to IL-2 even after lower doses and fewer administrations compared to plain interleukin-2 (IL-2). Enhanced cellular translocation of TAT peptide (transactivator of transcription of human immunodeficiency virus) was reported by incorporating it into the liposomes (Carsten et al. 2004). Dioleoyl phosphatidylethanolamine (DOPE) and guanidinium-cholesterol cationic lipid liposomes successfully entrapped the β-galactosidase enzyme and the anti-cytokeratin[8] (K8) antibody and enhanced cellular delivery of β-galactosidase devoid of any negotiation in its activity (Chatin et al. 2015). In recent time, the applicability of bilosomes as a viable approach for oral administration of larger peptides and proteins along with associated biopharmaceutical challenges has been reviewed by Ahmad et al. (2017).

Niosomes are spontaneously gathered non-ionic surfactants vesicles generally made of alkyl or dialkyl polyglycerol ether class surfactants are (Malhotra and Jain 1994). Lamellar structures fashioned by mingling cholesterol and non-ionic surfactant and followed by hydration which leads to closed bilayer vesicle upon providing some energy such as physical agitation. Such bilayer structures are spontaneous arrangement of surfactant in such a way that allows hydrophilic heads to remain exposed to the aqueous phase while lipophilic tails acquaint themselves with the lipid phase. Niosomes are extensively investigated as a cheaper substitute of liposomes however is of synthetic origin. These vesicles as a carrier systems are closely having similar physical characteristics, in vitro release behavior, as well as in vivo behavior to the liposomes. These carriers not only have maximum advantages of liposomes, but also the economy, enhanced stability, and flexibility in storage conditions make them as ideal alternatives to phospholipids vesicles. Cholesterol is used to facilitate vesicle preparation, which are less leaky. Additionally, stabilizers could also be added to avoid aggregation of niosomes by repulsive, steric, or electrostatic effect. Theoretically niosome formation necessitates the existence of a specific type of amphiphile which possesses an aqueous head and a liphophilic tail and aqueous solvent. The lipophilic unit may consist of one or two alkyl or perfluoroalkyl moieties or, in certain cases, a single steroidal group usually from C12 to C18. Unit chain alkyl as lipophilic tail is more toxic compared to respective dialkylether chain although the ester-based surfactants are chemically somewhat unstable compared to ether-type surfactants. $C_{16}EO_5$ (polyoxyethylene cetyl ether) or $C_{18}EO_5$ (polyoxyethylene stearyl ether) surfactants are mainly utilized for synthesis of polyhedral vesicles. Various aspects manipulating the formation of niosomes are as follows: nature of surfactants, membrane composition, nature of encapsulated drug, and hydration temperature which is responsible for shape and size of the niosomes. The perfect condition for hydration is favored to be above the gel-to-liquid phase transition temperature of the system. The following are applications of niosomes: delivery of anti-cancer drugs, i.e., doxorubicin, paclitaxel, methotrexate; delivery of peptide drugs, i.e., insulin, oligonucleotide; ophthalmic drug delivery; and also transdermal drug delivery. Niosomes demonstrated greater stability to both temperature and oxidation in contrast to the major components of liposomes, i.e., phospholipids, hence are easy to handle without any serious precautions to be taken in storage (Fang et al. 2001). The vasoactive intestinal peptide (VIP) is therapeutically utilized for the management of Alzheimer's disease however unable to cross the blood–brain barrier (BBB) similar to other endogenous peptides; hence, its parenteral delivery is restricted. The entrapment of VIP in glucose decorated niosomes significantly enhances brain transport of peptide compared to control niosomes (up to 86%, in 5 min) (Biswal et al. 2008). Niosomes containing ethanol showed elastic behavior, hence increasing the intracellular translocation and greater stability of the Tat-GFP fusion protein against chemical degradation (Manosroi et al. 2011). Niosomes were used as a transporter of insulin and demonstrated noteworthy plasma glucose-lowering effect which continued for prolonged period of time (Khaksa et al. 2000). Further niosomes are also used as adjuvant for oral vaccination which demonstrated significant enhancement of antibody levels after encapsulation of ovalbumin into niosomes and shown better immune response compared to Freund's complete adjuvant (FCA), in the BALB/c mice (Rentel et al. 1999).

19.3 Particulate Carriers

19.3.1 Polymeric Nanoparticles

Nanoparticles are known as particulate suspensions or dried particles with a size in the range of 10–1000 nm. As per the method used for their preparation, nanoparticles, nanospheres, or nanocapsules could be produced. Contrasting to nanospheres in which drugs are dispersed in polymer matrix, nanocapsules are vesicular systems in which the drug exits as a core in an aqueous or oily cavity which is covered by coating of polymer; hence, they are "reservoir" system. Nanoparticles not necessarily exhibit size-dependent characteristics which alter greatly in fine particles or bulk materials. The drug/bioactive may be solubilized, dispersed, encapsulated, or adhered to a nanoparticle matrix. A supplementary trustworthy technique to increase the stability of bioactives is encapsulation into a nanoparticle, which protects it from the non-friendly atmosphere inside the biological system and improves the delivery at the site of action (Niven et al. 1994). Encapsulated bioactives were released from the nanoparticles by different mechanisms, i.e., diffusion, erosion, swelling, or polymer degradation. Nanoparticles provide an appropriate way of transporting low molecular weight therapeutics, plus larger bioactives, i.e., proteins, peptides, or genes through localized or targeted delivery to the desired organ. Nanotechnology concentrates on encapsulating therapeutics in biocompatible nanocomposites, i.e., nanoparticles, nanocapsules, micellar systems, and nanoconjugates. These carriers could be utilized to offer targeted drug delivery (cellular/tissue) in order to enhance oral bioavailability, to prolong drug/gene response at the desired site, to dissolve drugs for IV administration, and to stabilize therapeutic agents against enzymatic degradation. Additionally, the release of an entrapped material from nanoparticles to attain wanted therapeutic response in target site for necessary extent should be managed. Nanoparticles have comparatively higher intracellular uptake than microparticles. Polymeric nanoparticles hold remarkable assurance for the efficient cure of disorders as they have marvelous physicochemical characteristics, i.e., size, surface charge, hydrophilicity, and lipophilicity, hence considered as prospective carriers for bioactives, i.e., vaccines, peptides, anticancer drugs, genes, etc. Eventually, nanoparticles offer numerous advantages compared to free drug, such as protection from unwanted reactions with biological moieties and breakdown, improving the absorption into a desired organ (tumors), and escalating the pharmacokinetics of the therapeutics. Additionally, rate of medicament release from nanoparticles can be amended effortlessly to match up desired therapeutic levels into target organ for desired period of time. Once designed properly, nanoparticles can serve as a model delivery vehicle by preferentially picked up by cancer cells or tumor mass and also avoids early degradation of medicament during its transport. Further, intracellular delivery of the associated agents could be accomplished by engulfment through endocytosis/phagocytosis of nanoparticles. Moreover, characteristics of a polymer could be easily altered; nanoparticles comprise of a versatile carrier system that can be modified to synthesize the nanoparticles which are capable to cross through the biological obstacles and transport the load into the cells and/or intracellular space. As per the rate and extent of prolonged and controlled release of the incorporated protein, various natural and/or artificial polymers have been explored for the formulation of nanoparticles, i.e., chitosan, dextran, starch, albumin, gelatin, 2-methoxyethyl vinyl ether, copolymer of maleic anhydride and cyanoacrylates, poly(lactic acid) (PLA), poly(lactic acid-co-glycolic acid) (PLGA), PEG-PLA block copolymers, and poly(n-hexadecyl cyanoacrylate) (Soppimath et al. 2001; Solaro and Chiellini 2006; Rytting et al. 2008; Solaro 2008; Duan et al. 2009). A list of polymers which are frequently used for development of nanoparticles as protein carriers is shown in Table 19.2. Out of different techniques reported for the development of polymeric nanoparticles, majority of them reports

Table 19.2 Most widely used polymers for development of nanoparticles as drug carriers (Vauthier and Bouchemal 2009; Vyas and Khar 2002)

Synthetic homopolymers	Copolymers	Natural polymers	Colloid stabilizers
Poly(lactic acid)	Polylactide-poly(ethylene glycol)	Chitosan	Dextran
Poly(lactide-*co*-glycolide)	Poly(lactide-co-glycolide)-poly(ethylene glycol)	Alginate	Pluronic F68
Poly(ε-caprolactone)	Poly(ε-caprolactone)-poly(ethylene glycol)	Gelatin	Poly(vinyl alcohol)
Poly(isobutylcyanoacrylate)	Poly(hexadecylcyanoacrylate-co-poly(ethylene glycol) cyanoacrylate)	Albumin	Tween® 20 or Tween® 80
Poly(isohexylcyanoacrylate)		Agarose	
Poly(n-butylcyanoacrylate)		Pullulan	
Polyacrylates and polymethacrylates			
Polystyrene			

two key steps: the first step is emulsion formation, and the second is solvent evaporation or gelation/precipitation of polymer or polymerization. A schematic demonstration of commonly used techniques for development of polymeric nanoparticles is shown in Fig. 19.2. Great attention has been given to biodegradable polymers which are self-eliminating, hence overcoming the concern of surgical removal of carrier system. Permeation of insulin and tetanus toxoid were increased through oral and nasal mucosa by encapsulating it into PEG and chitosan-coated PLA/PLGA nanoparticles, respectively (Vila et al. 2002). Nanoparticles composed of biodegradable polymers poly(ε-caprolactone) (PCL) and blend of PLA and PLGA not only protected insulin from proteolytic degradation into the GI tract but also fruitfully decrease plasma sugar concentration after oral administration to diabetic rats (Damge et al. 2007). Higher toxoid levels were observed in the blood stream and lymph nodes with PEGylated nanoparticles containing tetanus toxoid after

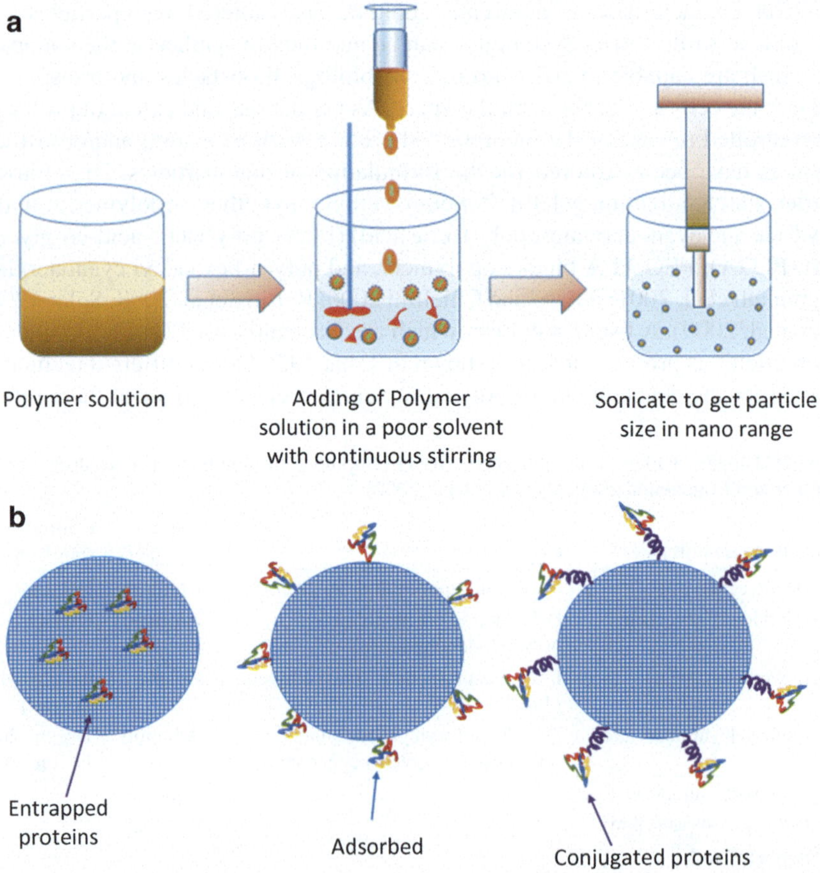

Fig. 19.2 (**a**) Schematic representation of the commonly used method for the preparation of polymeric nanoparticles. (**b**) Schematic of the entrapment of protein in nanocarrier composites

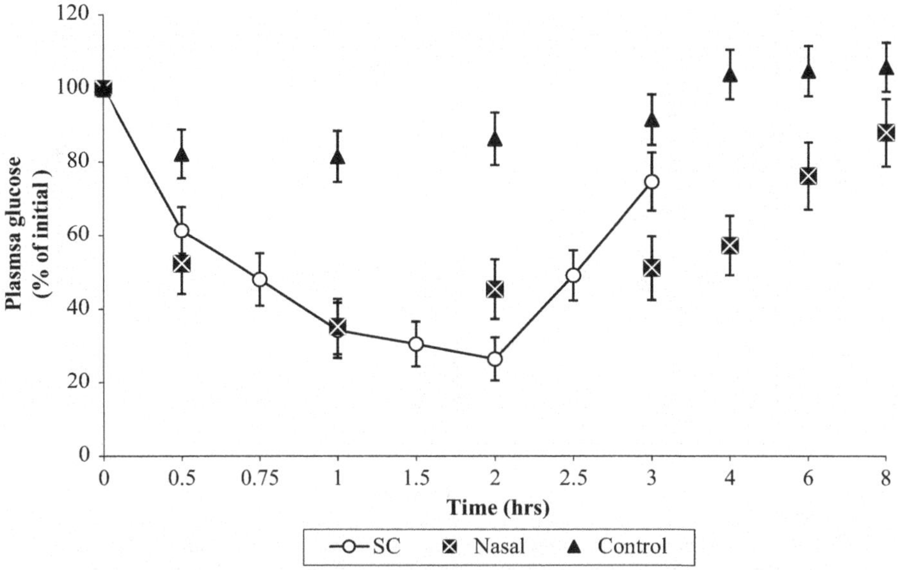

Fig. 19.3 Hypoglycemic effect of EE NPs in the presence of Na glycocholate after nasal administration and plain SC insulin to STZ-induced diabetic rats (mean ± SE, $n = 5$). *Hypoglycemic action of SC insulin is higher ($P < 0.05$) at 2 h compared to nasal NPs. **Hypoglycemic action of nasal insulin is significantly higher compared to SC and control. EE NPs = epichlorohydrin cross-linked starch NPs made via emulsion method. (With kind permission from Elsevier Jain et al. (2008))

nasal and oral administration to rats (Vila et al. 2005). Previously, our research group has established the usefulness of starch nanoparticles for nasal delivery of insulin (Jain et al. 2008) which showed superior hypoglcemic response compared to nasal insulin solution (Fig. 19.3). One of the approaches which is widely used for protein delivery by conjugation with various nanocarriers is receptor-mediated endocytosis (vide infra). In this regard, our research group previously tested the in vivo efficacy of vitamin B12 (VB12) coupled dextran nanoparticles after oral administration of VB12 coupled dextran NPs containing insulin (Fig. 19.4). We have also demonstrated the prevalence of VB12-mediated RME uptake of NPs by co-administering the large excess of VB12 to saturate the receptors 1 h prior to dosing of VB12-conjugated nanocarriers (Chalasani et al. 2007). Carboxymethyl-β-cyclodextrin-decorated chitosan (CMCD-g-CS) nanoparticles showed higher in vitro release profile of BSA in simulated intestinal fluid compared to gastric fluid (Song et al. 2017). Conjugation of cell-penetrating peptide (CPP) to the surface of the 15 nm chitosan nanoparticles (prepared by nanoemulsion method) was carried out and tested in Caco-2 cell line for translocation of insulin across the cell monolayer and found to be effective with 15–19% increase in insulin levels (Barbari et al. 2017). An exhaustive inventory of nanomedicines which are granted FDA approval and are classified as per type of carrier/material utilized in production of the nano formulation is shown in Table 19.1.

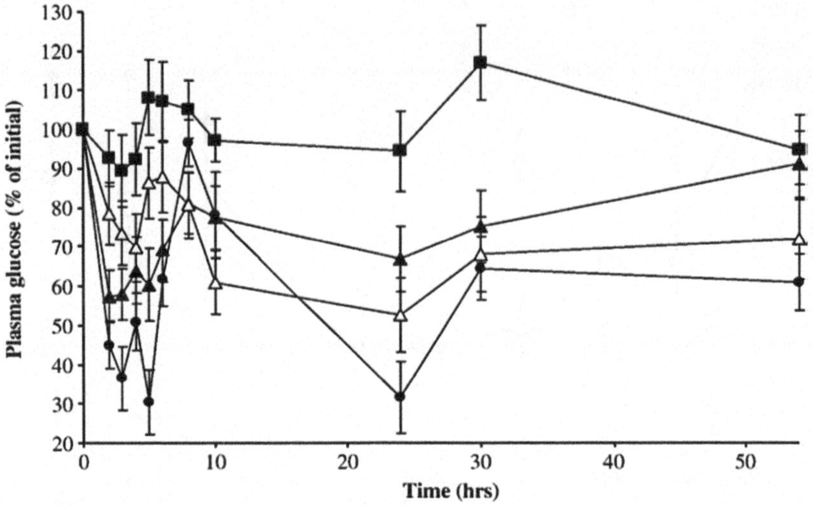

Fig. 19.4 VB12-mediated uptake study of dextran NPs of medium crosslinking density conjugated with O'Hexyl derivative of vitamin B12 (NPs-M35'OH) in terms of plasma glucose reduction after oral administration to STZ diabetic rats ($n = 3$, mean ± S.E.) (Δ) VB12 + 20 IU/kg. (▲) 10 IU/kg. (●) 20 IU/kg.∗ (■) Plain insulin control. ∗ $n = 6$. (With kind permission from Elsevier Chalasani et al. (2007))

19.3.2 Solid Lipid Nanoparticles (SLNs)

SLNs are composed of solid lipids which are stabilized by means of surfactants in an aqueous suspension. They hold close resemblance to nanoemulsion; the only difference is that liquid lipid is substituted with a solid lipid. This replacement of oils with solid lipids controls drug release in an outstanding style, since solid lipid lowers mobility of drug significantly, as compared to oil phase (Martins et al. 2007). SLNs have gained attention in the last few years as carrier system for macromolecules (Marcato and Duran 2008). SLNs are coupled with few advantages and at the same time circumvent some drawbacks of many different carriers such as nanoparticles, liposome, ethosomes, and lipid emulsion.

The advantages offered by SLNs include:

(i) They have drug targeting and controlled drug release ability.
(ii) Protection of labile drugs against photochemical, chemical, and oxidative degradation.
(iii) Limited toxicity compared to polymeric nanoparticles, as SLNs are made of physiological and biocompatible lipids.
(iv) Equally suitable for both hydrophilic and lipophilic drugs.
(v) Bypass exposure of macromolecules to non-friendly hydrophobic solvents.
(vi) SLNs are useful for delivery of macromolecules through various routes such as oral, pulmonary, intravenous, ophthalmic, and dermal (Garcia-Fuentes et al. 2003; Hou et al. 2003; Liu et al. 2008; Joshi and Muller 2009).

Delivery of macromolecules particularly proteins, peptides, and genes might face problems pertaining to their entrapment, as most of them have hydrophobic moieties that show adsorbing behavior onto surfaces (e.g., plastic and glass). This tendency could explain the ways to discrete losses in the amount of bioactives which reaches to the site of action (Duncan et al. 1995).

Mainly researchers described three models for incorporation of drugs or bioactives into the SLNs (Muller et al. 2000; Mehnert and Mader 2001), namely, (i) the drug-enriched shell model, (ii) the homogeneous matrix model, and (iii) the drug-enriched core model. These three models have different formulation composition such as chemical composition of the lipid, bioactive, and surfactants, along with the production method. Contrasting to most of the microsphere and nanoparticle which are made of polymers, SLN preparation techniques avoid potentially toxic organic solvents which may deteriorate bioactives. In addition, under prominent conditions, SLNs can be produced to entrap a huge variety of drugs and emerge to accomplish the desires of optimum nanocarriers (Muller et al. 1995; Wissing et al. 2004). The synthesis of SLN depends on solidification of dispersed phase. As a result, due to hydrophilic nature of bioactives, they have shown poor entrapment into the lipophilic material of SLN that tends to distribute into the aqueous phase during the course, which is promoted by the utilization of emulsifier (surfactants) and stabilizers. Solid lipid nanoparticles are usually prepared by solid lipid nanoparticles by hot homogenization technique (Fig. 19.5). Also, SLN can demonstrate partial

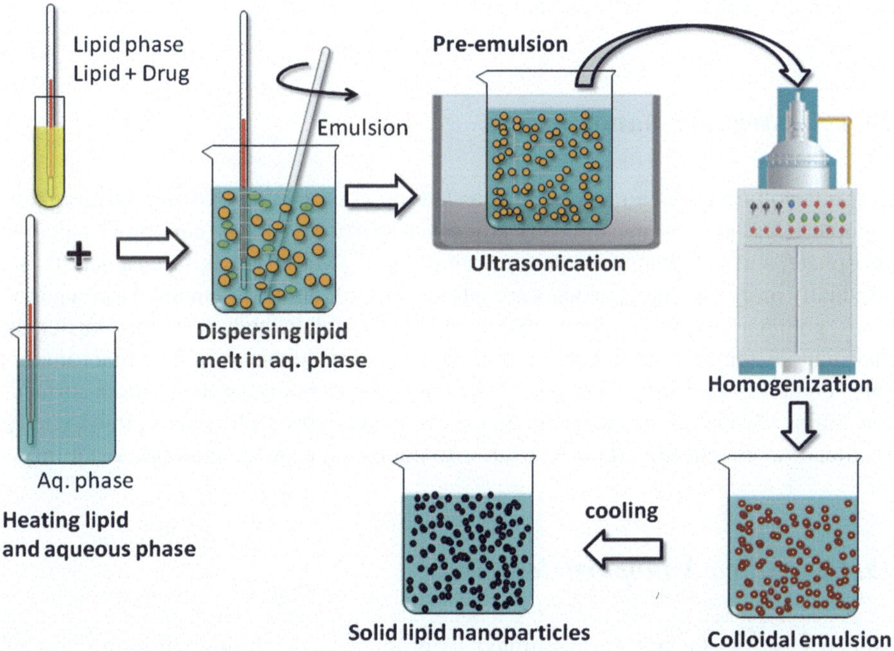

Fig. 19.5 Schematic depiction of various steps involved in the preparation of solid lipid nanoparticles by hot homogenization technique

entrapment owing to drug leakage during storage. In the last few decades, researchers have continuously publicized encouraging results regarding encapsulation of bioactives in SLNs. Numerous peptides such as somatostatin, lysozyme, LHRH, insulin, malaria antigens, calcitonin, and HBsAg have been encapsulated into the SLNs and studied for their stability, in vitro drug release kinetics, and in vivo performance (Table 19.3).

Various strategies have been explored to load macromolecules in ligand-decorated SLNs to enhance oral bioavailability and showed protein activity via both caveolae- and clathrin-mediated endocytosis (Fan et al. 2014). However, mucus was an obstacle to the translocation of SLNs. The absolute bioavailability of peptides was improved by 2.45 to 1.98 times compared to unmodified SLNs, suggesting their effectiveness in oral bioavailability enhancement of proteins. In another report, the gonadorelin (a model peptide) was incorporated into the SLN developed by solvent diffusion technique and evaluated for various parameters. (Hu et al. 2004). Lysozyme (a model peptide) was solubilized into the melted lipid phase, and it remained intact during the course of the process without diminishing its activity and found solubility-dependent entrapment efficiency of the peptide in the hydrophobic phase (Almeida et al. 1997). In another study, physiochemical stability, intracellular uptake by Caco-2 cells, and in vitro cytotoxicity of beta-carotene (BC) SLNs were evaluated and found minimum cytotoxicity of SLNs particularly after dilution ($\geq$10 times) (Yi et al. 2014). Goppert and Muller (2005) studied adsorption pattern of human plasma protein on the outermost layer of SLNs after IV injection and found SLN as a potential targeted drug delivery formulation.

19.4 Inorganic Nanocarriers

Current evolution in nanotechnology has led to the utilization of different inorganic nanoparticles such as mesoporous silica nanoparticle, carbon nanotube, calcium phosphate, and gold nanoparticle for drug delivery (Malmsten and Zauscher 2013). Gradually inorganic nanoparticles are adding weight; among them are carbon nanotubes, gold nanoparticles, and nanospheres, hence widely studied as drug carrier, as their nanometer size enables them to move easily inside the body. The macromolecules could be either placed inside of the nanotube or bound to the particle surface. The major advantages of inorganic nanocarriers are hydrophilic nature, low toxicity profile, biocompatibility, their resistance to microbial growth, and higher stability.

19.4.1 Silica Nanoparticles

Silicates and surfactant co-assembled to produce mesoporous silica, a class of surfactant-templated inorganic compound. Biocompatible nature of mesoporous silica nanoparticles (MSNs) is ideal for biological uses, hence used for delivery of

Table 19.3 Peptide and protein molecules successfully incorporated into the solid lipid nanoparticles

S. no.	Protein/peptide entrapped into solid lipid nanoparticles	Method of development of solid lipid nanoparticles	Entrapment efficiency	References
1.	BSA	Adsorption onto SLN	n.a.	Gualbert et al. (2003)
2.	Calcitonin	Solvent evaporation (w/o/w)	>90%	Garcia-Fuentes et al. (2005a, b)
3.	CyA	HPH hot dispersion HPH hot dispersion HPH hot dispersion HPH hot dispersion Warm microemulsion (o/w)	96.6–97.8% 88.4% n.a. 96.1% 13% protein content	Olbrich et al. (2000) Zhang et al. (2000) Radtke and Muller (2001a, b) Muller et al. (2006) Ugazio et al. (2002)
4.	Gonadorelin	Solvent displacement	50.4–69.4%	Hu et al. (2004)
5.	HAS	Adsorption onto SLN	12.4–32.4% (stealth) 7–13.5%	Cavalli et al. (1999)
6.	Insulin	Solvent displacement Solvent evaporation (w/o/w) Solvent evaporation (w/o/w) Supercritical CO_2 (PGSS)	26.8% n.a. 67.9% 75%	Zhang et al. (2006) Garcia-Fuentes et al. (2003) Zhang et al. (2006) Caliceti et al. (2006)
7.	[D-Trp-6] LHRH	Warm microemulsion (w/o/w)	90%	Morel et al. (1994)
8.	Lysozyme	HPH cold dispersion	43.2–59.2%	Almeida et al. (1997)
9.	Ovalbumin	Adsorption onto SLN Melt dispersion (o/w)	70–97% >80%	Videira et al. (2002) Videira et al. (1998)
10.	Thymopentin	Warm microemulsion (o/w with ionic pair or w/o/w)	5.2% or 1.7%	Morel et al. (1996)
11.	Yak interferon-α	Double emulsion solvent evaporation (w/o/w)	83.7%	Li et al. (2010)

BSA bovine serum albumin, *CyA* cyclosporine A, *HPH* high-pressure homogenization, *LHRH* luteinizing hormone-releasing hormone, *n.a.* not available, *PGSS* particles from gas-saturated solution technique therapeutics (Radin et al. 2002; Lai et al. 2003); however, they are not bioresorbable. Pore size and pore structure could be altered by ease by selection of surfactant and co-assembly conditions. Higher porous surface along with huge effective surface area of Mesoporus silica nanoparticles (MSN), hence MSN are being exploited as potential nanocarriers for bioactive molecules. Dimension, shape, and surface functionalization of MSN control the release rate of encapsulated bioactives. FITC-cytochrome C was effectively delivered into the HeLa cells after encapsulation into the MSN through a diffusion-mediated process. Also, it was demonstrated by confocal images that FITC-cytochrome C effectively runs away from the endosomal degradation pathways (Tan et al. 2004; Lu et al. 2007). These cellular uptake and endosomal breakout processes of MSNs are energy dependent, whereas surface modification of MSNs by amine and guanidinium makes them able to infiltrate into the cells by a clathrin- and covalent-free pathway (Xing et al. 2005; Slowing et al. 2006). Prasetyanto et al. (2016) published a straightforward approach to encapsulate a highly cytotoxic protein by encapsulating into the organosilica matrices which disintegrate while contact to a chemical response and demonstrated successful cellular delivery into the C6 glioma cells. Alternatively, ligands and antibodies govern the cellular translocation of MSNs by receptor-mediated endocytosis. Moreover, MSNs have been coated with release retarding agents such as chitosan-PEG copolymers, to modulate the release rate of entrapped material (Tan et al. 2004; Lu et al. 2007). Encapsulation efficiency of OVA (ovalbumin) was multiplied by 2.5-fold after surface modification of NPs with positively charged amine because of electrostatic interaction. Amino group modified the zeta potential of the particle from negativity toward positivity (Mahony et al. 2013). Amino-modified silica nanospheres along with photoluminescent CaF2:Tm,Yb nanocrystals have shown higher protein loading and prolonged release of the encapsulated cargo and showed usefulness in cell uptake of bioactives (Li et al. 2017). Besides its use as a delivery vehicle, MSNs were investigated as an immune booster which elicit immune response even at smaller OVA levels. These NPs are synthesized under controlled conditions allowing the modification of pore shape and surface groups (Vallet-Regi et al. 2007), which could be later on conjugated with various mAb structures. Larger pores offer chances for higher loading of mAbs. Surface-functionalized MSN has been encapsulated with a mAb against tumor cell surface protein and administered to mice with melanoma (Lei et al. 2010). The administration of particle-encapsulated mAb demonstrated enhanced suppression of tumor enlargement when compared to pure mAb injection. Findings suggest that encapsulation of mAb into MSN did not change the therapeutics effect or immunological potential, however and by assisting prolonged release of antibody results in enhancement of the half-life of the mAb at the tumor tissue. Mice with malignant mesotheliomas treated with intraperitoneal injection of doxorubicin (DOX) encapsulated into the MSNs which were attached to mesothelin-specific antibody (Macura et al. 2013). The surface-decorated MSN were shown higher effectiveness compared to plain DOX at lower dose level hence able to significantly diminish the undesired effects of DOX. MSN were also attached to antibody directed against epidermal growth factor in order to treat lung cancer cells, with cytotoxic drug pyrrolidone-2, and demonstrated a 38% reduction in tumor growth with low systemic toxicity (Sundarraj et al. 2014). Also, silica nanospheres composed of vacant core pooled with pH-sensitive chitosan transported encapsulated protein into breast cancer cells through targeting antibody ErbB2 and found higher levels in tumors because of lower local pH at tumors (pH 4.0 Vs physiological pH) upon administration to mouse model (Deng et al. 2011)

19.4.2 Gold Nanoparticles

Gold nanoparticles proved their applicability in biomedical field since they are bioinert, are biocompatible, have low toxicity, have flexibility of surface modifications, and have cellular imaging ability. Delivery of β-galactosidase into the cells using gold nanoparticles of 2.5 nm in size are explored as useful transporter for and found that β-galactosidase was effectively reached inside the cell membrane of HeLa cells (Ghosh et al. 2010). The mechanism through which gold NPs internalized depend on the surface characteristics, i.e., surface charge and/or size, mainly taken up by endocytosis, whereas antibody-decorated GNPs followed the receptor-mediated endocytic pathway (Tkachenko et al. 2003; Connor et al. 2005). Also, gold NPs were covalently conjugated with protein antigens without any chemical interactions and acted as adjuvant in order to produce a vaccine for cancer immunotherapy. Five nm gold nanoparticles were utilized for protein delivery by treating Balb/c mouse with OVA and AUNPs and were found to be highly effective in producing anti-OVA IgG antibodies (Tang et al. 2013). Gold nanocarriers, including nanorods, multifunctional nanocarriers made of a gold nanoshell, and conventional nanospheres, were conjugated with remedial mAbs and proven to be extremely selective to the target cells without losing any antibody functionality (Shao et al. 2011; Bisker et al. 2012; Cho et al. 2014; Lee et al. 2014; Shen et al. 2014).

19.4.3 Calcium Phosphate Nanoparticles

Calcium phosphate nanoparticles (CaP) having a diameter of 40–50 nm were developed, and their surface was altered by PEG; hence, these modified nanoparticles had zeta potential very close to zero and are used for protein delivery. Coating of pH-responsive material which solubilizes intestinal pH protected encapsulated insulin against the gastric degradation. In vitro release of insulin was negligible in acidic atmosphere whereas insulin was released for a time span of 8 h in intestinal pH (Ramachandran et al. 2008). Zinc is reported to be used for retarding insulin release (long-acting insulins); thus, calcium phosphates, zinc calcium, and zinc phosphates look attractive contenders for developing ceramic-based insulin carriers. Oral delivery is not only the most favored way of administration of drugs but also provides advantage of patient compliance. Preferably, the absorption of nanoparticles takes place via Peyer's patches region which arrived at the lymphatic system hence bypasses the first-pass metabolism during which insulin degraded significantly. BioSante Pharmaceuticals, a US-based company, successfully synthesized calcium phosphate nanoparticles and entered into the first stage of toxicity studies.

19.5 Concluding Remarks

Delivery of macromolecules is required to be more creative approach than smaller molecular weight therapeutics and demand safer administration, deserve enormous assurance for the treatment of complex diseases. In vivo delivery of macromolecules is emerging importance day by day particularly after latest advancement in recombinant technology which leads to commercial supply of huge variety of therapeutically effective macromolecules. The survival of these agents in the biological environment is of vital importance due to possible denaturation or enzymatic degradation in the absence of an ideal carrier. The selected carrier must provide the complete protection and should be able to translocate the cargo as per desired needs. Hence, it is critically important to design a suitable carrier of appropriate size, composition, and surface behavior as well as biocompatible. Out of the huge range of the carriers studied, nanocarriers emerged as an outstanding choice during the last decades for their successful therapeutic effects through oral, pulmonary, buccal, nasal, as well as parenteral routes.

Because of the biocompatible nature of lipids utilized in synthesis of nanoliposomes and solid lipid nanoparticles, these carriers are proved to be more safe than other types of nanocarriers. Moreover, lipid-based carriers serve as strong immunological adjuvants, capable of eliciting cellular and humoral response against a range of infectious agents related to human disease. However, generally they show considerable instability because of inadequate shelf-life and shorter half time. Indeed, because of high stability of polymeric nanoparticles over other nanocarriers, they emerged as promising carriers of bioactives to meet specific requirements. In addition, surface decoration of these nanoparticles with specific ligands RES uptake can be drastically decreased which is a key constraint with these carriers. On the other hand, synthesis of polymeric nanocarriers generally involves use of organic solvents which could have deleterious effect on macromolecules. Therefore, these drawbacks could be beaten by the picking up protein friendly method for particle synthesis by employment of self-assembling water-soluble polymers.

References

Ahmad J, Singhal M, Amin S, Rizwanullah M, Akhter S, Kamal MA et al (2017) Bile salt stabilized vesicles (Bilosomes): a novel nano-pharmaceutical design for oral delivery of proteins and peptides. Curr Pharm Des 23(11):1575–1588

Almeida AJ, Runge S, Muller RH (1997) Peptide-loaded solid lipid nanoparticles (SLN): influence of production parameters. Int J Pharm Sci 149(2):255–265

Anderson KE, Stevenson BR, Rogers JA (1999) Folic acid-PEO-labeled liposomes to improve gastrointestinal absorption of encapsulated agents. J Control Release 60:189–198

Barbari GR, Dorkoosh FA, Amini M, Sharifzadeh M, Atyabi F, Balalaie S et al (2017) A novel nanoemulsion-based method to produce ultrasmall, water-dispersible nanoparticles from chitosan, surface modified with cell-penetrating peptide for oral delivery of proteins and peptides. Int J Nanomedicine 12:3471–3483

Bi R, Shao W, Wang Q, Zhang N (2008) Spray-freeze-dried dry powder inhalation of insulin loaded liposomes for enhanced pulmonary delivery. J Drug Target 16:639–648

Bisker G, Yeheskely-Hayon D, Minai L, Yelin D (2012) Controlled release of Rituximab from gold nanoparticles for phototherapy of malignant cells. J Control Release 162:303–309

Biswal S, Murthy PN, Sahu J, Sahoo P, Amir F (2008) Vesicles of non-ionic surfactants niosomes and drug delivery potential. Int J Pharm Sci Nanotechnol 1:1–10

Caliceti P, Brossa A, Salmaso S, Bersani S, Elvassore N, Bertucco A (2006) Preparation of protein loaded solid lipid nano-particles by compressed fluid process. Proc Int Symp Control Release Bioact Mater 33:383

Carsten R, Ulrike S, Aurora O et al (2004) Application of novel solid lipid nanoparticle (SLN)-gene vector formulations based on a dimeric HIV-1 TAT-peptide in vitro and in vivo. Pharm Res 21:1662–1669

Cavalli R, Bocca C, Miglietta A, Caputo O, Gasco M (1999) Albumin adsorption on stealth and non-stealth solid lipid nanoparticles. STP Pharm Sci 9(2):183–189

Chalasani KB, Russel-Jones GJ, Jain AK, Jain SK, Diwan PV (2007) Effective oral delivery of insulin in animal models using vitamin B12-coated dextran nanoparticles. J Control Release 122:141–150

Chatin B, Mevel M, Devalliere J, Dallet L, Haudebourg T, Peuziat P et al (2015) Liposome-based formulation for intracellular delivery of functional proteins. Mol Ther Nucleic Acid 4:e244

Cho SK, Emoto K, Su LJ, Yang X, Flaig TW, Park W (2014) Functionalized gold nanorods for thermal ablation treatment of bladder cancer. J Biomed Nanotechnol 10:1267–1276

Connor EE, Mwamuka J, Gole A, Murphy CJ, Wyatt MD (2005) Gold nanoparticles are taken up by human cells but do not cause acute cytotoxicity. Small 1:325–327

Damge C, Maincent P, Ubrich N (2007) Oral delivery of insulin associated to polymeric nanoparticles in diabetic rats. J Control Release 117:163–170

Deng Z, Zhen Z, Hu X, Wu S, Xu Z, Chu PK (2011) Hollow chitosan-silica nanospheres as pH-sensitive targeted delivery carriers in breast cancer therapy. Biomaterials 32:4976–4986

Desai MP, Labhasetwar V, Amidon GL, Levy RJ (1996) Gastrointestinal uptake of biodegradable microparticles effect of particle size. Pharm Res 13:1838–1845

Duan J, Zhang Y, Chen W, Shen C, Liao M, Pan Y et al (2009) Cationic polybutyl cyanoacrylate nanoparticles for DNA delivery. J Biomed Biotechnol 2009:149254

Duncan MR, Lee JM, Warchol MP (1995) Influence of surfactants upon protein/peptide adsorption to glass and polypropylene. Int J Pharm Sci 120(2):179–188

Fan T, Chen C, Guo H, Xu J, Zhang J, Zhu X et al (2014) Design and evaluation of solid lipid nanoparticles modified with peptide ligand for oral delivery of protein drugs. Eur J Pharm Biopharm 88(2):518–528

Fang JY, Hong CT, Chiu WT, Wang YY (2001) Effect of liposomes and niosomes on skin permeation of enoxacin. Int J Pharm 219:61–72

Garcia-Fuentes M, Torres D, Alonso M (2003) Design of lipid nanoparticles for the oral delivery of hydrophilic macromolecules. Colloids Sur B Interfaces 27(2):159–168

Garcia-Fuentes M, Prego C, Torres D, Alonso MJ (2005a) A comparative study of the potential of solid triglyceride nanostructures coated with chitosan or poly(ethylene glycol) as carriers for oral calcitonin delivery. Eur J Pharm Sci 25(1):133–143

Garcia-Fuentes M, Torres D, Alonso MJ (2005b) New surface-modified lipid nanoparticles as delivery vehicles for salmon calcitonin. Int J Pharm Sci 296(1–2):122–132

Ghosh XP, Yang R, Arvizo ZJ, Zhu SS, Agasti Z, Rotello VJ (2010) Intracellular delivery of a membrane-impermeable enzyme in active form using functionalized gold nanoparticles. Am Chem Soc 132:2642–2645

Goppert TM, Muller RH (2005) Protein adsorption patterns on poloxamer- and poloxamine-stabilized solid lipid nanoparticles (SLN). Eur J Pharm Biopharm 3:361–372

Gualbert J, Shahgaldian P, Coleman AW (2003) Interactions of amphiphilic calix[4]arene-based Solid Lipid Nanoparticles with bovine serum albumin. Int J Pharm Sci 257(1–2):69–73

Hadjipanayis CG, Machaidze R, Kaluzova M, Wang L, Schuette AJ, Chen H et al (2010) EGFRvIII antibody-conjugated iron oxide nanoparticles for magnetic resonance imaging- guided convection-enhanced delivery and targeted therapy of glioblastoma. Cancer Res 70(15):6303–6312

Hou D, Xie C, Huang K, Zhu C (2003) The production and characteristics of solid lipid nanoparticles (SLNs). Biomaterials 24(10):1781–1785

Hu FQ, Hong Y, Yuan H (2004) Preparation and characterization of solid lipid nanoparticles containing peptide. Int J Pharm Sci 273(1–2):29–35

Ishida T, Harashim H, Kiwada H (2002) Liposome clearance. Biosci Rep 22:197–224

Jain AK, Khar RK, Ahmed FJ, Diwan PV (2008) Effective insulin delivery using starch nanoparticles as a potential trans-nasal mucoadhesive carrier. Eur J Pharm Biopharm 69(2):226–435

Jong WHD, Borm PJA (2008) Drug delivery and nanoparticle applications and hazards. Int J Nanomedicine 3(2):133–149

Joshi MD, Muller RH (2009) Lipid nanoparticles for parenteral delivery of actives. Eur J Pharm Biopharm 71(2):161–172

Kedar E, Braun E, Rutkowski Y, Emanuel N, Barenholz Y (1994) Delivery of cytokines by liposomes II. Interleukin-2 encapsulated in long-circulating sterically stabilized liposomes: Immunomodulatory and anti-tumor activity in mice. J Immunother Emphasis Tumor Immunol 16:115–124

Khaksa G, D'Souza R, Lewis S, Udupa N (2000) Pharmacokinetic study of niosome encapsulated insulin. Indian J Exp Biol 38:901–905

Lai CY, Trewyn BG, Jeftinija DM (2003) A mesoporous silica nanosphere-based carrier system with chemically removable CdS nanoparticle caps for stimuli-responsive controlled release of neurotransmitters and drug molecules. J Am Chem Soc 125:4451–4459

Lee H, Lee MY, Bhang SH (2014) Hyaluronate-gold nanoparticle/Tocilizumab complex for the treatment of rheumatoid arthritis. ACS Nano 8:4790–4798

Lei C, Liu P, Chen B (2010) Local release of highly loaded antibodies from functionalized nanoporous support for cancer immunotherapy. J Am Chem Soc 132:6906–6907

Li S, Zhao B, Wang F, Wang M, Xie S, Wang S et al (2010) Yak interferon-alpha loaded solid lipid nanoparticles for controlled release. Res Vet Sci 88:48–153

Li Y, Chen X, Liu H, Mou X, Ren Z, Ahmad Z et al (2017) Silica nanospheres entrapped with ultra-small luminescent crystals for protein delivery. Chem Eng J 330:166–174

Liguori L, Marques B, Villegas-Mendez A, Rothe R, Lenormand JL (2008) Liposomes mediated delivery of pro-apoptotic therapeutic membrane proteins. J Control Release 126:217–227

Liu W, He Z, Liang J, Zhu Y, Xu H, Yang X (2008) Preparation and characterization of novel fluorescent nanocomposite particles: CdSe/ZnS core-shell quantum dots loaded solid lipid nanoparticles. J Biomed Mater Res 84(4):1018–1025

Lu J, Liong M, Sherman S, Xia T, Kovochich M, Nel AE et al (2007) Mesoporous silica nanoparticles for cancer therapy: energy-dependent cellular uptake and delivery of paclitaxel to cancer cells. Nanobiotechnology 3:89–95

Macura SL, Steinbacher JL, MacPherson MB (2013) Microspheres targeted with a mesothelin antibody and loaded with doxorubicin reduce tumor volume of human mesotheliomas in xenografts. BMC Cancer 13:400

Mahony D, Cavallaro AS, Stahr F, Mahony TJ, Qiao SZ, Mitter N (2013) Mesoporous silica nanoparticles act as a self-adjuvant for ovalbumin model antigen in mice. Small 9(18):3138–3146

Malhotra M, Jain NK (1994) Niosomes as drug carriers. Indian Drugs 31:81–86

Malmsten M, Zauscher S (2013) Colloids and surfaces in biology. Curr Opin Colloid Sci 18:468

Manosroi A, Lohcharoenkal W, Gotz F, Werner RG, Manosroi W, Manosroi J (2011) Cellular uptake enhancement of Tat-GFP fusion protein loaded in elastic niosomes. J Biomed Nanotechnol 7:366–376

Marcato PD, Duran N (2008) New aspects of nanopharmaceutical delivery systems. J Nanosci Nanotechnol 8(5):2216–2229

Martins S, Sarmento B, Ferreira DC, Souto EB (2007) Lipid-based colloidal carriers for peptide and protein delivery--liposomes versus lipid nanoparticles. Int J Nanomedicine 2(4):595–607

Mehnert W, Mader K (2001) Solid lipid nanoparticles: production, characterization and applications. Adv Drug Deliv Rev 47(2–3):165–196

Morel S, Rosa Gasco M, Cavalli R (1994) Incorporation in lipospheres of [d-Trp-6]LHRH. Int J Pharm Sci 105(2):R1–R3

Morel S, Ugazio E, Cavalli R, Gasco MR (1996) Thymopentin in solid lipid nanoparticles, Int J Pharm 132(1–2):259–261

Muller R, Mehnert W, Lucks J-S, Schwarz C, Zur Mühlen A, Meyhers H et al (1995) Solid lipid nanoparticles (SLN): an alternative colloidal carrier system for controlled drug delivery. Eur J Pharm Biopharm 41(1):62–69

Muller RH, Mader K, Gohla S (2000) Solid lipid nanoparticles (SLN) for controlled drug delivery – a review of the state of the art. Eur J Pharm Biopharm 50(1):161–177

Muller RH, Runge S, Ravelli V, Mehnert W, Thunemann AF, Souto EB (2006) Oral bioavailability of cyclosporine: solid lipid nanoparticles (SLN) versus drug nanocrystals. Int J Pharm Sci 317(1):82–89

Niven RW, Lott FD, Ip AY, Cribbs JM (1994) Pulmonary delivery of powders and solutions containing recombinant human granulocyte colony-stimulating factor (rhG-CSF) to the rabbit. Pharm Res 11:1101–1109

Olbrich C, Runge SA, Mehnert W, Thunemann AF, Muller RH (2000) Entrapment efficiency and biodegradation of cyclosporine loaded solid lipid nanoparticles (SLN) for peroral administration. In: Proc. 3rd world meeting APV/APGI, Berlin, pp 425–426

Patel RP, Patel H, Baria AH (2009) Formulation and evaluation of liposomes of ketokonazole. Int J Drug Deliv Technol 1(1):16–23

Patel S, Bhirde AA, Rusling JF, Chen X, Gutkind JS, Patel V (2011) Nano delivers big: designing molecular missiles for cancer therapeutics. Pharmaceutics 3(1):34–52

Prasetyanto EA, Bertucci A, Septiadi D, Corradini R, Castro-Hartmann P, De Cola L (2016) Breakable hybrid organosilica nanocapsules for protein delivery. Angew Chem Int Ed Engl 55(10):3323–3327

Radin S, Falaize S, Lee MH, Ducheyne P (2002) In vitro bioactivity and degradation behavior of silica xerogels intended as controlled release materials. Biomaterials 23:3113–3122

Radtke M, Muller RH (2001a) Novel concept of topical cyclosporine delivery with supersaturated SLN creams. Proc Int Symp Control Release Bioact Mater 28:470–471

Radtke M, Muller RH (2001b) Stability study of creams containing cyclosporine SLN. Proc Int Symp Control Release Bioact Mater 28:472–473

Ramachandran R, Paul W, Sharma CP (2008) Synthesis and characterization of. J Biomed Mater Res B 18:699–703

Rentel CO, Bouwstra JA, Naisbett B, Junginger HE (1999) Niosomes as a novel peroral vaccine delivery system. Int J Pharm 186:161–167

Rezler EM, Khan DR, Lauer-Fields J, Cudic M, Baronas-Lowell D, Fields GB (2007) Targeted drug delivery utilizing protein-like molecular architecture. J Am Chem Soc 129:4961–4972

Rytting E, Nguyen J, Wang X, Kissel T (2008) Biodegradable polymeric nanocarriers for pulmonary drug delivery. Expert Opin Drug Deliv 5:629–639

Shao X, Zhang J, Rajian R (2011) ^{125}I-labeled gold nanorods for targeted imaging of inflammation. ACS Nano 5:8967–8973

Shen J, Li K, Cheng L, Liu Z, Lee ST, Liu J (2014) Specific detection and simultaneously localized photothermal treatment of cancer cells using layer-by-layer assembled multifunctional nanoparticles. ACS Appl Mater Interfaces 6:6443–6452

Slowing I, Trewyn BG, Lin VSY (2006) Effect of surface functionalization of MCM-41-type mesoporous silica nanoparticles on the endocytosis by human cancer cells. J Am Chem Soc 128:14792–14793

Solaro R (2008) Targeted delivery of proteins by nanosized carriers. J Polym Sci 46:1–11

Solaro R, Chiellini F (2006) Nanoparticles for the targeted delivery of peptides and proteins. In: Ravi Kumar MNV (ed) Handbook of particulate drug delivery, vol 2. American Scientific Publisher, New York, pp 193–222

Song YK, Liu D, Maruyama K, Takizawa T (1996) Antibody mediated lung targeting of long circulating emulsions. J Pharm Sci Tech 50:372–377

Song M, Zhang Y, Chen K, Wang H, Gong R (2017) Carboxymethyl-β-cyclodextrin grafted chitosan nanoparticles as oral delivery carrier of protein drugs. React Funct Polym 117:10–15

Soppimath KS, Aminabhavi TM, Kulkarni AR, Rudzinski WE (2001) Biodegradable polymeric nanoparticles as drug delivery devices. J Control Release 70:1–20

Sundarraj S, Thangam R, Sujitha MV, Vimala K, Kannan S (2014) Ligand-conjugated mesoporous silica nanorattles based on enzyme targeted prodrug delivery system for effective lung cancer therapy. Toxicol Appl Pharmacol 275:232–243

Tan W, Wang K, He X, Zhao XJ, Drake T, Wang L et al (2004) Bionanotechnology based on silica nanoparticles. Med Res Rev 24:621–638

Tang H, Kobayashi H, Niidome Y, Mori T, Katayama Y, Niidome T (2013) CW/pulsed NIR irradiation of gold nanorods: effect on transdermal protein delivery mediated by photothermal ablation. J Control Release 171(2):178–183

Tkachenko AG, Xie H, Coleman D, Glomm W, Ryan J, Anderson MF et al (2003) Multifunctional gold nanoparticle-peptide complexes for nuclear targeting. J Am Chem Soc 125:4700–4701

Ugazio E, Cavalli R, Gasco MR (2002) Incorporation of cyclosporin A in solid lipid nanoparticles (SLN). Int J Pharm Sci 241:341–344

Vallet-Regi M, Balas F, Arcos D (2007) Mesoporous materials for drug delivery. Angew Chem 46:7548–7558

Vauthier C, Bouchemal K (2009) Methods for the preparation and manufacture of polymeric nanoparticles. Pharm Res 26:1025–1058

Videira M, Azevedo AF, Almeida AJ (1998) Entrapment of a high molecular weight protein into solid lipid nanoparticles. APV/APGI:629–630

Videira M, Florin do H, Almeida AJ (2002) Preparation of solid lipid nanoparticles (SLN): a potential protein delivery system. V Spanish–Portuguese Con Control Drug Del:69–70

Vila A, Sanchez A, Tobio M, Calvo P, Alonso MJ (2002) Design of biodegradable particles for protein delivery. Bouchemal J Control Release 78:15–24

Vila A, Sanchez A, Evora CI, Soriano I, McCallion O, Alonso MJ (2005) PLA-PEG particlesas nasal protein carriers: the influence of the particle size. Int J Pharm 292:43–52

Vyas SP, Khar RK (2002) Targeted and controlled drug delivery, 1st edn. CBS Publishers and Distributors, New Delhi, pp 332–334

Wissing SA, Kayser O, Muller RH (2004) Solid lipid nanoparticles for parenteral drug delivery. Adv Drug Deliv Rev 56(9):1257–1272

Xing X, He X, Peng J, Wang K, Tan W (2005) Uptake of silica coated nanoparticles by HeLa cells. J Nanosci Nanotechnol 5:1688–1693

Yi J, Lam TI, Yokoyama W, Cheng LW, Zhong F (2014) Cellular uptake of beta-carotene from protein stabilized solid lipid nanoparticles prepared by homogenization-evaporation method. J Agri Food Chem 62(5):1096–1104

Yousefi A, Esmaeili F, Rahimian S, Atyabi F, Dinarvand R (2009) Preparation and *in vitro* evaluation of a pegylated nano-liposomal formulation containing docetaxel. Sci Pharm 77:453–464

Yu M, Wu J, Shi J, Farokhzad OC (2016) Nanotechnology for protein delivery: Overview and perspectives. J Control Release 240:24–37

Zhang Q, Yie G, Li Y, Yang Q, Na gai T (2000) Studies on the cyclosporin A loaded stearic acid nanoparticles. Int J Pharm Sci 200:153–159

Zhang N, Ping Q, Huang G, Xu W, Cheng Y, Han X (2006) Lectin-modified solid lipid nanoparticles as carriers for oral administration of insulin. Int J Pharm 327(1–2):153–159

Chapter 20
Efficacy of Nano-phytochemicals Over Pure Phytochemicals Against Various Cancers: Current Trends and Future Prospects

Asif Jafri, Saima Amjad, Shabana Bano, Sudhir Kumar, M. Serajuddin, and Md Arshad

20.1 Phytochemicals and Nano-phytochemicals as Potent Anticancer Agents

Cancer remains a universal destructive disease and the second most common cause of death in humans. Approximately 7.6 million mortalities per year are attributed to cancer. Developing nations are facing more cancer incidence and now represent about 60% of the deaths (Siegel et al. 2017). The World Health Organization has predicted that about 15 million new incidences of cancer will emerge globally in 2020. Anticancer activity is the consequence of natural or synthetic compounds to reverse, repress, or check the development of cancer. As the present radiotherapy and chemotherapy treatments do not distinguish normal cells from cancer cells and thus can cause severe side effects, researchers and the pharmaceutical industry are concerned to develop a smart drug that can cure this destructive disease without affecting normal healthy cells (Jafri et al. 2018). Fortunately, India is a vast resource of medicinal plants and natural products. Epidemiological observation suggests that Indian herbs, spices, and their isolated phytochemicals possess significant anticancer potential and that their consumption is associated with reduction of the progression of some cancers. The anticancer properties of several phytochemicals isolated from different medicinal plants have been reported. Table 20.1 depicts the anticancer properties of some potent phytochemicals and their anticancer mechanisms.

Asif Jafri and Saima Amjad contributed equally with all other contributors.

A. Jafri · M. Arshad (✉)
Molecular Endocrinology Lab, Department of Zoology, University of Lucknow, Lucknow, India

S. Amjad · S. Bano · S. Kumar · M. Serajuddin
Department of Zoology, University of Lucknow, Lucknow, India

I. Bhushan et al. (eds.), *Nanomaterials and Environmental Biotechnology*, Nanotechnology in the Life Sciences,
https://doi.org/10.1007/978-3-030-34544-0_20

Table 20.1 Potent phytochemicals and their anticancer mode of action against some cancers

Sample no.	Name of phytochemicals	Nature of phytochemical and the plant source	Mode of action in the induction of apoptosis	References
1.	Apigenin	A flavonoid present in the flowers of chamomile plants but also found in plenty in parsley, celery, celeriac, and chamomile tea	Encourages apoptosis, inhibits oxidative stress by scavenging free radicals, promotes anti-angiogenesis	Shukla and Gupta (2010)
2.	Berberin	An alkaloid present in *Hydrastis canadensis* Common name: Goldenseal or orangeroot	Apoptosis induction, blocks cell-cycle progression, inhibits signal transduction pathway, impairs cell adhesion and invasion	Tan et al. (2011)
3.	Combretastatin A4	Phenolic compound found in *Combretum caffrum* Common name: Cape bushwillow	Inhibits angiogenesis, encourages cell-cycle arrest via inhibition of tubulin polymerization	Tozer et al. (2002)
4.	Cucurbitacin B	A terpenoid present in *Trichosanthes cucumerina* Common name: Chichinda, snake gourd or Padwal	Apoptosis induction, inhibits cell-cycle progression, impairs signal transduction pathways, prevents angiogenesis	Kaushik et al. (2015)
5.	Curcumin	*Curcuma longa* Common name: Turmeric or Haldi	Enhances the expression of tumor suppressor genes and downregulation of oncogenes, promotes anti-angiogenesis, inhibits signal transduction pathway, damages cell adhesion and invasion properties	Ahmad et al. (2016)
6.	Epigallocatechin-3-gallate (EGCG)	*Camellia sinensis* Common name: Green tea	Initiates apoptosis, reduces oxidative stress by scavenging free radicals, inhibits angiogenesis, enhances the expression of the tumor suppressor gene, inhibits signal transduction pathways	Singh et al. (2011)
7.	Ellagic acid	A predominant phenol abundant in numerous fruits and vegetables	Triggers apoptosis, reduces oxidative stress and scavenges free radicals, inhibits angiogenesis, damages cell adhesion and invasion properties	Zhang et al. (2014)

(continued)

Table 20.1 (continued)

Sample no.	Name of phytochemicals	Nature of phytochemical and the plant source	Mode of action in the induction of apoptosis	References
8.	Emodin	A glycoside present in *Rhamnus frangula* Common name: Alder buckthorn	Encourages apoptosis, hinders tyrosine kinase enzyme action, reduces oxidative stress by scavenging free radicals, augments anti-angiogenesis, inhibits signal transduction pathway, upregulates the expression of tumor suppressor genes	Wei et al. (2013)
9.	Ferulic acid or hydroxycinnamic acid	A phenolic phyto-compound present in the cell walls of plants	Induces apoptosis, reduces oxidative stress by scavenging free radicals	Srinivasan et al. (2007)
10.	Gambogic acid	A glycoside present in *Garcinia hanburyi* (resins) Common name: Gamboge	Induces apoptosis, promotes anti-angiogenesis, inhibits telomerase enzyme action, inhibits signal transduction pathway	Kashyap et al. (2016)
11.	Genistein	Found in *Genista tinctoria* Common name: Dyer's greenweed or Dyer's broom	Encourages apoptosis, inhibits angiogenesis, inhibits cell-cycle progression, induces the expression of tumor suppressor gene	Li et al. (2012)
12.	Hesperitin	Found abundantly in citrus fruits	Induces apoptosis, promotes anti-angiogenesis, encourages the expression of tumor suppressor gene	Roohbakhsh et al. (2015)
13.	Honokiol	A lignin present in *Magnolia grandiflora* Common name: Southern magnolia, lily tree, laurel magnolia	Induces apoptosis, promotes anti-angiogenesis, and inhibits tumor growth; impairs signal transduction pathways	Fried and Arbiser (2009)
14.	Isoliquiritigenin	A terpenoid present in *Glycyrrhiza uralensis* Common name: Liquorice root, liquorice, sweetroot, Gan Cao	Enhances apoptosis, impairs signal transduction pathways, inhibits cell-cycle progression, obstructs angiogenesis, damages cell adhesion and invasion properties	Orlikova et al. (2011)
15.	Kaempferol	A flavonoid present in various plants and plant-derived foods	Encourages apoptosis, inhibits angiogenesis	Kim and Choi (2013)

(continued)

Table 20.1 (continued)

Sample no.	Name of phytochemicals	Nature of phytochemical and the plant source	Mode of action in the induction of apoptosis	References
16.	Luteolin	Found in *Salvia tomentosa* and also found in some dietary sources including carrots, green pepper, chamomile tea, broccoli, parsley, olive oil	Induces apoptosis, inhibits oxidative stress by scavenging free radicals, prevents the expression of the oncogene, impairs signal transduction pathway, encourages the expression of a tumor-suppressing gene, promotes cell-cycle arrest, prevent angiogenesis	Lin et al. (2008)
17.	Morin	*Morus alba* Common name: White mulberry	Promotes apoptosis, encourages anti-angiogenesis	Park et al. (2015)
18.	Naringenin	A predominant flavanone, abundant in grapefruit and also found in a variety of fruits and herbs	Promotes apoptosis, inhibits angiogenesis, encourages tumor suppressor gene expression, impairs signal transduction pathway	Ahamad et al. (2014)
19.	Nobiletin	A flavonoid isolated from citrus peels	Encourages apoptosis, inhibits angiogenesis, induces the expression of tumor suppressor gene, scavenging free radicals, and reduces oxidative stress	Rawson et al. (2014)
20.	Piperine	A predominant alkaloid of *Piper nigrum* Common name: Black pepper	Induction of apoptosis, upregulates tumor suppressor genes, activation of caspase-3, damages signal transduction pathways	Jafri et al. (2019)
21.	Quercetin	A major flavonoid present in several fruits and vegetables: apples, berries, red onions, green tea, etc.	Encourages apoptosis, reduce oxidative stress by scavenging free radicals, inhibits angiogenesis, impairs signal transduction pathway	Khan et al. (2016)
22.	Resveratrol	A phenol found in cranberry (*Vaccinium macrocarpon*), grapes (*Vitis vinifera*), and peanut (*Arachis hypogaea*)	Encourages apoptosis, reduces oxidative stress by scavenging free radicals, promotes cell-cycle arrest, inhibits expression of oncogene	Varoni et al. (2016)

(continued)

Table 20.1 (continued)

Sample no.	Name of phytochemicals	Nature of phytochemical and the plant source	Mode of action in the induction of apoptosis	References
23.	Silibinin	*Silybum marianum* Common names: *Carduus marianus*, milk thistle, blessed milk thistle, Scotch thistle	Induces apoptosis, promotes anti-angiogenesis, reduces oxidative stress by scavenging free radicals, hinders cell-cycle progression, enhances tumor suppressor gene expression, impairs cell adhesion and invasion properties	Ting et al. (2013)
24.	Tetrandrine	An alkaloid present in *Stephania tetrandra* Common name: Fang Ji	Apoptosis induction, anti-inflammatory and anti-carcinogenic effectiveness, anti-angiogenic agent	Chen (2002)
25.	Thymoquinone	A predominant quinone found abundantly in *Nigella sativa* Common name: Caraway, black cumin, or Kalonji	Promotes apoptosis, reduces oxidative stress by scavenging free radicals, arrests cell cycle, impairs cell adhesion and invasion, encourages enzyme induction and thus enhances detoxification, hinders angiogenesis	Woo et al. (2012)
26.	Vincristine	A potent alkaloid existing in *Catharanthus roseus* Common name: Sadabahar or periwinkle	Encourages apoptosis in cancerous cells, inhibition of cell-cycle progression via inhibiting polymerization of tubulin	Moudi et al. (2013)
27.	Vinblastine	An alkaloid existing in *Catharanthus roseus* Common name: Sadabahar or periwinkle	Encourages apoptosis in cancerous cells, inhibition of cell-cycle progression via inhibiting polymerization of tubulin	Moudi et al. (2013)
28.	Zerumbone	A predominant terpenoid found abundantly in *Zingiber zerumbet* Common name: Awapuhi, bitter ginger, pinecone ginger	Encourages cell-cycle arrest, inhibits expression of the oncogene, reduces oxidative stress by scavenging free radicals	Rahman et al. (2014)

(continued)

Table 20.1 (continued)

Sample no.	Name of phytochemicals	Nature of phytochemical and the plant source	Mode of action in the induction of apoptosis	References
29.	α-Mangostin	A xanthonoid obtained from *Garcinia mangostana* Common name: Mangosteen	Triggers apoptosis, reduces oxidative stress by scavenging free radicals, encourages cell-cycle arrest, enhances caspase-3 activity	Jafri et al. (2018)
30.	β-Lapachone	A quinone found in *Tabebuia avellanedae* Common name: Pink tabebuia	Induces apoptosis, impairs signal transduction pathway, reduces oxidative stress by scavenging free radicals	Kung et al. (2014)

20.2 The Advantage of Nano-phytochemicals Over Pure Phytochemicals

Phytochemicals have potential ability to protect against many chronic diseases such as heart disease, diabetes, neuronal degeneration, and cancer (Subramanian et al. 2016). Phytochemicals possess significant antiproliferative and apoptotic effects against various cancer cells and are widely used in anticancer research (Fig. 20.1). These phytochemicals induce apoptotic effects on various cancerous cells through different apoptotic signaling pathways and check cell-cycle progression, regulating the antioxidant system and detoxification.

Although the phytochemicals possess marvelous antiproliferative and apoptotic potential, they also have some limitations to be target-effective drugs because of their lower solubility in water, poor penetration power for entering targeted cells, restricted therapeutic potential, hepatic disposition, and prompt absorption by normal cells (Bhadoriya et al. 2011). Hence, to overcome these issues, scientists have shifted their interests toward a nano-based targeted drug delivery system of some promising phytochemicals to enhance their aqueous solubility, target specific to cancerous cells, improving cellular uptake, bioavailability and reducing the quantities of phytochemicals to achieve a secure therapeutic level of these potent phytochemicals for future targeted drug delivery systems (Fig. 20.2). The nano-phytochemicals also have various advantages over the pure phytochemicals in having excellent stability in the blood, a multi-efficient design, and hence improved anticancer effectiveness compared to the pure phytochemicals.

20.2.1 Role of Nanoform Phytochemicals in Cancer Research

Nanotechnology has improved the bioavailability, solubility, and targeted delivery of active phytochemicals by using nanoparticles as a carrier and enhancing the bioactivity of potent phytochemicals (Nishiyama 2007). There are some previous

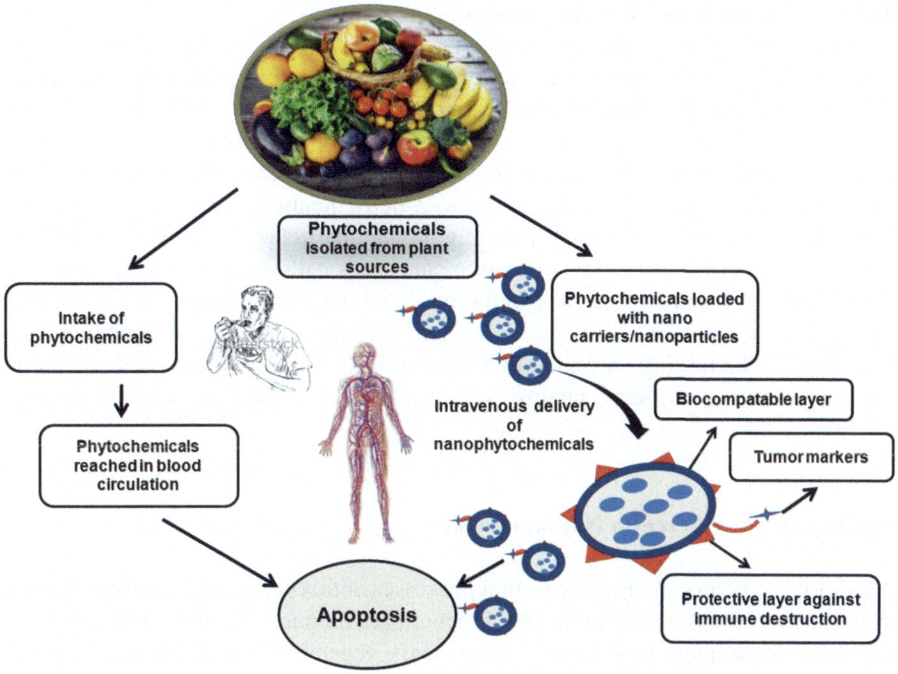

Fig. 20.1 Apoptotic induction by phytochemicals and nano-phytochemicals in cancer cells

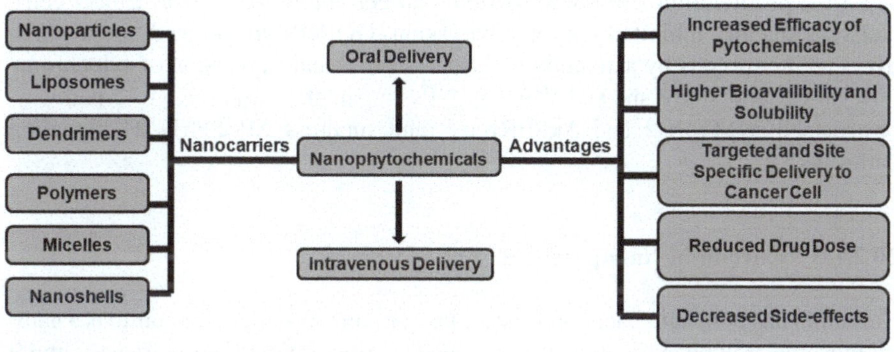

Fig. 20.2 Schematic presentation of nanocarrier nano-phytochemicals and their advantages in cancer research

reports of the active phytochemical-loaded nanoparticles targeting against cancer cells with reduced side effects. Table 20.2 shows some promising nano-phytochemicals and their anticancer potential against some cancers.

20.2.1.1 Broccoli Gold Nanoparticles

Broccoli (*Brassica oleracea*) is a cruciferous vegetable that is potentially protective against some cancers (Steinmetz and Potter 1991a, b). Broccoli phytochemicals are loaded with polyphenols, glucosinolates (GLs), vitamin C, carotene, folic acid, and fibers (Conaway et al. 2001; Mahn and Reyes 2011). These constituents enhance antioxidant activities and stimulate phase-2 detoxification enzymes (Tang et al. 2006; Qazi et al. 2010). Broccoli-loaded gold nanoparticles (B-Au NPs) were synthesized and were analyzed against some cancerous cells, such as MDA-MB-231, SkBr3, and T47D cells of breast carcinoma, PC-3 cells of prostate carcinoma, and U266 cells of myeloma to evaluate the anti-cancerous activity (Khoobchandani et al. 2013). The findings revealed that the conjugated B-Au NPs possessed excellent anti-cancerous effects against breast carcinoma, prostate carcinoma and myeloma cells.

20.2.1.2 Gold Quercetin Nanoparticles

Quercetin is a predominant flavone that possesses antioxidant and antiinflammatory properties and provides valuable health benefits to humans (Aguirre et al. 2011). It is isolated from many food items such as leafy vegetables, citrus fruits, red wine, green tea, apples, berries, onions, and *Ginkgo biloba* (Davis et al. 2009). A previous study reported the synthesis of gold-quercetin nanoparticles (GQ NPs) and its anticancer activity against liver carcinoma (Ren et al. 2017). The GQ NPs regulated abnormal proliferation, colony formation, and cell migration in liver cancer cells, and subsequently inhibited cancer progression. GQ NPs encouraged apoptosis in liver carcinoma cells by activation of caspases (3, 9) and expression of cytochrome *c* (Cyto-c). The study showed that GQ NPs impair the caspase/Cyto-c pathway, obstruct NF-κB/COX-2 and Akt/ERK1/2, and suppress AP-2β/hTERT signaling pathways.

20.2.1.3 Curcumin Nanoparticles

Curcumin has both anticancer and cancer prevention properties. Curcumin is a natural bioactive polyphenolic constituent, found in turmeric, with properties to inhibit proliferation and metastasis in a wide range of cancers (Joe et al. 2004). However, poor solubility in water has limited the systemic bioavailability of curcumin when administered orally. To overcome the solubility and bioavailability restrictions of curcumin, it was loaded on various types of nanoparticles. Methoxy polyethylene glycol (mPEG) and polycaprolactone (PCL) co-polymers enhanced the anticancerous activity of curcumin with efficient loading capability and sustainable release in A549 cells of lung epithelial carcinoma. This curcumin-loaded nanoparticle effectively enters into lung carcinoma A549 cells and interacts with the nucleus to encourage apoptosis (Yin et al. 2013). Poly lactic-co-glycolic acid curcumin

Table 20.2 Promising nano-phytochemicals and their anticancer potential against some cancers

Sample no.	Nano-phytoconstituents	Source of phytochemical	Mode of action	References
1.	Broccoli nanoparticle	Broccoli	Anticancer action against MDA-MB-231, SkBr3, and T47D cells of breast carcinoma, PC-3 cells of prostate carcinoma, and U266 cells of myeloma	Khoobchandani et al. (2013)
2.	Quercetin nanoparticle	Leafy vegetables, citrus fruits, red wine, green tea, apples, berries, onions, *Ginkgo biloba*	Inhibits growth in hepatic carcinoma cells, modulates caspase activities, upregulates cytochrome *c*, obstructs NFκB/COX-2 and Akt/ERK1/2 expression	Davis et al. (2009), Ren et al. (2017)
3.	Curcumin nanoparticle	Turmeric (*Curcuma longa*)	Inhibits phosphoinositide 3-kinase pathway, suppresses tumor burden, induced apoptosis,, enhanced therapeutic ability	Yin et al. (2013), Yallapu et al. (2014), Esfandiarpour-Boroujeni et al. (2017)
4.	*Selaginella doederleinii* nanoparticle	*Selaginella doederleinii* Hieron	Inhibits growth in A549 cells of human lung carcinoma	Syaefudin et al. (2016)
5.	*Nigella sativa* nanoparticles	*Nigella*	Upregulates miR-34 a,d, represses Rac1 mRNA protein expression, enhances antiproliferative and apoptotic properties	Bhattacharya et al. (2015), Ganea et al. (2010), Dehghani et al. (2015)
6.	Honokiol nanoparticle	*Magnolia officinalis*	Antiproliferative effect against osteosarcoma cells, induces apoptosis, inhibits proliferation of tumors and prevents angiogenesis, micro-vessel generation. and cell-cycle arrest	Cheng et al. (2016), Li et al. (2008)
7.	Silibinin nanoparticle	*Silybum marianum*	Suppresses cell proliferation, inhibits hTERT expression, prevents metastasis and angiogenesis in breast cancer cells	Ebrahimnezhad et al. (2013), Amirsaadat et al. (2017), Xu et al. (2013)
8.	Ursolic acid nanoparticles	Epicuticular waxes of apples, peels of fruits, herbs and spices such as rosemary and thyme	Encourages anticancer activity, promotes lysosomal destruction, enhances ROS augmentation, promotes apoptosis in MCF-7 cells of breast carcinoma	Baishya et al. (2016), Jiang et al. (2017)

(continued)

Table 20.2 (continued)

Sample no.	Nano-phytoconstituents	Source of phytochemical	Mode of action	References
9.	β-Lapachone nanoparticle	Bark of the Lapacho tree	Activates anti-cancerous and radiotherapeutic efficacy	Jeong et al. (2009)
10.	Ferulic acid nanoparticles	Vegetable sources, oat flours, wheat, rice	Reduces cancer cell proliferation, promotes intracellular ROS production, enhances DNA damage, lipid peroxidation, and encourages apoptosis	Zhao and Moghadasian (2008), Kumar and Pruthi (2014)

nanoparticles (PLGA-CUR NPs) showed tumor regression properties against prostate carcinoma cells by inhibiting colony formation and cell growth in prostate cancer cell xenografted mice (Yallapu et al. 2014). PLGA-CUR NPs conjugated with epidermoid growth factor receptor were targeted to GE11 peptides in breast carcinoma cells in vitro and in a tumorous mice model. It extended the epidermal growth factor receptor (EGFR) expression in MCF-7 cells of breast carcinoma and reduced the phosphoinositide-3 kinase pathway, suppressed tumor growth, suppressed cancer cell proliferation, and improved drug clearance from the circulation (Jin et al. 2017). Curcumin-loaded magnetic nanoparticles were also significantly taken up in MDA-MB-231 cells of human epithelial breast carcinoma. These nanoparticles showed strong anticancer activity via strong penetration power into the cell by endocytosis (Yallapu et al. 2012). The uploaded curcumin magnetic nanoparticles exhibited strong magnetic resonance properties and potentially enhanced the anticancer targeting capacity of curcumin.

Nano-curcumin effectively decreased the level of miRNA-21, β-catenin, and the expression of E6/E7 HPV oncoproteins in a mouse cervical cancer model (Zaman et al. 2016). Also, curcumin-loaded silica nanoparticles (Dinda et al. 2012) and curcumin-loaded folate-modified chitosan nanoparticles (Esfandiarpour-Boroujeni et al. 2017) have significant anticancer effect against HepG2 cells of hepatocellular carcinoma and MCF-7 cells of breast carcinoma, respectively.

20.2.1.4 *Selaginella doederleinii* Leaf Nanoparticles

Selaginella doederleinii Hieron possesses several bioactive phytochemicals and has been extensively used in traditional herbal medicine for curing rheumatoid arthritis and various cancers, specifically lung cancer, cervical cancer, choriocarcinoma, and nasopharyngeal carcinoma (Abdille et al. 2005; Ishii et al. 2005; Wang et al. 2015). *Selaginella* leaf extract nanoparticles (SDNPs) reduced the viability of lung carcinoma A549 cells and were least toxic to normal Chang liver cells (Syaefudin et al. 2016). The findings revealed the apoptotic potential of SDNPs and the targeted drug delivery system without affecting the normal Chang cells.

20.2.1.5 *Nigella sativa* Nanoformulation

Nigella sativa, commonly known as nigella, kalonji, black caraway, and black cumin, is broadly used as a traditional medicine universally (Heiss and Oeggl 2005). Thymoquinone is one of the major active phyto-constituents present in *Nigella sativa* and possesses various pharmacological properties: antimicrobial, antiinflammatory, antipyretic, hypoglycemic, analgesic, and antioxidant (Ali and Blunden 2003). Previous studies reported the anticancer and antimutagenic activity of thymoquinone in Caco-2, HT-29, HCT-116, DLD-1, and LoVo cancerous cells (Gali-Muhtasib et al. 2006). The encapsulated thymoquinone (TQ) nanoparticles (TQ NPs) were synthesized with the help of biodegradable polyvinyl pyrrolidone (PVP) and polyethylene glycol (PEG) (less than 50 nm in size) and significantly enhanced anticancer efficacy by increasing bioavailability and solubility (Bhattacharya et al. 2015) of TQ. In vitro study in MCF-7 cells of breast carcinoma and in vivo tumor-bearing mice demonstrated that PEG-TQ NPs enhance miR-34a, repress Rac1 mRNA protein expression, prevent cell migration, and hinder angiogenesis. TQ loaded with modified molecular micelles, PLGA nanoparticles, and chitosan myristic acid nanogel exhibited more antioxidant and antiproliferative effectiveness at low doses with controlled release as compared with pure TQ against MCF-7 cells and MDA-MB-231 of breast carcinoma (Ganea et al. 2010; Dehghani et al. 2015). These findings showed that TQ NP have a more potent apoptotic effect than pure TQ.

20.2.1.6 Honokiol Nanoparticle

Honokiol is a polyphenol, a lignan obtained from the medicinal plant *Magnolia officinalis* (Li et al. 2008). It possesses multifactorial pharmacological properties and is widely used to treat stomach upset, inflammation, nervous disturbance, and anxiety (Chiang et al. 2006; Kim and Cho 2008; Deng et al. 2008; Fried and Arbiser 2009). It also showed antineoplastic and apoptotic effectiveness against various cancerous cells (Yuan et al. 2009; Steinmann et al. 2012; Avtanski et al. 2013; Cheng et al. 2014; Subramaniam et al. 2015). Honokiol-loaded synthesized nanoformulation co-polymer micelles with monomethoxy polyethylene glycol and polycaprolactone (HK-MPEG/PCL) increased bioavailability and solubility. The anticancer activity of HK-MPEG/PCL showed reduced tumor growth and apoptotic induction in a cancer nude mouse model and prevented angiogenesis (Cheng et al. 2016). Another study on honokiol-loaded co-polymer synthesized with folate and polyethyleneimine nanoparticles (HK NPs) significantly prevented metastasis, proliferation, tumor growth, angiogenesis, and cell-cycle progression (Gou et al. 2010). The developed HK NPs potentially encouraged apoptosis of HNE-1 cells of nasopharyngeal carcinoma as compared to pure honokiol.

20.2.1.7 Silibinin-Loaded Nanoparticle

Silibinin is a natural antioxidant and a polyphenolic flavonoid found in *Silybum marianum*, also known as silybin or milk thistle (Davis-Searles et al. 2005). It possesses multifactorial pharmacological effects such as hepato-protective, antioxidant, and anticarcinogenic activities. Silibinin encourages apoptosis via inhibiting cell-cycle progression, preventing proliferation, impairs angiogenesis, and enhancing immune stimulation in several cancers (Mateen et al. 2010; Nejati-Koshki et al. 2012; Surai 2015). Some previous studies demonstrated that silibinin arrested the cancer cycle and inhibited the growth, progression, and angiogenesis of tumorous cells (Singh et al. 2006; Liang et al. 2014). Silibinin-loaded nanoparticles with PLGA-PEG co-polymer were synthesized to evaluate the anticancer potential and expression of the hTERT gene in T47D cells of breast carcinoma and A549 cells of lung carcinoma. The findings showed that hTERT expression more efficiently reduced the cell viability of both cancerous cells with the increasing concentrations of nano-silibinin as compared to the pure silibinin (Ebrahimnezhad et al. 2013; Amirsaadat et al. 2017). The silibinin-loaded phosphatidylcholine lipid nanoparticles significantly reduced the growth and angiogenesis of breast carcinoma (Xu et al. 2013). These findings suggested that nano-form silibinin was a more efficient anticancer agent than pure silibinin.

20.2.1.8 Ursolic Acid Nanoparticle

Ursolic acid is a plant-derived triterpenoid that is present in epicuticular waxes of apples, fruit peels, herbs, and spices such as rosemary and thyme (Shanmugam et al. 2013). Ursolic acid has multifactorial pharmacological properties, such as antiinflammatory (Chattopadhyay et al. 2002), antidiabetic (Jang et al. 2010), antiepileptic, anticancer (Tannock 2011), and liver protective (Shanmugam et al. 2011; Prasad et al. 2011) elements. Ursolic acid-based mesoporous silica nanospheres (UAMSN) were synthesized with the help of pH-sensitive chitosan and folic acid for the targeted drug delivery of tumor cells (Jiang et al. 2017). Enhanced apoptotic effects of ursolic acid-encapsulated PLGA nanoparticles (UA-NPs) were found in B16F10 mouse melanoma cells as compared with pure ursolic acid (Baishya et al. 2016). Folate and chitosan-based nanocarriers for the delivery of ursolic acid (FA-CS-UA-NPs) were studied for in vitro MCF-7 cells of breast carcinoma and in an in vivo tumor xenograft mouse model (Jin et al. 2016). The findings revealed FA-CS-UA-NPs penetrated into the cells via endocytosis and encouraged apoptosis. The FA-CS-UA-NPs had the potential to localize into the mitochondria, induce excessive reactive oxygen species (ROS), and enhance apoptosis in cancerous cells.

20.2.1.9 β-Lapachone Nanoparticle

Beta-lapachone is a quinone that is abundant in the bark of *Tabebuia avellanedae* (Lapacho tree) and has potent anti-cancerous properties (Pardee et al. 2002). It generates apoptosis via modulating NADPH quinone oxidoreductase-1 (NQO1) in cancerous cells (Blanco et al. 2007). Lapachone-loaded silver nanoparticles (Lap Au NPs) were synthesized and targeted on anti-EGFR antibody through intravenous injection of Lap Au NPs to tumor-bearing xenografted mice, showing an enhanced tumor suppression effect as well as enhanced radiotherapeutic efficacy (Jeong et al. 2009).

20.2.1.10 Ferulic Acid Nanoparticles

Ferulic acid is a phenolic phytochemical found in vegetable sources, oat flours, whole grain wheat, and rice (Zhao and Moghadasian 2008; Kumar and Pruthi 2014). A ferulic acid-loaded synthesized PLGA nanoparticle (FA-PLGA-NP) enhances apoptosis in NCI-H460 cells of lung carcinoma as compared to its pure form (Merlin et al. 2012). The FA-PLGA-NPs encourage apoptosis via decreasing cell proliferation, generating excessive intracellular ROS, enhanced DNA damage, and lipid peroxidation.

20.3 Conclusion

Numerous studies on phytochemicals have shown significant anticancer potential but some limitations. Studies on nano-phytochemicals, both in vitro and in animal models, showed enhanced bioavailability, improved cellular uptake, reduced doses, and enhanced solubility, overcoming the limitations of pure phytochemicals. Further, the potent anticancer nano-phytochemicals without side effects that are not destructive to normal cells are still to be investigated. The ongoing research on nano-phytochemicals opens a new avenue for cancer cure by following promising phytochemicals with the advanced approaches of nanotechnology.

Acknowledgments The authors Asif Jafri and Md Arshad acknowledge the support of Uttar Pradesh Council of Science and Technology, Lucknow, India (File No. CST/SERPD/D-299). Asif Jafri is also grateful to the Council of Scientific and Industrial Research, India for the award of Senior Research Fellowship (File No. 09/107 (0393)/2018- EMR-I).

References

Abdille MH, Singh RP, Jayaprakasha GK (2005) Antioxidant activity of the extracts from *Dillenia indica* fruits. Food Chem 90:891–896

Aguirre L, Arias N, Macarulla MT et al (2011) Beneficial effects of quercetin on obesity and diabetes. Open Nutraceuticals J 4:189–198

Ahamad MS, Siddiqui S, Jafri A et al (2014) Induction of apoptosis and antiproliferative activity of naringenin in human epidermoid carcinoma cell through ROS generation and cell cycle arrest. PLoS One 9:e110003

Ahmad MZ, Alkahtani SA, Akhter S et al (2016) Progress in nanotechnology-based drug carrier in designing of curcumin nanomedicines for cancer therapy: current state-of-the-art. J Drug Target 24:273–293

Ali BH, Blunden G (2003) Pharmacological and toxicological properties of *Nigella sativa*. Phytother Res 17:299–305

Amirsaadat S, Pilehvar-Soltanahmadi Y, Zarghami F et al (2017) Silibinin-loaded magnetic nanoparticles inhibit hTERT gene expression and proliferation of lung cancer cells. Artif Cells Nanomed Biotechnol 45:1649–1656

Avtanski DB, Nagalingam A, Kupusamy P et al (2013) Targeting epithelial-mesenchymal transition in breast cancer cells using Honokiol, a natural phenolic compound. Cancer Res 73:299

Baishya R, Nayak DK, Kumar D et al (2016) Ursolic acid loaded PLGA nanoparticles: in vitro and in vivo evaluation to explore tumor targeting ability on B16F10 melanoma cell lines. Pharm Res 33:2691–2703

Bhadoriya SS, Mangal A, Madoriya N et al (2011) Bioavailability and bioactivity enhancement of herbal drugs by "Nanotechnology": a review. J Curr Pharm Res 8:1–7

Bhattacharya S, Ahir M, Patra P et al (2015) PEGylated-thymoquinone-nanoparticle mediated retardation of breast cancer cell migration by deregulation of cytoskeletal actin polymerization through miR-34a. Biomaterials 51:91–107

Blanco E, Bey EA, Dong Y et al (2007) β-Lapachone-containing PEG–PLA polymer micelles as novel nanotherapeutics against NQO1-overexpressing tumor cells. J Control Release 122:365–374

Chattopadhyay D, Arunachalam G, Mandal AB et al (2002) Antimicrobial and anti-inflammatory activity of folklore: *Mallotus peltatus* leaf extract. J Ethnopharmacol 82:229–237

Chen YJ (2002) Potential role of tetrandrine in cancer therapy. Acta Pharmacol Sin 23:1102–1106

Cheng S, Welty M, Eliaz I et al (2014) Honokiol inhibits growth and migration of renal cell carcinoma. Cancer Res 74:3203

Cheng Y, Zheng S, Teng Y et al (2016) Preparation of honokiol with biodegradable nanoparticles for treatment of osteosarcoma. RSC Adv 6:94278–94286

Chiang CK, Sheu ML, Hung KY et al (2006) Honokiol, a small molecular weight natural product, alleviates experimental mesangial proliferative glomerulonephritis. Kidney Int 70:682–689

Conaway CC, Getachun SM, Liebes LL et al (2001) Disposition of glucosinolates and sulphoraphanes in human after ingestion of steam and fresh broccoli. Nutr Cancer 38:168–178

Davis JM, Murphy EA, Carmichael MD (2009) Effects of the dietary flavonoid quercetin upon performance and health. Curr Sports Med Rep 8:206–213

Davis-Searles PR, Nakanishi Y, Kim NC et al (2005) Milk thistle and prostate cancer: differential effects of pure flavonolignans from *Silybum marianum* on antiproliferative end points in human prostate carcinoma cells. Cancer Res 65:4448–4457

Dehghani H, Hashemi M, Entezari M et al (2015) The comparison of anticancer activity of thymoquinone and nanothymoquinone on human breast adenocarcinoma. Iran J Pharm Res 14:539

Deng J, Qian Y, Geng L et al (2008) Involvement of p38 mitogen-activated protein kinase pathway in honokiol-induced apoptosis in a human hepatoma cell line (hepG2). Liver Int 28:1458–1464

Dinda AK, Prashant CK, Naqvi S et al (2012) Curcumin loaded organically modified silica (ORMOSIL) nanoparticle; a novel agent for cancer therapy. Int J Nanotechnol 9:862–871

Ebrahimnezhad Z, Zarghami N, Keyhani M et al (2013) Inhibition of hTERT gene expression by silibinin-loaded PLGA-PEG-Fe3O4 in T47D breast cancer cell line. Bioimpacts 3:67

Esfandiarpour-Boroujeni S, Bagheri-Khoulenjani S, Mirzadeh H et al (2017) Fabrication and study of curcumin loaded nanoparticles based on folate-chitosan for breast cancer therapy application. Carbohydr Polym 168:14–21

Fried LE, Arbiser JL (2009) Honokiol, a multifunctional antiangiogenic and antitumor agent. Antioxid Redox Signal 11:1139–1148

Gali-Muhtasib H, Roessner A, Schneider-Stock R (2006) Thymoquinone: a promising anti-cancer drug from natural sources. Int J Biochem Cell Biol 38:1249–1253

Ganea GM, Fakayode SO, Losso JN et al (2010) Delivery of phytochemical thymoquinone using molecular micelle modified poly (D, L-lactide-co-glycolide) (PLGA) nanoparticles. Nanotechnology 21:285104

Gou M, Gong C, Zhang J et al (2010) Polymeric matrix for drug delivery: Honokiol-loaded PCL-PEG-PCL nanoparticles in PEG-PCL-PEG thermosensitive hydrogel. J Biomed Mater Res Part A 93:219–226

Heiss AG, Oeggl K (2005) The oldest evidence of *Nigella damascena* L. (Ranunculaceae) and its possible introduction to central Europe. Veg Hist Archaeobot 14:562–570

Ishii T, Yasuda K, Akatsuka A (2005) A mutation in the SDHC gene of complex increases oxidative stress resulting in apoptosis and tumorigenesis. Cancer Res 65:203–209

Jafri A, Siddiqui S, Deep A et al (2018) Induction of apoptosis by alpha-mangostin in human hepatocellular carcinoma cells via nuclear fragmentation and ROS dependent mitochondrial pathway. J Clin Exp Hepatol 8:S116

Jafri A, Siddiqui S, Rais J et al (2019) Induction of apoptosis by piperine in human cervical adenocarcinoma via ROS mediated mitochondrial pathway and caspase-3 activation. EXCLI J 18:154–164

Jang SM, Kim MJ, Choi MS et al (2010) Inhibitory effects of ursolic acid on hepatic polyol pathway and glucose production in streptozotocin-induced diabetic mice. Metabolism 59:512–519

Jeong SY, Park SJ, Yoon SM et al (2009) Systemic delivery and preclinical evaluation of Au nanoparticle containing β-lapachone for radiosensitization. J Control Release 139:239–245

Jiang K, Chi T, Li T et al (2017) A smart pH-responsive nano-carrier as a drug delivery system for the targeted delivery of ursolic acid: suppresses cancer growth and metastasis by modulating P53/MMP-9/PTEN/CD44 mediated multiple signaling pathways. Nanoscale 9:9428–9439

Jin H, Pi J, Yang F et al (2016) Folate-chitosan nanoparticles loaded with ursolic acid confer anti-breast cancer activities in vitro and in vivo. Sci Rep 6:30782

Jin H, Pi J, Zhao Y et al (2017) EGFR-targeting PLGA-PEG nanoparticles as a curcumin delivery system for breast cancer therapy. Nanoscale 9:16365–16374

Joe B, Vijaykumar M, Lokesh BR (2004) Biological properties of curcumin cellular and molecular mechanisms of action. Crit Rev Food Sci Nutr 44:97–111

Kashyap D, Mondal R, Tuli HS et al (2016) Molecular targets of gambogic acid in cancer: recent trends and advancements. Tumor Biol 37:12915–12925

Kaushik U, Aeri V, Mir SR (2015) Cucurbitacins – an insight into medicinal leads from nature. Pharmacogn Rev 9:12–18

Khan F, Niaz K, Maqbool F et al (2016) Molecular targets underlying the anticancer effects of quercetin: an update. Nutrients 8:529

Khoobchandani M, Zambre A, Katti K et al (2013) Green nanotechnology from Brassicaceae: development of broccoli phytochemicals-encapsulated gold nanoparticles and their applications in nanomedicine. Int J Green Nanotechnol 5(1):1–15

Kim BH, Cho JY (2008) Anti-inflammatory effect of honokiol is mediated by PI3K/Akt pathway suppression 1. Acta Pharmacol Sin 29:113–122

Kim SH, Choi KC (2013) Anti-cancer effect and underlying mechanism(s) of kaempferol, a phytoestrogen, on the regulation of apoptosis in diverse cancer cell models. Toxicol Res 29:229–234

Kumar N, Pruthi V (2014) Potential applications of ferulic acid from natural sources. Biotechnol Rep 4:86–93

Kung HN, Lu KS, Chau YP (2014) The chemotherapeutic effects of lapacho tree extract: β-lapachone. Chemotherapy 3:1–5

Li Z, Liu Y, Zhao X et al (2008) Honokiol, a natural therapeutic candidate, induces apoptosis and inhibits angiogenesis of ovarian tumor cells. Eur J Obstet Gynecol Reprod Biol 140:95–102

Li QS, Li CY, Li ZL et al (2012) Genistein and its synthetic analogs as anticancer agents. Anti-Cancer Agents Med Chem 12:271–281

Liang Z, Yang Y, Wang H et al (2014) Inhibition of SIRT1 signaling sensitizes the antitumor activity of silybin against human lung adenocarcinoma cells in vitro and in vivo. Mol Cancer Ther 13:1860–1872

Lin Y, Shi R, Wang X et al (2008) Luteolin, a flavonoid with potentials for cancer prevention and therapy. Curr Cancer Drug Targets 8:634–646

Mahn A, Reyes A (2011) An overview of health-promoting compounds of broccoli (*Brassica oleracea* var. *italica*) and the effect of processing. Food Sci Tech Int 18:503–514

Mateen S, Tyagi A, Agarwal C et al (2010) Silibinin inhibits human nonsmall cell lung cancer cell growth through cell-cycle arrest by modulating expression and function of key cell-cycle regulators. Mol Carcinog 49:247–258

Merlin JJ, Prasad NR, Shibli SM (2012) Ferulic acid loaded poly-*d*,*l*-lactide-co-glycolide nanoparticles: systematic study of particle size, drug encapsulation efficiency and anticancer effect in non-small cell lung carcinoma cell line in vitro. Biomed Prev Nutr 2:69–76

Moudi M, Go R, Yien CYS et al (2013) Vinca alkaloids. Int J Prev Med 4:1231–1235

Nejati-Koshki K, Zarghami N, Pourhassan-Moghaddam M et al (2012) Inhibition of leptin gene expression and secretion by silibinin: possible role of estrogen receptors. Cytotechnology 64:719–726

Nishiyama N (2007) Nanomedicine: nanocarriers shape up for long life. Nat Nanotechnol 2:203–204

Orlikova B, Tasdemir D, Golais F et al (2011) Dietary chalcones with chemopreventive and chemotherapeutic potential. Genes Nutr 6:125–147

Pardee AB, Li Y, Li CJ (2002) Cancer therapy with β-lapachone. Curr Cancer Drug Targets 2:227–242

Park C, Lee WS, Go SI et al (2015) Morin, a flavonoid from Moraceae, induces apoptosis by induction of BAD protein in human leukemic cells. Int J Mol Sci 16:645–659

Prasad S, Yadav VR, Kannappan R et al (2011) Ursolic acid, a pentacyclin triterpene, potentiates TRAIL-induced apoptosis through p53-independent up-regulation of death receptors evidence for the role of reactive oxygen species and JNK. J Biol Chem 286:5546–5557

Qazi A, Pal J, Maitah MI et al (2010) Anticancer activity of a broccoli derivative, sulforaphane, in Barrett adenocarcinoma: potential use in chemoprevention and as adjuvant in chemotherapy. Transl Oncol 3:389–399

Rahman HS, Rasedee A, Yeap SK et al (2014) Biomedical properties of a natural dietary plant metabolite, zerumbone, in cancer therapy and chemoprevention trials. Biomed Res Int 2014:920742

Rawson NE, Ho CT, Li S (2014) Efficacious anti-cancer property of flavonoids from citrus peels. Food Sci Human Wellness 3:104–109

Ren KW, Li YH, Wu G et al (2017) Quercetin nanoparticles display antitumor activity via proliferation inhibition and apoptosis induction in liver cancer cells. Int J Oncol 50:1299–1311

Roohbakhsh A, Parhiz H, Soltani F et al (2015) Molecular mechanisms behind the biological effects of hesperidin and hesperetin for the prevention of cancer and cardiovascular diseases. Life Sci 124:64–74

Shanmugam MK, Rajendran P, Li F et al (2011) Ursolic acid inhibits multiple cell survival pathways leading to suppression of growth of prostate cancer xenograft in nude mice. J Mol Med 89:713–727

Shanmugam MK, Dai X, Kumar AP et al (2013) Ursolic acid in cancer prevention and treatment: molecular targets, pharmacokinetics and clinical studies. Biochem Pharmacol 85:1579–1587

Shukla S, Gupta S (2010) Apigenin: a promising molecule for cancer prevention. Pharm Res 27:962–978

Siegel RL, Miller KD, Jemal A (2017) Cancer statistics, 2017. CA Cancer J Clin 67:7–30

Singh RP, Deep G, Chittezhath M et al (2006) Effect of silibinin on the growth and progression of primary lung tumors in mice. J Natl Cancer Inst 98:846–855

Singh BN, Shankar S, Srivastava RK (2011) Green tea catechin, epigallocatechin-3-gallate (EGCG): mechanisms, perspectives and clinical applications. Biochem Pharmacol 82:1807–1821

Srinivasan M, Sudheer AR, Menon VP (2007) Ferulic acid: therapeutic potential through its antioxidant property. J Clin Biochem Nutr 40:92–100

Steinmann P, Walters DK, Arlt MJ et al (2012) Antimetastatic activity of honokiol in osteosarcoma. Cancer 118:2117–2127

Steinmetz KA, Potter JD (1991a) Vegetables, fruit, and cancer. I. Epidemiology. Cancer Causes Control 2:325–357

Steinmetz KA, Potter JD (1991b) Vegetables, fruit, and cancer. II. Mechanisms. Cancer Causes Control 2:427–442

Subramaniam D, Ponnurangam S, Kwatra D et al (2015) Honokiol prevents colonic tumorigenesis and affects stem cell viability by affecting oncogenic YAP1 function. Cancer Res 75:1893

Subramanian AP, Jaganathan SK, Manikandan A et al (2016) Recent trends in nano-based drug delivery systems for efficient delivery of phytochemicals in chemotherapy. RSC Adv 6:48294–48314

Surai PF (2015) Silymarin as a natural antioxidant: an overview of the current evidence and perspectives. Antioxidants 4:204–247

Syaefudin JA, Rosiyana L, Setyani A et al (2016) Nanoparticles of *Selaginella doederleinii* leaf extract inhibit human lung cancer cells A549. Earth Environ Sci 31:1–5

Tan W, Li Y, Chen M et al (2011) Berberine hydrochloride: anticancer activity and nanoparticulate delivery system. Int J Nanomedicine 6:1773–1777

Tang L, Zhang Y, Jobson EH et al (2006) Potent activation of mitochondria-mediated apoptosis and arrest in S and M phases of cancer cells by a broccoli sprout extract. Mol Cancer Ther 5:935–944

Tannock LR (2011) Ursolic acid effect on atherosclerosis: apples and apples, or apples and oranges? Atherosclerosis 219:397–398

Ting H, Deep G, Agarwal R (2013) Molecular mechanisms of silibinin-mediated cancer chemoprevention with major emphasis on prostate cancer. AAPS J 15:707–716

Tozer GM, Kanthou C, Parkins CS et al (2002) The biology of the combretastatins as tumour vascular targeting agents. Int J Exp Pathol 83:21–38

Varoni EM, Faro AFL, Sharifi-Rad J et al (2016) Anticancer molecular mechanisms of resveratrol. Front Nutr 3:1–15

Wang G, Yao S, Cheng L et al (2015) Antioxidant and anticancer effection of the volatile oil from various habitats of *Selaginella doederleinii* Hieron. Technol Health Care 23:S21–S27

Wei WT, Lin SZ, Liu DL et al (2013) The distinct mechanisms of the antitumor activity of emodin in different types of cancer. Oncol Rep 30:2555–2562

Woo CC, Kumar AP, Sethi G et al (2012) Thymoquinone: potential cure for inflammatory disorders and cancer. Biochem Pharmacol 83:443–451

Xu P, Yin Q, Shen J et al (2013) Synergistic inhibition of breast cancer metastasis by silibinin-loaded lipid nanoparticles containing TPGS. Int J Pharm 454:21–30

Yallapu MM, Othman SF, Curtis ET, Bauer NA, Chauhan N, Kumar D, Jaggi M, Chauhan SC (2012) Curcumin-loaded magnetic nanoparticles for breast cancer therapeutics and imaging applications. Int J Nanomedicine 7:1761

Yallapu MM, Khan S, Maher DM et al (2014) Anti-cancer activity of curcumin loaded nanoparticles in prostate cancer. Biomaterials 35:8635–8648

Yin H, Zhang H, Liu B (2013) Superior anticancer efficacy of curcumin-loaded nanoparticles against lung cancer. Acta Biochim Biophys Sin 45:634–640

Yuan Z, Liu H, Yan F et al (2009) Improved therapeutic efficacy against murine carcinoma by combining honokiol with gene therapy of PNAS-4, a novel pro-apoptotic gene. Cancer Sci 100:1757–1766

Zaman MS, Chauhan N, Yallapu MM et al (2016) Curcumin nanoformulation for cervical cancer treatment. Sci Rep 6:20051

Zhang HM, Zhao L, Li H et al (2014) Research progress on the anticarcinogenic actions and mechanisms of ellagic acid. Cancer Biol Med 11:92–100

Zhao Z, Moghadasian MH (2008) Chemistry, natural sources, dietary intake and pharmacokinetic properties of ferulic acid: a review. Food Chem 109:691–702

Index

I. Bhushan et al. (eds.), *Nanomaterials and Environmental Biotechnology*, Nanotechnology in the Life Sciences,
https://doi.org/10.1007/978-3-030-34544-0

J

L

M

N

O

P

The manufacturer's authorised representative in the EU is Springer
Nature Customer Service Centre GmbH, Europaplatz 3, 69115 Heidelberg,
Germany. If you have any concerns regarding our products, please
contact ProductSafety@springernature.com

Printed and bound by CPI Group (UK) Ltd, Croydon, CR0 4YY
05/10/2026
02228145-0013